Probability and Mathematical Statistics (Continued)

WILLIAMS • Diffusions, Markov Processes, and Martingales, Volume I: Foundations

ZACKS • Theory of Statistical Inference

Applied Probability and Statistics

ANDERSON, AUQUIER, HAUCK, OAKES, VANDAELE, and WEISBERG • Statistical Methods for Comparative Studies

ARTHANARI and DODGE • Mathematical Programming in Statistics

BAILEY • The Elements of Stochastic Processes with Applications to the Natural Sciences

BAILEY • Mathematics, Statistics and Systems for Health

BARNETT • Interpreting Multivariate Data

BARNETT and LEWIS • Outliers in Statistical Data

BARTHOLOMEW • Stochastic Models for Social Processes, *Third Edition*

BARTHOLOMEW and FORBES • Statistical Techniques for Manpower Planning

BECK and ARNOLD • Parameter Estimation in Engineering and Science

BELSLEY, KUH, and WELSCH • Regression Diagnostics: Identifying Influential Data and Sources of Collinearity

BENNETT and FRANKLIN • Statistical Analysis in Chemistry and the Chemical Industry

BHAT • Elements of Applied Stochastic Processes

BLOOMFIELD • Fourier Analysis of Time Series: An Introduction

BOX • R. A. Fisher, The Life of a Scientist

BOX and DRAPER • Evolutionary Operation: A Statistical Method for Process Improvement

BOX, HUNTER, and HUNTER • Statistics for Experimenters: An Introduction to Design, Data Analysis, and Model Building

BROWN and HOLLANDER • Statistics: A Biomedical Introduction

BROWNLEE • Statistical Theory and Methodology in Science and Engineering, *Second Edition*

BURY • Statistical Models in Applied Science

CHAMBERS • Computational Methods for Data Analysis

CHATTERJEE and PRICE • Regression Analysis by Example

CHERNOFF and MOSES • Elementary Decision Theory

CHOW • Analysis and Control of Dynamic Economic Systems

CHOW • Econometric Analysis by Control Methods

CLELLAND, BROWN, and deCANI • Basic Statistics with Business Applications, *Second Edition*

COCHRAN • Sampling Techniques, *Third Edition*

COCHRAN and COX • Experimental Designs, *Second Edition*

CONOVER • Practical Nonparametric Statistics, *Second Edition*

CORNELL • Experiments with Mixtures: Designs, Models and The Analysis of Mixture Data

COX • Planning of Experiments

DANIEL • Biostatistics: A Foundation for Analysis in the Health Sciences, *Second Edition*

DANIEL • Applications of Statistics to Industrial Experimentation

DANIEL and WOOD • Fitting Equations to Data: Computer Analysis of Multifactor Data, *Second Edition*

DAVID • Order Statistics, *Second Edition*

DEMING • Sample Design in Business Research

DODGE and ROMIG • Sampling Inspection Tables, *Second Edition*

DRAPER and SMITH • Applied Regression Analysis, *Second Edition*

DUNN • Basic Statistics: A Primer for the Biomedical Sciences, *Second Edition*

DUNN and CLARK • Applied Statistics: Analysis of Variance and Regression

ELANDT-JOHNSON • Probability Models and Statistical Methods in Genetics

continued on back

Introduction to
Probability Theory
and Statistical Inference

Introduction to Probability Theory And Statistical Inference

THIRD EDITION

Harold J. Larson

Naval Postgraduate School
Monterey, California

1807 1982

175 YEARS OF PUBLISHING

JOHN WILEY & SONS
New York • Chichester • Brisbane • Toronto • Singapore

Library of Congress Cataloging in Publication Data

Larson, Harold J., 1934-
    Introduction to probability theory and statistical inference.

    (Wiley series in probability and mathematical statistics. Probability and statistics section
ISBN 0271-6232).
    Bibliography: p.
    Includes index.
    1. Probabilities. 2. Mathematical statistics.
I. Title. II. Series.

QA273.L352	1982	519.2	81-16246
ISBN 0-471-05909-9			AACR2

Printed in the United States of America

10 9 8 7 6 5 4

For Marie

Preface

This text provides an introduction to probability theory and to many methods used in problems of statistical inference. Some stress is laid on the concept of a probabilistic model for the mechanism generating a set of observed data, leading to the natural application of probability theory to answer questions of interest. As with the previous editions, a one-year course in calculus should allow the student easy access to the material presented. I believe that this method of presentation is ideal for students in the physical and social sciences, as well as in engineering; the examples and exercises are not restricted to any discipline.

This edition contains new topics, including the bivariate normal probability law and the F distribution, and discussions of descriptive statistics (Chapter 6), Cramér-Rao bounds for variances of estimators (Chapter 7), two-sample inference procedures (Chapters 7 and 8), the analysis of variance (Chapter 9), and nonparametric procedures (Chapter 10). Most of the material from Chapter 3 on has been totally rewritten and includes new examples and exercises.

The first five chapters are again devoted to probability theory, the last chapters devoted to problems of statistical inference. The probability topics, and order of coverage, are much the same as in the earlier editions. Chapter 1 again discusses useful concepts from set theory, whereas Chapters 2 and 3 introduce probability, random variables (both discrete and continuous), and some descriptions of the probability law for a random variable. Chapter 4 presents physical arguments leading to the most commonly encountered (one-dimensional) probability laws, including the negative binomial, gamma, and beta distributions. Chapter 5, which discusses jointly distributed random variables, has been tightened up in some senses and is more directly aimed toward results that are important (and frequently used) in problems of statistical inference. This chapter ends with a discussion of some of the probability laws associated with random samples of a normal random variable (χ^2, T, and F distributions).

Chapter 6 is completely new, starting with a section on descriptive statistics. Among other things, this provides a vehicle for discussing subscripted variables and manipulations of such variables, subjects that are (incorrectly) frequently assumed transparent to students. The rest of Chapter 6 describes the concept of a random sample and its use in making inferences about an underlying probability law. Chapter 7 discusses point and interval estimation (including two-sample procedures), and Chapter 8 introduces the Neyman–Pearson theory of tests (including two-sample situations); some discussion is also devoted to the test of significance approach because, in practice, this is a commonly adopted stance. Chapter 9 considers least squares estimation and inference and the analysis of variance for the one- and two-way models. Chapter 10 presents an introduction to nonparametric procedures, and Chapter 11 discusses Bayesian approaches; these last three chapters are independent of one another and can be mixed or matched as desired.

With the current widespread availability of hand-held calculators (and more powerful equipment), I feel that binomial and Poisson probability calculations are reasonably within the capability of today's students; tables of these two probability laws have been dropped for this edition. The appendix provides a table of the standard normal distribution function, as well as selected quantiles of certain χ^2, T, and F distributions. Answers to all exercises (except those of a "show-that" or "prove-that" nature) are also presented in the appendix; a solutions manual for all exercises is available from Wiley.

The material in this edition is ample for a one-year course in probability theory and statistical inference. Chapters 1 to 5 give a useful one-semester introduction to probability theory (aimed toward statistical inference, not stochastic processes); for a one-quarter introduction to probability, Sections 2.5, 3.4, 3.5, the negative binomial, gamma, and beta distributions (Chapter 4), and Section 5.7 can be omitted in part or in whole. For a two-quarter course on probability and inference, the probability material previously listed (except Section 5.7) and the two-sample procedures in Chapters 7 and 8 could be dropped, and the individual instructor can then choose as he or she wishes (or as time dictates) from Chapters 9, 10, and 11.

I am grateful to all those who used the earlier two editions of this book, especially those who had comments to share and look forward to receiving comments on this third edition. None of the material treated is original with the author; hopefully, the order and method of treatment will help to minimize difficulties in its use in practical problems. My thanks to Bob Lande for typing the manuscript.

Harold J. Larson

Contents

Introduction to
Probability Theory
and Statistical Inference

1

Set Theory

In the study of probability theory and statistics an exact medium of communication is extremely important; if the meaning of the question that is asked is confused by semantics, the solution is all the more difficult, if not impossible, to find. The usual exact language employed to state and solve probability problems is that of set theory. The amount of set theory that is required for relative ease and comfort in probability manipulations is easily acquired. We will look briefly at some of the simpler definitions, operations, and concepts of set theory, not because these ideas are necessarily a part of probability theory but because the time needed to master them is more than compensated for by later simplifications in the study of probability.

1.1   Set Notation, Equality, and Subsets

A *set* is a collection of objects. The objects themselves can be anything from numbers to battleships. An object that belongs to a particular set is called an *element* of that set. We will commonly use uppercase letters from the beginning of the alphabet to denote sets)A, B, C, etc.) and lowercase letters from the end of the alphabet to denote elements of sets (x, y, z, etc.).

To specify that certain objects belong to a given set, we will use braces { } (commonly called set builders) and either the *roster* (complete listing of

1

all elements) or the *rule* method. For example, if we want to write that the set A consists of the letters a, b, c and that the set B consists of the first 10 integers, we may write

$$A = \{a, b, c\} \;(\textit{roster} \text{ method of specification})$$
$$B = \{x : x = 1, 2, 3, \ldots, 10\} \;(\textit{rule} \text{ method of specification}).$$

These two sets can easily be read as "A is the set of elements a, b, c" and "B is the set of elements x such that $x = 1$ or $x = 2$ or $x = 3$ and so on up to $x = 10$." We will use the symbol $\in$ as shorthand for "belongs to" and thus can write for the two sets defined here that $a \in A$, $7 \in B$. Just as a line drawn through an equals sign is taken as negation of the equality, we will use $\notin$ to mean "does not belong to"; thus $a \notin B$, $9 \notin A$, $f \notin A$, $102 \notin B$, and so on, where A and B are the sets previously defined.

DEFINITION 1.1.1. Two sets A and B are *equal* if and only if every element that belongs to A also belongs to B and every element that belongs to B also belongs to A. ∎

Then two sets are equal only if both contain exactly the same elements. Notice in particular that the *order* of listing of elements of a set is of no importance and that the sets

$$A = \{1, 2, a, 3\}, \qquad B = \{a, 1, 2, 3\}$$

are equal. Also, the number of times that an element is listed in the roster specifying the set is of no concern; the two sets

$$C = \{1, 2, 3\}, \qquad D = \{1, 2, 2, 1, 3, 1\}$$

are equal. We will have no use for the redundancy exhibited in the roster of D and thus will always assume that if the roster of a particular set contains n elements, the elements are distinguishable and that the same element does not occur more than once.

DEFINITION 1.1.2. A is a *subset* of B (written $A \subset B$) if and only if every element that belongs to A also belongs to B. ∎

(Actually, two different types of subsets can be distinguished, proper and improper subsets. A is a *proper* subset of B if it is a subset of B and there exists at least one element of B that does not belong to A; otherwise A is an *improper* subset of B. We will have no need to make this distinction.) In a sense, A is a subset of B if A is contained in B. For example, $\{a\}$ is a subset of $\{a, z\}$ and both are subsets of $\{a, b, z\}$. The set of all people residing in California is a subset of the set of all people residing in the United States; the set of all pine trees is a subset of the set of all trees.

A word regarding the difference between something belonging to a set A and being a subset of A may be in order. For example, define

$$A = \{1, 2, 3\}, \qquad B = \{1, 3\}, \qquad C = \{1\}.$$

Then it is correct to say $B \subset A$ and $C \subset A$ since every element that belongs to B also belongs to A, as does every element that belongs to C. However, it is not correct to say that $B \in A$ or $C \in A$ since B and C are not specified in the roster of elements that belong to A. Similarly, it is correct to say that $1 \in A$ and $1 \in C$, and not that $1 \subset A$ and $1 \subset C$ since 1 is not a set. Thus we can say that the set of all married people is a subset of the set of all people, but we do not say that the set of all married people belongs to the set of all people. A particular married person belongs to the set of all married people (as well as to the set of all people), but this person is not a subset of either (since a person is not a set).

An alternative definition of set equality can be given by saying that $A = B$ if and only if $A \subset B$ and $B \subset A$. This, in fact, provides a very useful way of demonstrating that two sets are equal, as we will see. Most simple set equalities can be rigorously and easily proved in this way.

DEFINITION 1.1.3. The *null set* $\varnothing$ is the set with no elements; that is, $\varnothing = \{\ \}$. $\varnothing$ is also called the *empty* set. ∎

The null set $\varnothing$ serves a purpose in set manipulations that is rather similar to the role played by 0 (zero) in manipulating numbers. It is a subset of any set A ($\varnothing \subset A$) since every element that belongs to $\varnothing$ (there are none) also belongs to A; that is, "every x belonging to $\varnothing$ also belongs to A" is a true statement for any set A (vacuously) since there is no $x \in \varnothing$ to make the statement false. Picture a set, which has some elements belonging to it, as being a paper bag containing objects; the paper bag represents the set builders $\{\ \}$ and the objects are the elements of the set. Then $\varnothing$ is simply an empty paper bag. What is $\{\varnothing\}$ and is it different from $\varnothing$?

Suppose A is a subset of U ($A \subset U$); then the set

$$\bar{A} = \{x : x \notin A \text{ and } x \in U\}$$

contains all the elements that do not belong to A (but do belong to U). This set $\bar{A}$ is called the *complement* of A with respect to U. If, in a particular discussion, all the sets used (A, B, C, etc.) are subsets of the same set U, it is usual to call U the *universal set* for the discussion. It is also common in such cases to then refer to complements ($\bar{A}, \bar{B}, \bar{C}$, etc.) without specific mention of the fact that the complements are all taken with respect to the same universal set U. No confusion will arise with respect to the set referred to when complements are used.

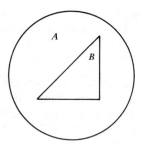

Figure 1.1.1

Venn (or *Euler*) diagrams are frequently useful in picturing sets and relationships between sets. These diagrams use geometric shapes to represent sets (the actual shape used has no real bearing). For example, in Figure 1.1.1, two sets are represented and labeled A and B; A is the set of all the points on the circle (and its interior) and B is the set of points on the triangle (and its interior). As pictured, note that $B \subset A$ since every point that belongs to B also belongs to A. If, however, the relationship is disturbed, as in Figure 1.1.2, B is no longer a subset of A, since a portion of B lies outside A (there are elements that belong to B that do not belong to A).

From the definitions it is easily seen that every set is equal to itself and that every set is a subset of itself. That is, we can say

$$A = A, \qquad \text{for all } A,$$
$$A \subset A, \qquad \text{for all } A.$$

In textbooks on algebra, these two statements are summarized by saying that both the equality and the subset relations are *reflexive*.

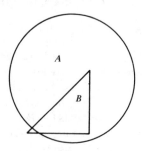

Figure 1.1.2

EXERCISE 1.1

1. Show that the subset relationship is *transitive*; that is, $A \subset B$ and $B \subset C$ imply $A \subset C$.
2. Show that the subset relationship is not *symmetric*; that is, $A \subset B$ and $B \subset A$ are logically different statements and one is not implied by the other. Give an example illustrating this nonsymmetry.
3. Give an example showing that if $E \subset F$ and $D \subset F$, we do not necessarily have $D \subset E$ or $E \subset D$.
4. Show that set equality is transitive. (Transitivity is defined in Exercise 1.1.1.)
5. If $A = \{x : x = 0, 1\}$, $B = \{-1, 0, 1\}$, $C = \{0, 1, -1\}$, $D = \{-1, 1\}$ mark each of the following either true or false.

    (a) $A = B$ (h) $C \subset B$
    (b) $A = C$ (i) $0 \in B$
    (c) $B = C$ (j) $0 \subset C$
    (d) $A \subset B$ (k) $A = D$
    (e) $A \subset C$ (l) $D \subset A$
    (f) $B \subset C$ (m) $B \subset D$
    (g) $C \subset A$ (n) $D \in C$.

6. Define $E = \{x : 0 \le x \le 1\}$, $F = \{y : 0 \le y \le 1\}$, $G = \{x : 0 < x < 1\}$, $H = \{1\}$. Mark each of the following either true or false.

    (a) $E = F$ (e) $H \subset G$
    (b) $F = G$ (f) $H \subset E$
    (c) $F \subset G$ (g) $H \in F$
    (d) $G \subset F$

7. Is the set of all students a subset of the set of all people? Is it a subset of the set of all people under 36 years of age? Under 50 years of age?
8. Consider the set of all people in your family.

    (a) Is this set well defined? What sort of questions must be considered to make it well defined?

    (b) Are you an element of this set or a subset of this set?
9. Is the set of all automobiles produced in Detroit in a calendar year well defined? What sort of questions must be considered to make it well defined?
10. Is the set of all living U.S. citizens, at a given instant of time, equal to the set of all people residing in the United States at the same instant of time?
11. Define

$$A = \{1, 2, 3, \dots, 8\}, \qquad B = \{2, 3, 4\}, \qquad C = \{1, 3, 5, 7, 9\}$$

and let the universal set be $U = \{x: x = 1, 2, ..., 10\}$. Define the sets $\bar{A}, \bar{B}, \bar{C}$. Is $\bar{B} \subset \bar{A}$?

12. Draw a picture like Figure 1.1.1, except put a big rectangle U enclosing the circle A, which itself encloses the triangle B. Now shade the set $\bar{A}$ and crosshatch the set $\bar{B}$. Is $\bar{A} \subset \bar{B}$?

13. Provide an analytic argument that if $B \subset A$, then $\bar{A} \subset \bar{B}$. (*Hint*: If $x \in \bar{A}$, is it possible that $x \in B$ when $B \subset A$?)

14. Show that $\bar{\bar{A}}$ (the complement of $\bar{A}$) must be A.

15. Granted a well-defined universal set U, what is $\bar{U}$? What is $\varnothing$?

1.2 Set Operations

You are quite used to performing various operations with numbers, such as addition, subtraction, multiplication, and division. Each of these operations takes a pair of numbers (say, 3 and 7) and the result of the operation (let us use addition as an example) is another number (10 in this case). In this section we will study 3 set operations: the union, intersection, and Cartesian product of two sets. Given two sets, A and B, the result of each of these set operations will be another set (generally a different set, depending on which operation was performed). We start with the union of two sets.

DEFINITION 1.2.1.   The *union* of A and B (written $A \cup B$) is the set that consists of all the elements that belong to A *or* to B or to both; that is, $A \cup B = \{x: x \in A \text{ or } x \in B\}$.   ■

For example, if $A = \{1, 2\}$, $B = \{1, 3\}$, $C = \{0\}$, then $A \cup B = \{1, 2, 3\}$, $A \cup C = \{0, 1, 2\}$, and $B \cup C = \{1, 3, 0\}$. Notice from the definition of the union that $A \cup B$ and $B \cup A$ are identical sets (the operation is *commutative*), because the collection of objects that belong to A or to B is the same; it

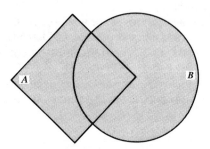

Figure 1.2.1. $A \cup B$ is shaded.

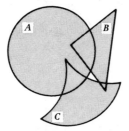

Figure 1.2.2. $A \cup B \cup C$ is shaded.

does not matter whether we first list those that belong to A or those that belong to B.

Figure 1.2.1 shows two sets, A and B, and their union $A \cup B$. Notice that we necessarily must have $A \subset A \cup B$ and $B \subset A \cup B$. Draw a Venn diagram with $B \subset A$ to prove to yourself that $A \cup B = A$ for this case. Given three sets A, B, and C we can form the set $A \cup B$ and then form the union of this set with C, which we would write $(A \cup B) \cup C$. Or with the same three sets we could first form the union $B \cup C$ and then form the union of A with this set, written $A \cup (B \cup C)$. A glance at Figure 1.2.2 should be convincing that these two sets must be equal, thus there is no need to use parentheses to indicate which union is formed first since

$$(A \cup B) \cup C = A \cup (B \cup C) = A \cup B \cup C.$$

The union operation is *associative*.

Next we define the intersection of two sets.

DEFINITION 1.2.2. The *intersection* of A and B (written $A \cap B$) is the set that consists of all elements that belong *both* to A and to B; that is, $A \cap B = \{x : x \in A \text{ and } x \in B\}$. ∎

For example, if $A = \{0, 1\}$, $B = \{1, 3\}$, $C = \{0, 1, 3\}$, then $A \cap B = \{1\}$, $A \cap C = \{0, 1\}$, $B \cap C = \{1, 3\}$. If A is the set of married people and B is the set of people living in California, then $A \cap B$ is the set of married people living in California.

The intersection operation is also commutative since $A \cap B$ and $B \cap A$ must be the same. Figure 1.2.3 pictures two sets A and B and their intersection $A \cap B$. Note that $A \cap B \subset A$ and $A \cap B \subset B$. Draw a Venn diagram with $B \subset A$ to show that $A \cap B = B$ in this case. It is possible, of course, that there are no elements that belong to both A and B, in which case $A \cap B = \varnothing$; we will say that two such sets are *disjoint* since they do not

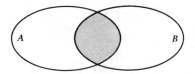

Figure 1.2.3. $A \cap B$ is shaded.

overlap. Figure 1.2.4 shows A and B disjoint; this is the only way to picture $\varnothing$ in a Venn diagram (by the intersection when they are disjoint). Note that any set A and its complement $\bar{A}$ will always be disjoint.

Given three sets A, B, and C the intersection of all three could be written $(A \cap B) \cap C$ or $A \cap (B \cap C)$. By looking at Figure 1.2.5, we can easily see that both of these represent the same set. The intersection operation is associative and, in a string of intersections, which intersection is formed first is not significant. If any two of A, B, and C are disjoint, then the intersection $A \cap B \cap C$ is necessarily null, as is easily seen by picturing a third set C with A and B in Figure 1.2.4. This condition is not necessary though; $A \cap B \cap C$ can be empty even when $A \cap B \neq \varnothing$, $A \cap C \neq \varnothing$, and $B \cap C \neq \varnothing$, as a glance back at Figure 1.2.2 will show.

Although both the union and intersection operations are associative and one arrives at the same set regardless of the order in which the operations are performed, the same is *not* necessarily true if both operations appear in the same expression. For example, $(A \cup B) \cap C$ and $A \cup (B \cap C)$ are not equal, as illustrated in Figure 1.2.6.

As discussed earlier, a set of elements is unchanged if we merely rearrange the ordering of its elements. To have at our disposal collections of n elements where order is of importance, we define an n-tuple as follows.

DEFINITION 1.2.3. An n-tuple is an *ordered* array of n components written $(x_1, x_2, \ldots, x_n)$. ∎

For example, $(1, 2)$, $(2, 1)$, $(0, 100)$, (a, b) are all 2-tuples; (a, b, c), (b, a, c), $(1, 1, 1)$, $(2, 1, 2)$ are all 3-tuples. Two n-tuples are different, even if they contain the same components, if they are written in a different order and,

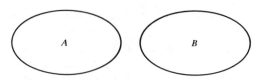

Figure 1.2.4. $\varnothing$ is shaded.

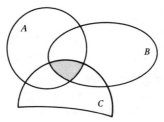

Figure 1.2.5. $A \cap B \cap C$ is shaded.

of course, an n-tuple and an $(n + 1)$-tuple cannot be equal because they have different numbers of components. Thus

$$(1, 2) \neq (2, 1)$$
$$(1, 3, 1) \neq (1, 1, 3)$$
$$(1, 1) \neq (1, 1, 1)$$

For those who may be acquainted with the term, an n-tuple whose components are real numbers is the same as an n-dimensional vector.

We will have use for *sets* of n-tuples as we proceed. For example, if we want to discuss the set of married couples living in Texas, then we want each couple to be an element of the set; for clarity we list the husband's name first when specifying the couple. Each element of the set then is a couple or a 2-tuple. Note that a married individual living in Texas does not belong to the set; he is a component of one of the 2-tuples that belongs to the set. As a second example, the set of points lying in the first quadrant of the usual Cartesian plane is a set of 2-tuples. Each point is represented by a 2-tuple (x, y) where x is the horizontal coordinate and y the vertical coordinate.

Remember that order counts in forming n-tuples but not in listing

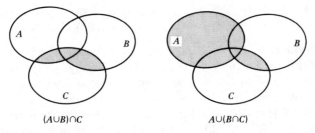

$(A \cup B) \cap C$ $A \cup (B \cap C)$

Figure 1.2.6

the roster of a set. Thus

$$\{(1, 2), (1, 3)\} \neq \{(2, 1), (1, 3)\}$$
$$\{(4, 2), (2, 5)\} = \{(2, 5), (4, 2)\}.$$

One further set operation will be useful—the Cartesian product. This operation again results in a set, but this set has elements whose nature differs from that of the elements of A and B.

DEFINITION 1.2.4.   The *Cartesian product* $A \times B$ of A and B is the set of all possible 2-tuples (x_1, x_2) where $x_1 \in A$, $x_2 \in B$. That is,

$$A \times B = \{(x_1, x_2): x_1 \in A, x_2 \in B\} \qquad \blacksquare$$

Let us consider a number of examples of Cartesian products. Suppose that $A = \{0, 1, 2\}$, $B = \{3, 5\}$, $C = \{0\}$. Then

$A \times B = \{(0, 3), (0, 5), (1, 3), (1, 5), (2, 3), (2, 5)\}$
$A \times C = \{(0, 0), (1, 0), (2, 0)\}$
$B \times C = \{(3, 0), (5, 0)\}$
$B \times A = \{(3, 0), (3, 1), (3, 2), (5, 0), (5, 1), (5, 2)\}$
$C \times A = \{(0, 0), (0, 1), (0, 2)\}$
$C \times B = \{(0, 3), (0, 5)\}$
$A \times A = \{(0, 0), (0, 1), (0, 2), (1, 0), (1, 1), (1, 2), (2, 0), (2, 1), (2, 2)\}$
$B \times B = \{(3, 3), (3, 5), (5, 3), (5, 5)\}$
$C \times C = \{(0, 0)\}.$

If D is the set of all positive real numbers, then $D \times D$ is the set of points in the first quadrant of the plane.

Notice that $A \times B$ and $B \times A$ are not equal in general. The operation is *not commutative*. (Compare $A \times B$ and $B \times A$ above.) We can form the Cartesian product of any number of sets; $A \times B \times C$ is a set of 3-tuples, $A_1 \times A_2 \times \cdots \times A_n$ is a set of n-tuples. Taking A, B, and C as previously defined,

$A \times B \times C = \{(0, 3, 0), (0, 5, 0), (1, 3, 0), (1, 5, 0), (2, 3, 0), (2, 5, 0)\}$
$A \times C \times B = \{(0, 0, 3), (0, 0, 5), (1, 0, 3), (1, 0, 5), (2, 0, 3), (2, 0, 5)\}.$

Strictly speaking, we should denote the first element listed in $A \times B \times C$ as $[(0, 3), 0]$ if we are thinking of $(A \times B) \times C$, or as $[0, (3, 0)]$ if we are thinking of $A \times (B \times C)$. We have no reason to distinguish between $[(0, 3), 0]$, $[0, (3, 0)]$, and $(0, 3, 0)$, so we simply define $A \times B \times C$, as noted, to be the collection of 3-tuples that results when the innermost pair of parentheses is deleted. Thus Cartesian products are *associative*, with this convention.

EXERCISE 1.2

1. Show that $A \cap B \subset A \subset A \cup B$ and that $A \cap B \subset B \subset A \cup B$.
2. If $A = \{1, 0\}$, $B = \{x: 0 < x < 1\}$, $C = \{\frac{1}{2}\}$, compute $A \cup B$, $A \cup C$, $B \cup C$, $A \cap B$, $A \cap C$, $B \cap C$.
3. Show that $B \cup B = B \cap B = B$ for any B.
4. What is the implication of the equation $E \cap F = F$?
5. What is the implication of the equation $E \cup F = E$?
6. Define:

$$A = \{x: x = 1, 2, 3, \ldots, 10\}$$
$$B = \{x: 1 \leq x \leq 10\}$$
$$C = \{x: x = 0, 1, 2, 3, 4, 5, 6\}$$
$$D = \{0, 10, 20, 30\}$$

and compute:

(a) $A \cup B$ (g) $B \cup C$ (m) $A \cup B \cup C$

(b) $A \cap B$ (h) $B \cap C$ (n) $A \cap (B \cup C)$

(c) $A \cup C$ (i) $B \cup D$ (o) $A \cup (B \cap C)$

(d) $A \cap C$ (j) $B \cap D$ (p) $A \cap B \cap C$

(e) $A \cup D$ (k) $C \cup D$ (q) $C \cup (A \cap D)$

(f) $A \cap D$ (l) $C \cap D$ (r) $(A \cup B) \cap (C \cup B)$.

7. (a) Construct a Venn diagram to show that

$$A \cap (B \cup C) = (A \cap B) \cup (A \cap C)$$
$$A \cup (B \cap C) = (A \cup B) \cap (A \cup C)$$

   (b) Show these two equations are true analytically by demonstrating that $x \in A \cap (B \cup C)$ implies $x \in (A \cap B) \cup (A \cap C)$ and vice versa. These two equations are called the *distributive* laws linking unions and intersections.

8. Define A to be the set of safecrackers in the United States and define B to be the set of prison inmates in the United States. What, in words, is $A \cap B$?

9. Let A be the set of female sociologists in the United States and let B be the set of male sociologists in the United States. What, in words, is $A \cup B$?

10. Mark each of the following true, if always true, or false.

   (a) $\bar{A} \cup B = A \cap \bar{B}$
   (b) $B \cap C \subset (A \cup B) \cap (A \cup C)$
   (c) $A \subset (A \cap C) \cup (A \cup B)$

11. Show that $\overline{A \cup B} = \bar{A} \cap \bar{B}$. (This is one of De Morgan's laws.)

12. Show that $\overline{A \cap B} = \bar{A} \cup \bar{B}$. (This is also one of De Morgan's laws.) *Hint.* In Exercise 1.2.11, replace A and B by their complements and then take the complement of both sides.

13. Draw Venn diagrams illustrating the sets $A \cap \bar{B}$, $\bar{A} \cap \bar{B}$, $A \cup B \cup C$, $A \cap B \cap C$, $(A \cup B) \cap C$, $(A \cap C) \cup (B \cap C)$.

14. Define $A = \{1, 2\}$, $B = \{2, 1\}$, $C = \{10, 12\}$, and form the Cartesian products $A \times B$, $A \times C$, $B \times C$, $B \times A$, $C \times A$, $C \times B$, $A \times A$, $B \times B$, $C \times C$, $A \times B \times C$, $C \times B \times A$, $C \times A \times B$.

15. What are the conditions under which $A \times B = B \times A$?

16. If A is the set of married men living in Texas and B is the set of married women living in Texas, is $A \times B$ the set of married couples living in Texas?

17. If A is the set of people with U.S. citizenship and B is the set of people with Canadian citizenship, does $A \cap B = \varnothing$?

18. Show that $(A \cup B) \times C = (A \times C) \cup (B \times C)$.

19. Show that $(A \times B) \cap (A \times C) = A \times (B \cap C)$.

1.3   Functions

The idea of a function is quite basic to many branches of mathematics, not least of all to probability theory. As we will see, there are several different types of functions that are useful. You are probably already acquainted with *real functions* of *real variables*, such as $y = 3x + 5$, say, for x between 0 and 10. As we proceed, we will use real functions of real variables to express either relations between real variables or the association of values of one variable (y) with values of another (x). We will also have need to express rules that associate real numbers with sets (probability functions) and rules that associate real numbers with elements of a set (random variables). The following definition of a function is general enough to cover all these cases.

DEFINITION 1.3.1.   A *function* is a set of pairs (2-tuples), such that no two pairs have equal first components. The set that has as elements the *first* components of all the 2-tuples is called the *domain of definition* of the function. The set that has as elements the *second* components of all the 2-tuples is called the *range* of the function. ■

This method of defining a function may not be familiar to you. For example, the function $y = 3x + 5$, for $0 \le x \le 10$, would be expressed as the set of pairs $\{(x, 3x + 5): 0 \le x \le 10\}$. The independent variable (x) is the first element of the pair and the second element gives the associated

y value ($3x + 5$ in this case). No two pairs can have equal first components to ensure that each x value has a single unique y value associated with it. The same y value could be associated with more than one x value. The domain of definition for this function is the set of values for x, $D = \{x: 0 \le x \le 10\}$, and the range for this function is the set of values for y, $R = \{y: 5 \le y \le 35\}$.

We will commonly express a function by using $f(x)$, say, $f(x) = 3x + 5$, $0 \le x \le 10$, for the preceding case. When explicitly using the set of pairs notation for a function, we will also use f to represent the set of pairs, say, $f = \{(x, 3x + 5): 0 \le x \le 10\}$ for the function just discussed. The first usage will occur far more frequently; no confusion will result from this dual usage of the same letter. Example 1.3.1 discusses several real functions of a real variable.

EXAMPLE 1.3.1. For simplicity of exposition, let us consider a real variable x that is constrained to lie between -5 and 5; thus we will consider several functions of a real variable x where $-5 < x < 5$. First consider the *constant* function:

$$y = f_1(x) = 2, \qquad -5 < x < 5.$$

From Definition 1.3.1 we would write this function as

$$f_1 = \{(x, 2): -5 < x < 5\}$$

to indicate that the same constant value, 2, is associated with each x. As a second example, consider the *identity* function

$$y = f_2(x) = x, \qquad -5 < x < 5,$$

which associates each value of x with itself. This would be written

$$f_2 = \{(x, x): -5 < x < 5\},$$

to indicate that f_2 associates each value of x with itself (the second, associated value for each pair is equal to the first value). The *square* function

$$y = f_3(x) = x^2, \qquad -5 < x < 5,$$

would be written

$$f_3 = \{(x, x^2): -5 < x < 5\}.$$

The *absolute value* function

$$y = f_4(x) = |x|, \qquad -5 < x < 5$$

would be written

$$f_4 = \{(x, |x|): -5 < x < 5\}.$$

Figure 1.3.1 presents the graphs of f_1, f_2, f_3, and f_4. Since in every case we restricted x to lie between -5 and 5, the domain of definition for each of the 4 functions is

$$D_{f_1} = D_{f_2} = D_{f_3} = D_{f_4} = \{x: -5 < x < 5\}.$$

The ranges, however, are not the same. It is easily verified that the ranges are

$$R_{f_1} = \{2\}$$
$$R_{f_2} = \{y: -5 < y < 5\}$$
$$R_{f_3} = \{y: 0 \le y < 25\}$$
$$R_{f_4} = \{y: 0 \le y < 5\}. \qquad \blacksquare$$

Using Definition 1.3.1, we see that the definition of a function is appropriate for more than just real functions of a real variable. There is no requirement that the first entry in each pair should necessarily be a real number (nor that the second should be either, for that matter). In succeeding chapters we will study *probability functions* and *random variables*; for each of these the first component (the quantity that will have a real number associated with it) in each pair is not a real number, although the second component will be. A probability function is a rule or function that associates a real number with every *subset* of a given set (and satisfies certain rules as well; see Section 2.2). Thus, for a probability function, the first component in each pair will be a subset of a given set; the second component will be a real number (called a probability). A random variable is a rule that associates a real number with every *element* of a given set. Thus the first component of each pair will be an element of the given set and the second component will be a real number (the value of the random variable).

The following examples illustrate these two types of functions:

EXAMPLE 1.3.2.   Define

$$U = \{1, 2, 3, 4, 5\}$$

and let $n(A) = $ number of elements in A, for $A \subset U$. Then n is a function that associates a number with each subset $A \subset U$; the number associated with each A is the number of elements belonging to A. Thus as a set of pairs we could write

$$n = \{(A, n(A)): A \subset U, n(A) = \text{number of elements in } A\},$$

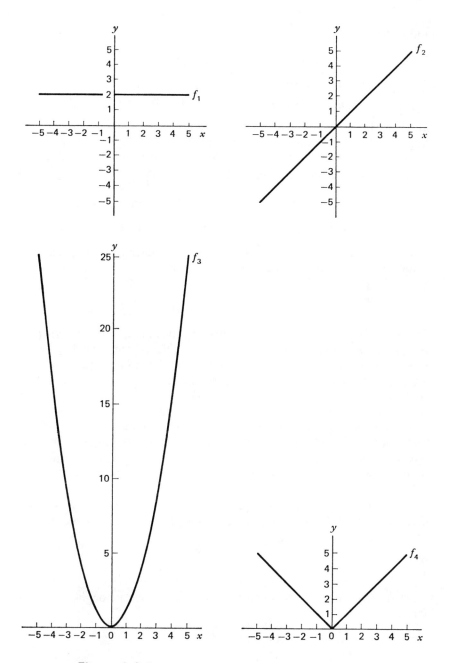

Figure 1.3.1

and we have

$$n(\{1\}) = n(\{2\}) = n(\{4\}) = 1$$
$$n(\{1, 2\}) = n(\{1, 5\}) = n(\{2, 3\}) = 2$$
$$n(\{1, 2, 3\}) = n(\{1, 3, 5\}) = n(\{2, 3, 4\}) = 3$$
$$n(\{1, 2, 3, 4, 5\}) = 5, \qquad n(\varnothing) = 0$$

for example. n is called a set function defined on the class of subsets of U. It does not satisfy the axioms for a probability function, which is simply a special type of set function. Note that the domain of definition of n is the class of subsets of U and its range is $R_n = \{0, 1, 2, 3, 4, 5\}$. ∎

EXAMPLE 1.3.3.   Let $S = \{2, 4, 6, 8\}$ and define the function

$$I = \{(A, y): A \subset S, y = 1 \quad \text{if} \quad 4 \in A \quad \text{and} \quad y = 0 \quad \text{if} \quad 4 \notin A\}.$$

This function associates 1 with every subset of S which has 4 as an element, and associates 0 with all other subsets of S. Thus

$$I(\{2\}) = 0 = I(\{6\}) = I(\{8\})$$
$$I(\{4\}) = 1 = I(\{2, 4\}) = I(\{4, 6, 8\})$$

and so on. This is a special case of what is called an *indicator function* (for the particular element 4). It indicates (or differentiates between) which subsets have 4 as an element versus which subsets do not. As we will see, it also satisfies the rules for a probability function. The range for I is $R_I = \{0, 1\}$ and its domain of definition is the class of all subsets of S. ∎

The functions discussed in Examples 1.3.2 and 1.3.3 are called *real-valued set functions* because they associate real numbers with sets. Probability functions are particular kinds of real-valued set functions. We will also use real-valued *element* functions. The following two examples are of this type. Random variables are real-valued element functions.

EXAMPLE 1.3.4.   Let $A = \{1, 2, 3, 4\}$ and define $U = A \times A$. Then each element of U is a 2-tuple (u_1, u_2). Now define the function X by

$$X(u_1, u_2) = u_1 + u_2$$

or

$$X = \{((u_1, u_2), u_1 + u_2): (u_1, u_2) \in U\}$$

Then X associates a real number with each *element* belonging to U. In fact,

$$X((1, 1)) = 2$$
$$X((2, 2)) = 4$$
$$X((1, 4)) = 5$$
$$X((3, 4)) = 7, \text{and so on.}$$

The domain of definition for X is the set U and the range of X can be verified to be

$$R_X = \{2, 3, 4, 5, 6, 7, 8\}.$$ ∎

EXAMPLE 1.3.5.   Let A and U be as defined in Example 1.3.4, and now define

$$Y = \{((u_1, u_2), u_1/u_2): (u_1, u_2) \in U\}.$$

Then

$$Y((1, 1)) = Y((2, 2)) = 1$$
$$Y((1, 3)) = \tfrac{1}{3}$$
$$Y((3, 1)) = 3, \text{and so on.}$$

Again, the domain of definition for Y is $D_Y = U$ and its range is

$$R_Y = \{\tfrac{1}{4}, \tfrac{1}{3}, \tfrac{1}{2}, \tfrac{2}{3}, \tfrac{3}{4}, 1, \tfrac{4}{3}, \tfrac{3}{2}, 2, 3, 4\}.$$ ∎

EXERCISE 1.3

1. Let A be the set of individuals in your family and define the function $f(\omega) = $ age of ω for $\omega \in A$. Specify the range of f.

2. Let

$$B = \{1, 2, 3, 4, 5, 6\}$$
$$C = B \times B$$

and define the function

$$g((x_1, x_2)) = x_1 + x_2 \qquad \text{for} \qquad (x_1, x_2) \in C.$$

What is the range of g?

3. Let $A = \{0, 1\}$ and define $B = A \times A \times A$. For each element $(x_1, x_2, x_3) \in B$, let

$$h((x_1, x_2, x_3)) = x_1 + x_2 + x_3.$$

What is the range of h? How many elements have the value 0 assigned to them? How many have the value 1 assigned to them?

4. For the set C defined in Exercise 1.3.2, let $d((x_1, x_2)) = x_1 - x_2$. What is the range of d?

5. In Exercise 1.3.4 let

$$a((x_1, x_2)) = |d((x_1, x_2))|.$$

What is the range of a?

6. Refer to the set U and the set function $n(A)$, defined in Example 1.3.2. Let $P(A) = n(A)/5$ for all $A \subset U$. What is the range of P? (This is also a probability function.)

7. Let $U = \{1, 2, 3, 4\}$ and define

$$q(\varnothing) = 2$$
$$q(A) = \text{reciprocal of the number of elements in } A$$

for every nonempty subset $A \subset U$. What is the range of q?

8. Let U and n be as defined in Example 1.3.2. Show that

$$n(A) \leq n(B) \qquad\qquad \text{if } A \subset B$$
$$n(A \cup B) = n(A) + n(B) \qquad \text{if } A \cap B = \varnothing.$$

(The first of these says that n is *monotonic*, the second that n is *finitely additive*.)

9. Show that the set function I defined in Example 1.3.3 is monotonic and finitely additive. (See Exercise 1.3.8.)

10. Show that the set function q defined in Exercise 1.3.7 is not monotonic and is not finitely additive. (These terms are defined in Exercise 1.3.8.)

1.4   Summary

Union: $A \cup B = \{x : x \in A \text{ or } x \in B \text{ or both}\}$
Intersection: $A \cap B = \{x : x \in A \text{ and } x \in B\}$
Complement: $\bar{A} = \{x : x \notin A\}$
Subset: $A \subset B$ if and only if $x \in A$ implies $x \in B$.
n-tuple: Ordered array of n components.
Cartesian product: $A \times B = \{(x_1, x_2) : x_1 \in A, x_2 \in B\}$
Function: Set of 2-tuples such that no two pairs have equal first components.
Domain of function: Set of first components of all the 2-tuples of the function.
Range of function: Set of second components of all the 2-tuples of the function.

2

Probability

We are now ready to begin our study of probability theory itself. Some of the earliest known applications of probability theory occurred in the seventeenth century. A French nobleman of that time was interested in several games then played at Monte Carlo; he tried unsuccessfully to describe mathematically the relative proportion of the time that certain bets would be won. He was acquainted with two of the best mathematicians of the day, Pascal and Fermat, and mentioned his difficulties to them. This began a famous exchange of letters between the two mathematicians concerning the correct application of mathematics to the measurement of relative frequencies of occurrences in simple gambling games. Historians generally agree that this exchange of letters was the beginning of probability theory as we now know it.

For many years a simple *relative frequency* definition of probability was all that was both known and that many felt was necessary. This definition proceeds roughly as follows. Suppose that a chance experiment is to be performed (some operation whose outcome cannot be predicted in advance); thus there are several possible outcomes that can occur when the experiment is performed. If an event A occurs with m of these outcomes, then the probability of A occurring is the ratio m/n, where n is the total number of outcomes possible. Thus if the experiment consists of one roll of a fair die and A is the occurrence of an even number, $m = 3, n = 6$ and the probability of A is $\frac{3}{6}$.

There are many problems for which this definition is appropriate, but such a heuristic approach is not conducive to a mathematical treatment of the theory of probability. The mathematical advances in probability theory were relatively limited and difficult to establish on a firm basis until the Russian mathematician A. N. Kolmogorov gave a simple set of three *axioms* or rules that probabilities are assumed to obey. Since the establishment of this firm axiomatic basis, great strides have been made in the theory of probability and in the number of practical problems to which it is applied.

In this chapter we will see what these three axioms are and why we might reasonably adopt these rules for probabilities to follow. The axioms do not give any unique value that the probability of an event must equal; rather, they express internal rules that ensure consistency in our arbitrary assignment of probabilities. The relative frequency definition of probability, already mentioned, is only one way in which probabilities can be computed; it is discussed more fully in Section 2.3 and, as shown, does satisfy the axioms. This relative frequency definition has built within it certain assumptions about the outcomes of the experiment that are not always appropriate. Thus this arbitrary way of assigning probabilities is not always applicable.

It is useful to discuss briefly the notion of a probability model. An experiment is a physical operation that, in the real world, can result in one of many possible outcomes. For example, if we roll a pair of dice one time the two numbers we might observe can range anywhere from a pair of ones to a pair of sixes. Or, if a particular individual is going to run one hundred yards as fast as he can, the elapsed time from when he starts until when he finishes might be anywhere from 9 to 30 seconds. Or if a particular person lives in a given way in terms of rest and habits of various kinds, his total life span might lie anywhere from zero years to one hundred years. In each of these cases the particular outcome we might observe cannot be predicted in advance, but the total collection of outcomes can be assumed to be known (one of which will be observed when the experiment is completed). When building a probability model for an experiment, we are concerned with specifying: (1) what the total collection of outcomes could be and (2) the relative frequency of occurrence of these outcomes, based on an analysis of the experiment. We are in many senses idealizing the physical situation by restricting the set of possible outcomes; but to the extent that this idealization does not affect the relative frequency of events of interest, we can profitably use the results of our computations as descriptions of the actual physical experiment. The probability model then consists of the assumed collection of possible outcomes and the assigned relative frequencies or probabilities of these outcomes. The axioms are used to assure consistency in this assignment of probabilities.

In the latter half of this book we will study problems and methods of statistical inference. Probability theory is essentially deductive in nature; from the facts known about an experiment one wants to deduce the probabilities of occurrence of events of interest. As we will see, statistics is more inductive in nature. By observing what is true in a few cases, the statistician is interested in making generalizations or inferences about what may also be true in cases not observed. Statisticians may, for example, observe the outcome of an experiment one or more times and then ask whether these observations are consistent with one or another type of probability distribution. Or they may observe what has happened through time to a particular quantity, such as the price of a stock or the yearly production of a fruit tree, and then ask what will happen to this quantity in the future. Or, in a science such as genetics or physics, a "scientific law" may predict what should be observed when a certain experiment is performed; by observing the actual experiment statisticians then are interested in deciding whether or not the "law" appears to be correct. In each case an inference is to be made; probability theory proves very useful to statisticians in comparing different ways or methods of making the inference.

2.1 Sample Space; Events

As mentioned, modern probability theory has its roots in the study of games of chance. We will also rely on this type of mechanism for some illustrations, because simple game examples are easy both to understand and to analyze. We will also, however, try to indicate examples of applications that may be of wider practical interest.

We will use the word *experiment* to represent generically any sort of operation whose outcome cannot be predicted in advance with certainty. Thus flipping a coin and observing which face lands on top, planting a particular hybrid corn on a given plot of ground and observing its yield, firing a space craft past Saturn to observe its rings, all fall within our definition of an experiment. Set theory proves useful in defining the *sample space* for an experiment, which we now define.

DEFINITION 2.1.1. The sample space for an experiment is the set of all possible outcomes that might be observed. ∎

Sample spaces are not unique for the same experiment and may or may not be simple to describe. Let us examine some examples.

EXAMPLE 2.1.1. We roll a die one time. Then the experiment is the roll of the die. A sample space for this experiment could be

$$S = \{1, 2, 3, 4, 5, 6\}$$

where each of the integers 1 through 6 is meant to represent the face having that many spots being uppermost when the die stops rolling.   ■

EXAMPLE 2.1.2.   Suppose a small town has 500 registered voters. We select one of these at random (the experiment); by "at random" we mean each of the 500 voters has the same chance of being the one we will select. If the roster of names and addresses of the registered voters is available, this could clearly be taken as our sample space. Or if these people have been numbered from 1 to 500, we could equivalently use

$$S = \{1, 2, 3, \ldots, 500\}.$$ ■

EXAMPLE 2.1.3.   Rachel purchases an electronic digital watch that has a counter for seconds. She sets her watch with radio station WWV at 12 noon on one day and, on the same day of the following week she again listens to WWV at 12 noon and observes the digits given by the second counter. A sample space for this experiment is simply the listing of values her second counter might display when WWV gives the times back.

$$S = \{00, 01, 02, \ldots, 59\}.$$ ■

Frequently the performance of an experiment naturally gives rise to more than one piece of information that we may want to record. If we observe two pieces of information when the experiment is performed, we might reasonably want a sample space that is a collection of 2-tuples, the two positions corresponding to the two pieces of information. Or if we observe three pieces of information, we might want a sample space of 3-tuples. Or, more generally, if we observe r pieces of information, we might want a sample space of r-tuples. In each of the three examples given, a single piece of information was generated when performing the experiment; thus each of these sample spaces had 1-tuples (or scalars) as elements. The next three examples discuss experiments in which more than one piece of information is observed.

EXAMPLE 2.1.4.   Suppose our experiment consists of one roll of two dice, one red and the other green. A reasonable sample space for the experiment would be the collection of all possible 2-tuples (x_1, x_2) that could occur where the number in the first position of any 2-tuple corresponds to the number on the red die and the number in the second position corresponds to the number on the green die. Thus we might use as our sample space

$$S = D \times D$$

where $D = \{1, 2, 3, 4, 5, 6\}$: that is,

$$S = \{(x_1, x_2): x_1 = 1, 2, 3, \ldots, 6; \quad x_2 = 1, 2, 3, \ldots, 6\}.$$ ■

EXAMPLE 2.1.5. Doug, Joe, and Hugh match coins. The experiment they perform is one flip of three coins. A reasonable sample space for this experiment is the set of 3-tuples, each of which has H (head) or T (tail) in every position. The first position in the 3-tuple corresponds to the face on Doug's coin, the second position to the face on Joe's coin, and the third position to the face on Hugh's coin. This sample space S can be written

$$S = C \times C \times C,$$

where $C = \{H, T\}$:

$$S = \{(x_1, x_2, x_3): x_1 = H \text{ or } T, x_2 = H \text{ or } T, x_3 = H \text{ or } T\}. \quad ■$$

EXAMPLE 2.1.6. If 10 distinct voters are selected from the 500 registered in the small town mentioned in Example 2.1.2, we could use a sample space S whose elements are 10-tuples, the first component of each 10-tuple representing the first person selected, the second component representing the second person selected, and so on. With the numbering of voters mentioned earlier, then,

$$S = \{(x_1, x_2, \ldots, x_{10}): x_i = 1, 2, \ldots, 500, x_i \neq x_j \text{ for } i \neq j\}.$$

This set is not the Cartesian product of the same set with itself 10 times, because we have not allowed selection of the same voter two or more times. ■

The sample space is assumed to list all possible outcomes that might be observed when the experiment is performed. After the experiment is over, we have observed the occurring outcome (element of S). As we know, there are many subsets of S, each of which contains any particular element of S. In the language of probability theory, these subsets of S are called *events*. Granted that a particular element of S, call it a, was observed to occur, we will say that *any* event (subset) that has a as an element occurred. Because this concept is quite basic to the sequel, we will summarize this discussion with a definition.

DEFINITION 2.1.2. An event A is a subset of a sample space S, $A \subset S$. An event is said to have *occurred* if any one of its elements is the outcome observed. ■

The sample space used for Example 2.1.1 was

$$S = \{1, 2, 3, 4, 5, 6\}.$$

Then each of the sets

$$A = \{1\}, \quad B = \{1, 3, 5\}, \quad C = \{2, 4, 6\}, \quad D = \{4, 5, 6\}, \quad E = \{1, 3, 4, 6\},$$

is an event (these are not the only events since they are not the only subsets of S). These are all distinct (different) events because no two of these subsets are equal. If we actually were to perform the experiment (roll the die) and got a 1, then events A, B, and E are said to have occurred since each of these has 1 as an element. Events C and D did not occur since $1 \notin C$ and $1 \notin D$. If we got a 4 when the die was rolled, then we would say that events C, D, and E occurred since 4 is an element of each of them. Notice that no matter which outcome we observe when the experiment is performed, *many* different events have each occurred (as we will see, exactly half the possible events occur for any particular outcome).

In solving probability problems we need generally take a word description of an event and translate it into a subset of the sample space. With experience, this step is not particularly difficult, but it is frequently bothersome when first studying probability. For example, in Example 2.1.4 a pair of dice is rolled one time. The sample space S is the set of 2-tuples

$$S = \{(x_1, x_2): x_1 = 1, 2, \ldots, 6,\ x = 1, 2, \ldots, 6\}.$$

The events:

> A : The sum of the two dice is 3.
> B : The sum of the two dice is 7.
> C : The two dice show the same number.

are, as subsets,

> $A = \{(1, 2), (2, 1)\}$
> $B = \{(1, 6), (2, 5), (3, 4), (4, 3), (5, 2), (6, 1)\}$
> $C = \{(1, 1), (2, 2), (3, 3), (4, 4), (5, 5), (6, 6)\}.$

Suppose that in Example 2.1.5 Doug, Joe and Hugh are playing a game called "odd man loses." That is, if two of the coins' faces match and the third person's does not, the third person loses. The sample space is

$$S = \{(x_1, x_2, x_3): x_1 = H \text{ or } T, x_2 = H \text{ or } T, x_3 = H \text{ or } T\},$$

where the first position corresponds to Doug's coin; the second, to Joe's; the third, to Hugh's. Define the events:

> A : Doug loses.
> B : Doug does not lose.
> C : Joe loses.
> D : No one loses.

Then, written as subsets, we have

$$A = \{(H, T, T), (T, H, H)\}$$
$$B = \{(H, H, T), (H, T, H), (H, H, H), (T, T, H), (T, H, T), (T, T, T)\}$$
$$\quad = \bar{A}$$
$$C = \{(T, H, T), (H, T, H)\}$$
$$D = \{(H, H, H), (T, T, T)\}.$$

EXAMPLE 2.1.7.   Suppose our experiment consists of a hundred-yard dash involving four college-age sprinters. It is clear that there are many facets of the experiment that we might be interested in, such as the name of the winner, the winning time, the order in which the four cross the finish line, the time of the second-place man, or the time of the third-place man. Which facets were of interest would determine the sample space to be used. For example, if we were going to refer only to the winning time, we could use

$$S_1 = \{t: 0 \leq t \leq 15\}$$

(measuring time in seconds). Or if we are interested in the times of both the first- and second-place men, we could use

$$S_2 = \{(t_1, t_2): 0 \leq t_1 \leq 15, t_1 < t_2 \leq 20\}.$$

In this latter case t_1 is the time of the winner and t_2 is the time of the second-place man; thus the requirement $t_2 > t_1$. For either of these sample spaces we could define the event A : winning time is between 9.45 and 9.65 seconds; using S_1 we would have

$$A = \{t: 9.45 \leq t \leq 9.65\}.$$

And using S_2, we would have

$$A = \{(t_1, t_2): 9.45 \leq t_1 \leq 9.65, t_1 < t_2 \leq 20\}.$$

Again, then, the sample space is not unique. If A were the only event of interest, we would undoubtedly decide to use S_1 rather than S_2 since this would require keeping track of only the single time of the first-place finisher. ■

EXERCISE 2.1

1. Specify a sample space for the experiment that consists of drawing 1 ball from an urn containing 10 balls of which 4 are white and 6 are red. (Assume that the balls are numbered 1 through 10.)
2. Specify a sample space for the experiment that consists of drawing 2

balls with replacement from the urn containing 10 balls (that is, the first ball removed is replaced in the urn before the second is drawn out). Again, assume that they are numbered.

3. Specify a sample space for the experiment that consists of drawing 2 balls without replacement from the urn containing 10 balls (that is, the first ball removed is not replaced in the urn before the second is drawn out). Assume that they are numbered.

4. For the sample space given in Exercise 2.1.1, define the events (as subsets):

A : A white ball is drawn.
B : A red ball is drawn.

5. For the sample space given in Exercise 2.1.2, define the events (as subsets):

C : The first ball is white.
D : The second ball is white.
E : Both balls are white.

Does $C \cap D = E$?

6. A cigarette company packs one of five different slips, labeled a, b, c, d, e, respectively, with each pack it produces. Suppose that you buy two packs of cigarettes of this brand. What is a good sample space for the experiment whose outcome is the pair of slips you receive with the two packs?

7. Suppose that all of the residents of a particular town are bald or have brown hair or have black hair. Furthermore, each resident has blue eyes or brown eyes. We select one resident at random. Give a sample space S for this experiment and define, as subsets, these events:

A : The selected resident is bald.
B : The selected resident has blue eyes.
C : The selected resident has brown hair and brown eyes.

8. Three girls, Marie, Sandy, and Tina, enter a beauty contest. Prizes are awarded for first and second place. Specify a sample space for the experiment that consists of the choice of the two winners. Define as subsets, the events:

A : Marie wins.
B : Marie gets second prize.
C : Tina and Sandy get the prizes.

9. Three cards are selected at random without replacement from a deck that contains 3 red, 3 blue, 3 green, and 3 black cards. Give a sample space for this experiment and define the events:

   A : All the selected cards are red.
   B : One card is red, 1 green, and 1 blue.
   C : Three different colors occur.
   D : All 4 colors occur.

10. A small town contains three grocery stores (call them 1, 2, 3). Four ladies living in this town each randomly and independently pick a store in which to shop (in this town). Give a sample space for the experiment that consists of the selection of stores by the ladies and define the events:

   A : All the ladies choose store 1.
   B : Half the ladies choose store 1 and half choose store 2.
   C : All the stores are chosen (by at least one lady).

11. Nine horses are entered in a race (numbered 1 through 9). Prizes are given for win, place, and show (first, second, and third place, respectively). Give a reasonable sample space for this experiment and, as subsets, define the events:

   A : Horse number 5 wins.
   B : Horse number 5 shows.
   C : Horse number 5 does not finish in the money.

12. Two light bulbs are placed on test until both fail. Assume that each will burn no more than 1600 hours. Define a reasonable sample space for this experiment and describe, as subsets, the events:

   A : Both bulbs fail in less than 1000 hours.
   B : Neither bulb fails in less than 1000 hours.
   C : The shortest time to failure (of the two) is 1000 hours.
   D : The longest time to failure (of the two) is 1000 hours.

13. Each day a newspaper boy sells papers on the same corner; he is supplied with 30 papers. On any particular day, the number he will sell is not known. Define a reasonable sample space for the experiment that consists of the number of sales he will make on a given day. Define the events:

   A : He sells at least 5 papers.
   B : He sells at most 5 papers.
   C : He sells exactly 5 papers.

14. (Refer to Exercise 2.1.13.) Consider the experiment that consists of the number of sales this newspaper boy will make on two successive days. Give a reasonable sample space for the experiment and define the events:

A : He sells at least 5 papers the first day.
B : He sells at least 5 papers the second day.
C : He sells at least 5 papers both days.

15. A large retail radio and television store receives a shipment of 50 radios, all of the same make and model. Each radio either works correctly or it does not. A receiving clerk must verify that each radio accepted does work correctly (or not, in which case it is returned). What is a reasonable sample space for this experiment? Define the events:

A : All work correctly.
B : None work correctly.
C : The first two do not work correctly, but the rest do.

2.2   Probability Axioms

The word "probability" is a relative newcomer to the English language. Run-of-the-mill U. S. dictionaries of just 50 years ago did not consider the word worthy of a regular entry, although some included short phrases describing probability under the entry "probable". Then, as now, the most frequently mentioned synonym for probability was likelihood, which in turn was defined roughly as the character of being likely. The current technical usage of the word probability is essentially the same. If the weather forecaster in your area says that the probability of rain in your locale tomorrow is .4, her intended meaning is as follows: If you conceive of a very large number of days in the past with conditions like today, then 40 percent of the days were followed by days with rain, 60 percent were followed by days without rain. Thus the probability or likelihood for rain tomorrow is measured to be 40 percent or .4.

Given an experiment that could result in any one of many different outcomes, probability is used to measure the likelihood of occurrence of those different outcomes or, more generally, it is used to measure the likelihood of occurrence of the different possible events (subsets of the sample space). Probabilities are (dimensionless) real numbers that are associated with events. Thus a *probability function* (rule giving assignments of probabilities) is an example of a real-valued set function, as mentioned in Chapter 1. In this section we will study the Kolmogorov axioms for a probability function, rules that probabilities must follow to ensure that they behave like relative frequencies and to ensure consistency in their assignment.

Let us first discuss the reasons behind the axioms before we list them. They all arise from the properties of relative frequencies. First, the relative frequency of an event that is certain to occur must be 1, because it would occur 100 percent of the time; probability is called a *bounded* measure because of this requirement, as we will see. Second, the relative frequency of occurrence of any event can never be negative; thus we will insist that probabilities must never be negative. The only remaining property required is that of additivity. This says that if two events cannot occur simultaneously (because they are disjoint subsets), the probability of the event defined by their union must be equal to the sum of the probabilities of the two events. In the language of probability, two events that cannot occur simultaneously are called *mutually exclusive* (the subsets are called disjoint, the events are called mutually exclusive).

Formally, a probability function is a real-valued set function defined on the class of all subsets of the sample space S: the value that is associated with a subset A is denoted by $P(A)$. The assignment of probabilities must satisfy the following three rules (in order that the set function may be called a probability function):

1. $P(S) = 1$.
2. $P(A) \geq 0$ for all $A \subset S$.
3. $P(A \cup B) = P(A) + P(B)$ if $A \cap B = \varnothing$.

These are the three requirements just discussed. If S really does list all the possible outcomes for the experiment, some one of its elements is certain to occur. Probabilities must be nonnegative and the probability of the union of two mutually exclusive events is the sum of their probabilities. The sample space S plays the role of the universal set; any complements taken are with respect to S. In working with probabilities, remember they are real numbers. Thus they must follow the rules for real numbers.

Many consequences or theorems can be derived for probability functions. Let us take a look at some of these now. Recall that $\varnothing \subset S$. Thus our probability measure must assign some number to this event. The number that must always be assigned is 0, as is proved in Theorem 2.2.1.

Theorem 2.2.1. $P(\varnothing) = 0$ for any S.

Proof: $S \cup \varnothing = S$ and thus $P(S \cup \varnothing) = P(S) = 1$ by Axiom 1. But $S \cap \varnothing = \varnothing$ so that $P(S \cup \varnothing) = P(S) + P(\varnothing) = 1 + P(\varnothing)$ by Axiom 3. Thus $1 + P(\varnothing) = 1$; that is, $P(\varnothing) = 0$.   ■

A second consequence of the assumed axioms is given as Theorem 2.2.2. There are many instances in which it saves a great deal of effort in computing probabilities.

Theorem 2.2.2. $P(\bar{A}) = 1 - P(A)$, where $\bar{A}$ is the complement of A with respect to S.

Proof: $A \cup \bar{A} = S$ so $P(A \cup \bar{A}) = P(S) = 1$ by Axiom 1. But $A \cap \bar{A} = \varnothing$ and thus $P(A \cup \bar{A}) = P(A) + P(\bar{A})$ by Axiom 3. Thus we have established that $P(A) + P(\bar{A}) = 1$, from which the result follows immediately. ∎

Axiom 3 tells us that if two events A and B have no elements in common, then the probability of their union is the sum of their individual probabilities. Theorem 2.2.3 derives a preliminary result that is used to establish Theorem 2.2.4, regarding $P(A \cup B)$ when A and B have elements in common.

Theorem 2.2.3. $P(\bar{A} \cap B) = P(B) - P(A \cap B)$.

Proof: By referring to Figure 2.2.1, we can see that

$$B = (\bar{A} \cap B) \cup (A \cap B).$$

Then $P(B) = P\big((\bar{A} \cap B) \cup (A \cap B)\big)$. Furthermore, $(\bar{A} \cap B) \cap (A \cap B) = \varnothing$; thus $P\big((\bar{A} \cap B) \cup (A \cap B)\big) = P(\bar{A} \cap B) + P(A \cap B)$, and we have established that

$$P(B) = P(\bar{A} \cap B) + P(A \cap B)$$

from which the desired result follows immediately. ∎

Theorem 2.2.4. $P(A \cup B) = P(A) + P(B) - P(A \cap B)$.

Proof: By referring to Figure 2.2.1, we can also see that

$$A \cup B = A \cup (\bar{A} \cap B),$$

and thus $P(A \cup B) = P\big(A \cup (\bar{A} \cap B)\big)$. Furthermore, $A \cap (\bar{A} \cap B) = \varnothing$ so

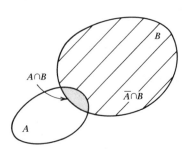

Figure 2.2.1

that

$$P(A \cup (\bar{A} \cap B)) = P(A) + P(\bar{A} \cap B)$$
$$= P(A) + P(B) - P(A \cap B),$$

by Theorem 2.2.3. Thus

$$P(A \cup B) = P(A) + P(B) - P(A \cap B). \qquad \blacksquare$$

Axiom 3 extends immediately to the union of any number of mutually exclusive events. Suppose we have an experiment with sample space S and $A_1, A_2, \ldots, A_k$ are mutually exclusive events (thus $A_i \cap A_j = \varnothing$ for all $i \neq j$). Then, clearly, $A_1 \cup A_2 \cup \ldots \cup A_{k-1}$ and A_k are mutually exclusive so

$$P(A_1 \cup A_2 \cup \ldots \cup A_{k-1} \cup A_k) = P(A_1 \cup A_2 \cup \ldots \cup A_{k-1}) + P(A_k)$$

by Axiom 3. Repeating this reasoning $k - 2$ more times establishes the following theorem.

Theorem 2.2.5. If $A_1, A_2, \ldots, A_k$ are mutually exclusive events, then

$$P(A_1 \cup A_2 \cup \ldots \cup A_k) = P(A_1) + P(A_2) + \cdots + P(A_k),$$

for any $k = 2, 3, 4, \ldots$.

It is important to realize that the axioms will not give a unique assignment of probabilities to events; rather, the axioms simply clarify relationships between probabilities that we assign so that we will be consistent with our intuitive notions of probability. For example, if a rocket has been designed to take a man to the moon, then the experiment that consists of firing the rocket and the man at the moon can be thought of as having two outcomes: success and failure. Success would be the safe arrival of the man on the moon, failure anything else that might occur. Then the axioms do not imply that the probability of the event {success} must be $\frac{1}{2}$ or $\frac{3}{4}$ or .99 or any other particular value. If we denote this probability by p, they do imply that $0 \leq p \leq 1$ and that the probability of the event {failure} must be $1 - p$. Beyond this, p is still unspecified.

Actual specification of the value of p must come from analytical considerations of the experiment performed and the mechanism behind it. For the rocket example just mentioned, this would consist of detailed examination of the rocket design and conditions under which it is to be fired, in addition to any prior test firings or performance data available. Generally, considerations of prior data and their implications regarding the value of p fall into the realm of statistics, the topic for the latter half of this volume.

EXERCISE 2.2

1. Given $S = \{1, 2, 3\}$, $A = \{1\}$, $B = \{3\}$, $C = \{2\}$, $P(A) = \frac{1}{3}$, $P(B) = \frac{1}{3}$, find

   (a) $P(C)$ (d) $P(\bar{A} \cap \bar{B})$
   (b) $P(A \cup B)$ (e) $P(\bar{A} \cup \bar{B})$
   (c) $P(\bar{A})$ (f) $P(B \cup C)$.

2. Let S, A, B, C be defined as in Exercise 2.2.1, but now let $P(A) = \frac{1}{2}$, $P(B) = \frac{1}{5}$. Compute the probabilities asked for in (a) through (f).

3. Let S, A, B, C be defined as in Exercise 2.2.1 and let $P(A) = 1$. Compute the probabilities asked for in (a) through (f). Could we let $P(A)$ be 2?

4. Define $S = \{a, b\}$, $B = \{b\}$. Give three different assignments of probabilities to the subsets of S.

5. Prove, from the axioms, that probabilities are monotonic; that is, $P(A) \leq P(B)$ if $A \subset B$.

6. Prove, from the axioms, that $P(A) \leq 1$ for all A.

7. Given an experiment such that $P(A) = \frac{1}{2}$, $P(B) = \frac{1}{2}$, $P(A \cup B) = \frac{2}{3}$, compute

   (a) $P(\bar{A})$ (e) $P(\bar{A} \cup \bar{B})$
   (b) $P(\bar{B})$ (f) $P(A \cap \bar{B})$
   (c) $P(A \cap B)$ (g) $P(\bar{A} \cap B)$
   (d) $P(\bar{A} \cap \bar{B})$ (h) $P(\bar{A} \cup B)$.

8. Given an experiment such that $P(A) = \frac{1}{2}$, $P(B) = \frac{1}{3}$, $P(A \cap B) = \frac{1}{4}$, compute

   (a) $P(A \cup B)$ (d) $P(A \cap \bar{B})$
   (b) $P(\bar{A} \cup B)$ (e) $P(\bar{A} \cap \bar{B})$
   (c) $P(\bar{A} \cap B)$ (f) $P(\bar{A} \cup \bar{B})$.

9. Is it possible to have an assignment of probabilities such that $P(A) = \frac{1}{2}$, $P(A \cap B) = \frac{1}{3}$, $P(B) = \frac{1}{4}$?

10. If we know that $P(A \cup B) = \frac{2}{3}$ and $P(A \cap B) = \frac{1}{3}$, can we determine $P(A)$ and $P(B)$?

11. The sample space S for an experiment is $S = \{a, b, c\}$. Is it possible for a probability measure to have values

$$P(\{a, b\}) = \frac{2}{3}$$
$$P(\{a, c\}) = \frac{1}{3}$$
$$P(\{b, c\}) = \frac{1}{3}?$$

   Why or why not?

12. A memo (addressed to two people) is circulated by the originator. Assume it will eventually be received by 0 or 1 or 2 of the addressees, and the probability that it will be received by at least one addressee

is .9, whereas the probability that it will be received by at most 1 addressee is .8. What is the probability that the number of addressees to receive
it is

(a) 0   (b) 1   (c) 2?

13.   At the end of its fiscal year, the total book value of a corporation may
       be smaller than it was the preceding fiscal year (call this -1), it may
       be equal to what it was the preceding year (call this 0) or it may be
       greater than what it was the preceding year (call this 1). We might
       then look at the balance sheet of this corporation for the latest year
       as being an experiment with

$$S = \{-1, 0, 1\}.$$

Is it possible to have a probability measure on this sample space such
that

$$P(\{-1, 1\}) = .6$$
$$P(\{0, 1\}) = .9$$
$$P(\{-1, 0\}) = .5?$$

Why or why not?

14.   Given an experiment such that $P(\bar{A} \cap B) = .1$, $P(A \cap \bar{B}) = .4$,
       $P(\bar{A} \cap B) = .6$, compute

(a) $P(A)$ (c) $P(A \cup B)$
(b) $P(B)$ (d) $P(\bar{A} \cup B)$

15.   A family drives a camper across the United States. Assume that the
       number of flat tires they will have during the trip is either 0 or 1 or 2.
       Also, assume the probability is .9 that they will have at most 1 flat
       and the probability is .2 that they will have at least 1 flat. Compute
       the probability that the number of flat tires they have is

(a) 0   (b) 1   (c) 2.

2.3 Finite Sample Spaces

In many cases the sample space we will use for an experiment will
have a finite number of elements. By finite number we mean there is some
integer that is larger than the number of elements in S; put another way,
S is finite if it is meaningful to use the roster method in specifying the elements
that belong to S. The list of possible outcomes does have an end, although
it might be very time consuming to construct the full list.

EXAMPLE 2.3.1.   Suppose, as in Example 2.1.3, that Rachel purchases an electronic digital watch with a counter for seconds and, as before, she sets the watch with WWV at 12 noon and then observes the value on the second counter at 12 noon seven days later. The value she will observe must be one of the elements of

$$S = \{00, 01, 02, ..., 59\},$$

which is a sample space with 60 distinct elements, and thus is finite. ■

EXAMPLE 2.3.2.   The standard deck of playing cards has 52 cards with 4 suits and 13 cards within each suit. In the game of bridge each player receives 13 of these cards. Suppose that you are playing bridge and perform the "experiment" of observing the 13 cards you receive in one hand. The sample space for the experiment then consists of the list of possible collections of 13 cards that you might receive. The number of elements in this sample space is very large, in fact, it is 635,013,559,600, approaching the federal debt (in dollars). This again is a finite sample space since there is an end to the roster of elements. (See Example 2.4.5 to verify the number of elements in this sample space.) ■

The specification of the probability function for any sample space with a finite number of elements is accomplished simply by defining the probability of occurrence for every *single-element* event, those events (subsets) that have only one element belonging to them. Clearly, if S has k distinct elements, there are exactly k distinct (and mutually exclusive) single-element events. Any event with two or more elements (subset of S) must be a union of single-element events and immediately, from Theorem 2.2.5, the probability for such a union is given by the sum of the probabilities for the single-element events in the union. Thus if a rule is described that gives the probabilities for the single-element events, the probabilities for all other events are also specified. Of course, if $A_1, A_2, ..., A_k$ are the distinct single-element events, where S has k elements, then $A_1 \cup A_2 \cup ... \cup A_k = S$ so $\sum_{i=1}^{k} P(A_i) = 1$. Thus it is really only necessary to specify the values for $k - 1$ of the single-element events; the probability for the remaining one must be the value to make this sum equal 1.

EXAMPLE 2.3.3.   A die has been loaded in a manner such that the probability of face i being uppermost, when it stops rolling, is proportional to i, $i = 1, 2, 3, ..., 6$. (If you think about it, this would not be easy to accomplish.) The sample space for the experiment then is

$$S = \{1, 2, 3, 4, 5, 6\},$$

the elements representing the number of spots on the face uppermost when the die has stopped. There are, of course, six distinct single-element events,

$$A_i = \{i\}, \qquad i = 1, 2, 3, 4, 5, 6.$$

Let $p = P(A_1)$. Then the fact that the probability of occurrence should be proportional to the number of spots on the uppermost face says we must have

$$P(A_2) = 2p, \qquad P(A_3) = 3p, \ldots, P(A_6) = 6p.$$

The requirement that these probabilities for single-element events must sum to 1 then gives

$$p + 2p + 3p + 4p + 5p + 6p = 1,$$

from which we have $p = \frac{1}{21}$ and the probability function is completely specified. ∎

The probability functions for fair gambling games are generally based on the assumption that all single-element events are equally likely to occur (if the right sample space is used). Indeed, this sort of assumption was tacitly made in the early correspondence between Fermat and Pascal, mentioned earlier. If S has k elements and it is assumed the single-element events are equally likely, their common probability must equal $\frac{1}{k}$. Furthermore, since any event is the union of single-element events, then the probability of any event $A \subset S$ is given by the ratio of the number of elements in A (the number of single-element events whose union is A) to the number of elements in S. That is, we use the rule

$$P(A) = \frac{n(A)}{n(S)} \qquad \text{for} \qquad A \subset S,$$

where $n(A)$ is the number of elements in A (see Example 1.3.2 in Chapter 1). That this rule will satisfy the three axioms given in Section 2.2 is proved in the next theorem.

Theorem 2.3.1. If S has k elements, the rule

$$P(A) = \frac{n(A)}{n(S)}$$

satisfies the three axioms for a probability function.

Proof: If S has k elements, then $n(S) = k$.

$$P(S) = \frac{n(S)}{n(S)} = \frac{k}{k} = 1$$

so Axiom 1 is satisfied. If A is any subset of S, it contains a nonnegative number of elements; that is, $n(A) \geq 0$ for all $A \subset S$. Then

$$\frac{n(A)}{k} = \frac{n(A)}{n(S)} = P(A) \geq 0 \qquad \text{for all} \qquad A \subset S$$

and Axiom 2 is satisfied. If $A \cap B = \varnothing$, then A and B have no elements in common and $n(A \cup B) = n(A) + n(B)$. Thus

$$\frac{n(A \cup B)}{n(S)} = \frac{n(A)}{n(S)} + \frac{n(B)}{n(S)},$$

that is,

$$P(A \cup B) = P(A) + P(B). \qquad \blacksquare$$

Thus, for any problem in which we are justified in assuming equally likely single-element events, we now have a rule that enables us to compute the probability of occurrence of any event.

EXAMPLE 2.3.4. Suppose we roll a fair die one time. What is the probability of getting an even number? What is the probability of getting a number that is greater than 4?

Our sample space is $S = \{1, 2, 3, 4, 5, 6\}$. Since the die is fair, we assume the single-element events to be equally likely to occur; each then has probability $\frac{1}{6}$ of occurrence. Let A be the event that an even number occurs and let B be the event that we get a number greater than 4.

$$A = \{2, 4, 6\}, \qquad B = \{5, 6\}, \qquad n(A) = 3, \qquad n(B) = 2, \qquad n(S) = 6,$$

and we have $P(A) = \frac{3}{6}$, $P(B) = \frac{2}{6}$. $\qquad \blacksquare$

EXAMPLE 2.3.5. We roll a pair of fair dice one time. What is the probability that the sum of the two numbers is 2? That it is 7? That it is 11?

Our sample space is $S = \{(x_1, x_2) : x_1 = 1, 2, ..., 6; x_2 = 1, 2, ..., 6\}$. Since the first die can have a number from 1 through 6 on it and, quite independently, the second die can also have any number 1 through 6 on it, we reason that there are $6 \cdot 6 = 36$ elements belonging to S (or we simply list them all and count them). Thus $n(S) = 36$. Let A be the event that the sum is 2, B the event that the sum is 7, C the event that the sum is 11. Then

$$A = \{(1, 1)\}$$
$$B = \{(1, 6), (2, 5), (3, 4), (4, 3), (5, 2), (6, 1)\}$$
$$C = \{(5, 6), (6, 5)\},$$

and we see that $n(A) = 1$, $n(B) = 6$, $n(C) = 2$. Then since we are assuming that the dice are fair, $P(A) = \frac{1}{36}$, $P(B) = \frac{6}{36}$, $P(C) = \frac{2}{36}$. ∎

EXAMPLE 2.3.6. Assume four people enter a tournament; the game to be played is one that cannot end in a tie (such as tennis, cribbage, backgammon, etc.). For concreteness, assume the four people are named Doug, Joe, Hugh, and Ray. The tournament will be single elimination; this means that Doug and Joe will play each other (game 1) and the loser will be eliminated. Simultaneously or subsequently, Hugh and Ray will play each other (game 2) and the loser will be eliminated. The two winners then play each other (game 3); the winner of this game wins the tournament. What is the probability that Ray wins the tournament? Let us first find a reasonable sample space for this experiment. Since three games will be played, and the winners of the games are the important quantities to record, let us adopt a sample space whose elements are 3-tuples. The first position records the winner of game 1, the second position records the winner of game 2, and the third position records the winner of game 3 (the winner of the tournament). Thus we will use

$$S = \{(\text{Doug, Hugh, Doug}), (\text{Doug, Hugh, Hugh}), (\text{Joe, Hugh, Joe}),$$
$$(\text{Joe, Hugh, Hugh}), (\text{Doug, Ray, Doug}), (\text{Doug, Ray, Ray}),$$
$$(\text{Joe, Ray, Joe}), (\text{Joe, Ray, Ray})\}.$$

To answer questions about who will win the tournament, we must assign values to the probabilities of occurrence of the single-element events. *If the players are all of equal ability,* each single-element event would be assigned probability $\frac{1}{8}$. Under this assumption

$$P(\text{Ray wins}) = P(\{(\text{Doug, Ray, Ray}), (\text{Joe, Ray, Ray})\})$$
$$= P(\{(\text{Doug, Ray, Ray})\}) + P(\{(\text{Joe, Ray, Ray})\})$$
$$= \tfrac{2}{8} = \tfrac{1}{4}.$$

Not surprisingly, with this assumption, we also have

$$P(\text{Doug wins}) = P(\text{Joe wins}) = P(\text{Hugh wins}) = \tfrac{1}{4}. \qquad ∎$$

Many important problems have equally likely single-element events. As seen, computations of probabilities in these cases reduce to counting the number of elements in S and the number of elements in the events of interest. Probabilities are then given by the ratio of these quantities. Because counting the number of elements belonging to an event plays a role in many practical problems, the next two sections are devoted to counting techniques.

EXERCISE 2.3

1. For the die discussed in Example 2.3.3 compute the probability that the face uppermost will be (with the die rolled once)
    (a) An even number.
    (b) Greater than 4.
    (c) Between 2 and 5 inclusive.
2. A trick coin is to be flipped one time. The probability of getting a head is three times as large as the probability of getting a tail. What are the probabilities for the two single-element events?
3. A TV repairman gets calls to repair TV sets made by four different manufacturers. Call them M_1, M_2, M_3, and M_4: thus the next repair call he receives will request that he repair a set of one of these four brands. From his previous records and experience it is twice as likely that the request will be for a set made by M_1 as that the request will be for a set made by M_2, whereas this latter probability is three times as large as the probability that the request is for a set made by M_3. There have been equal numbers of requests for repair for sets made by M_1 and M_4. What are the probabilities that the next repair requested is for a set made by M_1? M_2? M_3?
4. If two fair coins are flipped, what is the probability that the two faces are alike?
5. If we draw 1 card at random from a standard deck of 52, what is the probability that it is red? That it is a diamond? That it is an ace? That it is the ace of diamonds?
6. Five different colored rubber bowls with the same identical dog food in them are laid out in a row. If a dog chooses a bowl at random from which to eat, what is the probability that he selects the blue one? If a second dog is used, what is the probability that he selects the blue one? What is the probability that both choose the blue one? What is the probability that they both choose the same color?
7. A pair of fair dice is rolled once. Compute the probability that the sum is equal to each of the integers 2 through 12.
8. A one is painted on the head side and a two is painted on the tail side of each of 3 fifty-cent pieces. If the 3 coins are all tossed once (together), compute the probability that the sum of the three numbers occurring is each of the integers 3 through 6.
9. Forty people are riding on the same railroad car. Of this number, 5 are Irish ladies with blue coats, 2 are Irish men with green coats, 1 is an Irish man with a black coat, 7 are Norwegian ladies with brown coats, 2 are Norwegian ladies with blue coats, 6 are Norwegian men with black coats, 4 are German men with green coats, 3 are German ladies with black coats, 5 are German ladies with blue coats, and 5

are German men with black coats. If we select one person at random from this car, what is the probability that the selected person is a man? Is wearing a green coat? Is wearing a brown coat? Is a Norwegian? Is a German? Is a German wearing a green coat?

10. For the tournament example (Example 2.3.6), keep the elements of the sample space in the order listed. Instead of assuming they are all equally likely, assume the probabilities assigned to the single-element events are, respectively, 4, 8, 1, 5, 4, 2, 1, 2, each divided by 27. Compute the probabilities that each of the four contestants will win the tournament.

11. Using the assignment of probabilities in Exercise 2.3.10, compute the probability that each of the four contestants loses his first game.

12. Assume eight players of equal ability enter a single-elimination tournament (no ties allowed). What is the probability that each will be the winner of the tournament? What is the probability that player 1 wins his first two games and loses the final?

2.4 Counting Techniques

When the equally likely assumption is made for a finite sample space, the probability of occurrence of any event A is given by the ratio of the number of elements belonging to A to the number of elements belonging to S. For such cases it is useful to be able to count the number of elements belonging to given sets. We will study some simple methods appropriate to this type of counting problem in this section. The reader should be warned that counting problems are easy to define but can be difficult to solve. No great expertise in solving counting problems is required for the successful study of the material in this text.

A very simple technique that is frequently useful in counting problems is called the *multiplication principle*. This can be stated as follows.

DEFINITION 2.4.1.   If a first operation can be performed in any of n_1 ways and a second operation can then be performed in any of n_2 ways, both operations can be performed (the second immediately following the first) in $n_1 \cdot n_2$ ways.   ∎

For example, if we can travel from town A to town B in 3 ways and from town B to town C in 4 ways, then we can travel from A to C in a total of $3 \cdot 4 = 12$ ways. Or if the operation of tossing a die gives rise to 1 of 6 possible outcomes and the operation of tossing a second die gives rise to 1 of 6 possible outcomes, then the operation of tossing a pair of dice gives rise to $6 \cdot 6 = 36$ possible outcomes.

EXAMPLE 2.4.1.   Suppose that a set A has n_1 elements and a second set B has n_2 elements. Then the Cartesian product $A \times B$ (see Definition 1.2.6) has $n_1 n_2$ elements. The Cartesian product $A \times A$ has n_1^2 elements, $B \times B$ has n_2^2 elements.   ■

This definition can immediately be extended to any number of operations. The operation of tossing three coins gives rise to $2^3 = 8$ possible outcomes. The operation of rolling 5 six-sided dice gives rise to $6^5 = 7776$ possible outcomes. If the set A_i has n_i elements, $i = 1, 2, ..., k$, the Cartesian product $A_1 \times A_2 \times \cdots \times A_k$ has $n_1 n_2 \cdots n_k$ different elements.

Tree diagrams can frequently be useful. These are essentially graphic expressions of the multiplication principle. The operations to be performed are represented by nodes (small circles) and the number of ways of performing the operation are represented by "branches" (or lines). The final number of branch tips is the total number of ways the collection of operations can be performed. Figure 2.4.1 presents a tree diagram to count the total number of possible outcomes when three coins are flipped, operations corresponding to separate coins.

DEFINITION 2.4.2.   An arrangement of n symbols in a definite order is called a *permutation* of the n symbols.   ■

We will frequently want to know how many different n-tuples can be made using n different symbols each only once (that is, how many permutations are possible). The multiplication principle will immediately give us the answer. We can count the number of n-tuples by reasoning as follows. In listing all the possible n-tuples, we would perform n natural operations. First we must fill the leftmost position of the n-tuple. Then we must fill the second leftmost position, the third leftmost position, and so on. Since we could put any of the n elements available into the leftmost position, this operation can be performed in n ways. After we fill the leftmost position, we can use any of the remaining $n - 1$ elements to fill the second leftmost position; we can use any of the remaining $n - 2$ elements to fill the third leftmost position and so on until we finally arrive at the rightmost position and want to count the number of ways it can be filled. We have used $n - 1$ elements at this point to fill the first $n - 1$ positions and thus have left only one element that must be used to fill the final position. Then the total number of ways we can perform all n operations (which is also the total number of n-tuples we could make with the given n symbols) is given by the product of the numbers of ways of doing the individual operations. Thus the number of different n-tuples is

$$n(n - 1)(n - 2) \cdots 2 \cdot 1,$$

which we write $n!$ (read n-factorial). Thus $4! = 4 \cdot 3 \cdot 2 \cdot 1 = 24$, and $8! = 8 \cdot 7 \cdot 6 \cdot 5 \cdot 4 \cdot 3 \cdot 2 \cdot 1 = 40,320$. $n!$ gets large very fast as n increases in size.

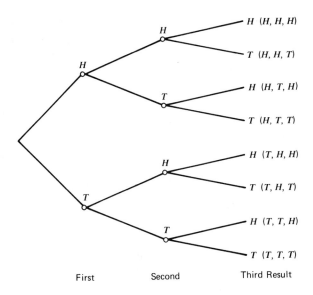

First   Second   Third Result

Figure 2.4.1

Since $n(n - 1)! = n!$, note that for $n = 1$ this equation says $1 \cdot 0! = 1!$ and $1!$ is of course 1. Thus we will define $0! = 1$.

EXAMPLE 2.4.2   Suppose that the same 5 people park their cars on the same side of the street in the same block every night. How many different orderings of the 5 cars parked on the street are possible? The different orderings of the cars parked on the street could be represented by 5-tuples with 5 distinct elements; then if we can count the number of different 5-tuples that are possible, we also know the number of different orderings on the street. The number of 5-tuples possible is of course $5! = 5 \cdot 4 \cdot 3 \cdot 2 \cdot 1 = 120$; thus these 5 people could park their cars on the street in a different order every night for 4 months without repeating an ordering they had already used.   ■

DEFINITION 2.4.3.   The number of r-tuples we can make ($r \leq n$), using n different symbols (each only once), is called the *number of permutations of n things r at a time* and is denoted by $_nP_r$.   ■

How might we compute the value of $_nP_r$? Each r-tuple has exactly r positions. The leftmost position could be filled by any of the n symbols; the second leftmost position could then be filled by any of the remaining $n - 1$

symbols, and so on. By the time we are ready to fill the rth position, we have used $(r - 1)$ symbols already and any of the remaining $n - (r - 1)$ symbols could be used in the rth position. Thus the total number of r-tuples we could construct is $n(n - 1) \cdots (n - r + 1)$ and we have

$$_nP_r = n(n - 1) \cdots (n - r + 1).$$

If we multiply this number by $(n - r)!/(n - r)!$, we certainly do not change its value; thus we also have

$$_nP_r = n(n - 1)(n - 2) \cdots (n - r + 1) \frac{(n - r)!}{(n - r)!}$$

$$= \frac{n!}{(n - r)!}.$$

EXAMPLE 2.4.3.   (a) Fifteen cars enter a race. In how many different ways could trophies for first, second, and third place be awarded? This answer is simply $_{15}P_3 = 15!/12! = 2730$ since the question is equivalent to asking how many permutations are there of 15 objects, 3 at a time.

(b) How many of the 3-tuples just counted have car number 15 in the first position? This can be computed in two ways. First, we might reason that there are $_{14}P_2 = 14!/12! = 182$ ways in which the last two positions could be filled, having already put 15 into the first position of the 3-tuple. Alternatively, there must obviously be equal numbers of 3-tuples (in the totality of all possible) having car 15 in first place as there are having car 14 in first place, 13 in first place, and so on. Thus if we divide the total number of 3-tuples by 15, we should get the number that have car 15 in first place; this gives $2730/15 = 182$—the same answer. ∎

EXAMPLE 2.4.4.   (a) How many three-letter "words" can we make using the letters w, i, n, t, e, r (allowing no repetition)? (A "word" is any arrangement of letters, regardless of whether it is in actual fact a word listed in the dictionary for some language.) This is, of course, just $_6P_3 = 6!/3!$ $= 120$. The number of four-letter words is $_6P_4 = 360$, and so on.

(b) Suppose that repetition of a letter is allowed in making three-letter words using w, i, n, t, e, r. How many three-letter words can we make? The answer is $6 \cdot 6 \cdot 6 = 6^3 = 216$, since we now would be able to fill each position with six letters. The number of four-letter words we could make, allowing repetition, then is $6 \cdot 6 \cdot 6 \cdot 6 = 6^4 = 1296$.

(c) How many three-letter words are there with one or more repeated letters? How many four-letter words are there with one or more repeated letters? We know that there are 216 three-letter words if repetitions are allowed and 120 three-letter words if repetition is not allowed. Thus there are $216 - 120 = 96$ three-letter words with one or more repeated letters. Analogously, there are $1296 - 360 = 936$ four-letter words with one or more repeated letters. ∎

DEFINITION 2.4.4.   The number of distinct subsets, each of size r, that can be constructed from a set with n elements is called the number of *combinations* of n things r at a time; this number is represented by $\begin{pmatrix} n \\ r \end{pmatrix}$. ∎

To evaluate $\begin{pmatrix} n \\ r \end{pmatrix}$ remember that sets are not ordered and that r different symbols can be used to construct $r!$ different r-tuples. Thus since $\begin{pmatrix} n \\ r \end{pmatrix}$ is the number of different collections (subsets) of size r that we could get from a set with n elements, each of which leads to $r!$ r-tuples, it must be true that

$$\begin{pmatrix} n \\ r \end{pmatrix} r! = {}_nP_r = \frac{n!}{(n-r)!};$$

that is,

$$\begin{pmatrix} n \\ r \end{pmatrix} = \frac{n!}{r!(n-r)!}.$$

Thus if a set has, say, $n = 4$ elements (let the set be $\{1, 2, 3, 4\}$), it has $\begin{pmatrix} 4 \\ 1 \end{pmatrix} = \frac{4!}{1!\,3!} = 4$ subsets of size 1 ($\{1\}, \{2\}, \{3\}, \{4\}$) and $\begin{pmatrix} 4 \\ 2 \end{pmatrix} = \frac{4!}{2!2!} = 6$ subsets of size 2 ($\{1, 2\}, \{1, 3\}, \{1, 4\}, \{2, 3\}, \{2, 4\}, \{3, 4\}$).

Remember that order is ignored in counting the number of combinations that can be constructed and that order counts in determining the number of permutations. The most difficult part of many counting problems is deciding whether or not order should be considered.

EXAMPLE 2.4.5.   (a) How many distinct 5-card hands can be dealt from a standard 52-card deck? Since the 5-card hand remains unchanged if you received the same 5 cards, but in a different order, the answer is

$$\begin{pmatrix} 52 \\ 5 \end{pmatrix} = \frac{52!}{5!\,47!} = 2{,}598{,}960.$$

(b) How many distinct 13-card hands can be dealt from a standard deck?

$$\binom{52}{13} = 635{,}013{,}559{,}600.$$

(c) Suppose that 10 boys go out for basketball at a particular school. How many different teams could be fielded from this school?

$$\binom{10}{5} = 252.$$

This is the number of different collections or subsets of 5 boys that could be chosen from the 10. If different allocations of boys to the positions are to be counted as different teams, then there are

$$5!(252) = 30{,}240 = {_{10}P_5}$$

different teams.

(d) One of the 10 boys out for basketball is named Joe. How many of the 252 teams include Joe as a member? If we want to count only those teams that include Joe, then we need count only how many ways we might select 4 additional individuals to be on the team. This is

$$\binom{9}{4} = 126.$$

∎

If you compare parts (c) and (d) of Example 2.4.5, notice that Joe is on exactly half (which equals $r = 5$ divided by $n = 10$) the total possible number of teams that the school could field. This result is true in general. Suppose we define

$$S = \{1, 2, \ldots, n\};$$

then we know immediately that the total number of subsets of S, each with r elements, is $\binom{n}{r} = \dfrac{n!}{r!\,(n-r)!}$. Reasoning as in Example 2.4.5, we find that the number of these subsets that have the element 1 as a member is given by the number of ways we can select $r - 1$ other elements from the $n - 1$ available: $\binom{n-1}{r-1} = \dfrac{(n-1)!}{(r-1)!\,(n-r)!}$. Thus the proportion of subsets of size r, each of which has 1 (or any other specific element) as a member, is $\binom{r-1}{n-1} \Big/ \binom{n}{r} = \dfrac{r}{n}$.

The number $\binom{n}{r}$ occurs in many applications of mathematics and is frequently called a combinatorial coefficient or a binomial coefficient. It is quite straightforward to evaluate $\binom{n}{r}$ on hand-held calculators. The best formulation for this purpose is $\binom{n}{r} = \dfrac{n(n-1)\cdots(n-r+1)}{r(r-1)\cdots 2 \cdot 1}$, possibly intermixing the multiplication and division operations (depending on the sizes of n and r) to avoid overflow. It is easy to see that $\binom{n}{0} = \binom{n}{n} = 1$ for any n; if a set S has n elements, it has exactly one subset with no elements ($\emptyset$) and one with n elements (S). It is equally easy to see that $\binom{n}{1} = \binom{n}{n-1} = n$ and that $\binom{n}{r} = \binom{n}{n-r}$, as is easily verified by writing out the factorial representations of the two; every selection of a subset of size n leaves behind a subset of size $n - r$, so there must be equal numbers of subsets of these two sizes.

Again, assume S has n elements and we are to count the number of subsets of S that have r elements (which is of course $\binom{n}{r}$). The number of these subsets that have 1 (or any specific element) belonging to them is $\binom{n-1}{r-1}$ and the number of these subsets that *do not* have 1 belonging to them is $\binom{n-1}{r}$; thus it must be true that $\binom{n-1}{r-1} + \binom{n-1}{r} = \binom{n}{r}$, as is easily verified by writing the combinatorial coefficients out in factorial form and then adding $\binom{n-1}{r-1}$ and $\binom{n-1}{r}$. This identity is the basis for what is called *Pascal's* triangle, a simple scheme whereby the values for the combinatorial coefficients can be built up recursively from the knowledge that $\binom{1}{0} = \binom{1}{1} = 1$. Figure 2.4.2 gives a short example of Pascal's triangle. The two entries in the first row are the values for $\binom{1}{0}$ and $\binom{1}{1}$;

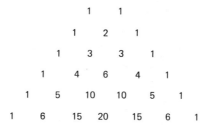

Figure 2.4.2

the three entries in the second row are the values for $\binom{2}{0}, \binom{2}{1}, \binom{2}{2}$, and
so on. Each subsequent row is started and ended with a 1, and each interior
entry is the sum of the two values above its position. We could conceivably
evaluate $\binom{50}{30}$, say, in this way but the modern hand-held calculator is
faster and easier.

As is evident from Figure 2.4.2, with n fixed, the value for $\binom{n}{r}$ starts
at 1 for $r = 0$, increases to a maximum at $r = n/2$, if n is even, or at
$r = (n - 1)/2$, if n is odd, and then decreases symmetrically back to 1. When
n is even, there is a unique maximum for $\binom{n}{r}$, given by $r = n/2$, and
when n is odd, the two middle terms, with $r = (n - 1)/2$ and $r = (n + 1)/2$,
are both equal and both maximize $\binom{n}{r}$.

Let us now state the *binomial theorem*, which may already be familiar
from high school algebra, at least for some values of n. This result is useful
in a surprising number of different problems, not the least of which, for
our purposes, will be in describing the binomial random variable. It is
also the reason that $\binom{n}{r}$ is called a binomial coefficient.

Theorem 2.4.1.   If x and y are any two real numbers and n is a positive
integer, then

$$(x + y)^n = \sum_{i=0}^{n} \binom{n}{i} x^i y^{n-i}, \qquad \text{where} \binom{n}{i} = \frac{n!}{(n - i)!\, i!}.$$

This result is useful in evaluating certain types of sums, for example,

$$2^4 + (4)(2)^3(3) + (6)(2)^2(3)^2 + (4)(2)(3)^3 + 3^4$$

$$= \sum_{i=0}^{4} \binom{4}{i} 2^i 3^{4-i} = (2+3)^4 = 5^4 = 625.$$

The binomial theorem is also useful in approximating the values for certain numbers (this type of problem was perhaps of more interest before the advent of the hand-held calculator, but it is still helpful at times). For example, from Theorem 2.4.1, we have

$$(1 - x)^n = 1 - nx + \binom{n}{2}x^2 - \binom{n}{3}x^3 + \cdots + (-1)^n x^n.$$

If x is "very small" then x raised to higher powers gets much smaller and, especially since the terms in the preceding sum alternate in value, it would seem reasonable to use $1 - nx$ as an approximation for $(1 - x)^n$. Thus, for example, $(.995)^5$ would be approximated by $1 - 5(.005) = .975$ and $(.99)^{10}$ is approximated by $1 - 10(.01) = .90$. The binomial theorem also is useful in solving counting problems. Example 2.4.6 discusses two such cases.

EXAMPLE 2.4.6.   Let $S = \{1, 2, \ldots, n\}$ and let us evaluate the total number of subsets S has. We know the number of subsets of size r is $\binom{n}{r}$ and thus the total number of subsets S has is

$$\sum_{r=0}^{n} \binom{n}{r} = \sum_{r=0}^{n} \binom{n}{r} 1^r 1^{n-r} = (1+1)^n = 2^n.$$

A set with $n = 5$ elements has $2^5 = 32$ subsets, whereas a set with $n = 10$ elements has $2^{10} = 1024$ subsets. The total number of subsets increases very quickly with n. We can also easily evaluate the number of these subsets that have any particular element, say, 1, belonging to them: There are $\binom{n-1}{r-1}$, $r = 1, 2, \ldots, n$, subsets of size r that have 1 as an element, and therefore the total number of subsets of S that have 1 as an element is

$$\sum_{r=1}^{n} \binom{n-1}{r-1} = \sum_{j=0}^{n-1} \binom{n-1}{j} = (1+1)^{n-1} = 2^{n-1}.$$

Thus any particular element belongs to exactly $\frac{1}{2}$ of all the subsets of S. ∎

Thus far in our discussion we have assumed that the objects we are working with are all distinguishable or different from one another. If this is not the case, our formulas that involve order (permutations) have to be modified appropriately. Suppose we have n objects, n_1 of which are alike, n_2 of which are alike, but different from all others, and so on, out to n_k which are of a kth kind. Note then that $\sum n_i = n$. Suppose we want to know the number of different permutations, or arrangements in a row, that are possible with these n objects. The answer is no longer $n!$, since those items that are alike are indistinguishable and could be permuted among themselves without changing the arrangement. Thus the answer must be less than $n!$. We can choose any n_1 of the n positions into which to put the first kind of object. This operation can be performed in $\binom{n}{n_1}$ ways. Once the objects of the first type have been assigned their positions, any n_2 of the remaining $n - n_1$ positions may be chosen for the second type of object; this operation can be performed in $\binom{n - n_1}{n_2}$ ways; similarly, the third type of object may be assigned its positions in $\binom{n - n_1 - n_2}{n_3}$ different ways. Thus we can see that the total number of permutations is

$$\binom{n}{n_1}\binom{n - n_1}{n_2}\binom{n - n_1 - n_2}{n_3}\cdots\binom{n - n_1 - n_2 - \cdots - n_{k-1}}{n_k}$$

$$= \frac{n!}{n_1!(n - n_1)!} \cdot \frac{(n - n_1)!}{n_2!(n - n_1 - n_2)!} \cdot \frac{(n - n_1 - n_2)!}{n_3!(n - n_1 - n_2 - n_3)!} \cdots$$

$$\frac{\left(n - \sum_{1}^{k-1} n_i\right)!}{n_k!\left(n - \sum_{1}^{k} n_i\right)!}$$

$$= \frac{n!}{n_1!\, n_2!\, n_3!\, \cdots n_k!}$$

because of the cancellation that occurs from one term to the next (and recall that $n = \sum_{1}^{k} n_i$). This quantity is frequently called a multinomial coefficient and is denoted $\binom{n}{n_1\, n_2\, n_3\, \cdots n_k}$. That is,

$$\binom{n}{n_1\, n_2\, \cdots n_k} = \frac{n!}{\prod_{i=1}^{k} n_i}.$$

EXAMPLE 2.4.7.   The word statistics has 10 letters, of which 3 are s, 3 are t, 2 are i, 1 is a and 1 is c. The number of different 10-letter sequences that can be made with these letters then is

$$\binom{10}{3\,3\,2\,1\,1} = \frac{10!}{3!\,3!\,2!\,1!\,1!\,1!} = 50{,}400,$$

considerably smaller than $10! = 3{,}628{,}800$, the number possible if all the letters were different.

EXERCISE 2.4

1. How many ways can 3 different books be arranged side by side on a shelf?
2. If an item sold by a vending machine costs 40 cents, and the money deposited into the machine must consist of a quarter, a dime, and a nickel, in a how many different orders could the money be inserted into the machine?
3. Six people are about to enter a cave in single file. In how many ways could they arrange themselves in a row to go through the entrance?
4. A bag contains 1 red, 1 black, and 1 green marble. I randomly select 1 of the marbles and record its color. I then replace it in the bag, shake the bag, and randomly select a second marble, again recording its color. The second marble is then replaced and a third marble is randomly selected and its color recorded. How many different samples of 3 colors could occur?
5. An ant farm contains both red and black ants. A particular passage in the farm is so narrow that only 1 ant can get through at a time. If 4 ants follow each other through the passage, how many different color patterns (having 4 elements) could be produced (assuming that red ants are indistinguishable from one another, as are black ants)?
6. A particular city is going to give 3 awards to outstanding residents. If 4 people are eligible to receive them, in how many different ways could they be distributed among the 4 people (assuming that no person may receive more than 1 award)?
7. If a set has 3 elements, how many subsets does it have?
8. Could we define a set that has exactly 9 subsets?
9. How many selections of 5 dominoes can be made from a regular 28-domino double-6 set?
10. In how many ways can 2 teams be chosen from an 8-team league? Thus how many games are necessary if each team is to play every

other team one time? How many games are necessary if each team is to play every other team six times?

11. How many committees of 3 people could be chosen from a group of 10?

12. How many 5-man squads could be chosen from a company of 20 men?

13. Given a set of 15 points in a plane, how many lines would be necessary to connect all possible pairs of points?

14. A "complete graph of order 3" is given by connecting 3 points in all possible ways. If 15 points are joined in all possible ways, how many complete graphs of order 3 would be included? Of order $k = 4, 5, 6$, ... 15?

15. Given a box with two 25-watt, three 40-watt, and four 100-watt bulbs, in how many ways could 3 bulbs be selected from the box, assuming all bulbs are distinguishable?

16. Referring to the light bulbs in Exercise 2.4.15, how many of these selections of three bulbs would include both 25-watt bulbs? How many would include no 25-watt bulbs?

17. How many bulb selections defined in Exercise 2.4.15 would include exactly 1 of each of the 3 different wattages?

18. Many states use automobile license plates that have three letters followed by three digits.
    (a) Assuming all possible combinations can be used, how many different plates can be made?
    (b) If 196 three-letter combinations have a bad connotation, and will not be used, how many plates can be made?
    (c) Assume that each license plate again is to have a total of six positions, three of which must be letters and three of which must be digits. How many different plates could be made (ignore the restriction in (b))?

19. A small town has 100 registered voters, 60 of whom favor a school bond issue. The town treasurer is going to take a poll of 10 registered voters.
    (a) How many different selections of 10 people can be made from the 100 registered voters?
    (b) How many of these will include 6 or more people who favor the bond issue?
    (c) How many will not include 6 or more people who favor the bond issue?

20. If n is any even number, show that

$$\binom{n}{1} + \binom{n}{3} + \cdots + \binom{n}{n-1} = \binom{n}{0} + \binom{n}{2} + \cdots + \binom{n}{n}.$$

(*Hint.* Consider the binomial expansion of $(1 - 1)^n$.)

21. Approximate the value for
    (a) $(.999)^{10}$
    (b) $(.95)^5$
    (c) $(1.99)^6$ (Remember $2(.995) = 1.99$).
22. A committee of three people is to be chosen from four married couples.
    (a) How many different committees are there?
    (b) How many committees in (a) contain two women and one man?
    (c) How many committees are there such that no two committee members are married to each other?
23. Twenty automobiles enter a race. Eight are made by manufacturer A, seven by manufacturer B, and the rest by manufacturer C. Keeping track of only the manufacturer of the cars, in how many ways can the cars cross the finish line?
24. (a) How many of the arrangements listed in Exercise 2.4.23 have one of A's cars in first position?
    (b) How many of the arrangements listed in Exercise 2.4.23 have A's cars in both of the first two positions?
25. How many different 11-letter sequences can be made using the letters in Mississippi? How many of these begin with M and end with i?
26. Count the number of different 4-letter sequences that can be made using the letters in Mississippi.

2.5 Some Particular Probability Problems

In this section we will take up a few problems that should help acquaint the reader with counting techniques and their applications to probability problems. The number of elements in $A \subset S$ is denoted by $n(A)$.

EXAMPLE 2.5.1.   A bag contains 4 red and 2 white marbles. If these are randomly laid out in a row, what is the probability that the 2 end marbles are white? That they are not both white? That the 2 white marbles are side by side? For convenience we assume that the white marbles are numbered 1 and 2 and that the red marbles are numbered 3 through 6. Then we might adopt as our sample space S the collection of $6! = 720$ permutations of 6 things; that is,

$$S = \{(x_1, x_2, \ldots, x_6): x_i = 1, 2, 3, \ldots, 6, \text{ for all } i \text{ and } x_i \neq x_j \text{ for } i \neq j\}.$$

If the marbles are randomly laid out in a row, then each of these 6-tuples is equally likely to occur and we can use our equally likely formula for computing probabilities. Define A to be the event that the first and last marbles are white (the collection of 6-tuples with marbles 1 and 2 on the ends) and

B to be the event that marbles 1 and 2 are side by side. Then the number of elements belonging to *A* is

$$n(A) = 2 \cdot 4! = 48$$

(the white marbles could be on the ends in 2 ways and, for either of these, the red marbles could be arranged between in 4! ways). We also find that

$$n(B) = 5 \cdot 2 \cdot 4! = 240.$$

(There are 5 side-by-side positions for the white marbles to occupy, namely, 12, 23, 34, 45, 56; whichever of these is the one to occur, the white marbles can occupy the selected pair of positions in 2 ways and the red marbles can be permuted in the remaining positions in 4! ways.) As we previously noted

$$n(S) = 6! = 720,$$

and we have

$$P(A) = \tfrac{48}{720} = \tfrac{1}{15}$$

and

$$P(B) = \tfrac{240}{720} = \tfrac{1}{3}.$$

$\bar{A}$ is the event that the white marbles are not on the 2 end positions; we immediately have

$$P(\bar{A}) = 1 - P(A) = \tfrac{14}{15}.$$

An alternative sample space for this problem can be derived as follows. If we pretend that the marbles are going to be put into numbered spots, then all possible outcomes of the experiment can be recorded by keeping track of just the 2 positions that the white marbles occupy; all remaining positions are, of course, filled with the red marbles. All possible pairs of positions are equally likely to occur if the marbles are put down randomly. Thus

$$S = \{(x_1, x_2): x_1 = 1, 2, \ldots, 5; \; x_2 = 2, \ldots, 6; \; x_1 < x_2\}.$$

Note that *S* does list every possible pair of position numbers that we could select for the white marbles and that it lists each one only once. The number of elements belonging to *S* is equal to the number of subsets of size 2 that a set with 6 elements has, that is

$$n(S) = \binom{6}{2} = 15.$$

If we define *A* and *B* as before, exactly 1 of these subsets consists of the largest and the smallest elements of *S* and exactly 5 of them consist of consecutive pairs. Thus

$$n(A) = 1, \qquad n(B) = 5$$

and, as shown,

$$P(A) = \tfrac{1}{15}, \qquad P(\bar{A}) = \tfrac{14}{15}, \qquad P(B) = \tfrac{1}{3}. \qquad \blacksquare$$

In many examples more than one equally likely sample space is possible; used correctly any one of them will give answers to problems of interest.

EXAMPLE 2.5.2.   Suppose that we select a whole number at random between 100 and 999, inclusive. What is the probability that it has at least one 1 in it? What is the probability that it has exactly two 3's in it? For a sample space, we choose

$$S = \{x : x = 100, 101, ..., 999\}.$$

Then $n(S) = 900$ and, since the number is chosen at random, we assume that all single-element events are equally likely to occur. Define the events:

  A: The selected number has at least one 1 in it.
  B: The selected number has exactly two 3's in it.

We will find it easy to compute $n(\bar{A})$, then use this to get $P(\bar{A})$, and finally use Theorem 2.2.2 to compute $P(A) = 1 - P(\bar{A})$. We will compute $n(B)$ directly. The event $\bar{A}$ would be the collection of 3-digit numbers, each of which contains no 1's. The first position of any such number can be filled in 8 ways (since the first digit can be neither 1 nor 0) and each of the succeeding 2 positions can be filled in 9 ways (since 1 cannot occur in either). Thus

$$n(\bar{A}) = 8 \cdot 9 \cdot 9 = 648$$

and

$$P(\bar{A}) = \frac{n(\bar{A})}{n(S)} = .72$$

$$P(A) = 1 - P(\bar{A}) = .28.$$

To compute $n(B)$, we can reason as follows. If the first digit is a 3, then one of the succeeding digits must be a 3 and the other can be any of the remaining nine digits. These two succeeding digits can occur in either of 2 orders, so there are $9 \cdot 2 = 18$ 3-digit numbers having a 3 in the first position and each containing exactly two 3's. If the first digit is not a 3, then this position can be filled in 8 ways (neither 0 nor 3 can be used). The last two positions must then both be filled with 3's. Thus

$$n(B) = 18 + 8 = 26$$

and

$$P(B) = \frac{n(B)}{n(S)} \doteq .029.$$

∎

EXAMPLE 2.5.3.   Suppose that n people are in a room. If we make a list of their birthdates (month and day of the month), what is the probability that there will be one or more repetitions in the list? (What is the probability that two or more people have the same birthday?) We will make the assumption that there are only 365 days available for each birthday (ignoring February 29) and that each of these days is equally likely to occur for any individual's birthday. (It can, in fact, be shown that this is the worst possible assumption we might make relative to this probability; that is, if days in March or some other month are more likely to occur as birthdays, then the probability of one or more repeated birthdays is larger than if all days are equally likely.) Our sample space is the collection of all possible n-tuples that could occur for the birthdays, numbering the days of the year sequentially from 1 to 365. Thus

$$S = \{(x_1, x_2, \ldots, x_n): x_i = 1, 2, \ldots, 365; \qquad i = 1, 2, \ldots, n\}.$$

The first position in each n-tuple gives the first person's birthday; the second position gives the second person's birthday, and so on. Assuming that all days of the year are equally likely for each person's birthday implies that each of these n-tuples is equally likely to occur. By using the counting techniques presented in Section 2.4, we see that

$$n(S) = 365^n.$$

Define A to be the event that there is one or more repetitions of the same number in the n-tuple that occurs. Then $\bar{A}$ is the collection of n-tuples that have no repetitions; we can see easily that (with $n \leq 365$)

$$n(\bar{A}) = 365 \cdot 364 \cdot 363 \cdots (365 - n + 1) = {}_{365}P_n,$$

which gives us

$$P(\bar{A}) = \frac{n(\bar{A})}{n(S)} = \frac{365 \cdot 364 \cdot 363 \cdots (365 - n + 1)}{365^n}.$$

Again, from Theorem 2.2.2,

$$P(A) = 1 - P(\bar{A}).$$

Table 2.5.1 gives the values of $P(\bar{A})$ and $P(A)$ for various values of n. It is somewhat surprising that the probability of a repetition exceeds $\frac{1}{2}$ for as few as 23 people in the room and that for 60 people it is a virtual certainty.

∎

Table 2.5.1

n	$P(\bar{A})$	$P(A)$
10	.871	.129
20	.589	.411
21	.556	.444
22	.524	.476
23	.493	.507
24	.462	.538
25	.431	.569
30	.294	.706
40	.109	.891
50	.030	.970
60	.006	.994

EXAMPLE 2.5.4. Suppose that Mrs. Riley claims to be a clairvoyant. Specifically, she claims that if she is presented with 8 cards, 4 of which are red and 4 of which are black, she will correctly identify the color of at least 6 of the cards without being able to see their colors. If she is guessing and has no special ability, what is the probability that she would correctly identify at least 6 out of the 8 cards? (She will identify 4 of the cards as red and 4 of the cards as black.) We arbitrarily decide to present her with the 4 red cards first, one by one, and then present her with the 4 black cards. The sample space for the experiment is the set of all possible guesses she might give for the colors of the cards; that is,

$$S = \{(x_1, x_2, \ldots, x_8) : x_i = R \text{ or } B, \text{ for all } i; \text{ exactly 4 } x_i\text{'s are } R\}.$$

If she is guessing, then each of the single-element events is equally likely to occur. Since each 8-tuple belonging to S contains exactly 4 R's and exactly 4 B's, we can compute $n(S)$ by counting how many ways we might select 4 positions from the 8 in which to place the R's; thus

$$n(S) = \binom{8}{4} = 70.$$

Define

> *A* : She identifies at least 6 cards correctly.
> *B* : She identifies exactly 6 cards correctly.
> *C* : She identifies all 8 cards correctly.

Since she will call 4 cards red and 4 cards black, it is not possible for her to be correct on exactly 7 cards; thus

$$A = B \cup C$$

and since

$$B \cap C = \emptyset,$$
$$P(A) = P(B) + P(C).$$

Clearly, $n(C) = 1$ so $P(C) = \frac{1}{70}$. If B is to occur, she must identify exactly 3 of the 4 red cards correctly and exactly 3 of the 4 black cards correctly. The 1 red card on which she is wrong could be any of the 4 and the 1 black card on which she is wrong could be any one of the 4. Thus the number of 8-tuples having 1 B in the first 4 positions and 1 R in the last 4 positions is

$$n(B) = 4 \cdot 4 = 16$$

and we have

$$P(B) = \tfrac{16}{70}.$$

Thus

$$P(A) = \tfrac{1}{70} + \tfrac{16}{70} = \tfrac{17}{70} \doteq .243;$$

if she only guesses, there is slightly less than 1 chance in 4 that she will do as well as she claims.   ■

Equally likely sample spaces must be chosen with care. Occasionally, what might appear to be a reasonable "equally likely" sample space is not. Example 2.5.5 presents a simple case of this type.

EXAMPLE 2.5.5.   Inadvertently, two cold tablets and two aspirin tablets were placed in the same box; to the eye, all four tablets appear identical. Douglas and Hugh are each going to select a tablet from the box (at random, Hugh after Douglas). Because there are two tablets of each kind in the box, it is possible for both of them to select a cold tablet or for both of them to select an aspirin tablet. Thus we might use the sample space

$$S = \{aa, ac, ca, cc\},$$

where *a* represents aspirin tablet, *c* represents cold tablet and Douglas's selection is stated first; because the selections are made at random, we could also make the equally likely assumption. Let D be the event that Douglas selects a cold tablet, H be the event that Hugh selects a cold tablet and then $D \cap H$ is the event that both select a cold tablet. Thus $D = \{ca, cc\}$, $H = \{ac, cc\}$, $D \cap H = \{cc\}$, and with the equally likely assumption we have $P(D) = P(H) = \frac{1}{2}$, $P(D \cap H) = \frac{1}{4}$. Having read the preceding discussion and analysis, consider the following alternative approach. Assume the tablets are numbered 1 through 4 with the numbers 1 and 2 assigned to the aspirin tablets. The sample space is

$$S = \{(x_1, x_2) : x_i = 1, 2, 3, 4, i = 1, 2, \ x_1 \neq x_2\}$$

and the equally likely assumption is made (again, the first position corresponds to Douglas's selection, the second, to Hugh's). Thus the number of elements in S is $4 \cdot 3 = 12$ and defining events D and H as before,

$$n(D) = 2 \cdot 3 = 6, \qquad n(H) = 3 \cdot 2 = 6, \qquad n(D \cap H) = 2 \cdot 1 = 2,$$

and we have

$$P(D) = P(H) = \tfrac{6}{12} = \tfrac{1}{2}, \qquad P(D \cap H) = \tfrac{2}{12} = \tfrac{1}{6},$$

not $\frac{1}{4}$ as previously. Why are different answers obtained for $P(D \cap H)$ and which of the two approaches appears correct to you? ∎

EXERCISE 2.5

1. Five white and three red balls are laid out in a row at random. What is the probability that both end balls are white? That one is red and the other is white?

2. For the situation discussed in Exercise 2.5.1, what is the probability that all the red balls are together? That all the white balls are together?

3. A 5-digit number is selected at random. What is the probability that it contains no 5's? That it contains exactly one 5?

4. For the selection discussed in Exercise 2.5.3, what is the probability that all the digits are the same? That all the digits are different?

5. Ten people in total are nominated for a slate of 3 offices. If every group of 3 people has the same probability of winning, what is the probability that a particular person will be on the winning slate? That a particular pair of people will be on the winning slate?

6. Two people are to be selected at random to be set free from a prison with a population of 100. What is the probability that the oldest prisoner is 1 of the 2 selected? That the oldest and youngest are the pair selected?

7. A particular item is stocked in three different sizes. An order is received for two items, but the sizes are not specified. The clerk filling the order

arbitrarily (or by guessing) simply selects two items and ships them out.

   (a) What is the probability that the two items shipped were the ones desired (they were the right sizes)?

   (b) What is the probability that neither item shipped was correct?

8.   A, B, and C are going to race. What is the probability that A will finish ahead of C, given that all are of equal ability (and no ties can occur)? What is the probability that A will finish ahead of both B and C?

9.   Each of 5 people is asked to distinguish between vanilla ice cream and French vanilla custard (each is given a small sample of both and asked to identify which is ice cream). If all 5 people are guessing, what is the probability that all will correctly identify the ice cream? If all 5 are guessing, what is the probability that at least 4 will identify the ice cream correctly?

10.   Compute the probability that a group of 5 cards drawn at random from a 52-card deck will contain

   (a) Exactly 2 pair.

   (b) A full house (3 of one denomination and 2 of another).

   (c) A flush (all 5 from the same suit).

   (d) A straight (5 in sequence, beginning with ace or deuce or trey, ..., or ten).

11.   n people are in a room. Compute the probability that at least 2 have the same birth month. Evaluate this probability for $n = 3, 4, 5, 6$.

12.   The same 3 pilots use the same 3 aircraft for a flight, on each of two successive days. Which pilot uses which plane is determined by chance, on both days.

   (a) What is the probability that each pilot gets the same plane both days?

   (b) What is the probability that the plane each pilot gets the second day is different from the plane he had the first day?

13.   Ten men go salmon fishing on a commercial boat. When the boat returns, you see that three salmon were caught. Assume all possibilities of who caught each fish are equally likely.

   (a) What is the probability that the three fish were caught by three different men?

   (b) What is the probability that all three were caught by the same man?

14.   A person is to be presented with 3 red and 3 white cards in a random sequence. He knows that there will be 3 of each color; thus he will identify 3 cards as being of each color. If he is guessing, what is his probability of correctly identifying all 6 cards? Of identifying exactly 5 correctly? Exactly 4?

15.   In Example 2.5.4, what is the probability that Mrs. Riley identifies exactly 4 cards correctly?

2.6 Conditional Probability

There are applications in which we will be given that an event B has occurred and we want to know the probability that A occurred, conditioned on, or given the occurrence of the event B. For example, a manufacturer of programmable calculators may order silicon chips from any of three different suppliers; given that an arriving lot was produced by supplier 1 (event B has occurred), he would be interested in the probability that the proportion of defective chips is less than, say, 3 percent (event A). Or given that a person selected at random in an opinion poll is registered as a Republican (event B), we could be interested in the probability that she will vote for the Republican candidate in the next election. Or granted that a person selected at random has a family history of diabetes (event B), it is of interest to know the probability that he also has diabetes. This conditional probability for A, given that B has occurred, will be denoted by $P(A|B)$ to distinguish it from the unconditional probability of the occurrence of A.

How should $P(A|B)$ be defined? Clearly, if event B has occurred, it must have occurred either with event A (thus $A \cap B$ occurred) or it must have occurred without event A occurring (thus $\bar{A} \cap B$ occurred). It would seem rational to let $P(A|B)$ equal

$$\frac{P(A \cap B)}{P(A \cap B) + P(\bar{A} \cap B)},$$

which in the relative frequency sense gives the proportion of the time that A occurs with B relative to the proportion of the time that B occurs (either with or without A). Recognizing that $A \cap B$ and $\bar{A} \cap B$ are mutually exclusive and that $(A \cap B) \cup (\bar{A} \cap B) = B$, the preceding denominator is simply $P(B)$ and we have the following definition for $P(A|B)$.

DEFINITION 2.6.1. The *conditional* probability of A occurring, given that B has occurred, is

$$P(A|B) = \frac{P(A \cap B)}{P(B)}$$

where $P(B) > 0$. ■

The conditional probability for A may be smaller than, equal to, or greater than the unconditional probability for A.

EXAMPLE 2.6.1.   We roll a pair of fair dice 1 time and are given that the 2 numbers that occur are not the same. Compute the probability that the sum is 7 or that the sum is 4 or that the sum is 12. Define the event:

A: The two numbers that occur are different.

Then we are given that A has occurred. Also define the events:

B: The sum is 7.
C: The sum is 4.
D: The sum is 12.

Then assuming equally likely single-element events, we find that

$$P(A) = \tfrac{5}{6}, \qquad P(B) = \tfrac{1}{6}, \qquad P(C) = \tfrac{1}{12}, \qquad P(D) = \tfrac{1}{36},$$
$$P(A \cap B) = \tfrac{1}{6}, \qquad P(A \cap C) = \tfrac{1}{18}, \qquad P(A \cap D) = 0$$

and thus we have

$$P(B \mid A) = \frac{\tfrac{1}{6}}{\tfrac{5}{6}} = \tfrac{1}{5}$$

$$P(C \mid A) = \frac{\tfrac{1}{18}}{\tfrac{5}{6}} = \tfrac{1}{15}$$

$$P(D \mid A) = \frac{0}{\tfrac{5}{6}} = 0.$$

∎

EXAMPLE 2.6.2.   A car manufacturer produced 500,000 copies of one of his popular models in a given year. To hold costs down, he produced the car in only four colors: red, yellow, blue, and green. The numbers of cars painted these colors were 100,000, 150,000, 175,000, and 75,000, respectively. Six months after this model went out of production it was found that the steering linkage installed in 75,000 of these models had a defect that could prove serious and the cars affected were to be recalled and repaired. Of the 75,000 with the steering defect, 10,000 were painted red; 10,000, blue; 30,000, yellow; and the remainder, green. Suppose one of the green cars of this model is selected at random (I purchased it), what is the probability it will be recalled for the steering repair? To answer this question, we will use a sample S whose elements are 2-tuples. The first element of each 2-tuple identifies the color (r, y, b, g) and the second element identifies whether or not the steering linkage is defective (d for defective, n for nondefective). Thus

$$S = \{(r, d), (r, n), (y, d), (y, n), (b, d), (b, n), (g, d), (g, n)\}.$$

The information given then specifies that

$$P(\{(r,d),(r,n)\}) = \frac{100{,}000}{500{,}000} = .2$$

$$P(\{(y,d),(y,n)\}) = \frac{150{,}000}{500{,}000} = .3$$

$$P(\{(b,d),(b,n)\}) = \frac{175{,}000}{500{,}000} = .35$$

$$P(\{(g,d),(g,n)\}) = \frac{75{,}000}{500{,}000} = .15$$

and that

$$P(\{(r,d)\}) = \frac{10{,}000}{500{,}000} = .02$$

$$P(\{(b,d)\}) = \frac{10{,}000}{500{,}000} = .02$$

$$P(\{(y,d)\}) = \frac{30{,}000}{500{,}000} = .06$$

$$P(\{(g,d)\}) = \frac{25{,}000}{500{,}000} = .05$$

from which it is possible to completely specify the probability function for S. Now define B to be the event a green car is selected and let A be the event the car has the steering linkage defect. Then $P(A|B) = \frac{.05}{.15} = \frac{1}{3}$, and it is straightforward to compute the probability of recall for any other color as well. ∎

One of the most frequent uses for conditional probability is to give an easy procedure for assigning probabilities to intersections of events. Since

$$P(A|B) = \frac{P(A \cap B)}{P(B)},$$

it must be true that

$$P(A \cap B) = P(B)\,P(A|B),$$

and reversing the roles of A and B, we have

$$P(A \cap B) = P(A)\,P(B|A).$$

This type of result is easily extended by induction. For example,

$$P(A|B \cap C) = \frac{P(A \cap B \cap C)}{P(B \cap C)}$$

immediately gives

$$P(A \cap B \cap C) = P(A|B \cap C) P(B \cap C),$$

and since $P(B \cap C) = P(B|C) P(C)$,

$$P(A \cap B \cap C) = P(C) P(B|C) P(A|B \cap C).$$

Thus probabilities for intersections of events can be built up recursively from unconditional and conditional probabilities.

EXAMPLE 2.6.3. We select 2 balls at random without replacement from an urn that contains 4 white and 8 black balls. (a) Compute the probability that both balls are white. (b) Compute the probability that the second ball is white.

(a) Define

A: The first ball is white.
B: The second ball is white.
C: Both balls are white.

Then

$$A \cap B = C \quad \text{and} \quad P(C) = P(A \cap B) = P(A) P(B|A) = \tfrac{4}{12} \cdot \tfrac{3}{11} = \tfrac{1}{11}.$$

(b) Clearly, $B = (A \cap B) \cup (\bar{A} \cap B)$ and

$$(A \cap B) \cap (\bar{A} \cap B) = \varnothing.$$

Then

$$\begin{aligned}
P(B) &= P(A \cap B) + P(\bar{A} \cap B) \\
&= P(A) P(B|A) + P(\bar{A}) P(B|\bar{A}) \\
&= \tfrac{4}{12} \cdot \tfrac{3}{11} + \tfrac{8}{12} \cdot \tfrac{4}{11} \\
&= \tfrac{1}{3}.
\end{aligned}$$

Notice that the probability of drawing a white ball the second draw (even though the first one is not replaced) is $\tfrac{1}{3}$, the same as the probability that the first ball drawn is white. This can be shown to be the case generally. $P(B)$ is an unconditional probability and, as indicated by the way it was previously computed, it is an average of the two conditional probabilities $P(B|A)$ and $P(B|\bar{A})$. ∎

EXAMPLE 2.6.4. Box 1 contains 4 defective and 16 nondefective light bulbs. Box 2 contains 1 defective and 1 nondefective light bulb. We roll a fair die 1 time. If we get a 1 or a 2, then we select a bulb at random from box 1. Otherwise we select a bulb from box 2. What is the probability that the selected bulb will be defective? Define

A : We select a bulb from box 1.
B : The selected bulb is defective.

Then $P(A) = \frac{1}{3}$, $P(\bar{A}) = \frac{2}{3}$, $P(B|A) = \frac{1}{5}$, and $P(B|\bar{A}) = \frac{1}{2}$. Since

$$B = (B \cap A) \cup (B \cap \bar{A})$$

and

$$(B \cap A) \cap (B \cap \bar{A}) = \varnothing$$

we have

$$
\begin{aligned}
P(B) &= P(B \cap A) + P(B \cap \bar{A}) \\
&= P(A)\,P(B|A) + P(\bar{A})\,P(B|\bar{A}) \\
&= \tfrac{1}{3} \cdot \tfrac{1}{5} + \tfrac{2}{3} \cdot \tfrac{1}{2} = \tfrac{6}{15} = \tfrac{2}{5}.
\end{aligned}
$$ ∎

The events $E_1, E_2, \ldots, E_n$ are called a *partition* of the sample space S if $E_i \cap E_j = \varnothing$ for all $i \ne j$ and $E_1 \cup E_2 \cup \cdots \cup E_n = S$. Thus a partition cuts the whole sample space into mutually exclusive pieces. Figure 2.6.1 gives a Venn diagram with $n = 7$ events in the partition.

If $A \subset S$ is any event and $E_1, E_2, \ldots, E_n$ is a partition of S, then $E_1, E_2, \ldots, E_n$ also partition A; that is,

$$A = (A \cap E_1) \cup (A \cap E_2) \cup \cdots \cup (A \cap E_n),$$

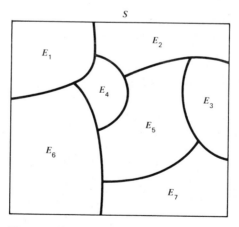

Figure 2.6.1

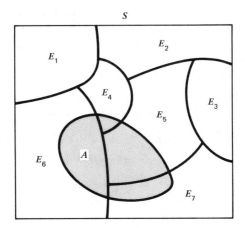

S

Figure 2.6.2

and, of course, $(A \cap E_i) \cap (A \cap E_j) = \varnothing$ for all $i \neq j$. Figure 2.6.2 pictures this partitioning of the event A, again with $n = 7$ events in the partition. It then follows that we can write

$$P(A) = P(A \cap E_1) + P(A \cap E_2) + \cdots + P(A \cap E_n),$$

a result known as the *theorem of total probability*. This result was actually used earlier in Example 2.6.4 and is illustrated once more in the following example.

EXAMPLE 2.6.5.   A calculator manufacturer buys the same integrated circuit from three different suppliers, call them I, II, and III. From past experience, 1 percent of the circuits supplied by I have been defective, 3 percent of those supplied by II have been defective, and 4 percent of those supplied by III have been defective. Granted that this manufacturer buys 30 percent of his circuits from I, 50 percent from II, and the rest from III, we can use the theorem of total probability to compute the probability that an integrated circuit, checked just before final assembly into a calculator, is found to be defective. Thus let A be the event that the chip is found defective and let E_1, E_2, E_3 be the events that the chip selected was manufactured by I, II, III, respectively, and we have $P(E_1) = .3$, $P(E_2) = .5$, $P(E_3) = .2$, $P(A|E_1) = .01$, $P(A|E_2) = .03$, $P(A|E_3) = .04$. Thus $P(A \cap E_1) = .003$, $P(A \cap E_2) = .015$, $P(A \cap E_3) = .008$ so

$$P(A) = .003 + .015 + .008 = .021. \qquad \blacksquare$$

The theorem of total probability can be used to easily establish *Bayes' Theorem,* named after the Reverend T. Bayes; the result is commonly credited with first being published in 1764 in Bayes' posthumous memoirs. It is used extensively in Bayesian methods of statistical inference, some of which are discussed in Chapter 11.

Theorem 2.6.1. Let $E_1, E_2, ..., E_n$ be a partition of S. Then for any event $A \subset S$

$$P(E_i|A) = \frac{P(E_i) P(A|E_i)}{\sum\limits_{j=1}^{n} P(E_j) P(A|E_j)}, \qquad i = 1, 2, ..., n.$$

Proof: By definition

$$P(E_i|A) = \frac{P(E_i \cap A)}{P(A)},$$

and since $P(E_i \cap A) = P(E_i) P(A|E_i)$,

$$P(A) = \sum\limits_{j=1}^{n} P(A \cap E_j)$$

$$= \sum\limits_{j=1}^{n} P(E_j) P(A|E_j),$$

the result follows immediately. ■

EXAMPLE 2.6.6. Assume that the probability is .95 that the jury selected to try a criminal case will arrive at the appropriate verdict. That is, given a guilty defendant on trial, the probability is .95 that the jury will find him guilty and, conversely, given an innocent man on trial, the probability is .95 that the jury will find him innocent. Suppose that the local police force is quite diligent in its duties and that 99 percent of the people brought before the court are actually guilty. Let us compute the probability that a defendant is innocent, given that the jury finds him innocent. Let G be the event that the defendant is guilty and let J be the event that the jury finds him guilty. Then we are given that $P(J|G) = P(\bar{J}|\bar{G}) = .95$, $P(G) = .99$, and we want to compute $P(\bar{G}|\bar{J})$. G and $\bar{G}$ form a partition of the sample space, so from Bayes' Theorem

$$P(\bar{G}|\bar{J}) = \frac{P(\bar{J}|\bar{G}) P(\bar{G})}{P(\bar{J}|\bar{G}) P(\bar{G}) + P(\bar{J}|G) P(G)}$$

$$= \frac{(.95)(.01)}{(.95)(.01) + (.05)(.99)} = .161.$$

Thus there is about 1 chance in 6 he really is innocent, if found innocent, and about 5 chances in 6 he is guilty, although found innocent. Similarly, the probability is .0005 that he is innocent, if found guilty, and is .9995 that he is in fact guilty if found guilty. ∎

It is not, at first glance, easy to grasp the meaning of Bayes' Theorem. Note that it gives the probability of occurrence of E_i, one of the events in a partition, given an event A has occurred, which seems in a sense backward. Some of its original uses were concerned with $E_1, E_2, ..., E_n$, which represented various possible mutually exclusive theories of how the world was created or how it had reached its current state; the values $P(E_i)$ were called *prior* probabilities. The event A represents some event that is known to have occurred, such as a recorded history to a given point in time. Bayes' Theorem then shows how to evaluate $P(E_i|A)$, the conditional probability theory i is correct (called the *posterior* probability), given the occurrence of A. It is well suited for adaptive schemes of using data or experience to modify prior beliefs. Criticism of Bayesian procedures generally centers on the assumption that $P(E_i)$ and $P(A|E_i)$ are necessarily known. If they are not, of course, it is not possible to employ the theorem.

EXAMPLE 2.6.7.   Let us assume the same situation as Example 2.6.5: 30 percent of the integrated circuits are supplied by I, 50 percent, by II, and 20 percent, by III, and the probabilities of defects for these suppliers are $P(A|E_1) = .01$, $P(A|E_2) = .03$, $P(A|E_3) = .04$. Now suppose that an unlabeled box of these integrated circuits is found; it is known only that all the integrated circuits in the box came from the same supplier, which particular one is unknown. Without testing any of the circuits, it would seem reasonable to assume that the probabilities the box came from each of th three are $P(E_1) = .3$, $P(E_2) = .5$, $P(E_3) = .2$, respectively, because these ? c the proportions of circuits purchased from the three suppliers. If one cir.uit is selected from the box and tested, Bayes' Theorem can be used to compute new probabilities that the box came from each of the three suppliers, given the result of testing the circuit. If, for example, the circuit is tested and found defective, then the event A, defective circuit, has occurred and we have

$$P(E_1|A) = \frac{P(E_1)\,P(A|E_1)}{P(E_1)\,P(A|E_1) + P(E_2)\,P(A|E_2) + P(E_3)\,P(A|E_3)}$$

$$= \frac{(.3)(.01)}{(.3)(.01) + (.5)(.03) + (.2)(.04)}$$

$$= .115$$

$$P(E_2|A) = \frac{(.5)(.03)}{(.3)(.01) + (.5)(.03) + (.2)(.04)}$$

$$= .577$$

$$P(E_3|A) = \frac{(.2)(.04)}{(.3)(.01) + (.5)(.03) + (.2)(.04)}$$

$$= .308.$$

On the other hand, if the circuit tested is found to be nondefective then the event $\bar{A}$, nondefective circuit, has occurred and we would have

$$P(E_1|\bar{A}) = \frac{P(E_1)P(\bar{A}|E_1)}{P(E_1)P(\bar{A}|E_1) + P(E_2)P(\bar{A}|E_2) + P(E_3)P(\bar{A}|E_3)}$$

$$= \frac{(.3)(.99)}{(.3)(.99) + (.5)(.97) + (.2)(.96)}$$

$$= .305$$

$$P(E_2|\bar{A}) = \frac{(.5)(.97)}{(.3)(.99) + (.5)(.97) + (.2)(.96)}$$

$$= .498$$

$$P(E_3|\bar{A}) = \frac{(.2)(.96)}{(.3)(.99) + (.5)(.97) + (.2)(.96)}$$

$$= .197.$$

Thus the posterior probabilities that the box was supplied by I, II, or III are changed more from their initial unconditional values by finding the circuit defective than they are by finding the circuit nondefective.

Bayes' Theorem or formula can be applied more than once. If one circuit has already been selected, tested, and found defective the probabilities that the box came from suppliers I, II, and III are (with a slight shift in notation so we can apply Bayes' Theorem again without confusion) $P(E_1) = .115$, $P(E_2) = .577$, and $P(E_3) = .308$. If a second item is selected and found defective, we have

$$P(E_1|A) = \frac{(.115)(.01)}{(.115)(.01) + (.577)(.03) + (.308)(.04)}$$

$$= .037$$

$$P(E_2|A) = \frac{(.577)(.03)}{(.115)(.01) + (.577)(.03) + (.308)(.04)}$$

$$= .562$$

$$P(E_3|A) = \frac{(.308)(.04)}{(.115)(.01) + (.577)(.03) + (.308)(.04)}$$

$$= .400$$

which would be the probabilities that the box came from the three suppliers, given two circuits were tested and both found defective. ∎

EXERCISE 2.6

1. An urn contains 4 balls numbered 1, 2, 3, 4, respectively. Two balls are drawn without replacement. Let A be the event that the sum is 5 and let B_i be the event that the first ball drawn has an i on it, $i = 1, 2, 3, 4$. Compute $P(A|B_i)$, $i = 1, 2, 3, 4$, and $P(B_i|A)$, $i = 1, 2, 3, 4$.
2. Suppose that the two balls of Exercise 2.6.1 are drawn with replacement. Let A and B_i be as previously defined and compute $P(A|B_i)$ and $P(B_i|A)$, $i = 1, 2, 3, 4$.
3. A fair coin is flipped 4 times. What is the probability that the fourth flip is a head, given that each of the first 3 flips resulted in heads?
4. A fair coin is flipped 4 times. What is the probability that the fourth flip is a head, given that 3 heads occurred in the 4 flips? Given that 2 heads occurred in the 4 flips?
5. Urn 1 contains 2 red and 4 blue balls, urn 2 contains 10 red and 2 blue balls. If an urn is chosen at random and a ball is removed from the chosen urn, what is the probability that the selected ball is blue? That it is red?
6. Suppose that in Exercise 2.6.5, instead of selecting an urn at random, we roll a die and select the ball from urn 1 if a 1 occurs on the die and otherwise select the ball from urn 2. What is the probability that the selected ball is blue? That it is red?
7. Five cards are selected at random without replacement from a 52-card

deck. What is the probability that they are all red? That they are all diamonds?

8.   An urn contains 2 red, 2 white, and 2 blue balls. Two balls are selected at random without replacement from the urn. Compute the probability that the second ball drawn is red.

9.   An urn contains 2 black and 5 brown balls. A ball is selected at random. If the ball drawn is brown, it is replaced and 2 additional brown balls are also put into the urn. If the ball drawn is black, it is not replaced in the urn and no additional balls are added. A ball is then drawn from the urn the second time. What is the probability that it is brown?

10.   The two-stage experiment described in Exercise 2.6.9 was performed and we are given that the ball selected at the second stage was brown. What is the probability that the ball selected at the first stage was also brown?

11.   Suppose that medical science has a cancer-diagnostic test that is 95 percent accurate on both those who do and those who do not have cancer. If .005 of the population actually does have cancer, compute the probability that a particular individual has cancer, given that the test says he has cancer.

12.   In a large midwestern school 1 percent of the student body participates in the intercollegiate athletic program; 10 percent of these people have a grade point of 3 or more (out of 4), whereas 20 percent of the remainder of the student body have a grade point of 3 or more. What proportion of the total student body has a grade point of 3 or more? Suppose we select 1 student at random from this student body and find that he has a grade point of 3.12. What is the probability that he participates in the intercollegiate athletic program?

13.   Two different suppliers, A and B, provide a manufacturer with the same part. All supplies of this part are kept in a large bin. In the past, 5 percent of the parts supplied by A and 9 percent of the parts supplied by B have been defective. A supplies four times as many parts as B. Suppose you reach into the bin and select a part, and find it is nondefective. What is the probability that it was supplied by A?

14.   Two fair dice are rolled one time. Given that the sum of the two numbers that occurred was at least 7, compute the probability that it was equal to i, for $i = 7, 8, 9, 10, 11, 12$.

15.   Assume that, among families with two children, there are equal numbers of families with (boy, boy), (boy, girl), (girl, boy) and (girl, girl), where the ordering in the 2-tuple indicates the order of birth. We select a family with two children at random.

(a) What is the probability that the family has 2 boys, given that it has at least one boy?

(b) Suppose you are married and currently have one child, a boy. You are expecting your second child. Clearly, once you have two children, you will have at least one boy. Is the answer to (a) the probability that your second child will be a boy?

16. Sixteen teams enter a single-elimination tournament, playing a game that cannot end in a tie. Thus a total of four rounds will be required to determine the winner. Assume that the probability your team wins its first game is .9 and that the probabilities it will win its succeeding games are .8, .7, .6, respectively, given that it won its preceding games.
   (a) What is the probability that your team will win the tournament?
   (b) What is the probability that your team is eliminated in the third round? The second round? The first round?

17. Every 25th person to enter an amusement park is given a prize.
   (a) What is the probability that you receive a prize when you enter?
   (b) Given that you have waited, and observed that none of the 15 people to enter before you received a prize, what is the probability that you receive a prize?

18. Assume that two circuits are selected from the box, discussed in Example 2.6.7, and both are found nondefective. Compute the probabilities that the box came from each of the three suppliers.

2.7   Independent Events

The concept of independent events is very important and will be used frequently in the material to follow. By saying events A and B are independent, we mean that the occurrence or nonoccurrence of event B provides no information about whether event A has also occurred; stated more precisely, we want to call events A and B independent if

$$P(A) = P(A|B) = P(A|\bar{B}).$$

If the unconditional probability of A occurring is equal to $P(A|B)$ and $P(A|\bar{B})$, then we do not know any more about the occurrence of A after we are given that B occurred (or did not occur) than we did before we had this knowledge. Granted

$$P(A) = P(A|B) = \frac{P(A \cap B)}{P(B)}$$

it then follows that $P(A \cap B) = P(A)P(B)$, if A and B are independent. We will take this equation as our definition of independence of events A and B, as given in the following definition.

DEFINITION 2.7.1. Two events, A and B, are independent if and only if $P(A \cap B) = P(A) P(B)$. ■

Independence is a symmetric relation. If $P(A \cap B) = P(A) P(B)$, then $P(A|B) = P(A)$ and $P(B|A) = P(B)$; you can also verify that if A and B are independent, then so are A and $\bar{B}$, $\bar{A}$ and B, as well as $\bar{A}$ and $\bar{B}$.

EXAMPLE 2.7.1. Assume that the numbers given in the cells of Table 2.7.1 give the probabilities of a randomly selected individual falling into the given cell.

Table 2.7.1

	Gets Cancer	Does Not Get Cancer
Smoker	.50	.20
Nonsmoker	.10	.20

That is, if we let A be the event that the selected individual is a smoker and let B be the event that the selected individual gets cancer, then

$$P(A \cap B) = .5, \qquad P(A \cap \bar{B}) = .2, \qquad P(\bar{A} \cap B) = .1$$

and

$$P(\bar{A} \cap \bar{B}) = .2.$$

Since

$$P(A) = P(A \cap B) + P(A \cap \bar{B}) = .7$$
$$P(B) = P(A \cap B) + P(\bar{A} \cap B) = .6,$$

we see that $P(A \cap B) = .5 \neq (.7)(.6)$, so A and B are not independent. ■

EXAMPLE 2.7.2. If 2 fair dice are rolled 1 time, show that the 2 events

A : The sum of the 2 dice is 7.
B : The 2 dice have the same number.

are not independent. As seen before, $P(A) = P(B) = \frac{1}{6}$; $A \cap B = \varnothing$, so $P(A \cap B) = 0$, which is not $\frac{1}{6} \cdot \frac{1}{6}$. Thus the 2 events are not independent. ■
The definitions of independence and mutually exclusive are often confused; this is probably caused by the fact that in common English usage the word independent is frequently used to mean "having nothing to do with." The phrase "having nothing to do with" could be interpreted to mean that

two events could not happen together, which means the two events are mutually exclusive. The two concepts are certainly not the same in content, as we can see from the following theorem.

Theorem 2.7.1. Assume that $P(A) \neq 0$ and $P(B) \neq 0$. Then A and B independent implies that they are not mutually exclusive and A and B mutually exclusive implies that they are not independent.

Proof: Suppose that A and B are independent. Then $P(A \cap B)$ $= P(A) P(B) \neq 0$, since $P(A) \neq 0$ and $P(B) \neq 0$. Thus they are not mutually exclusive. Now suppose A and B are mutually exclusive. Then $A \cap B = \varnothing$ and $P(A \cap B) = 0$. But since $P(A) \neq 0$ and $P(B) \neq 0$, $P(A) P(B) \neq 0$ and they are then not independent. ∎

Independence of three events is defined as follows.

DEFINITION 2.7.2. A, B, and C are independent if and only if:

1. $P(A \cap B) = P(A) P(B)$.
2. $P(A \cap C) = P(A) P(C)$.
3. $P(B \cap C) = P(B) P(C)$.
4. $P(A \cap B \cap C) = P(A) P(B) P(C)$.   ∎

Many examples can be given that show that the first three of these conditions do not imply the fourth and vice versa. The following is an example in which the first three equations hold but the fourth does not.

EXAMPLE 2.7.3.   Assume a fair coin is flipped two times and define the three events.

A: Head on first flip.
B: Head on second flip.
C: Same face on both flips.

Then we can easily see that $P(A) = P(B) = P(C) = \frac{1}{2}$ and that $P(A \cap B)$ $= P(A \cap C) = P(B \cap C) = \frac{1}{4}$, so equations 1, 2, and 3 of Definition 2.7.2 are satisfied; A, B, and C are called *pairwise independent events*. But since

$$P(A \cap B \cap C) = \tfrac{1}{4} \neq (\tfrac{1}{2})^3,$$

they do not satisfy equation (4) and thus are not independent events. ∎

The n events $A_1, A_2, ..., A_n$ are completely or mutually independent if and only if $P(A_i \cap A_j) = P(A_i) P(A_j)$ for any two of the events and $P(A_i \cap A_j \cap A_k) = P(A_i) P(A_j) P(A_k)$ for any three of them, and so on, out to and including $P(A_1 \cap A_2 \cap ... \cap A_n) = P(A_1) P(A_2) ... P(A_n)$. Thus to prove the independence of n events we would have to verify $\binom{n}{2} + \binom{n}{3}$ $+ \cdots + \binom{n}{n} = 2^n - n - 1$ separate equations. The real use of independence of events is in the assignment of probabilities to intersections of events. Granted n events are independent, the probability of the intersection of any number of them is given by the product of the unconditional probabilities of the events in the intersection.

EXAMPLE 2.7.4. According to the 1958 standard ordinary mortality table, the probability a 20-year-old person will live to age 65 (or more) is about .704. If we assume three 20-year-old friends will reach age 65 (or not) independently, we then can compute the probabilities of any events defined by whether or not the individuals will live to age 65. Perhaps the clearest way to see this is to formally adopt a sample space S with 3-tuples as elements, the three positions corresponding to the 3 friends; each element of each 3-tuple is either y (for yes, the person is alive on her 65th birthday) or n (for no, she did not make it). Then we assume that the probability of y occurring for each person is .704 and the probability of n occurring is $1 - .704 = .296$; the independence assumption allows us to assign probabilities to all of the 3-tuples in S by multiplying together the probabilities for the appropriate components. Thus

$$P(\{(y, y, y)\}) = (.704)(.704)(.704) = .349$$
$$P(\{(y, n, y)\}) = (.704)(.296)(.704) = .147$$
$$P(\{(n, y, y)\}) = (.296)(.704)(.704) = .147$$

and so on. Note that several of the 3-tuples are assigned the same probability of occurrence. The probability that exactly two of the three are alive at their 65th birthdays is

$$P(\{(y, y, n), (y, n, y), (n, y, y)\}) = 3(.704)^2(.296) = .440$$

and the probability that exactly one of them is alive at her 65th birthday is

$$P(\{(y, n, n), (n, y, n), (n, n, y)\}) = 3(.704)(.296)^2 = .185. \quad \blacksquare$$

The method used in assigning probabilities in Example 2.7.4 is used frequently in many applications. Note that we could have defined $T_i = \{y, n\}$ as a sample space for each of the individuals, y indicating the person is alive at age 65, n that she is not. Our sample space for the three individuals is the Cartesian product of T_1, T_2, and T_3: $S = T_1 \times T_2 \times T_3$. Furthermore, we could just as well define the probability function $P_i(\{y\}) = .704$, $P_i(\{n\}) = .296$, $i = 1, 2, 3$ for each of the three individual sample spaces T_1, T_2, T_3 and then the probability function for $S = T_1 \times T_2 \times T_3$ is actually given by

$P(\{(x_1, x_2, x_3)\}) = P_1(\{x_1\}) \cdot P_2(\{x_2\}) \cdot P_3(\{x_3\})$ for all $(x_1, x_2, x_3) \in S$. The probability function for the single-element events belonging to S is the product of values of the probability functions for the single-element events of T_1, T_2, and T_3. This works, of course, because we assumed that the same sample space was appropriate for each of the three individuals, that the same probability function was appropriate for each of them, and that each of them would or would not reach age 65 independently or unaffected by whether the others would. This is an example of an experiment with *independent trials* with the same probability function. It is not necessary that the same probability function is used for each of the individuals, as stated in the following definition. The important aspect is that multiplication is appropriate in assigning probabilities to the single-element events.

DEFINITION 2.7.3.   An experiment with sample space S consists of n *independent trials* if and only if
    (a) S is a Cartesian product of n sets

$$S = T_1 \times T_2 \times \ldots \times T_n$$

    (b) The probability of every single-element event $\{(x_1, x_2, \ldots, x_n)\} \in S$
    is the product of the probabilities of the appropriate single-element events defined on T_1, T_2, $\ldots$, T_n; that is

$$P(\{(x_1, x_2, \ldots, x_n)\}) = P_1(\{x_1\}) \cdot P_2(\{x_2\}) \cdots P_n(\{x_n\})$$

    where $P_1, P_2, \ldots, P_n$ are the probability measures for $T_1, T_2, \ldots, T_n$, respectively.   ∎

Note immediately then that an experiment that consists of n independent trials has n-tuples as elements of its sample space. Furthermore, probabilities of single-element events are assigned in a special way; this special way in fact gives an easy method of computing many probabilities.

EXAMPLE 2.7.5.   A television-radio retail sales outlet receives a shipment of five radios from a manufacturer. In the past, 90 percent of the radios received from this manufacturer work correctly when first tried; 10 percent have required some adjustment or repair before working correctly. Let us discuss the experiment that consists of checking which of the five radios just received work correctly. We assume that each radio works correctly (nondefective, labeled n) or does not work correctly (defective, labeled d). Thus the simple experiments that consist of checking individual radios could all use the sample space

$$T_i = \{n, d\}, \qquad i = 1, 2, 3, 4, 5.$$

Based on the historical information given, we should also assume

$$P_i(\{n\}) = .9, \qquad P_i(\{d\}) = .1, \qquad i = 1, 2, 3, 4, 5.$$

The experiment that consists of checking all five radios could have the sample space

$$S = T_1 \times T_2 \times T_3 \times T_4 \times T_5.$$

Furthermore, because the individual radios are separate physical entities, it seems reasonable that the occurrence or nonoccurrence of defects would be independent. That is, we have an experiment that consists of independent trials. Then using the rule mentioned in Definition 2.7.3, we would assign probabilities to the single-element events of S by multiplying together the probabilities of observing the appropriate results in the individual trials. For example,

$$P(\{(n, n, n, n, n)\}) = (.9)(.9)(.9)(.9)(.9) = (.9)^5$$
$$P(\{(d, d, d, d, d)\}) = (.1)(.1)(.1)(.1)(.1) = (.1)^5$$
$$P(\{(n, d, n, d, n)\}) = (.9)(.1)(.9)(.1)(.9) = (.1)^2 (.9)^3$$
$$P(\{(d, d, n, n, n)\}) = (.1)(.1)(.9)(.9)(.9) = (.1)^2 (.9)^3$$

and so on. Once the probabilities of the single-element events are known, of course, we can compute the probabilities of any other events of interest. ∎

EXAMPLE 2.7.6. A "pop quiz" is given unannounced in the first 10 minutes of a psychology class. The quiz has 10 questions. The first six questions are true-false, so only two possible choices can be made. The remaining four questions are multiple-choice, each with four options. Assume that Sue comes to class completely unprepared, not having had time to study, and thus will simply guess at the correct answer to every question. If we consider the "experiment" whose outcome is defined by her performance on the test, we can define

$$T_i = \{r, w\}, \qquad i = 1, 2, ..., 10,$$

where r stands for right answer and w stands for wrong answer and the sample space then is simply

$$S = T_1 \times T_2 \times \cdots \times T_{10}.$$

Furthermore, if she is guessing at the right answer for every question, we would have

$$P_i(\{r\}) = P_i(\{w\}) = \tfrac{1}{2}, \qquad i = 1, 2, ..., 6$$
$$P_i(\{r\}) = \tfrac{1}{4}, \qquad P_i(\{w\}) = \tfrac{3}{4}, \qquad i = 7, 8, 9, 10$$

because she has an equal chance of selecting any of the possible responses.

Equally, if she is guessing, whether she is right or wrong on any question does not affect whether she is right or wrong on any other, so the probability of any single-element event in S is the product of 10 numbers, the probabilities of the appropriate responses in the 10-tuple considered. For example, the probability that she answers all 10 questions correctly is $(\frac{1}{2})^6 (\frac{1}{4})^4$, and the probability that she answers all 10 questions incorrectly is $(\frac{1}{2})^6 (\frac{3}{4})^4$. The probability that she is right on all the even-numbered questions and wrong on the odd-numbered ones is $(\frac{1}{2})^6 (\frac{1}{4}) (\frac{3}{4}) (\frac{1}{4}) (\frac{3}{4}) = (\frac{1}{2})^6 (\frac{1}{4})^2 (\frac{3}{4})^2$. Suppose she passes the quiz if and only if she is right on at least 5 of the true-false and on at least 3 of the multiple-choice questions. Then the probability that she passes the quiz is the union of four mutually exclusive events.

A_1: Right on 5 true-false and 3 multiple-choice.
A_2: Right on 6 true-false and 3 multiple-choice.
A_3: Right on 5 true-false and 4 multiple-choice.
A_4: Right on 6 true-false and 4 multiple-choice.

You can verify (use your counting skills) that

$$P(A_1) = \binom{6}{5}\left(\frac{1}{2}\right)^6\binom{4}{3}\left(\frac{1}{4}\right)^3\left(\frac{3}{4}\right) = .0044$$

$$P(A_2) = \binom{6}{6}\left(\frac{1}{2}\right)^6\binom{4}{3}\left(\frac{1}{4}\right)^3\left(\frac{3}{4}\right) = .0007$$

$$P(A_3) = \binom{6}{5}\left(\frac{1}{2}\right)^6\binom{4}{4}\left(\frac{1}{4}\right)^4 = .0004$$

$$P(A_4) = \binom{6}{6}\left(\frac{1}{2}\right)^6\binom{4}{4}\left(\frac{1}{4}\right)^4 = .0001$$

Thus the probability she passes is

$$P(A_1) + P(A_2) + P(A_3) + P(A_4) = .0056$$

She could expect to have a much higher probability of passing by keeping up her studies and being prepared for such quizzes. ■

EXERCISE 2.7

1.   One fair coin is flipped 2 times. Are the 2 events

A : A head occurs on the first flip.
B : A head occurs on the second flip.

independent?

2.   A fair coin is flipped two times. Let A be the event that a head occurs

on the first flip and let B be the event that the same face does not occur on both flips. Are A and B independent?

3. An urn contains 4 balls numbered 1, 2, 3, 4, respectively. Two balls are drawn without replacement. Let A be the event that the first ball drawn has a 1 on it and let B be the event that the second ball has a 1 on it. Are A and B independent?

4. If the drawing is done with replacement in Exercise 2.7.3, are A and B independent?

5. A pair of dice is rolled one time. Let A be the event that the first die has a 1 on it, B the event that the second die has a 6 on it, and C the event that the sum is 7. Are A, B, and C independent?

6. A fair coin is flipped three times. Let A be the event that a head occurs on the first flip, let B be the event that at least 2 tails occur, and let C be the event that we get exactly 1 head or that we get tail, head, head in that order. Show that these three events satisfy equation 4 of Definition 2.7.2, but do not satisfy equations 1, 2, or 3.

7. Prove that if A and B are independent, so are $\bar{A}$ and $\bar{B}$.

8. The probability that a certain basketball player scores on a free throw is .7. If in a game he gets 15 free throws, compute the probability that he makes them all. Compute the probability that he makes 14 of them. What assumptions have you made in deriving your answer?

9. Three teams, A, B, and C, enter a round-robin tournament. (Each team plays 2 games, 1 against each of the possible opponents. The winner of the tournament, if there is a winner, is the team winning both its games.) Assume that the game played is one in which a tie is not allowed. We assume the following probabilities

$$P(A \text{ beats } B) = .7$$
$$P(B \text{ beats } C) = .8$$
$$P(C \text{ beats } A) = .9.$$

Assuming independence, compute the probability that

(a) A wins.
(b) B wins.
(c) No one wins.

10. In Example 2.7.5, assume that 10 radios are received.
(a) What would you use as a sample space for the experiment?
(b) What is the probability that all 10 work correctly?
(c) What is the probability that at least 9 work correctly?

11. A department store has a sales display of 1000 light bulbs, 10 of which are defective.

(a) If you buy 20 of these bulbs, what is the probability you get no defectives?

(b) If you buy 20 of these bulbs, what is the probability you get exactly one defective bulb?

12. In one morning a Fuller brush salesman can call at 16 homes. If the probability of his making a sale is .1, at each home, what is the probability he makes at least one sale, during his morning?

13. Assume that a student is carrying 4 courses in a given quarter and that his probabilities of getting A's are .2, .5, .1, .7, respectively, in these courses.

(a) What is the probability that he gets all A's?

(b) What is the probability that he gets exactly 3 A's?

Can you find anything to criticize in the way you've answered this question?

14. Each of three different students, Joe, Hugh, and Rachel, are given the same problem to solve. They work on the problem independently and have probabilities .8, .7, and .6 of solving it, respectively.

(a) What is the probability that none of them solve the problem?

(b) What is the probability that the problem will be solved (by one or more of them)?

(c) Granted the problem was solved, what is the probability that the solution is due to Rachel alone?

2.8 Discrete and Continuous Sample Spaces

In many cases the elements of the sample space are real numbers. For example, if our experiment consists of a 100-meter dash with six runners competing, the elements of our sample space could be the possible times required by the winner to complete the dash. Or if we roll a pair of fair dice one time, the elements of S could be the possible values for the sum of the two uppermost numbers. If a student takes an exam in probability, the elements of the sample space could be the possible scores he will make.

We are forced to distinguish between two different types of real-element sample spaces S because the two require different methods for assigning probabilities to events. The first type we will discuss is called a discrete sample space.

DEFINITION 2.8.1. A real-element sample space S is called *discrete* if no finite-length interval on the real line contains an infinite number of elements of S. ∎

Examples of discrete sample spaces are

$S = \{1, 2, 3, ..., n\}$, the first n integers
$S = \{0, 1, 2, 3, ...\}$, the set of nonnegative integers
$S = \{0, 2, 4, 6, ...\}$, the set of even integers
$S = \{..., -1, 0, 1, ...\}$, the set of all integers

Discrete sample spaces may contain a finite or an infinite number of elements. The elements of a discrete sample space are isolated points on the real line; there are other points on the real line, between any two elements of S, which do not belong to S. There are, however, sets of points that have this property but which are not included in our definition of a discrete set, namely, sets that contain convergent sequences. Thus the set $\{1, \frac{1}{2}, \frac{1}{4}, \frac{1}{8}, \frac{1}{16}, ..., \frac{1}{2^n}, ...\}$ consists of isolated points on the line but is not discrete because the interval from 0 to 1 includes an infinite number of its elements.

It is possible for an event A to be given by the infinite union of mutually exclusive events $B_1, B_2, B_3, ...$ (generally in a variety of ways), when working with sample spaces that have an infinite number of elements. To avoid possible ambiguities in the assignment of probabilities to such sets, we must strengthen Axiom 3, by saying

3*. $\quad P(B_1 \cup B_2 \cup B_3 \cup ...) = P(B_1) + P(B_2) + P(B_3) + \cdots$
$\quad\quad$ if $\quad\quad B_i \cap B_j = \varnothing \quad\quad$ for all $\quad i \neq j.$

This called the *countable additivity* axiom, stating that if A is the union of an infinite number of mutually exclusive events, then $P(A) = \sum_{i=1}^{\infty} P(B_i)$, the value of $P(A)$ must be the limit of the infinite series of partial sums $P(A) = \lim_{n \to \infty} \sum_{i=1}^{n} P(B_i)$.

Discrete sample spaces are ones for which it is meaningful to specify the probabilities for single-element events. Probabilities for any other events (subsets of S) then are given by summing the probabilities of the single-element events of which they are comprised. Of course, if S has an infinite number of elements, we will be forced to use some type of rule that specifies the single-element event probabilities $P(\{x\})$, at least for most $x \in S$, because we cannot construct an infinite table listing the individual probabilities. It must still be true that $\sum_{x \in S} P(\{x\}) = 1$.

A frequently occurring type of assignment of probabilities for single-element events is one that employs a geometric series, or the derivatives of such series. Recall that

$$(1 + r + r^2 + \cdots + r^n)(1 - r) = 1 - r^{n+1},$$

and thus if $r \neq 1$,

$$1 + r + r^2 + \cdots + r^n = \frac{1 - r^{n+1}}{1 - r}.$$

Then, if $|r| < 1$,

$$\lim_{n \to \infty} \frac{1 - r^{n+1}}{1 - r} = \frac{1}{1 - r},$$

which we write as

$$\sum_{i=0}^{\infty} r^i = \frac{1}{1 - r}.$$

Subtracting the first term from this series (which is 1), we also have

$$\sum_{i=1}^{\infty} r^i = \frac{1}{1 - r} - 1 = \frac{r}{1 - r}, \qquad |r| < 1.$$

Now take the derivative of this series; that is,

$$\frac{d}{dr}(r + r^2 + r^3 + \cdots) = 1 + 2r + 3r^2 + \cdots = \sum_{i=0}^{\infty} (i + 1) r^i,$$

which must be the same as the derivative of $\dfrac{r}{1 - r}$, $\dfrac{d}{dr}\left(\dfrac{r}{1 - r}\right) = \dfrac{1}{(1 - r)^2}.$

Thus we also have

$$\sum_{i=0}^{\infty} (i + 1) r^i = \frac{1}{(1 - r)^2}.$$

Equating the kth derivative of $1 + r + r^2 + r^3 + \cdots$ to the kth derivative of $1/(1 - r)$ shows as well that

$$\sum_{i=0}^{\infty} \binom{k + i}{k} r^i = \sum_{j=k}^{\infty} \binom{j}{k} r^{j-k} = \frac{1}{(1 - r)^{k+1}}$$

as long as $|r| < 1$. These geometric series results all prove useful in our studies. The following example uses a geometric series assignment of probabilities to single-element events.

EXAMPLE 2.8.1.   A fair coin is flipped until a head occurs. As a sample space let us use

$$S = \{1, 2, 3, 4, \ldots\},$$

the set of positive integers, where $x \in S$ just counts the flip number on which the first head occurs (1 means the first flip was a head; 2 means the first head occurred on the second flip; 3, that the first head occurred on the third flip; etc.). Granted the coin is fair, it would seem reasonable to assign $P(\{1\}) = \frac{1}{2}$. The event $\{2\}$ occurs if and only if a tail occurs on the first flip *and* a head occurs on the second. Assuming the outcomes of the flips to be independent, we should assign $P(\{2\}) = \frac{1}{2} \cdot \frac{1}{2} = \frac{1}{4}$. Similarly, the event $\{3\}$ occurs if and only if we get tails on both of the first two flips and a head on the third; thus we should assign $P(\{3\}) = \frac{1}{2} \cdot \frac{1}{2} \cdot \frac{1}{2} = \frac{1}{8}$. The pattern should be clear. The facts that the coin is fair and that the flips are independent lead to the assignment

$$P(\{x\}) = \frac{1}{2^x}, \qquad x \in S.$$

Does this assignment satisfy Axiom 1? Recalling the geometric series previously discussed,

$$P(S) = \sum_{x \in S} P(\{x\}) = \sum_{x=1}^{\infty} \frac{1}{2^x} = \frac{1/2}{1 - 1/2} = 1,$$

and we see that Axiom 1 is satisfied. Probabilities of events are again given by summing probabilities of single-element events. Thus if A is the event that it takes no more than 4 flips to get the first head and B is the event that the first head occurs on an even-numbered flip, we have

$$P(A) = P(\{1\}) + P(\{2\}) + P(\{3\}) + P(\{4\})$$
$$= \frac{1}{2} + \frac{1}{4} + \frac{1}{8} + \frac{1}{16} = \frac{15}{16},$$
$$P(B) = P(\{2\}) + P(\{4\}) + P(\{6\}) + \cdots$$
$$= \sum_{j=1}^{\infty} \frac{1}{2^{2j}} = \frac{1/2^2}{1 - 1/2^2} = \frac{1}{4} \cdot \frac{4}{3} = \frac{1}{3}. \qquad \blacksquare$$

EXAMPLE 2.8.2. Suppose that we select a number at random from the positive integers. What is the probability that it is even? The sample space for this experiment is

$$S = \{1, 2, 3, \ldots\},$$

so S is discrete. If we select an integer at random, then each single-element event should have the same probability. However, it is not possible to assign the same probability to each of the single-element events. No matter how small the value of this common probability, the sum of the probabilities of the single-element events is infinite. That is, suppose we say

$$P(\{x\}) = p > 0 \qquad \text{for} \qquad x = 1, 2, 3, \ldots$$

Then

$$P(S) = \sum_{x=1}^{\infty} P(\{x\}) = \sum_{x=1}^{\infty} p$$

and this sum diverges, no matter how close to 0 p is. Thus it would appear that there is no way we could satisfactorily describe such an experiment. This conclusion is correct because, after a little reflection, we would have to admit that there is absolutely no way in which we could perform the experiment. It is impossible to select an integer at random from the set of positive integers. The experiment itself makes no sense; this is why we cannot describe it. In spite of the foregoing it would seem reasonable to give $\frac{1}{2}$ as the probability of selecting an even number since every other integer is even. This conclusion is actually based on reasoning such as the following. If we were to randomly select an integer from the set $\{1, 2, 3, \ldots, M\}$ where M is very large, the probability that the number is even is $\frac{1}{2}$ if M is even and $\frac{1}{2} - (1/2\,M)$ if M is odd. In either case it is about $\frac{1}{2}$ and, as M gets bigger and bigger, it gets closer and closer to $\frac{1}{2}$ (for odd M; for even M it is $\frac{1}{2}$). ■

The second type of real-element sample space that finds frequent use is called continuous. We will define it as follows.

DEFINITION 2.8.2.   A sample space S that has as elements all the points in an interval, or all the points in a union of intervals, on the real line is called *continuous*. ■

Such sets as

$$S = \{x: 0 \le x \le 10\}$$
$$S = \{x: 0 \le x \le 1 \text{ or } 2 \le x \le 3\}$$

are called continuous. A continuous sample space always has an infinite number of elements. Furthermore, if $x \in S$ and $y \in S$, then all the points between x and y also belong to S, if x and y are selected from one of the intervals belonging to S. There are "more" points in any continuous interval than there are integers. Thus a continuous sample space is said to contain an uncountable or nondenumerable number of points. Consequently, we are forced to assign zero as the probability of occurrence for essentially all the individual points in S; if we did not, it would not be possible to satisfy the first axiom, that $P(S) = 1$. Thus in continuous sample spaces we are not in general able to compute the probability of an event A by summing the probabilities of the single-element events (points) that are subsets of A. What, then, must we do to specify the probabilities for events when we have a continuous sample space?

The measure called length of a line used in geometry is a familiar example that is conceptually somewhat similar to defining probability functions for a continuous sample space S. Recall that the length of a line actually is a rule that associates a number (called length) with intervals on the real line. You will also recall that the length of any single point on the line is zero and thus the length of the interval from 0 to 1 (which is 1) is not obtained by summing the lengths of the individual points between 0 and 1. Just as with the measure length, the probability function for a continuous sample space S will be a rule that associates numbers (called probabilities) with intervals that are subsets of S.

In any discrete sample space, finite or infinite, all subsets of S are events and we can consistently compute the probability of any event by summing probabilities of single-element events. If S is continuous, there are technical reasons for not allowing *all* possible subsets of S to be events. All intervals that are subsets of S, and any unions and intersections of such intervals, are events (and we can assign probabilities to them) so these technical restrictions really will not limit our ability to solve practical problems.

There are, of course, many ways of assigning numbers to intervals that will satisfy the axioms and can thus be used as probability functions. One of the simplest of these makes use simply of the lengths of intervals. For example, suppose our sample space consists of all the points in the interval from a to b inclusive, where $a < b$; either or both of a and b could be negative or positive numbers. Thus we have

$$S = \{x: a \le x \le b\}.$$

Let $L(A)$ represent the length of the interval A where $A \subset S$ and if A is a union of nonoverlapping intervals that are subsets of S, let $L(A)$ be the sum of the lengths of the intervals. Then the rule

$$P(A) = \frac{L(A)}{L(S)}, \qquad A \subset S$$

satisfies the axioms for a probability function. To see that this is true is easy; this rule gives $P(S) = L(S)/L(S) = 1$ so that Axiom 1 is satisfied. Because the length of any interval is nonnegative, we have

$$P(A) = \frac{L(A)}{L(S)} \ge 0$$

so that Axiom 2 is satisfied. If $C_1, C_2, C_3, \ldots$ are nonoverlapping intervals and each is a subset of S, we have defined

$$L(C_1 \cup C_2 \cup C_3 \cup \ldots) = L(C_1) + L(C_2) + L(C_3) + \cdots$$

and thus

$$P(C_1 \cup C_2 \cup C_3 \cup \ldots) = \frac{L(C_1) + L(C_2) + L(C_3) + \cdots}{L(S)}$$

$$= P(C_1) + P(C_2) + P(C_3) + \cdots,$$

so Axiom 3 is also satisfied. Since the length of any point is zero, this rule also gives $P(A) = 0$ if $A = \{x\}$, $x \in S$, is a single-element event.

It is useful to think geometrically about the way this rule assigns probabilities to intervals. With

$$S = \{x : a \leq x \leq b\}$$

$L(S) = b - a$, so the divisor of each ratio is $b - a$. Define the function of a real variable

$$f(x) = \begin{cases} \dfrac{1}{b-a}, & a \leq x \leq b \\ 0, & \text{otherwise,} \end{cases}$$

which is pictured in Figure 2.8.1 with $0 < a < b$, although, as noted earlier, a or b or both could be negative. Then the total area between $f(x)$ and the x axis is $(b - a)(1/(b - a)) = 1$, which we describe by saying the total area under $f(x)$ is 1. Now let $A \subset S$ be any interval, such as the one pictured in Figure 2.8.2, where $a < c < d < b$. Then the area under $f(x)$ over the interval A is in fact $(d - c)/(b - a) = L(A)/L(S) = P(A)$. That is, this rule that assigns probabilities to intervals according to their lengths can also be

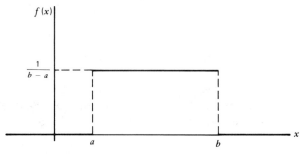

Figure 2.8.1

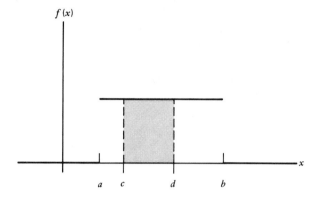

Figure 2.8.2

thought of as assigning probabilities by computing the *area* over the interval under the constant function $f(x) = 1/(b - a)$. (These are equivalent, of course, because the area under this constant function is $1/(b - a)$ times the length of the interval; that is, the area is proportional to the length of the interval.) The total area under $f(x)$ must be 1 so that $P(S) = (b - a)/(b - a) = 1$.

The rule that computes $P(A)$ by $L(A)/L(S)$, $A \subset S$, has an interesting property. If $B \subset S$ is twice as long as $A \subset S$, then $L(B) = 2L(A)$, which of course implies $P(B) = 2P(A)$; that is, the event B is twice as likely to occur (contains the observed outcome x) as is event A. This is also true of discrete equally likely sample spaces. If S has equally likely outcomes (and thus S must be finite) and event $B \subset S$ contains twice as many elements as event $A \subset S$, then it is also true for this rule that $P(B) = 2P(A)$. For this reason an experiment with a continuous sample space and probability function defined by

$$P(A) = \frac{L(A)}{L(S)}, \qquad A \subset S,$$

is frequently (rather sloppily) called one with "equally likely outcomes" (it is also called a uniform probability law). In the continuous case this would more aptly be called a sample space with equally likely intervals, meaning the probability assigned to the interval depends only on the length of the interval and not on where it is located within S.

EXAMPLE 2.8.3.   Doug is a 2-year-old boy. From his family history it seems plausible to assume that his adult height is equally likely to lie between 5 feet 9 inches and 6 feet 2 inches. Making this assumption, what is the probability that he will stand at least 6 feet high as an adult? What is the probability that his adult height will lie between 5 feet 10 inches and 5 feet 11 inches?

We use as our sample space

$$S = \{x: 69 \leq x \leq 74\},$$

where we are recording his achieved adult height in inches. Define

$$A = \{x: 72 \leq x \leq 74\}$$
$$B = \{x: 70 \leq x \leq 71\}.$$

Then

$$L(S) = 5, \qquad L(A) = 2, \qquad L(B) = 1,$$

and we have

$$P(A) = \tfrac{2}{5}, \qquad P(B) = \tfrac{1}{5}. \qquad \blacksquare$$

EXAMPLE 2.8.4.   Assume that you daily ride a commuter train from your home in Connecticut into Manhattan. The station you leave from has trains leaving for Manhattan at 7 A.M., 7:13 A.M., 7:20 A.M., 7:25 A.M., 7:32 A.M., 7:45 A.M., and 7:55 A.M. It is your practice to take the first train that leaves after your arrival at the station. Due to the vagaries of your rising time and the traffic you encounter driving to the station, you are equally likely to arrive at the station at any instant between 7:15 A.M. and 7:45 A.M. On a particular day, what is the probability that you have to wait less than 5 minutes at the station? Less than 10 minutes? Suppose the 7:25 A.M. and 7:45 A.M. trains are expresses. What is the probability that you catch an express on a given day?

Let us, for convenience, take as our sample space

$$S = \{x: 0 \leq x \leq 30\},$$

where the elements of S are actually meant to represent minutes after 7:15 A.M. that might occur as your arrival time. (Refer to Figure 2.8.3.) Define the events.

A: You wait less than 5 minutes.
B: You wait less than 10 minutes.
C: You catch an express.

Then

$$A = \{x: 0 \leq x < 10 \text{ or } 12 \leq x < 17 \text{ or } 25 \leq x < 30\}$$
$$B = \{x: 0 \leq x < 17 \text{ or } 20 \leq x < 30\}$$
$$C = \{x: 5 \leq x < 10 \text{ or } 17 \leq x < 30\}$$

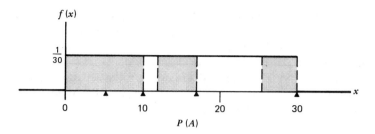

$P(A)$

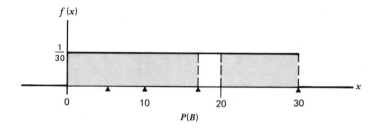

$P(B)$

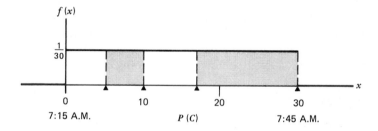

7:15 A.M. $P(C)$ 7:45 A.M.

Figure 2.8.3

and

$$L(S) = 30, \qquad L(A) = 20, \qquad L(B) = 27, \qquad L(C) = 18.$$

Thus

$$P(A) = \tfrac{2}{3}, \qquad P(B) = \tfrac{9}{10}, \qquad P(C) = \tfrac{3}{5}. \qquad ■$$

As we will see in Chapter 3, probabilities for events defined on continuous sample spaces are evaluated by computing areas under a function $f(x)$ over the interval whose probability is being evaluated. It is not necessary, of course, that $f(x)$ be constant for all $x \in S$. If it is not a constant, then probabilities of events (intervals) of equal length are not necessarily equal. The circumstances governing the experiment performed dictate the type of function that should be used to compute areas over intervals.

In reality, of course, all measuring instruments have finite accuracy. That is, it is truly not possible to distinguish all the separate points in an interval on the real line. For example, if we let S be the sample space for the winning time in, say, a 100-yard dash involving six college-age sprinters, then

$$S = \{x: 5 \le x \le 15\},$$

with measurements made in seconds, would surely contain the winning time. However, even with today's sophisticated time-measuring devices, winning times are quoted to the nearest hundredth of a second (it is debatable whether they are truly accurate even to that extent), and thus the winning time might be 9.47 seconds, say, or 9.48 seconds, but none of the times between these two numbers could in fact occur (as the recorded winning time). Therefore one might argue that the sample space should be $S = \{5.00, 5.01, 5.02, \dots, 14.98, 14.99, 15.00\}$ for this experiment, using the 1001 possible $1/100th$ of a second readings that lie between 5 and 15 seconds. This could in fact be done, but it is considerably simpler to idealize the situation— act as though all points in the interval could occur and to use continuous techniques (probabilities represented by areas) to define probability functions. In addition to being easier to use, the continuous approach can be made to be as accurate as one desires, compared to the discrete sample space implied by the limited accuracy of measuring devices.

EXERCISE 2.8

1. A fair die is rolled until a 1 occurs. Compute the probability that:
   (a) Ten rolls are needed.
   (b) Less than 4 rolls are needed.
   (c) An odd number of rolls is needed.
2. A fair pair of dice is rolled until a 7 occurs (as the sum of the 2 numbers on the dice). Compute the probability that
   (a) Two rolls are needed.
   (b) An even number of rolls is needed.
3. You fire a rifle at a target until you hit it. Assume the probability that you hit it is .9 for each shot and that the shots are independent. Compute the probability that:
   (a) It takes more than 2 shots.
   (b) The number of shots required is a multiple of 3.
4. Hugh takes a written driver's license test repeatedly until he passes it.

Assume the probability that he passes it any given time is .1 and that the tests are independent. Compute the probability that:
(a) It takes him more than 4 attempts.
(b) It takes him more than 10 attempts.

5. A traffic light on a route you travel every day turns red every 4 minutes, stays red 1 minute and then turns green again (thus it is green 3 minutes, red 1, etc.), with the red part of the signal starting on the hour, every hour.
   (a) If you arrive at the light at a random instant between 7:55 A.M. and 8:05 A.M., what is the probability that you have to stop at the light?
   (b) If you arrive at the light at a random instant between 7:54 A.M. and 8:04 A.M. what is the probability that you have to stop for the light?

6. The plug on an electric clock with a sweep second hand is pulled at a random instant of time within a certain minute. What is the probability that the second hand is between the 4 and the 5? Between the 1 and the 2? Between the 1 and the 6?

7. A point is chosen at random between 0 and 1 on the x axis in the (x, y) plane. A circle centered at the origin is then drawn in the plane, with radius determined by the chosen point. Compute the probability that the area of the circle is less than $\pi/2$.

8. A 12-inch ruler is broken into 2 pieces at a random point along its length. What is the probability that the longer piece is at least twice the length of the shorter piece?

9. The game of odd man out is played with 3 people, each flipping a single coin. All 3 flip their coins simultaneously; if one face is different from the other two, its owner is the odd man and he loses.
   (a) What is the probability that there is an odd man on a given turn, assuming all three coins are fair?
   (b) If there is no odd man on the first turn, the coins are all flipped again, until the odd man is determined. What is the probability that an even number of turns is required to determine the loser?

10. Answer Exercise 2.8.9 assuming that four people play (here three must match and one is different).

11. Answer Exercise 2.8.9 assuming that n people play ($n - 1$ must match). Does this seem like a feasible game as n gets large?

12. For the continuous sample space discussed in Example 2.8.3, assume that probabilities are assigned to intervals by the area between $f(x)$ and the x axis, where $f(x)$ is as pictured on the top of the next page. Answer the questions posed in the text for this case.

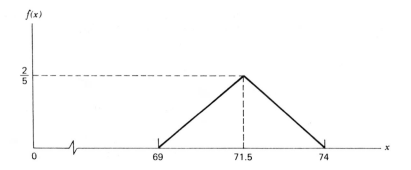

13. For the continuous sample space discussed in Example 2.8.4, assume
     that probabilities are assigned to intervals by the area between $f(x)$
     and the x axis, where $f(x)$ is as pictured below. Answer the questions
     posed in the text for this case.

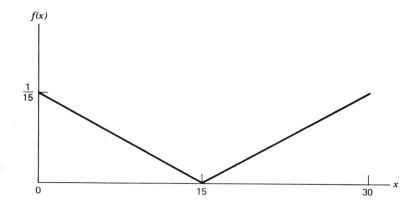

2.9 Summary

Sample space: Collection of all possible outcomes for an experiment
Event: A subset of a sample space
Mutually exclusive events: $A \cap B = \varnothing$.
Probability of an event: Real number associated with the event that repre-
    sents the relative frequency of occurrence for the event
Probability axioms:
    1. $P(S) = 1$
    2. $P(A) \geq 0$ for all events
    3. $P(A \cup B) = P(A) + P(B)$ if $A \cap B = \varnothing$

3*. $P(B_1 \cup B_2 \cup \cdots \cup B_n \cup \cdots) = \sum_{i=1}^{\infty} P(B_i)$ if $B_i \cap B_j = \varnothing$ for all $i \neq j$

Consequences:

$P(\varnothing) = 0$

$P(\bar{A}) = 1 - P(A)$

$P(A \cup B) = P(A) + P(B) - P(A \cap B)$

Single-element event: An event with only one element

Equally likely rule:

$$P(A) = \frac{\text{number of elements in } A}{\text{number of elements in } S}$$

Number of permutations of n things r at a time:

$_nP_r = n!/(n - r)! = $ number of different r-tuples that can be constructed from n elements.

Number of combinations of n things r at a time:

$$\binom{n}{r} = \frac{n!}{r!(n - r)!} = \text{number of subsets of size } r \text{ that can be constructed from a set with } n \text{ elements.}$$

Binomial theorem:

$$(x + y)^n = \sum_{k=0}^{n} \binom{n}{k} x^k y^{n-k}, \qquad n = 1, 2, 3, \ldots$$

Conditional probability: $P(A|B) = \dfrac{P(A \cap B)}{P(B)}$

$P(A \cap B \cap C) = P(A) P(B|A) P(C|A \cap B)$

Partition: $E_1 \cup E_2 \cup \cdots \cup E_n = S$ and $E_i \cap E_j = \varnothing$ if $i \neq j$

Bayes' Theorem: $P(E_i|A) = \dfrac{P(E_i) P(A|E_i)}{\sum_{j=1}^{n} P(E_j) P(A|E_j)}$ if $E_1, E_2, \ldots, E_n$ is a partition of S.

Independent events: $P(A \cap B) = P(A) P(B)$

Discrete sample space: Real-element sample space such that no finite-length interval contains an infinite number of elements

Geometric series: $\displaystyle\sum_{j=0}^{\infty} r^j = \dfrac{1}{1 - r}$ if $|r| < 1$

Derivatives of geometric series:

$$\sum_{i=0}^{\infty} \binom{k + i}{k} r^i = \sum_{j=k}^{\infty} \binom{j}{k} r^{j-k} = \frac{1}{(1 - r)^{k+1}}, \ |r| < 1, \ k = 1, 2, 3, \ldots$$

Continuous sample space: One that has as its elements all the points in an interval, or some union of intervals on the real line.

3

Random Variables and Distribution Functions

Heuristically, a random variable is a numerical quantity whose observed value is determined, at least in part, by some chance mechanism. For example, if a person takes a gambling weekend at a casino, the total amount of money he or she wins (or loses) is a random variable. The number of games a baseball team will win, in a given season, is a random variable. The number of bushels per acre that a farmer will harvest, having planted a hybrid corn, is a random variable. The number of sales a used-car salesman will make in a given month is a random variable. Examples abound of numerical quantities whose values are affected by a chance mechanism.

The most frequent use of probability theory is the description of random variables. Virtually all areas of modern science involve numerical measurements whose values are affected to some extent by chance mechanisms. Indeed, the method of scientific experimentation culminates in observing the outcome of an experiment and generally quantifying or describing the results observed with numbers, observed values of random variables. In this chapter we will study how random variables are defined and a number of ways of describing their behavior, from what is needed to specify their probability laws to various ways the probability law itself can be at least roughly described or summarized.

3.1 Random Variables

Granted we have an experiment and a sample space S for this experiment, any random variable defined on the experiment can be thought

92

of simply as a rule that associates a real number with every possible outcome (element of S). In the language of Chapter 1 a random variable is an element function whose domain of definition is the sample space S and whose range is a set of real numbers. Our formal definition follows.

DEFINITION 3.1.1. A *random variable* X is a real-valued function of the elements of a sample space S. The range of X will be denoted by R_X. ∎

Especially now, when we are first studying random variables, we will use a functional notation to stress the fact that a random variable is a special sort of mathematical function. The Greek letter ω will be used to represent a generic element of the sample space (regardless of whether the elements are 1-tuples or n-tuples) and $X(\omega)$ will be the functional representation of the random variable (the rule) X. We will reserve capital letters from the end of the alphabet (X, Y, Z, U, V, W, etc.) to represent random variables and lowercase letters (x, y, z, u, v, w, etc.) to stand for particular values in the range of the random variable.

EXAMPLE 3.1.1. We roll a pair of fair dice 1 time. Let X be the sum of the 2 numbers that occur. Then the sample space is

$$S = \{(x_1, x_2): x_1 = 1, 2, ..., 6; x_2 = 1, 2, ..., 6\}$$

and we have defined

$$X(\omega) = x_1 + x_2 \qquad \text{for} \qquad \omega = (x_1, x_2) \in S.$$

The range of X (the set of possible values that can occur for X) is $R_X = \{2, 3, ..., 12\}$. ∎

EXAMPLE 3.1.2. During a basketball game let us assume that player number 1 will get a total of 5 free throws. Then the sample space that records the outcomes of these 5 free throws can be

$$S = \{(x_1, x_2, x_3, x_4, x_5): x_i = h \text{ or } m, i = 1, 2, 3, 4, 5\},$$

the Cartesian product of $S_1 = \{h, m\}$ with itself 5 times, where h represents hit (he makes the free throw) and m represents a miss. Now define Y to be the number of free throws he makes during the game and Y is a random variable with

$$Y[(h, h, h, h, h)] = 5$$
$$Y[(h, m, m, h, h)] = 3$$
$$Y[(h, h, m, h, m)] = 3, \text{ and so on.}$$

The range of Y is clearly $R_Y = \{0, 1, 2, 3, 4, 5\}$. ∎

EXAMPLE 3.1.3.   The game of darts (popular in some pubs and other places) involves tossing a dart at a bull's-eye target. Suppose we super-impose an (x, y) coordinate system over the target, with its center located at the center of the bull's-eye. A single dart is tossed at the target by an experienced player; assume we are sure the dart will hit somewhere within the outer ring, whose diameter is 25 centimeters. Then the experiment de-fined by observing the impact point of one dart thrown by this player could have as a sample space

$$S = \{(x, y): x^2 + y^2 \leq (12.5)^2\}$$

where the units used are centimeters, and the impact point of the dart is recorded by its (x, y) coordinates (see Figure 3.1.1). Now define Z to be the distance between the dart's impact point and the center of the target (also the center of the coordinate system). Then Z is a random variable and associates the number $\sqrt{x^2 + y^2}$ with every element of S, that is,

$$Z[(x, y)] = \sqrt{x^2 + y^2}, \qquad (x, y) \in S$$

since the radial distance of any point (x, y) from the center is $\sqrt{x^2 + y^2}$. The range for Z is $R_Z = \{z: 0 \leq z \leq 12.5\}$. The impact point pictured in the figure has $x = 6$, $y = 3$, so the radial distance is 6.7.   ∎

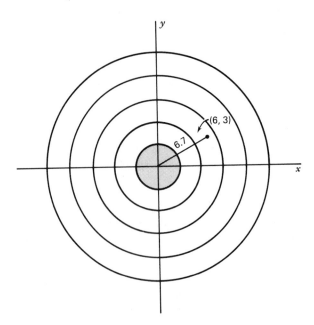

Figure 3.1.1

Random variables can be discrete (Examples 3.1.1, 3.1.2) or continuous (Example 3.1.3), depending on their range, the collection of values associated with elements of S. As we will see, the distinction made here between the two types of random variables is the same as the distinction made between discrete and continuous sample spaces in Section 2.8. In this section we will concentrate on discrete random variables and the specification of their probability laws. The definition of a discrete random variable follows.

DEFINITION 3.1.2.   A random variable X is called *discrete* if its range, R_X, is a discrete set.   ∎

The range of a discrete random variable, R_X, is a discrete set (of real numbers). Thus it is just like a discrete sample space S, which we studied in Chapter 2. That is, the probability law for a discrete random variable can be specified by simply defining $p_X(x) = P(X = x)$ for $x \in R_X$ and then the probability for any event A involving the observed value for X can be evaluated by summation. The function $p_X(x)$ is called the probability function for the discrete random variable X, as given in the following definition.

DEFINITION 3.1.3.   The *probability function* for a discrete random variable X is

$$p_X(x) = P(X = x) \qquad \text{for} \qquad x \in R_X.$$ ∎

How do we find the probability function for X? This is quite straightforward if X is defined on a sample space S and we know the probability function for S. To see how this is done, assume we are given an experiment with sample space S and know the probability function for S. That is, if $A \subset S$ is any event, then we already have the rule that tells us the value for $P(A)$. If X is a discrete random variable defined on S, then X has associated a real number $X(\omega)$ with each element $\omega \in S$ and R_X is a discrete set. For each $x \in R_X$ let $B(x) \subset S$ be the collection of all elements $\omega \in S$ such that $X(\omega) = x$: $B(x) = \{\omega : \omega \in S \text{ and } X(\omega) = x\}$. Then $B(x)$ is an event and we will observe $X = x$ if and only if $B(x)$ occurs when the experiment is performed. Thus we define

$$p_X(x) = P(X = x) = P\big(B(x)\big),$$

and the probability function for X is a direct descendant of the probability function specified for S. The following examples should help clarify this procedure.

EXAMPLE 3.1.4. Two fair dice are rolled one time, as in Example 3.1.1, and let X be the sum of the two numbers that occur. The sample space for the experiment is

$$S = S_1 \times S_1,$$

where $S_1 = \{1, 2, 3, 4, 5, 6\}$. Granted the dice are fair, the 36 2-tuples in S are equally likely to occur and the probability function for S is specified by

$$P(\{\omega\}) = \tfrac{1}{36}, \qquad \omega \in S.$$

We have defined X to be the sum of the two numbers that occur:

$$X[(x_1, x_2)] = x_1 + x_2, \qquad (x_1, x_2) \in S,$$

and the range of X then is $R_X = \{2, 3, \ldots, 12\}$. Thus as before, we can define the events

$$B(x) = \{\omega: \omega \in S \text{ and } X(\omega) = x\}, \qquad x \in R_X;$$

we have

$$B(2) = \{(1, 1)\}$$
$$B(3) = \{(1, 2), (2, 1)\}$$
$$B(4) = \{(1, 3), (2, 2), (3, 1)\}$$
$$B(5) = \{(1, 4), (2, 3), (3, 2), (4, 1)\}$$

and so on for $x = 6, 7, 8, \ldots, 12$. Because we already know the probability function for S, it is straightforward to evaluate the probabilities $P(B(x))$, $x = 2, 3, \ldots, 12$. In fact,

$$P(B(2)) = \tfrac{1}{36}$$
$$P(B(3)) = \tfrac{2}{36}$$
$$P(B(4)) = \tfrac{3}{36}, \text{ and so on,}$$

since the probability of any event is given by the number of elements it contains divided by 36, the number of elements in S. Then the probability function for the random variable X is given by the values for $P[B(x)]$, determined from the probability function for S. Thus

$$p_X(2) = P(B(2)) = \tfrac{1}{36}$$
$$p_X(3) = P(B(3)) = \tfrac{2}{36}$$
$$p_X(4) = P(B(4)) = \tfrac{3}{36}, \text{ and so on.}$$

The nonzero values for $p_X(x)$ are given in the following table.

x	2	3	4	5	6	7	8	9	10	11	12
$p_X(x)$	$\frac{1}{36}$	$\frac{2}{36}$	$\frac{3}{36}$	$\frac{4}{36}$	$\frac{5}{36}$	$\frac{6}{36}$	$\frac{5}{36}$	$\frac{4}{36}$	$\frac{3}{36}$	$\frac{2}{36}$	$\frac{1}{36}$

For any $x \notin R_X$ the event $B(x) = \varnothing$, and we have $p_X(x) = P(\varnothing) = 0$ for all such values. ∎

We will generally define $p_X(x)$ only for those x such that $p_X(x) > 0$; it is understood that $p_X(x) = 0$ for all other x.

EXAMPLE 3.1.5.   Assume the basketball player mentioned in Example 3.1.2 has the same probability .7 of making every one of his 5 free throws and that his attempts are independent. Then the experiment consists of 5 independent trials with probability of success .7 for each trial; thus the probabilities for the single-element events in S are given by multiplying together probabilities for the separate trials. For example,

$$P(\{(h, h, h, h, h)\}) = (.7)^5$$
$$P(\{(m, m, m, m, m)\}) = (.3)^5$$
$$P(\{(m, h, m, h, m)\}) = (.3)(.7)(.3)(.7)(.3) = (.3)^3 (.7)^2 \ ;$$

each of the $2^5 = 32$ different single-element events has a probability that is the product of five terms, each term being .3 or .7. Now let Y be the number of shots he makes in the five attempts; clearly, the range of Y is $R_Y = \{0, 1, 2, 3, 4, 5\}$ so its probability function is positive for each of these integers. Again, $p_Y(y)$ is obtained directly from the probability function defined on S. Again, with

$$B(y) = \{\omega: \omega \in S \text{ and } Y(\omega) = y\}$$

we have

$$B(0) = \{(m, m, m, m, m)\}$$
$$B(1) = \{(h, m, m, m, m), (m, h, m, m, m), (m, m, h, m, m), (m, m, m, h, m),$$
$$(m, m, m, m, h)\}$$

and so on. Given any element $\omega \in S$, $Y(\omega)$ is simply equal to the number of components in ω that are h, and it is easy to count the number of 5-tuples that belong to $B(y)$, $y = 0, 1, 2, 3, 4, 5$. All we must do is to count the number of ways that we can select y positions in the 5-tuples (each to be filled with h, the remaining positions are filled with m). But the number of ways that we can select y positions from the 5 available is $\binom{5}{y}$. Thus $B(0)$

has $\binom{5}{0} = 1$ element, $B(1)$ has $\binom{5}{1} = 5$ elements (already listed), $B(2)$ has $\binom{5}{2} = 10$ elements, $B(3)$ has $\binom{5}{3} = 10$, $B(4)$ has $\binom{5}{4} = 5$ and $B(5)$ has $\binom{5}{5} = 1$ element. Every element $\omega \in S$ must belong to (exactly) one $B(y)$ and thus the sum of the numbers of elements in these 6 sets should total 32, as you can verify that it does. Each 5-tuple belonging to $B(y)$ contains exactly y h's and 5-y m's (or it would not belong to $B(y)$), so the probability of each single element event $\{\omega\} \subset B(y)$ must be $(.7)^y(.3)^{5-y}$. Since every single-element event $\{\omega\} \subset B(y)$ has the same probability, $P[B(y)]$ is simply the product of the number of elements that belong to $B(y)$ $\left(\text{which is } \binom{5}{y}\right)$ times their common probability $(.7)^y(.3)^{5-y}$. This gives the probability function for Y

$$p_Y(y) = P[B(y)] = \binom{5}{y}(.7)^y(.3)^{5-y}, \qquad y = 0, 1, 2, 3, 4, 5.$$

The following table gives the numerical values for this probability function, which are easily verified on a hand-held calculator.

y	0	1	2	3	4	5
$p_Y(y)$	.0024	.0284	.1323	.3087	.3602	.1681

We can use this probability function for Y to evaluate the probability of any event defined in terms of Y. For example, the probability Y is an odd number (we really mean the observed value for Y is odd) is

$$P(Y \text{ is odd}) = p_Y(1) + p_Y(3) + p_Y(5) = .5052$$

and thus $P(Y \text{ is even}) = .4948$. The probability that Y is at least 3 is

$$P(Y \geq 3) = p_Y(3) + p_Y(4) + p_Y(5) = .8370,$$

whereas the probability that Y is no greater than 1 is

$$P(Y \leq 1) = p_Y(0) + p_Y(1) = .0308. \qquad \blacksquare$$

The probability function $p_X(x)$ gives the probabilities of occurrence for the different possible observed values x. Thinking of the range of X

$$R_X = \{x : p_X(x) > 0\}$$

as being a discrete sample space S whose probability function is $p_X(x)$, we easily see that it is necessary that $\sum_{x \in R_X} p_X(x) = 1$; the only other requirement that $p_X(x)$ must satisfy is $p_X(x) \geq 0$ for any real number, because probabilities cannot be negative. Any function of a real variable, $p_X(x)$, such that $p_X(x) \geq 0$ for all x and such that

$$\sum_{x \in R_X} p_X(x) = 1,$$

where R_X is the collection of values such that $p_X(x) > 0$, could be the probability function for a discrete random variable X. We will close this section with one further example.

EXAMPLE 3.1.6. The game of Chuck-a-Luck is played as follows. Three fair dice are rolled. You as the bettor are allowed to bet 1 dollar (or some other amount) on the occurrence of one of the integers 1, 2, 3, 4, 5, 6. Suppose you bet on the occurrence of a 5. Then if one 5 occurs (on the 3 dice) you win 1 dollar, if two 5's occur you win 2 dollars, and if three 5's occur you win 3 dollars. If no 5's occur you lose your 1 dollar. Let V be the net amount you win in one play of this game. The range of V is $R_V = \{-1, 1, 2, 3\}$ so V is a discrete random variable. We choose as our sample space

$$S = \{(x_1, x_2, x_3): x_i = 1, 2, \ldots, 6; i = 1, 2, 3\}.$$

Since we are assuming the dice to be fair, the 3 dice constitute 3 independent trials, each with probability $\frac{1}{6}$ of giving a 5. We immediately then are able to assign probabilities to all of the single-element events and can use these to find the probability function for V.

$$
\begin{aligned}
p_V(v) &= \tfrac{125}{216}, & \text{for } v &= -1 \\
&= \tfrac{75}{216}, & \text{for } v &= 1 \\
&= \tfrac{15}{216}, & \text{for } v &= 2 \\
&= \tfrac{1}{216}, & \text{for } v &= 3.
\end{aligned}
$$

As you might gather, it is quite significant that $p_V(-1) > \frac{1}{2}$. We will soon see exactly why this should play a role in Las Vegas casinos' offers to play this game. ∎

EXERCISE 3.1

1. An urn contains 4 balls numbered 1, 2, 3, 4, respectively. Let Y be the number that occurs if 1 ball is drawn at random from the urn. What is the probability function for Y?

2. Consider the urn defined in Exercise 3.1.1. Two balls are drawn from the urn without replacement. Let Z be the sum of the two numbers that occur. Derive the probability function for Z.

3. Assume that the sampling in Exercise 3.1.2 is done with replacement and define Z in the same way. Derive the new probability function for Z.

4. Two balls are drawn with replacement from the urn mentioned in Exercise 3.1.1. Let X be the sum of the squares of the two numbers drawn and derive the probability function for X.

5. A class in statistics contains 10 students, 3 of whom are 19, 4 are 20, 1 is 21, 1 is 24, and 1 is 26. Two students are selected at random without replacement from this class. Let X be the average age of the 2 selected students and derive the probability function for X.

6. A man has four keys in his pocket and, since it is dark, cannot see which is his door key. He will try each key in turn until he finds the right one. Let X be the number of keys tried (including the right one) to open the door. What is the probability function for X?

7. Five cards are dealt from a standard 52-card deck. Let Y be the number of red cards that are dealt. What is the probability function for Y?

8. For the experiment defined in Exercise 3.1.7, let Z be the number of spades dealt. What is the probability function for Z?

9. Assume two fair dice are rolled one time and let M be the maximum (larger) of the two numbers that occur. Derive the probability function for M.

10. Let W be the minimum (smaller) of the two numbers that occur when the dice in Exercise 3.1.9 are rolled. Find the probability function for W.

11. A package of four light bulbs contains one defective bulb. You select two bulbs at random from the package to go into the same fixture. Let X be the number of bulbs you select that work (are not defective) and derive the probability function for X.

*12. The game of Keeno involves selecting Ping-Pong balls at random from a large container. The container contains 100 balls numbered from 00 to 99. Twenty of these balls will be selected at random from the container. Before they are selected, you are allowed to choose any three numbers you like from 00 to 99. Let X be the number of your selections that occur in the 20 that are pulled from the container. What is the probability function for X?

3.2　Distribution Functions and Density Functions

The probability function $p_X(x)$ for a discrete random variable X gives the probability of occurrence of the elements in the range of X. It can then be used to compute the probability of occurrence for any event defined by the observed value of X. In this section we will study the *distribution function* (also frequently called the cumulative distribution function or cdf) for a random variable X. It is simply an alternative function that can be used to evaluate probabilities of events defined by the observed value for a random variable. It is very easy to use in certain theoretical developments and has the added advantage of giving a straightforward method for describing what we will call *continuous* random variables. Let us start with the definition of the distribution function for a random variable.

DEFINITION 3.2.1. The distribution function $F_X(t)$ for a random variable X gives the value of $P(X \leq t)$ for any real t; that is, $F_X(t) = P(X \leq t)$ for $-\infty < t < \infty$. ∎

We note, then, that the domain of definition for a distribution function is the whole real line and since $F_X(t)$ is a probability, we must have $0 \leq F_X(t) \leq 1$ for any t. Since the distribution function for X gives the probability that X is less than or equal to t, it is very simple to get $F_X(t)$, granted we know the probability function, $p_X(x)$, for X. The value of $F_X(2)$, say, is $P(X \leq 2)$. The event $\{X \leq 2\}$ occurs if the observed value for X is any number less than or equal to 2 and, if $p_X(x)$ is known, we would evaluate this probability by summing $p_X(x)$ over those $x \in R_X$ that are no bigger than 2. Thus in general we derive $F_X(t)$ from $p_X(x)$ by summing values of $p_X(x)$ and we will write

$$F_X(t) = \sum_{x \leq t} p_X(x),$$

where the symbol $\sum_{x \leq t}$ is shorthand for sum over all values of $x \in R_X$ such that $x \leq t$. The following examples illustrate this.

EXAMPLE 3.2.1. Suppose a hat contains four slips of paper; each slip bears the number 1, 2, 3, or 4 (each number occurs on one slip). One slip is drawn from the hat without looking. Let X be the number on the slip that is drawn. Then the values X could equal are 1, 2, 3, and 4 and because the slip is drawn without looking, these are equally likely to occur. Thus the probability function for X is

$$p_X(x) = \tfrac{1}{4}, \quad x = 1, 2, 3, 4$$
$$p_X(x) = 0, \quad \text{otherwise.}$$

Now let us find the distribution function for X. Remember $F_X(t)$ is defined for all real t so we must compute $P(X \leq t)$ for both positive and negative t. Clearly, the smallest number that can occur for X is 1; thus for any $t < 1$

the event $\{X \leq t\}$ is empty, $\{X \leq t\} = \emptyset$ for any $t < 1$, because the smallest number that could be drawn from the hat is 1. The event $\{X \leq 1\}$ could occur, namely, the observed value for X could be 1, which is less than or equal to 1, and

$$F_X(1) = p_X(1) = \tfrac{1}{4}.$$

Now consider any real number t bigger than 1 but less than 2 (for example, $t = \tfrac{3}{2}$); the event $\{X \leq t\}$ for $1 < t < 2$ occurs only if we observe $X = 1$; that is,

$$F_X(t) = p_X(1) = \tfrac{1}{4}, \qquad 1 < t < 2$$

and thus the distribution function is constant over this interval. The accumulation of probability cannot change over intervals of values that X cannot equal.

Now consider $F_X(2) = P(X \leq 2)$. The event $\{X \leq 2\}$ occurs if *either* $X = 1$ or $X = 2$ is observed and thus

$$F_X(2) = p_X(1) + p_X(2) = \tfrac{2}{4}.$$

There are no observed values for X in the interval from 2 to 3, thus $F_X(t)$, $2 < t < 3$ must remain constant and

$$F_X(t) = \tfrac{2}{4}, \qquad 2 < t < 3.$$

Continuing in this way, we would find that the event $\{X \leq 4\}$ occurs if $X = 1$ or 2 or 3 or 4 and thus

$$F_X(4) = p_X(1) + p_X(2) + p_X(3) = \tfrac{4}{4}.$$

Indeed, the event $\{X \leq t\}$, where t is *any* number larger than 4 (consider $t = 5$), occurs if $X = 1$ or 2 or 3 or 4 so

$$F_X(t) = 1 \qquad \text{for all} \qquad t \geq 4.$$

The distribution function $F_X(t)$ is pictured in Figure 3.2.1. Notice that $F_X(t)$ has a jump in it at each of the points $t = 1, 2, 3, 4$, exactly the values that X can equal. Thinking of $F_X(t)$ as giving the accumulation of probability (on the real line) up to and including the point t, it is obvious why $F_X(t)$ must jump at $t = 1, 2, 3, 4$. These are the (only) values that have positive probability of occurring as the observed value for X. Thus the accumulation of probability jumps at each of these points. ∎

EXAMPLE 3.2.2.   In Example 3.1.6 we found

$$\begin{aligned} p_V(v) &= \tfrac{125}{216}, & \text{for } v &= -1 \\ &= \tfrac{75}{216}, & \text{for } v &= 1 \\ &= \tfrac{15}{216}, & \text{for } v &= 2 \\ &= \tfrac{1}{216}, & \text{for } v &= 3. \end{aligned}$$

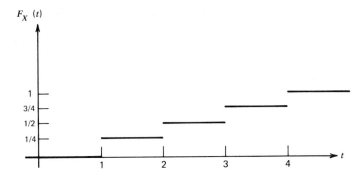

Figure 3.2.1

Then the distribution function for V is

$$
\begin{aligned}
F_V(t) &= 0, & t &< -1 \\
&= \tfrac{125}{216}, & -1 &\leq t < 1 \\
&= \tfrac{200}{216}, & 1 &\leq t < 2 \\
&= \tfrac{215}{216}, & 2 &\leq t < 3 \\
&= 1, & t &\geq 3.
\end{aligned}
$$

Figure 3.2.2 gives a plot of $F_V(t)$. ■

The distribution function F_X can be used to evaluate probability statements of any kind regarding X; in particular, if we define the two events

$$
\begin{aligned}
A&: X \leq a \\
B&: a < X \leq b,
\end{aligned}
$$

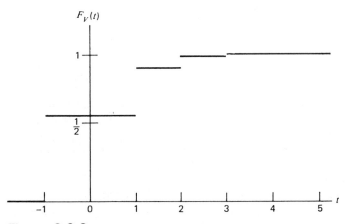

Figure 3.2.2

where a and b are any two real numbers such that $a < b$, we obviously have

$$A \cap B = \varnothing$$

so

$$P(A \cup B) = P(A) + P(B).$$

Since $A \cup B$ is simply the event that $X \leq b$,

$$P(X \leq b) = P(X \leq a) + P(a < X \leq b),$$

or

$$F_X(b) = F_X(a) + P(a < X \leq b).$$

Thus

$$P(a < X \leq b) = F_X(b) - F_X(a),$$

which shows us how to evaluate the probability that X lies in a particular interval, given F_X.

EXAMPLE 3.2.3.   Given the random variable V discussed in Example 3.2.2 (refer to Figure 3.2.2), use F_V to compute

$$P(0 < V \leq 3), \qquad P(V \leq 0), \qquad P(-1 < V \leq 0).$$

Using the preceding equation, we have

$$
\begin{aligned}
P(0 < V \leq 3) &= F_V(3) - F_V(0) \\
&= 1 - \tfrac{125}{216} = \tfrac{91}{216}, \\
P(V \leq 0) &= F_V(0) = \tfrac{125}{216}, \\
P(-1 < V \leq 0) &= F_V(0) - F_V(-1) \\
&= \tfrac{125}{216} - \tfrac{125}{216} = 0.
\end{aligned}
$$

This last probability is zero since the distribution function F_V is constant or level on the interval $-1 < t \leq 0$. This means that there are no values of t in the interval $-1 < t \leq 0$ that V will equal with positive probability. This is the case in general; if the distribution function of a random variable X is constant over any interval, then the probability of X taking on values in that interval is zero. ∎

We have seen how we might derive the distribution function for a discrete random variable X from the probability function for X. We might logically then ask how we could get the probability function for X from

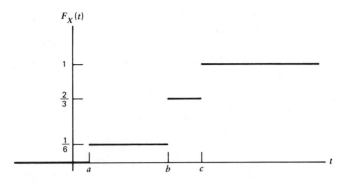

Figure 3.2.3

the distribution function for X. To answer this, consider the real numbers t in the interval $b - h < t \leq b$ where h is some positive number. As h tends to zero, this interval clearly tends to the single point $t = b$; that is, the limit of any such interval as $h \to 0$ is the single point $t = b$ since $b - h \to b$. Then we might expect, as is indeed the case, that

$$\lim_{h \to 0} P(b - h < X \leq b) = P(X = b).$$

But

$$P(b - h < X \leq b) = F_X(b) - F_X(b - h)$$

so

$$\lim_{h \to 0} P(b - h < X \leq b) = \lim_{h \to 0} \left[F_X(b) - F_X(b - h) \right]$$

$$= F_X(b) - \lim_{h \to 0} F_X(b - h)$$

$$= F_X(b) - F_X(b -),$$

where we have written $F_X(b-)$ for $\lim_{h \to 0} F_X(b - h)$. Thus if b is a point of discontinuity of F_X, then b is a value that X takes on with positive probability; the probability that $X = b$ is the size of the jump at $F_X(b)$.

If a random variable X has the distribution function plotted in Figure 3.2.3 then the range for X is $R_X = \{a, b, c\}$, the points at which the jumps occur. The value of the probability function at these points is given by the size of the jump:

$$p_X(a) = F_X(a) - F_X(a-) = \tfrac{1}{6} - 0 = \tfrac{1}{6}$$
$$p_X(b) = F_X(b) - F_X(b-) = \tfrac{2}{3} - \tfrac{1}{6} = \tfrac{1}{2}$$
$$p_X(c) = F_X(c) - F_X(c-) = 1 - \tfrac{2}{3} = \tfrac{1}{3},$$

and $p_X(x)$ is 0 for other values since there is no jump in the accumulation of probability at any other value.

EXAMPLE 3.2.4.   The distribution function for a random variable Y is specified to be

$$
\begin{aligned}
F_Y(t) &= 0, & t < -2 \\
&= \tfrac{1}{3}, & -2 \le t < 1 \\
&= \tfrac{7}{12}, & 1 \le t < 5 \\
&= \tfrac{47}{60}, & 5 \le t < 11 \\
&= \tfrac{57}{60}, & 11 \le t < 20 \\
&= 1, & 20 \le t.
\end{aligned}
$$

To find $p_Y(y)$, we must first locate the points of discontinuity for $F_Y(t)$. These are $-2, 1, 5, 11$, and 20 since for each of these numbers the value of F_Y at t [which is $F_Y(t)$] is different from the value F_Y approaches as its argument approaches t from smaller values $[F_Y(t-)]$. The value of the probability function at each of these numbers is given by the size of the jump in $F_Y(t)$.

$$
\begin{aligned}
p_Y(-2) &= \tfrac{1}{3} \\
p_Y(1) &= \tfrac{1}{4} \\
p_Y(5) &= \tfrac{1}{5} \\
p_Y(11) &= \tfrac{1}{6} \\
p_Y(20) &= \tfrac{1}{20}.
\end{aligned}
$$
∎

The distribution function $F_X(t)$ gives the value of $P(X \le t)$ where t is any real number. Because numbers called probabilities are required to satisfy the three axioms discussed in Section 2.2, there are rules that a function $H(t)$ of a real variable must satisfy if it is to be the distribution function for a random variable. These are:

1. $0 \le H(t) \le 1$ for all $-\infty < t < \infty$. This is necessary if $H(t)$ is to be the probability of an event.
2. $\lim_{t \to -\infty} H(t) = 0, \qquad \lim_{t \to \infty} H(t) = 1.$
    If a random variable associates a real number $X(\omega)$ with every element $\omega \in S$, then the event $A(t) = \{\omega : X(\omega) \le t\}$ must have the properties
    $\lim_{t \to -\infty} A(t) = \varnothing, \qquad \lim_{t \to \infty} A(t) = S.$
3. If $a < b$ are any two real numbers, then $H(a) \le H(b)$. This requirement follows from the fact that $A(a) = \{\omega : X(\omega) \le a\} \subset A(b) = \{\omega : X(\omega) \le b\}$ whenever $a < b$. Thus $H(a) = P(A(a)) \le P(A(b)) = H(b)$ (see Exercise 2.2.5).

4. $\lim_{h \to 0} H(t + h) = H(t)$ for any t and $h > 0$; that is, $H(t)$ must be continuous from the right.

This requirement is technical; let

$$C(t, h) = \{\omega : t < X(\omega) \le t + h\}, \quad -\infty < t < \infty, h > 0.$$

Then in the limit as $h \to 0$, $C(t, h) \to \varnothing$, so that

$$\lim P\big(C(t, h)\big) = \lim \big(H(t + h) - H(t)\big)$$
$$= P(\varnothing) = 0$$

which is ensured by (4).

Any function of a real variable satisfying (1) through (4) could be used as the distribution function for a random variable. You can easily verify that the distribution functions discussed in Examples 3.2.1 through 3.2.4 do in fact satisfy these requirements.

There is no requirement that a distribution function must have discontinuities, that its value must jump at certain points (all the distribution functions we have used so far did have discontinuities). That is, there are many different *continuous* functions $H(t)$ that satisfy (1) through (4). For each of these we would have $H(t) = H(t-)$, for all t; but this in turn would say that the probability function defined by $H(t) = H(t-)$ must be zero for all t and the random variable whose probability law is described by such a distribution function then has probability zero of equaling any particular real number. Such distribution functions give the probability laws for *continuous* random variables. These are defined as follows.

DEFINITION 3.2.2. If X is a random variable whose distribution function

$$F_X(t) = P(X \le t)$$

is a continuous function of t, $-\infty < t < \infty$, we will call X a *continuous* random variable. ∎

EXAMPLE 3.2.5. The function

$$H_1(t) = 0, \quad t < 0$$
$$= t, \quad 0 \le t \le 1$$
$$= 1, \quad 1 < t,$$

which is graphed in Figure 3.2.4, satisfies the requirements (1) through (4) for a distribution function. Because it is a continuous function of t, the random variable whose probability law is described by $H_1(t)$ is a *continuous*

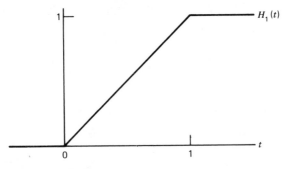

Figure 3.2.4

random variable. Similarly, the function

$$H_2(t) = 0, \qquad t < 0$$
$$= 1 - e^{-t}, \qquad t \geq 0,$$

which is graphed in Figure 3.2.5, also satisfies the requirements for a distribution function; because it is a continuous function, the corresponding random variable is called continuous. Let X be the random variable whose distribution function is $H_1(t)$ and let Y be the random variable whose distribution function is $H_2(t)$. Then

$$P(.1 < X \leq .4) = H_1(.4) - H_1(.1) = .4 - .1 = .3$$
$$P(.6 < X \leq .7) = H_1(.7) - H_1(.6) = .7 - .6 = .1$$
$$P(X \leq .5) = H_1(.5) = .5$$
$$P(.9 < X) = H_1(\infty) - H_1(.9) = 1 - .9 = .1$$
$$P(.1 < Y \leq .4) = H_2(.4) - H_2(.1) = (1 - e^{-.4}) - (1 - e^{-.1}) = .23$$
$$P(.6 < Y \leq .7) = H_2(.7) - H_2(.6) = (1 - e^{-.7}) - (1 - e^{-.6}) = .05$$
$$P(Y \leq .5) = 1 - e^{-.5} = .39$$
$$P(.9 < Y) = H_2(\infty) - H_2(.9) = 1 - (1 - e^{-.9}) = .41$$

and so on. The probability that a continuous random variable X is greater than a but no more than b is given by $F_X(b) - F_X(a)$, just as with a discrete random variable. ∎

Suppose X is a continuous random variable. Then, as already noted, $P(X = x) = 0$ for any x. Since the events

$$A: \{X = a\}$$
$$B: \{X = b\}$$
$$C: \{a < X < b\}$$

are clearly disjoint and $P(A) = P(B) = 0$, the following four probabilities

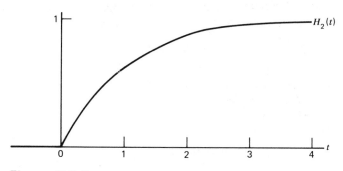

Figure 3.2.5

must all be equal for a continuous random variable:

$$
\begin{aligned}
P(a < X < b) &= P(C) \\
&= P(C) + P(A) = P(a \leq X < b) \\
&= P(C) + P(B) = P(a < X \leq b) \\
&= P(C) + P(A) + P(B) = P(a \leq X \leq b).
\end{aligned}
$$

Thus when X is continuous the probability assigned to any interval (a, b) is unchanged, regardless of whether either or both end points are included; this, of course, is not true for a discrete random variable.

All the continuous distribution functions that we will study actually will be differentiable and their derivatives will also be continuous (except possibly at a discrete set of points, having no effect on any probabilities computed). That is, if $F_X(t)$ is the distribution function for a continuous random variable X, then

$$
\frac{d}{dt} F_X(t) = f_X(t)
$$

exists and is continuous for all but at most a discrete set of values for t. Recalling that the derivative of a function $F_X(t)$ is defined by

$$
\lim_{\Delta t \to 0} \frac{F_X(t + \Delta t) - F_X(t)}{\Delta t} = f_X(t)
$$

we can see that for small Δt we should have

$$
F_X(t + \Delta t) - F_X(t) \doteq f_X(t)\, \Delta t,
$$

that is,

$$
P(t \leq X \leq t + \Delta t) \doteq f_X(t)\, \Delta t
$$

and, abusing this equation slightly, $f_X(t)\,\Delta t$ is approximately the probability that a continuous random variable will equal some value in an interval of length Δt centered at t. This derivative of a continuous distribution function is called the *density* function for the continuous random variable.

DEFINITION 3.2.3.   Let X be a continuous random variable with distribution function $F_X(t)$. The *density function* (also called the probability density function or pdf) for X is

$$f_X(t) = \frac{dF_X(t)}{dt}.$$

The *range* for a continuous random variable X is $R_X = \{t: f_X(t) > 0\}$, where $f_X(t)$ is the density function for X. ∎

Because the density function is the derivative of the distribution function, it gives the rate of change of the distribution function; in a sense it is the rate at which the probability is accumulated at the value t. Recall that

$$P\left(t - \frac{\Delta t}{2} < X < t + \frac{\Delta t}{2}\right) \doteq f_X(t)\,\Delta t.$$

If we subdivide the interval $a - \Delta t/2 < t < b + \Delta t/2$ into a large number n of intervals of length $\Delta t = (b - a)/(n - 1)$, so that $t_1 = a$, $t_n = b$ as in Figure 3.2.6, then the sum of the areas of the n rectangles pictured is approximately $P(a \le X \le b)$; that is,

$$P(a \le X \le b) \doteq \sum_{i=1}^{n} f_X(t_i)\,\Delta t$$

since the area of rectangle i is $f_X(t_i)\,\Delta t$. Taking the limit as $n \to \infty$ (and

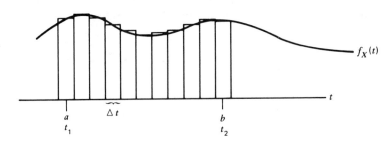

Figure 3.2.6

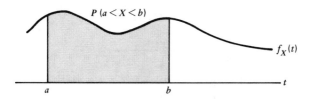

Figure 3.2.7

thus $\Delta t \to 0$), we have

$$\lim_{n \to \infty} \sum_{i=1}^{n} f_X(t_i) \, \Delta t = \int_a^b f_X(t) \, dt,$$

that is,

$$F_X(b) - F_X(a) = \int_a^b f_X(t) \, dt.$$

Because a definite integral gives the area under the integrand between the definite limits, the *area* under $f_X(t)$ between a and b gives $P(a \le X \le b)$ for a continuous random variable. (See Figure 3.2.7.) The *range* for X is the collection of values for which the density for X (derivative of $F_X(t)$) is positive. Thus the range again gives the collection of possible observed values for a continuous random variable, just as it does for a discrete random variable.

EXAMPLE 3.2.6.   Let X be the random variable whose distribution function is $H_1(t)$ from Example 3.2.5. That is,

$$F_X(t) = 0, \qquad t < 0$$
$$= t, \qquad 0 \le t \le 1$$
$$= 1, \qquad 1 < t;$$

the density function for X then is

$$f_X(t) = \frac{d}{dt} F_X(t) = 1, \qquad 0 < t < 1$$
$$= 0, \qquad t < 0 \text{ or } t > 1.$$

This density function is graphed in Figure 3.2.8.

Since the density function for X is equal to 1 for $0 < t < 1$, the area under the density over any interval between 0 and 1 is equal to the length of the interval; thus the probability that the observed value for X lies in an interval contained in $(0, 1)$ is proportional to the length of the interval.

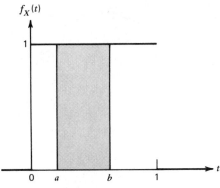

$f_X(t)$

1

0   a b 1 t

Figure 3.2.8

This is the same method for computing probabilities for continuous sample spaces as was discussed in Section 2.8. Because the density function is constant or uniform over (0, 1), the random variable X is called *uniform*; we will discuss uniform random variables in Chapter 4. We will somewhat inaccurately say the values from the range of a uniform random variable are "equally likely" to occur. It provides a continuous analog to the discrete equally likely case already discussed.

Now let Y be the continuous random variable whose distribution function is $H_2(t)$ defined in Example 3.2.5.

$$F_Y(t) = H_2(t) = 0, \qquad t < 0$$
$$= 1 - e^{-t}, \qquad t \geq 0.$$

The density function for Y is

$$f_Y(t) = \frac{d}{dt} F_Y(t) = e^{-t}, \qquad t > 0$$
$$= 0, \qquad t < 0,$$

which is graphed in Figure 3.2.9. As we will see in Chapter 4. Y is a particular example of an *exponential* random variable. ∎

Let us conclude this section by commenting that any function $h(t)$ can be considered the probability density function of some continuous random variable if

$$h(t) \geq 0, \qquad \text{for all } t,$$

and

$$\int_{-\infty}^{\infty} h(t)\, dt = 1.$$

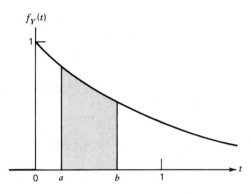

Figure 3.2.9

EXERCISE 3.2

1. Verify that

$$F_X(t) = 0, \qquad t < -3$$
$$= \tfrac{1}{3}, \qquad -3 \le t < -1$$
$$= \tfrac{2}{3}, \qquad -1 \le t < 0$$
$$= 1, \qquad t \ge 0$$

is a distribution function and derive the probability function for X.

2. Verify that

$$F_Z(t) = 0, \qquad t < -2$$
$$= \tfrac{1}{2}, \qquad -2 \le t < 0$$
$$= 1, \qquad t \ge 0$$

is a distribution function and specify the probability function for Z. Use it to compute $P(-1 \le Z \le 1)$.

3. Verify that

$$F_W(t) = 0, \qquad t < 3$$
$$= \tfrac{1}{3}, \qquad 3 \le t < 4$$
$$= \tfrac{1}{2}, \qquad 4 \le t < 5$$
$$= \tfrac{2}{3}, \qquad 5 \le t < 6$$
$$= 1, \qquad t \ge 6$$

is a distribution function and specify the probability function for W. Use it to compute $P(3 < W \le 5)$.

4. Verify that

$$
\begin{aligned}
F_Y(t) &= 0, & t &< 0 \\
&= \tfrac{1}{4}, & 0 &\le t < 5 \\
&= \tfrac{1}{3}, & 5 &\le t < 7 \\
&= \tfrac{1}{2}, & 7 &\le t < 100 \\
&= \tfrac{5}{6}, & 100 &\le t < 102 \\
&= 1, & t &\ge 102
\end{aligned}
$$

is a distribution function and specify the probability function for Y. Use it to compute $P(Y \le 100)$.

5. The random variable Z has the probability function

$$
\begin{aligned}
p_Z(x) &= \tfrac{1}{3}, & \text{for} \quad x &= 0, 1, 2 \\
&= 0, & \text{otherwise.}
\end{aligned}
$$

What is the distribution function for Z?

6. The random variable U has the probability function

$$
\begin{aligned}
p_U(-3) &= \tfrac{1}{2} \\
p_U(0) &= \tfrac{1}{6} \\
p_U(4) &= \tfrac{1}{3}.
\end{aligned}
$$

What is the distribution function for U?

7. Verify that

$$
\begin{aligned}
F_X(t) &= 0, & t &< -1 \\
&= \frac{t+1}{2}, & -1 &\le t \le 1 \\
&= 1, & t &> 1
\end{aligned}
$$

is a distribution function and specify the probability density function for X. Use it to compute $P(-\tfrac{1}{2} \le X \le \tfrac{1}{2})$.

8. Verify that

$$
\begin{aligned}
F_Y(t) &= 0, & t &< 0 \\
&= \sqrt{t}, & 0 &\le t \le 1 \\
&= 1, & t &> 1
\end{aligned}
$$

is a distribution function and specify the probability density function for Y. Use it to compute $P(\tfrac{1}{4} < Y < \tfrac{3}{4})$.

9. Verify that

$$
\begin{aligned}
F_Z(t) &= 0, & t &< 0 \\
&= t^2, & 0 &\le t < \tfrac{1}{2} \\
&= 1 - 3(1-t)^2, & \tfrac{1}{2} &\le t < 1 \\
&= 1, & t &\ge 1
\end{aligned}
$$

is a distribution function and derive the density function for Z.

10. Is

$$H(t) = 0, \qquad\qquad t < -1$$
$$= 1 - t^2, \qquad -1 \le t < \tfrac{1}{2}$$
$$= \tfrac{1}{2} + t^2, \qquad \tfrac{1}{2} \le t < 1$$
$$= 1, \qquad\qquad t \ge 1$$

a distribution function?

*11. Verify that

$$H(t) = 0, \qquad t < 0$$
$$= \tfrac{1}{2}, \qquad 0 \le t < \tfrac{1}{2}$$
$$= t, \qquad \tfrac{1}{2} \le t \le 1$$
$$= 1, \qquad t > 1$$

is a distribution function. This is the distribution function of what is called a *mixture random variable.* It has one point of discontinuity (at $t = 0$) and thus there is a positive probability ($\tfrac{1}{2}$) of the random variable equaling zero. Note that the distribution function is continuous and increasing for $\tfrac{1}{2} \le t \le 1$. Thus the random variable takes on particular values in this interval with probability zero; we could define a pseudo density function for the random variable lying in subintervals of this interval.

12. Given X has probability density function

$$f_X(x) = 1, \qquad 99 < x < 100$$
$$= 0, \qquad \text{otherwise,}$$

derive $F_X(t)$.

13. Y is a continuous random variable with

$$f_Y(y) = 2(1 - y), \qquad 0 < y < 1$$
$$= 0, \qquad\qquad \text{otherwise.}$$

Derive $F_Y(t)$.

14. Z is a continuous random variable with probability density function

$$f_Z(z) = 10e^{-10z}, \qquad z > 0$$
$$= 0, \qquad\qquad \text{otherwise.}$$

Derive $F_Z(t)$.

15. Let X be the random variable whose density function is defined in Exercise 2.8.12 and derive $F_X(t)$.

16. Let Y be the random variable whose density function is defined in Exercise 2.8.13 and derive $F_Y(t)$.

3.3 Expected Values and Summary Measures

The probability law for a random variable can be defined by its distribution function, $F_X(t)$, or its density function, $f_X(t)$, if X is continuous, or its probability function, $p_X(k)$, if X is discrete. Once we know the probability law for X, we are able to compute the probabilities of occurrence for any events of interest. In many applications we will be interested in describing various aspects of different probability laws, ways of describing certain properties of probability laws. For example, what is a "typical" value for the random variable to equal, where "typical" may be defined in various ways. How much variability is exhibited by the probability law, how spread out are the possible observed values for the random variable? In this section we will study some common measures of certain aspects of probability laws, concentrating on measures of the "middle" of the probability law (the typical observed value) and of the "variability" of the probability law (the spread of the possible observed values).

Let us discuss the expected or average value of a random variable. A specific gambling example is perhaps the easiest case to visualize first. The game of Chuck-a-Luck was discussed in Example 3.1.6. Briefly, $1 is bet and the amount won is a discrete random variable V with probability function

$$p_V(v) = \tfrac{125}{216} \quad \text{at } v = -1$$
$$= \tfrac{75}{216} \quad \text{at } v = 1$$
$$= \tfrac{15}{216} \quad \text{at } v = 2$$
$$= \tfrac{1}{216} \quad \text{at } v = 3.$$

Suppose you were to play this game $n > 1$ times, winning v_1 dollars the first time, v_2 dollars the second time, ..., v_n dollars the nth time (remember each v_i can be any of the values $-1, 1, 2, 3$). The average amount won over these n plays then is

$$\bar{v} = \frac{1}{n} \sum_{i=1}^{n} v_i.$$

What might we expect for the value of $\bar{v}$?

Since each v_i must equal $-1, 1, 2,$ or 3, let k_1 be the number of times in the n plays that you win -1 dollars (you lose your dollar), let k_2 be the number of times you win 1 dollar, k_3 the number of times you win 2 dollars, and k_4 the number of times you win 3 dollars. Then your average winnings over the n plays is

$$\bar{v} = \frac{1}{n} \sum_{i=1}^{n} v_i = (-1)\frac{k_1}{n} + (1)\frac{k_2}{n} + (2)\frac{k_3}{n} + (3)\frac{k_4}{n}.$$

The ratio k_1/n is the relative frequency in your n plays that you win -1 dollars; but this number should be roughly equal to $\frac{125}{216} = p_V(-1) = P(V = -1)$, because probabilities should equal relative frequencies of occurrence. Similarly, we would expect that

$$\frac{k_2}{n} \doteq p_V(1), \qquad \frac{k_3}{n} \doteq p_V(2), \qquad \frac{k_4}{n} \doteq p_V(3)$$

and thus we should find the average amount won to approximately be given by

$$(-1) p_V(-1) + (1) p_V(1) + (2) p_V(2) + (3) p_V(3)$$
$$= (-1) (\tfrac{125}{216}) + (1) (\tfrac{75}{216}) + (2) (\tfrac{15}{216}) + (3) (\tfrac{1}{216})$$

$$= \frac{-17}{216} = -.08.$$

You should expect, on the average, to lose about 8¢ for every dollar you bet playing this game. This number we have computed, $-17/216$, will be called the *expected* value or average value of the random variable V. It is a measure of the middle of the probability law for V. Notice that an expected value does not have to be a possible observed value for the random variable.

Very similar reasoning is appropriate in the continuous case. Suppose you agree to meet a friend (who is generally late) at a specified hour; you arrive on time (call this time zero on the time scale) and your friend will arrive at a random time $T > 0$ with density function $f_T(t)$ pictured in Figure 3.3.1,

$$f_T(t) = \frac{2}{15} - \frac{2t}{225}, \qquad 0 < t < 15,$$

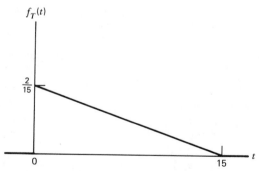

Figure 3.3.1

where we are measuring time in minutes. We could break the interval $0 < t < 15$ into a large number, n, of small pieces of length $\Delta t = 15/n$, and let $t_1, t_2, \ldots, t_n$ be the midpoints of these small intervals. Then the probability (relative frequency) your friend arrives at time t_i is about $f_T(t_i)\,\Delta t$ and, if this agreed-to meeting situation were to take place a large number of times, you could expect your average waiting time to be about $\sum_{i=1}^{n} t_i f_X(t_i)\,\Delta t$—the sum of products of the length of time you have to wait (t_i) times the probability of this time occurring $(f_X(t_i)\,\Delta t)$—just as previously discussed for the discrete case. If we take the limit of this sum as n, the number of pieces we subdivided the continuous range for T into, approaches ∞, the expected value converges to

$$\int_0^{15} t f_X(t)\, dt = \int_0^{15} t\left(\frac{2}{15} - \frac{2t}{225} \right) dt = 5 \text{ minutes.}$$

If the density for T truly describes the time you have to wait to meet your friend, your average or expected waiting time is 5 minutes. Again, this expected value is a measure of the middle of the probability law for T. The definition of the expected value for a random variable follows.

DEFINITION 3.3.1.   (a) If X is a discrete random variable with probability function $p_X(x)$, the *expected value* for X is

$$E[X] = \sum_{x \in R_X} x p_X(x),$$

as long as the sum is absolutely convergent.
(b) If X is a continuous random variable with density function $f_X(t)$, the *expected value* for X is

$$E[X] = \int_{-\infty}^{\infty} x f_X(x)\, dx,$$

as long as the integral is absolutely convergent. We will also use μ_X to represent the mean or expected value for X.   ∎

The student of mechanics will recognize the fact that as long as it exists, the expected value of a random variable locates the balance point of its probability function or density function. That is, if X is discrete with probability function $p_X(x)$ and we imagine points with mass $p_X(x)$ located on the real line at x, then $E[X]$ or μ_X locates the balance position of the real line. If X is continuous with density $f_X(x)$ and we imagine a rod whose thickness at x is given by the density $f_X(x)$, then again $E[X]$ locates the balance point for the rod. Figure 3.3.2 illustrates this balancing interpretation for the two random variables previously discussed.

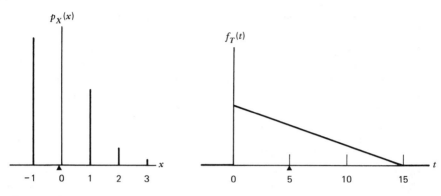

Figure 3.3.2

If the integral or sum defining $E[X]$ does not converge absolutely, we simply say the expected value does not exist. The following example discusses such a case.

EXAMPLE 3.3.1. As discussed, suppose you agree to meet a friend at a specified time and that your friend's arrival time is a random variable T. If you were to assume that the density for T is

$$f_T(t) = \frac{1}{t^2}, \qquad t > 1$$

then, from Definition 3.3.1, the time you should expect to wait is

$$E[T] = \int_1^\infty t \frac{1}{t^2} \, dt = \int_1^\infty \frac{1}{t} \, dt,$$

which diverges to ∞, so we say $E[T]$ does not exist. This would be, of course, a strange density to assume for the length of time you must wait. Note that the distribution function for T is

$$F_T(t) = \begin{cases} 0, & t < 1 \\ 1 - \dfrac{1}{t}, & t \geq 1 \end{cases}$$

so the probability your waiting time lies between t and $t + 1$ minutes is

$$\left(1 - \frac{1}{t+1}\right) - \left(1 - \frac{1}{t}\right) = \frac{1}{t(t+1)}.$$

Thus the probability you must wait between 1 and 2 minutes is $\frac{1}{2}$, between 2 and 3 minutes the probability is $\frac{1}{6}$, between 3 and 4 minutes the probability is $\frac{1}{12}$, and so on. Although these probabilities may not sound terribly unrealistic, they are not getting small very fast with t. In fact there is too much probability at the large observed values of T for the integral to converge. In terms of balance points this density has too heavy a tail out to the right for it to balance at any finite point. We will simply say $E[T]$ does not exist. The probability laws with which we will be most concerned will have expected values. ∎

The expected value of a random variable (granted it exists) locates the middle of a probability law, in the sense of a balance. To describe other aspects of probability laws (and for various other purposes), we shall define the expected value of a function of a random variable. Let $G(X)$ be a function of a random variable X. For example,

$$G_1(X) = X^2$$
$$G_2(X) = \ln X$$
$$G_3(X) = \frac{X}{1 - X}$$

are all functions of X. Anytime $X = x$, of course, we have $G(X) = G(x)$; the value of the function is given by simply evaluating it at the observed value x. Consequently, the expected value for any function $G(X)$ is very easy to define. Instead of multiplying the observed value x by its probability and summing, $G(x)$ is multiplied by the probability and summed.

DEFINITION 3.3.2.   Let $G(X)$ be a function of a random variable X. The expected value for $G(X)$ is

$$E[G(X)] = \sum_{x \in R_X} G(x)\, p_X(x) \qquad \text{if } X \text{ is discrete}$$
$$= \int_{-\infty}^{\infty} G(x)\, f_X(x)\, dx \quad \text{if } X \text{ is continuous,}$$

as long as the sum or integral is absolutely convergent. ∎

Suppose $G(X) = c$, a constant; that is, no matter what the observed value for X (let us assume X is discrete), the value for $G(X)$ is always c. It seems intuitive that the expected or average value for $G(X)$ then should also be c. That it must be so is easily verified,

$$E[G(X)] = \sum_{x \in R_X} c p_X(x) = c \sum_{x \in R_X} p_X(x) = c,$$

since the sum of the probability function for X must be 1. Similarly, if $G(X) = cH(X)$, where c is a constant,

$$E[G(X)] = \sum_{x \in R_X} G(x)\, p_X(x)$$

$$= \sum_{x \in R_X} cH(x)\, p_X(x)$$

$$= c \sum_{x \in R_X} H(x)\, p_X(x)$$

$$= cE[H(X)];$$

constants factor through the expectation. If $G(X)$ can be written as the sum of two functions, say, $G(X) = H(X) + J(X)$, then

$$E[G(X)] = \sum_{x \in R_X} G(x)\, p_X(x)$$

$$= \sum_{x \in R_X} [H(x) + J(x)]\, p_X(x)$$

$$= \sum_{x \in R_X} H(x)\, p_X(x) + \sum_{x \in R_X} J(x)\, p_X(x)$$

$$= E[H(X)] + E[J(X)];$$

the expected value of a sum is the sum of the expected values. We have then in fact established the following theorem.

Theorem 3.3.1. If X is any random variable, then
   (a) $E[c] = c$, where c is any constant
   (b) $E[cH(X)] = cE[H(X)]$, where c is any constant
   (c) $E[H(X) + J(X)] = E[H(X)] + E[J(X)]$
      as long as the expectations exist. ∎

You can easily verify these results are also true if X is any continuous random variable.

The variability of a random variable (or more accurately, of its probability law) can be measured in many ways. The two most commonly used measures of variability are the *variance* σ_X^2, and its positive square root, σ_X, called the standard deviation of X. These are now defined.

DEFINITION 3.3.3. The *variance* of a random variable X (denoted by σ_X^2) is defined to be

$$\sigma_X^2 = E[(X - \mu_X)^2];$$

its positive square root is denoted by σ_X and is called the *standard deviation* of X. Thus $\sigma_X = \sqrt{\sigma_X^2}$. ∎

If there are units attached to the random variable X, the variance would be measured in the squares of those units. The standard deviation is measured in the same units as the original random variable.

If we think of X as a point on the real line, the distance between X and μ_X is simply $X - \mu_X$. Since σ_X^2 is defined to be the expected value of the square of $X - \mu_X$, it gives the average squared distance between the observed values of X and the middle of the probability law, μ_X. Since σ_X^2 is an average of squares, it can never be negative and σ_X is always defined (as long as $E[X^2]$ exists). If X were a random variable with probability law defined by $P(X = 10) = 1$, $P(X = x) = 0$ for all $x \neq 10$, say, it is easy to see that the only possible observed value for X is 10 and that $\mu_X = 10$. If we were to observe values for X repeatedly, all the observed values would be 10; there is no variability exhibited in the observed values. We can also see that the variance for this random variable is

$$\sigma_X^2 = E[(X - 10)^2] = (10 - 10)^2 \cdot 1 = 0$$

and its standard deviation is also 0. A random variable that has probability 1 of equaling some specific number always has variance 0 (the converse of this is also true), since its probability law exhibits no variability.

Because the variance is the average squared distance between a random variable and its mean, it would seem that the variance should increase as the range of values with positive probability gets larger. That is, if the observed values for X must lie between -1 and 1, and the observed values for Y must lie between -10 and 10, then apparently one should have $\sigma_Y^2 > \sigma_X^2$ because Y has "more room" to vary. This does not necessarily prove true, since the distribution of probability has a great effect on σ_X^2; the random variable with more probability close to μ_X has the smaller variance. The following example should help make the point.

EXAMPLE 3.3.2.   Consider random variables U, V, and W with probability laws defined by the density functions

$$
\begin{aligned}
f_U(u) &= u + 1, & -1 \leq u \leq 0 \\
&= 1 - u, & 0 < u \leq 1 \\
&= 0, & \text{otherwise,} \\
f_V(v) &= \tfrac{1}{2}, & -1 \leq v \leq 1 \\
&= 0, & \text{otherwise,} \\
f_W(w) &= -w, & -1 \leq w \leq 0 \\
&= w, & 0 \leq w \leq 1 \\
&= 0, & \text{otherwise.}
\end{aligned}
$$

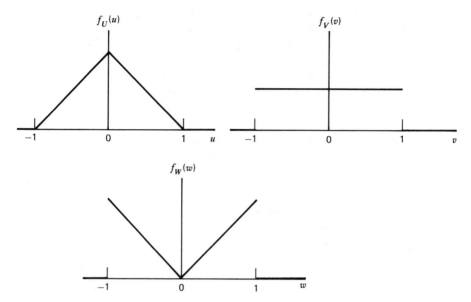

Figure 3.3.3

These are graphed in Figure 3.3.3. Each density is positive over the interval -1 to 1 so the observed values for all three random variables can vary over equal length intervals and all three mean values are zero. It is obvious from the picture that we should have $\sigma_U^2 < \sigma_V^2 < \sigma_W^2$, even though the ranges are equal. You can verify that in fact

$$\sigma_U^2 = \tfrac{1}{6}, \qquad \sigma_V^2 = \tfrac{1}{3}, \qquad \sigma_W^2 = \tfrac{1}{2};$$

the standard deviations are $1/\sqrt{6}$, $1/\sqrt{3}$, and $1/\sqrt{2}$, respectively.   ■

An alternative formula for computing the variance of a random variable is

$$\begin{aligned}
\sigma_X^2 &= E[(X - \mu_X)^2] = E[X^2 - 2X\mu_X + \mu_X^2] \\
&= E[X^2] - 2\mu_X E[X] + E[\mu_X^2] \\
&= E[X^2] - \mu_X^2,
\end{aligned}$$

so the variance can also be thought of as the average of the squares of the values of the random variable less the square of the average. This alternative formula is an easier way to evaluate σ_X^2 in almost all cases.

Linear transformations of random variables occur quite frequently in applications. Suppose T is the (random) time needed in minutes to complete a task and T has mean μ_T and variance σ_T^2; the time needed to complete the task in hours then is $H = T/60$. Intuitively, it would seem that the mean of H and variance of H should be rather simply related to μ_T and σ_T^2. Or if F is the maximum temperature (in degrees Fahrenheit) to be recorded in Washington, D.C., tomorrow, then the maximum temperature in degrees Celsius is $C = \frac{5}{9} F - \frac{160}{9}$, again a linear relation. Granted μ_F and σ_F are known, can we also evaluate μ_C and σ_C? Theorem 3.3.2, which you are asked to prove in Exercise 3.3.17, shows what these relationships are.

Theorem 3.3.2. Assume X is a random variable with mean μ_X and variance σ_X^2. If $Y = aX + b$, where a and b are any constants, then

$$\mu_Y = a\mu_X + b, \qquad \sigma_Y^2 = a^2\sigma_X^2, \qquad \sigma_Y = |a|\,\sigma_X. \qquad \blacksquare$$

The quantiles of a continuous distribution offer alternative (and perhaps more transparent) ways of summarizing or describing probability laws. These are now defined.

DEFINITION 3.3.4.　Let X be a continuous random variable with distribution function $F_X(t)$. The $(100k)th$ *quantile* for X is the number t_k such that

$$F_X(t_k) = k, \qquad 0 < k < 1. \qquad \blacksquare$$

The $(100k)$th quantile is that value t_k that makes the value of the distribution function equal to k; alternatively, it is the value such that the area under the density, to its left, is equal to k. Figure 3.3.4 illustrates both these statements for the random variable U whose density is defined in Example 3.3.2.

The 50th quantile, $t_{.5}$, is also called the *median* for the random variable (really its probability law). It is the value that cuts the density function into two pieces with equal area, .5 to the left and .5 to the right. The median is an alternative measure of the middle of a probability law. The 25th, 50th,

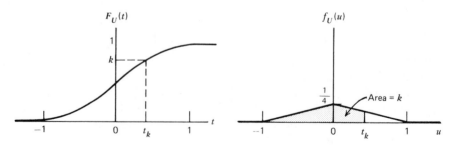

Figure 3.3.4

and 75th quantiles are also called the *quartiles* of the probability law, since they cut the density into four pieces, each of which has area $\frac{1}{4}$ under it. The *interquartile range*, $t_{.75} - t_{.25}$, gives an interval, "centered" at the median, which includes 50 percent of the probability. It is sometimes used as an alternative measure of the variability of a probability law. The larger the difference, $t_{.75} - t_{.25}$, is, the larger will be the interval needed to contain 50 percent probability and thus, the more variable will be the observed values. The following example compares the median, mean, interquartile range, and standard deviation for a random variable.

EXAMPLE 3.3.3. Suppose X is a random variable with distribution function

$$
\begin{aligned}
F_X(t) &= 0, && t < 0 \\
&= \sqrt{t}, && 0 \le t \le 1 \\
&= 1, && t > 1.
\end{aligned}
$$

Let us find the median of X and the interquartile range for X. Since

$$ F_X(t) = \sqrt{t}, $$

the median is defined by

$$ \sqrt{t_{.50}} = .5 $$

and thus $t_{.50} = (.5)^2 = .25$. Similarly,

$$
\begin{aligned}
t_{.25} &= (.25)^2 = .0625 \\
t_{.75} &= (.75)^2 = .5625.
\end{aligned}
$$

so the interquartile range is

$$ t_{.75} - t_{.25} = .5. $$

Let us also compute μ_X and σ_X. We find that the density function is

$$ f_X(t) = \frac{1}{2\sqrt{t}}, \qquad 0 < t < 1 $$

$$ = 0, \qquad\qquad \text{otherwise.} $$

Then

$$\mu_X = \int_0^1 x \frac{1}{2\sqrt{x}} \, dx = \tfrac{1}{3}$$

$$E[X^2] = \int_0^1 x^2 \frac{1}{2\sqrt{x}} \, dx = \tfrac{1}{5}$$

$$\sigma_X^2 = E[X^2] - \mu_X^2 = \tfrac{1}{5} - \tfrac{1}{9} = \tfrac{4}{45},$$

and

$$\sigma_X = \sqrt{\frac{4}{45}} = \frac{2}{3\sqrt{5}}.$$ ∎

Note the median and μ_X are not equal in this example. Unless the probability density function is symmetric with respect to some line perpendicular to the x axis, the point at which the density balances and the point that cuts it into two equal areas may not coincide. Exercise 3.3.15 shows a case in which the density function is not symmetric but the mean and median still coincide.

EXAMPLE 3.3.4.   It is possible, of course, for the distribution function $F_X(t)$ to remain constant over an interval, in which case certain quantiles may not be uniquely defined by Definition 3.3.4. For example, suppose X has distribution function

$$F_X(t) = 0, \qquad t \le 0$$

$$= \frac{t}{2}, \qquad 0 < t \le 1$$

$$= \frac{1}{2}, \qquad 1 < t \le 2$$

$$= \frac{t}{4}, \qquad 2 < t \le 4$$

$$= 1, \qquad 4 < t$$

pictured in Figure 3.3.5. Any number t between 1 and 2 has the property that the distribution function equals .5 and thus any number in this interval could be called the median for X. We will not encounter this case in the probability laws we will concentrate on, but just to make the quantiles

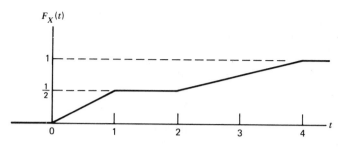

Figure 3.3.5

well defined, we will define them to be the smallest value that makes $F_X(t) \geq k$. Thus for the distribution function pictured in Figure 3.3.5, the median for X is 1 with this convention. ∎

EXERCISE 3.3

1. If
$$p_X(x) = \tfrac{1}{4}, \qquad x = 2, 4, 8, 16$$
$$= 0, \qquad \text{otherwise,}$$
   compute
   (a) $E[X]$ (c) $E[1/X]$ (e) σ_X^2 and σ_X.
   (b) $E[X^2]$ (d) $E[2^{X/2}]$

2. Suppose $f_X(x) = \tfrac{1}{2}$, $-1 < x < 1$, compute
   (a) $E[X]$ (c) $E[X + 2]$ (e) σ_X^2
   (b) $E[X^2]$ (d) $E[X/4 + 7]$ (f) σ_X.

3. Given
$$f_X(x) = 2(1 - x), \qquad 0 < x < 1$$
$$= 0, \qquad\qquad \text{otherwise,}$$
   compute
   (a) $E[X]$ (c) $E[(X + 10)^2]$ (e) σ_X^2
   (b) $E[X^2]$ (d) $E[1/(1 - X)]$ (f) σ_X.
4. Show that $E[X - \mu_X] = 0$.
5. Given
$$F_X(t) = 0, \qquad t < 0$$
$$= \frac{t}{100}, \qquad 0 \leq t \leq 100$$
$$= 1, \qquad t > 100,$$

find (a) $t_{.10}$, (b) $t_{.20}$, (c) $t_{.55}$, (d) $t_{.99}$, (e) $t_{.47}$, (f) $t_{.80}$, and (g) the inter-quartile range.

6. Suppose

$$
\begin{aligned}
F_U(t) &= 0, & t < 1 \\
&= \log_e t, & 1 \le t \le e \\
&= 1, & t > e,
\end{aligned}
$$

find the median, the interquartile range for U, μ_U, and σ_U^2.

7. If

$$
\begin{aligned}
F_Z(t) &= 0, & t < 0 \\
&= 2^t - 1, & 0 \le t \le 1 \\
&= 1, & t > 1,
\end{aligned}
$$

find the median, the interquartile range for Z, μ_Z, and σ_Z.

8. Show that $E[(X - a)^2]$ is minimized if $a = \mu_X$.

9. The adult height of a 3-year-old boy is equally likely to fall in the interval from 5 feet 6 inches to 5 feet 11 inches. What is his expected adult height?

10. A church lottery is going to give away a 3000-dollar car. They sell 10,000 tickets at 1 dollar apiece. If you buy 1 ticket, what is your expected gain? What is your expected gain if you buy 100 tickets? Compute the variance of your gain in these two instances.

11. Compute the mean and variance of the sum of the two numbers that occur when a pair of fair dice is rolled one time.

12. According to recent mortality tables, the probability an American citizen will die in his twentieth year is .00178. Assume that a life insurance company will sell a $1000 one-year term policy to a 19-year-old for $5.00. (This means that the company will pay the beneficiaries $1000 if the person dies during his twentieth year; if the person does not die, the company retains the $5.00.) What is the expected gain to the insurance company from selling one such policy, ignoring expenses of selling and administration? What is the standard deviation of the gain?

13. Suppose you are allowed to flip a fair coin until the first head appears. Let X be the total number of flips required. You will win 2^X dollars at the occurrence of the first head.
(a) What is the probability function for X?
(b) What is the expected value of your winnings?
(c) If the amount you pay to play a game is equal to the amount you expect to win, the game is called *fair*. How much should you pay to play this game to make it fair? This game is an example of the *Saint Petersburg paradox*.

14. Suppose, in Exercise 3.3.13, the game is truncated at 20 flips. That is, you flip the coin until the first head appears. You win 2^X dollars for $X \leq 19$ and you win 2^{20} dollars $= \$1,048,576$ if $X \geq 20$. How much should you pay to make this a fair game?

15. Assume X has density function

$$\begin{aligned} f_X(x) &= x, & 0 \leq x \leq 1 \\ &= \tfrac{3}{4}, & 1 < x \leq \tfrac{5}{3} \\ &= 0, & \text{otherwise.} \end{aligned}$$

Show that $\mu_X = t_{.5}$ and find their common value.

16. Find the median time you would wait for your friend if the density for T, the time you wait, is as given in Example 3.3.1. Evaluate the interquartile range for this probability law.

17. Prove Theorem 3.3.2.

3.4 Moments and Generating Functions

The mean and the variance describe two aspects of the probability law for a random variable (as long as they exist): the "middle" of the probability law and a measure of "how variable" the probability law is when measured from its middle. As seen in the last section, both the mean and variance are actually specified by the values of $E[X]$ and $E[X^2]$, the expected values of X raised to the first and second powers. Thus the values of $E[X]$ and $E[X^2]$ give rather natural measures of two particular aspects of a probability law. These two numbers, however, are not sufficient to specify the probability law for X completely. It can be shown that knowledge of $E[X^k]$, $k = 1, 2, 3, \ldots$, is sufficient to specify the probability law for X completely (granted all these expectations exist). Thus knowledge of the *moments* of X, $E[X^k]$ for $k = 1, 2, 3, \ldots$, provides another way of describing the probability law for X. In this section we will define the moments of a random variable, as well as various ways of evaluating these moments.

DEFINITION 3.4.1. The *k*th *moment* of a random variable X is

$$m_k = E[X^k], \qquad k = 1, 2, 3, \ldots,$$

as long as the expectation exists. ∎

The first moment m_1 is defined to be

$$m_1 = E[X^1] = E[X]$$

and thus $m_1 = \mu_X$. The second moment m_2 is

$$m_2 = E[X^2]$$

and thus $m_2 = \sigma_X^2 + \mu_X^2$. Thus knowledge of m_1 and m_2 for a random variable would immediately allow us to calculate the mean and the variance of the random variable. These two moments give information about the middle of the probability law and the relative variability about that middle value. The higher order moments $m_3, m_4, m_5, \ldots$, can be shown to give information about other facets of the probability law: the relative peakedness, how similar the two tails are on either side of the mean value, and so on. Let us look at some particular examples.

EXAMPLE 3.4.1.   Suppose we throw a dart at a target and have probability .9 of hitting the bull's-eye. Let X be 1 if we hit the bull's-eye and 0 if we miss it. Then

$$\begin{aligned} p_X(x) &= .1, & \text{at } x = 0 \\ &= .9, & \text{at } x = 1 \\ &= 0, & \text{otherwise.} \end{aligned}$$

The kth moment of X is

$$m_k = E[X^k] = 0^k(.1) + 1^k(.9) = .9 \qquad \text{for } k = 1, 2, 3, \ldots.$$

Thus this random variable X has constant moments of .9 for all k.   ∎

EXAMPLE 3.4.2.   Suppose we are going to drive a car to an athletic event. From past experience in making this trip we are willing to assume that our driving time is equally likely to be anywhere from 20 to 30 minutes. If we let X be the number of minutes it will take us to get there, then we have

$$\begin{aligned} f_X(x) &= \tfrac{1}{10}, & 20 < x < 30 \\ &= 0, & \text{otherwise} \end{aligned}$$

The kth moment of this random variable is

$$m_k = E[X^k] = \int_{20}^{30} \frac{x^k}{10}\, dx = \frac{30^{k+1} - 20^{k+1}}{10(k+1)}, \qquad k = 1, 2, 3, \ldots,$$

Thus

$$m_1 = \frac{(30)^2 - (20)^2}{10(2)} = 25$$

$$m_2 = \frac{(30)^3 - (20)^3}{10(3)} = 633\tfrac{1}{3},$$

and so on, and we have $\mu_X = 25$, $\sigma_X^2 = 8\tfrac{1}{3}$.   ∎

The moments of a random variable can be evaluated by expectations of powers, directly from the definition, as just illustrated. They can also be evaluated from the moment generating function of the probability law.

DEFINITION 3.4.2. The *moment generating function* for a random variable X is

$$m_X(t) = E[e^{tX}], \quad -\infty < t < \infty,$$

as long as the expectation exists. ∎

For many standard probability laws it is quite easy to evaluate $E[e^{tX}]$. How then can this function of a real variable t be used to evaluate the moments, $m_k = E[X^k]$? Recall that the Taylor series of e^{tx} about $x = 0$ gives

$$e^{tx} = 1 + tx + \frac{(tx)^2}{2!} + \frac{(tx)^3}{3!} + \cdots$$

for any real t and x. Thus for any fixed t we can write

$$
\begin{aligned}
m_X(t) &= E[e^{tX}] \\
&= E\left[1 + tX + \frac{t^2 X^2}{2!} + \frac{t^3 X^3}{3!} + \cdots \right] \\
&= 1 + tE[X] + \frac{t^2}{2!} E[X^2] + \frac{t^3}{3!} E[X^3] + \cdots \\
&= 1 + tm_1 + \frac{t^2}{2!} m_2 + \frac{t^3}{3!} m_3 + \cdots
\end{aligned}
$$

where $m_1, m_2, m_3, \ldots$ are the moments for X. Then

$$\frac{dm_X(t)}{dt} = m_1 + tm_2 + \frac{t^2}{2!} m_3 + \cdots$$

$$\frac{d^2 m_X(t)}{dt^2} = m_2 + tm_3 + \frac{t^2}{2!} m_4 + \cdots$$

and if we evaluate these derivatives at $t = 0$ all terms except the first disappear because they are multiplied by zero to some power. That is,

$$\frac{dm_X(t)}{dt}\bigg|_{t=0} = m_1$$

$$\frac{d^2 m_X(t)}{dt^2}\bigg|_{t=0} = m_2$$

and so on; as long as the moment generating function exists, its kth derivative evaluated at $t = 0$ gives m_k, the kth moment for X. Of course, $m_X(0) = E[e^{0 \cdot X}] = E[1] = 1$.

EXAMPLE 3.4.3.   Suppose the length of time a transistor will work (in a given circuit) is a random variable Y with density function

$$f_Y(y) = .001 e^{-.001y}, \quad y > 0$$
$$= 0, \qquad\qquad \text{otherwise.}$$

The moment generating function for Y is

$$m_Y(t) = E(e^{tY})$$

$$= \int_0^\infty e^{ty} .001 e^{-.001y} \, dy$$

$$= \int_0^\infty .001 e^{-y(.001 - t)} \, dy$$

$$= \frac{.001}{.001 - t}, \quad \text{for } t < .001.$$

From this we find that

$$\mu_Y = m_1 = m^{(1)}(0) = 1000$$
$$m_2 = m^{(2)}(0) = 2(1000)^2$$

so

$$\sigma_Y^2 = m_2 - \mu_Y^2 = (1000)^2$$

and

$$\sigma_Y = 1000. \qquad\qquad \blacksquare$$

EXAMPLE 3.4.4.   A fair coin is flipped twice. Let Z be the number of heads that occur. Then

$$p_Z(z) = \tfrac{1}{4}, \quad \text{at } z = 0$$
$$= \tfrac{1}{2}, \quad \text{at } z = 1$$
$$= \tfrac{1}{4}, \quad \text{at } z = 2$$
$$= 0, \quad \text{otherwise.}$$

The moment generating function for Z is

$$m_Z(t) = E[e^{tZ}]$$
$$= \tfrac{1}{4} + \tfrac{1}{2}e^t + \tfrac{1}{4}e^{2t}$$
$$= \tfrac{1}{4}(1 + e^t)^2,$$

from which we find $\mu_Z = m_1 = 1$, $\sigma_Z^2 = m_2 - \mu_Z^2 = \tfrac{1}{2}$. ∎

The moment generating function of a linear function of X is very simply related to the moment generating function for X. (Recall that Theorem 3.3.2 shows the relation between the means and variances of X and Y, when Y is a linear function of X.) That is, suppose X is a random variable with moment generating function $m_X(t)$ and let $Y = aX + b$, where a and b are any constants. Then

$$m_Y(t) = E[e^{tY}]$$
$$= E[e^{t(aX + b)}]$$
$$= E[e^{atX}e^{bt}]$$
$$= e^{bt}E[e^{atX}]$$
$$= e^{bt}m_X(at).$$

This establishes the following theorem.

Theorem 3.4.1. Let X be a random variable with moment generating function $m_X(t)$ and let $a \neq 0$ and b be constants. Then if $Y = aX + b$,

$$m_Y(t) = e^{bt}m_X(at).$$ ∎

For many random variables the *cumulant generating function* proves even easier to use in evaluating the mean and variance. This is defined to be the natural log of the moment generating function (assuming it exists). That is, if $m_X(t)$ is the moment generating function for X, then the *cumulant generating function* for X is

$$c_X(t) = \ln m_X(t),$$

and, of course,

$$m_X(t) = e^{c_X(t)},$$

so if $c_X(t)$ were known, it is easy to find $m_X(t)$. Then

$$\frac{d}{dt} c_X(t) = \frac{m'_X(t)}{m_X(t)}$$

$$\frac{d^2}{dt^2} c_X(t) = \frac{m''_X(t) m_X(t) - (m'_X(t))^2}{m_X^2(t)},$$

where

$$\frac{d}{dt} m_X(t) = m'_X(t), \qquad \frac{d^2}{dt^2} m_X(t) = m''_X(t).$$

Then, since $m_X(0) = 1$

$$\frac{d}{dt} c_X(t)\Big|_{t=0} = \frac{m'_X(0)}{m_X(0)} = \frac{m_1}{1} = \mu_X$$

$$\frac{d^2}{dt^2} c_X(t)\Big|_{t=0} = \frac{m''_X(0) m_X(0) - (m'_X(0))^2}{m_X^2(0)}$$

$$= \frac{m_2 \cdot (1) - (m_1)^2}{(1)^2} = \sigma_X^2.$$

The first two derivatives of $c_X(t)$ evaluated at $t = 0$ directly give the mean and variance for X. For many of the standard random variables that we will study, these two derivatives are very easy to evaluate.

EXAMPLE 3.4.5.   In Example 3.4.3 Y was a random variable with moment generating function

$$m_Y(t) = \frac{.001}{.001 - t}, \qquad t < .001.$$

The cumulant generating function for Y is

$$c_Y(t) = \ln m_Y(t) = \ln.001 - \ln(.001 - t)$$

from which we find

$$\frac{d}{dt} c_Y(t) = \frac{1}{.001 - t}$$

$$\frac{d^2}{dt^2} c_Y(t) = \frac{1}{(.001 - t)^2}$$

so

$$\mu_Y = \frac{d}{dt} c_Y(t)\big|_{t=0} = \frac{1}{.001} = 1000$$

$$\sigma_Y^2 = \frac{d^2}{dt^2} c_Y(t)\big|_{t=0} = \frac{1}{(.001)^2} = (1000)^2$$

as before. In Example 3.4.4 Z was a random variable with $m_Z(t) = \frac{1}{4}(1 + e^t)^2$. The cumulant generating function for Z then is

$$c_Z(t) = \ln m_Z(t) = \ln \tfrac{1}{4} + 2\ln(1 + e^t),$$

from which we find

$$\frac{d}{dt} c_Z(t) = \frac{2e^t}{1 + e^t}$$

$$\frac{d^2}{dt^2} c_Z(t) = \frac{2e^t}{(1 + e^t)^2}.$$

Thus

$$\frac{d}{dt} c_Z(t)\big|_{t=0} = \frac{2}{2} = 1 = \mu_Z$$

$$\frac{d^2}{dt^2} c_Z(t)\big|_{t=0} = \frac{2}{2^2} = \frac{1}{2} = \sigma_Z^2. \qquad \blacksquare$$

Either the cumulant generating function or the moment generating function can be used to evaluate means and variances (and other moments). Yet a third generating function also proves useful, especially with discrete random variables. This is the *factorial moment generating function*, defined by

$$\psi_X(t) = E[t^X] = E[e^{X \ln t}] = m_X(\ln t).$$

Expanding t^X in a Taylor series about the point $t = 1$ gives

$$t^X = 1 + X(t - 1) + X(X - 1)\frac{(t-1)^2}{2!} + X(X-1)(X-2)\frac{(t-1)^3}{3!} + \cdots$$

and thus

$$\psi_X(t) = 1 + (t - 1)E[X] + \frac{(t-1)^2}{2!} E[X(X-1)]$$

$$+ \frac{(t-1)^3}{3!} E[X(X-1)(X-2)] + \cdots$$

from which it is easy to see that

$$\frac{d}{dt}\psi_X(t)\big|_{t=1} = E[X]$$

$$\frac{d^2}{dt^2}\psi_X(t)\big|_{t=1} = E[X(X-1)] = E[X^2] - E[X]$$

$$\frac{d^3}{dt^3}\psi_X(t)\big|_{t=1} = E[X(X-1)(X-2)] = E[X^3] - 3E[X^2] + 2E[X],$$

and so on. $\psi_X(t)$ is called the factorial moment generating function because of the factorial-like structure of the expectations generated by its derivatives evaluated at $t = 1$. This same function is also called a *probability generating function*. To see why this name is used, assume X is a discrete random variable with probability function $p_X(x)$, nonzero for $x = 0, 1, \ldots, n$. Then

$$\psi_X(t) = E[t^X] = \sum_{x=0}^{n} t^x p_X(x)$$

$$= p_X(0) + t p_X(1) + t^2 p_X(2) + \cdots + t^n p_X(n)$$

and it is easy to see that

$$\frac{d^k}{dt^k}\psi_X(t)\big|_{t=0} = k! p_X(k), \qquad k = 0, 1, \ldots, n.$$

Thus the derivatives of $\psi_X(t)$ evaluated at $t = 0$ give a known constant times the probability function for X.

EXAMPLE 3.4.6.   A fair coin is flipped until a head occurs. If W is the number of flips required, the probability function for W is

$$p_W(w) = \frac{1}{2^w}, \qquad w = 1, 2, 3, \ldots$$

(see Example 2.8.1). The probability generating function for W then is

$$\psi_W(t) = E[t^W] = \sum_{w=1}^{\infty} \frac{t^w}{2^w} = \frac{t/2}{1 - t/2} = \frac{t}{2 - t},$$

for $|t/2| < 1$; that is, $-2 < t < 2$. Then

$$\frac{d}{dt}\psi_W(t) = \frac{2}{(2-t)^2}$$

$$\frac{d^2}{dt^2}\psi_W(t) = \frac{2 \cdot 2}{(2-t)^3}$$

$$\vdots$$

$$\frac{d^k}{dt^k}\psi_W(t) = \frac{2k!}{(2-t)^{k+1}}$$

and

$$\frac{d^k}{dt^k}\psi_W(t)\Big|_{t=0} = \frac{k!}{2^k} = p_W(k) \cdot k!, \qquad k = 1, 2, 3, \ldots$$

as it must. Note also that

$$\frac{d^k}{dt^k}\psi_W(t)\Big|_{t=1} = 2 \cdot k! = E[X(X-1)\ldots(X-k+1)],$$

so

$$E[X] = 2 \cdot 1! = 2$$
$$E[X(X-1)] = 2 \cdot 2! = 4 = E[X^2] - E[X]$$
$$E[X(X-1)(X-2)] = 2 \cdot 3! = 12$$
$$= E[X^3] - 3E[X^2] + 2E[X]. \qquad \blacksquare$$

EXERCISE 3.4

1. A single die is rolled one time. Let X be the number of 6's that occur. Compute $m_X(t)$ and use it to evaluate the first 3 moments of X.
2. Assume that Y, the number of minutes it takes you to eat lunch on an average day, is equally likely to lie in the interval from 30 to 40 minutes. Compute $m_Y(t)$.
3. The number of hours of satisfactory operation that a certain brand of TV set will give (without repair) is a random variable Z whose probability density function is

$$f_Z(z) = .0001e^{-.0001z}, \qquad z > 0$$
$$= 0, \qquad z \le 0.$$

Derive $m_Z(t)$ and use it to compute μ_Z and σ_Z^2.

4. A fair coin is flipped three times. Let U be the total number of heads that occur. Derive $\psi_U(t)$ and use it to compute μ_U and σ_U^2.

5. One integer is selected at random from the set $\{1, 2, 3, \ldots, n\}$. Let V be the selected integer and derive $\psi_V(t)$.

6. A discrete random variable X has moment generating function

$$m_X(t) = \exp[2(e^t - 1)].$$

Find the cumulant generating function for X and use it to evaluate μ_X and σ_X^2. (Note: $\exp[x] = e^x$.)

7. The factorial moment generating function for a discrete random variable Y is

$$\psi_Y(t) = (e^t - 1)/(e - 1)$$

Find the probability function for Y.

8. Evaluate the probability generating function for the random variable X discussed in Exercise 3.4.6.

9. A continuous random variable V has density function

$$f_V(v) = ve^{-v}, \qquad v > 0.$$

Find the moment generating function for V and the cumulant generating function for V.

10. A discrete random variable has probability generating function $(q + pt)^n$, where $0 \le p \le 1$, $q = 1 - p$. Find its probability function.

11. Find the moment and cumulant generating functions for the random variable discussed in Exercise 3.4.10.

12. Find the moment generating function for the discrete uniform random variable whose probability function is

$$p_U(u) = \frac{1}{n}, \qquad u = 1, 2, \ldots, n.$$

13. Find the moment generating function for the continuous uniform random variable whose density is

$$f_U(u) = \frac{1}{b - a}, \qquad a < u < b$$

$$= 0, \qquad \text{otherwise.}$$

14. Let X be a discrete random variable with factorial moment generating function $\psi_X(t)$ and define $Y = aX + b$, where a and b are constants. Express the factorial moment generating function for Y in terms of $\psi_X(t)$.

3.5 Functions of a Random Variable

In many applications one may know the probability law for a random variable X and then want to find the probability law for Y, a function of X. You undoubtedly already have some acquaintance with the metric system of units and know that the conversion from inches, feet, yards, to millimeters, centimeters, meters, and so on, involves a linear transformation; the metric measurement is a linear function of the measurement made in our more customary units. The same is true for conversions of weight, area, volume, temperature, and the various other types of scales that we employ for sundry purposes; the conversion from our customary units to the metric units is given by a linear function.

Now suppose a state survey party is used to measure the distance between two famous landmarks in California. The actual measurement they will arrive at is a random variable X; the units that they use are feet. Let us also assume an analysis of the procedures they employ makes it reasonable to assume we know the density function $f_X(x)$ for X. (As we will see in Chapter 4, there is good reason to assume $f_X(x)$ will be what we will call a normal density; these details need not concern us now.)

If we do know the density $f_X(x)$, then we can use it to evaluate probability statements about the accuracy of the survey party's measured distance (in feet). For example, if μ is the "true" distance between the landmarks, then the error in the measured distance is $|X - \mu|$ (in feet) and the density for X could be used to evaluate probability statements about the magnitude of the error. For the sake of illustration, assume we find the probability is .95 that the error in the survey is no more than 10 feet (the event $\{|X - \mu| \le 10\}$ has probability .95). Since 10 feet is the same as 3.048 meters, it would seem plausible that the probability is .95 that the survey error is no more than 3.048 meters. What we are really doing in making this conversion is defining $Y = aX$, where $a = .3048$ and $m = a\mu$; Y is the measurement made by the survey party in meters and m is the true distance in meters. The error made then is $|Y - m| = |aX - a\mu| = a|X - \mu|$ (in meters). Since

$$|X - \mu| \le 10 \qquad \text{is equivalent to}$$
$$a|X - \mu| \le 10a, \qquad \text{that is, } |Y - m| \le 3.048,$$

the probabilities of these equivalent statements must be equal. This is a very special example of using the probability law for a random variable X to derive a probability statement about the value of Y, a function of X. In this section we will study how to derive the probability law for Y, from the probability law for X, when Y is a function of X.

First we will discuss the discrete case. We assume X is a discrete random variable with probability function $p_X(x)$ and range R_X and let $Y = g(X)$ be a function of X. Then immediately, Y must also be a discrete random variable with range $R_Y = \{y: y = g(x), x \in R_X\}$. For example, suppose $R_X = \{-1, 1, 2, 3\}$ and we define $Y = X^2$. Then $R_Y = \{1, 4, 9\}$, since $(-1)^2 = 1^2 = 1$, $2^2 = 4$, and $3^2 = 9$; no other values could occur for Y. The probability function for Y is obtained simply by equating probabilities of equivalent events, as previously done in the survey party discussion and as originally done in deriving the probability function for a discrete random variable from the probability function for the sample space. That is, for $y \in R_Y$ we have

$$p_Y(y) = \sum_{x \in B_Y} p_X(x),$$

where $B_Y = \{x: x \in R_X \text{ and } g(x) = y\}$, since if there is more than one value of $x \in R_X$ such that $g(x) = y$, we get the total probability of Y equaling y by summing $p_X(x)$ over all x giving $g(x) = y$.

If, as previously, $R_X = \{-1, 1, 2, 3\}$

$$
\begin{aligned}
p_X(x) &= \tfrac{125}{216} && \text{for} && x = -1 \\
&= \tfrac{75}{216} && \text{for} && x = 1 \\
&= \tfrac{15}{216} && \text{for} && x = 2 \\
&= \tfrac{1}{216} && \text{for} && x = 3
\end{aligned}
$$

and we define $Y = X^2$, then $R_Y = \{1, 4, 9\}$ and

$$
\begin{aligned}
p_Y(1) &= p_X(-1) + p_X(1) = \tfrac{200}{216} \\
p_Y(4) &= p_X(2) = \tfrac{15}{216} \\
p_Y(9) &= p_X(3) = \tfrac{1}{216},
\end{aligned}
$$

which completely specifies the probability law for Y. When X is discrete and $Y = g(X)$, it is frequently easiest to first find R_Y and then to simply compute $p_Y(y)$ for each $y \in R_Y$.

EXAMPLE 3.5.1. A bakery owner has observed that he frequently receives calls for a fresh (made that day) birthday cake for a party. After observing this phenomenon for a while he decides that the probability distribution for D, the number of such daily requests, is

d	0	1	2	3	4
$p_D(d)$	.1	.2	.4	.2	.1

As is easily verified, the expected number of such calls per day is $E[D] = 2$. He then decided he would bake two birthday cakes each day for this purpose, in addition to any prior order cakes he might have for that day. Let Y be the number of these cakes he will sell each day. Then Y, the number sold, is a function of D; in fact $R_Y = \{0, 1, 2\}$ and

$$Y = D, \quad \text{for} \quad D = 0, 1$$
$$Y = 2, \quad \text{for} \quad D \geq 2.$$

Thus the probability law for Y is

y	0	1	2
$p_Y(y)$	.1	.2	.7

and the expected number of these special cakes he will sell is $E[Y] = 1.6$. ■

EXAMPLE 3.5.2. Let us continue the bakery example and discuss the marginal amount of income the bakery owner receives from the two daily birthday cakes he makes for last minute callers. Each birthday cake he makes costs \$3.50 and is sold for \$6.00 (if it is sold at all). Let us also assume that he donates any cake not sold to the local senior citizen club on the next day. This gives him a \$6.00 charitable deduction on his income tax form, which in turn translates into a \$2.00 reduction in his tax; thus he effectively loses \$1.50 for any cake he does not sell. Let X be the net marginal income, per day, which he realizes from baking two cakes per day for these special requests. This marginal income is \$2.50 for each cake sold and $-\$1.50$ for each cake not sold. Thus if he sells Y cakes on a given day (and does not sell $2 - Y$ cakes), we have $X = 2.5Y - 1.5(2 - Y) = 4Y - 3$ where the units are dollars. The range for X is $R_X = \{-3, 1, 5\}$ and the probability function for X is

x	-3	1	5
$p_X(x)$	.1	.2	.7

His expected marginal daily net income from this source then is $E[X] = \$3.40$. You are asked in Exercise 3.5.2 to consider other possible strategies that this baker might employ, in terms of the number of cakes he bakes daily for unexpected requests. ■

It is frequently quite straightforward to use the distribution function approach to find the probability law for $Y = g(X)$, especially when X is a continuous random variable and $g(x)$ is a continuous function for all $x \in R_X$. That is, the distribution function for Y is

$$F_Y(t) = P(Y \le t) \qquad \text{for all real } t;$$

but since $Y = g(X)$, the event $\{Y \le t\}$ is equivalent to $\{g(X) \le t\}$ so we have

$$F_Y(t) = P(g(X) \le t).$$

For many functions $g(X)$ the event $\{g(X) \le t\}$ can be expressed in a simple way as X itself lying in some interval or union of intervals. Then $P(g(X) \le t)$ can be evaluated directly from $F_Y(t)$.

EXAMPLE 3.5.3.   The meteorology station in Monterey, California, has gathered weather data for many years. Based on this previous experience, and today's current weather, they assume the following day's high recorded temperature for Monterey is a random variable X with density function

$$f_X(x) = \frac{x - 55}{100}, \qquad 55 \le x \le 65$$

$$= \frac{75 - x}{100}, \qquad 65 < x \le 75,$$

pictured in Figure 3.5.1; X is measured in degrees Fahrenheit. The dis-

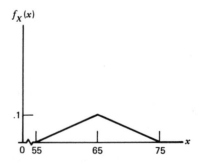

Figure 3.5.1

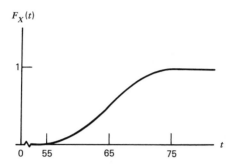

Figure 3.5.2

tribution function for X is

$$F_X(t) = P(X \le t) = \int_{55}^{t} \frac{(x - 55)}{100} \, dx = \frac{(t - 55)^2}{200}, \quad 55 \le t \le 65$$

$$= \frac{1}{2} + \int_{65}^{t} \frac{(75 - x)}{100} \, dx = 1 - \frac{(75 - t)^2}{200},$$

$$65 < t \le 75$$

$$= 1, \quad t > 75,$$

which is graphed in Figure 3.5.2. Now suppose that the Monterey weather station wishes to comply with the forces urging our conversion to the metric system. This system measures temperature on the centigrade or Celsius scale. If X is the temperature in degrees Fahrenheit and Y is the same temperature in degrees Celsius, then

$$Y = \tfrac{5}{9}(X - 32) = g(X),$$

a linear function of X. The distribution function approach previously mentioned is very easy to apply for a linear function. Indeed,

$$\begin{aligned} F_Y(t) = P(Y \le t) &= P\big(\tfrac{5}{9}(X - 32) \le t\big) \\ &= P(X \le \tfrac{9}{5}t + 32) \\ &= F_X(\tfrac{9}{5}t + 32). \end{aligned}$$

In other words, the events $\{Y \le t\}$ and $\{X \le \tfrac{9}{5}t + 32\}$ are equivalent (one happens if and only if the other does) and thus their probabilities of occurrence must be equal. Thus we have

$$F_Y(t) = F_X(\tfrac{9}{5}t + 32)$$

$$= \frac{(\tfrac{9}{5}t + 32 - 55)^2}{200} = \frac{(9t - 115)^2}{5000}, \qquad 55 \le \tfrac{9}{5}t + 32 \le 65,$$

that is,

$$\tfrac{115}{9} \le t \le \tfrac{165}{9}$$

and

$$F_Y(t) = 1 - \frac{(75 - \tfrac{9}{5}t - 32)^2}{200} = 1 - \frac{(215 - 9t)^2}{5000}, \qquad 65 < \tfrac{9}{5}t + 32 \le 75,$$

that is, for

$$\tfrac{165}{9} < t \le \tfrac{215}{9}.$$

This distribution function is graphed in Figure 3.5.3. Once we know the distribution function $F_Y(t)$ for Y (and we note it is continuous for all real t), we get the density for Y by differentiating $F_Y(t)$. Thus the density for the next day's high temperature, in degrees Celsius, is

$$f_Y(t) = \frac{d}{dt} F_Y(t) = \frac{9(9t - 115)}{2500}, \qquad \tfrac{115}{9} \le t \le \tfrac{165}{9},$$

$$= \frac{9(215 - 9t)}{2500}, \qquad \tfrac{165}{9} < t \le \tfrac{215}{9},$$

which is pictured in Figure 3.5.4. Notice that the density for Y has the same triangular shape as the density for X. This linear transformation shifts the origin and the length measure of the scale but leaves the triangular shape unchanged. ∎

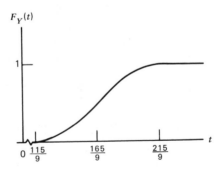

Figure 3.5.3

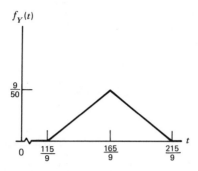

Figure 3.5.4

The linear transformation in Example 3.5.3 is a particular case of a monotonic (increasing) function. It is quite straightforward to derive the probability law for Y from the probability law for X when Y is a monotonic function of X. We will first discuss monotonic functions.

Let $y = g(x)$ be a real function of a real variable x, defined for all x in some set D. For simplicity we will also assume the derivative $dg(x)/dx$ exists for all $x \in D$. If for all $x_1 \in D$, $x_2 \in D$, where $x_1 < x_2$, we have $g(x_1) < g(x_2)$, we will call $g(x)$ *monotonic increasing*; if $g(x_1) > g(x_2)$ for all such x_1 and x_2, we will call $g(x)$ *monotonic decreasing*. Figure 3.5.5 illustrates a monotonic increasing function and Figure 3.5.6 illustrates a monotonic decreasing function. The important property these monotonic functions have is that in either case the inverse function $g^{-1}(\,)$ exists. That is, if $y = t$ the value of x that gives this value for y is $g^{-1}(t)$. Note as well from the figures that if $g(x)$ is increasing, then the set of y values such that $y \le t$

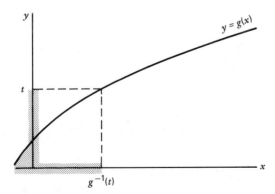

Figure 3.5.5

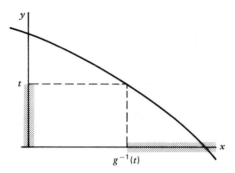

Figure 3.5.6

corresponds to the set of x values such that $x \leq g^{-1}(t)$, whereas if $g(x)$ is decreasing, $y \leq t$ corresponds to $x \geq g^{-1}(t)$. If $y = g(x)$ is monotonic increasing, then so is $x = g^{-1}(y)$, the inverse function. If $y = g(x)$ is monotonic decreasing, then so is $x = g^{-1}(y)$. Thus the slopes of the inverse functions, $dg^{-1}(y)/dy$, are positive or negative, depending on whether $g(x)$ is increasing or decreasing.

Now let X be a continuous random variable and define the random variable Y by $Y = g(X)$, where $y = g(x)$ is monotonic increasing for $x \in R_X$. In this case the event $\{Y \leq t\}$ is equivalent to $\{g(X) \leq t\}$, which in turn is equivalent to $\{X \leq g^{-1}(t)\}$ (see Figure 3.5.5). Two events that are equivalent have equal probabilities. Thus we have

$$
\begin{aligned}
F_Y(t) &= P(Y \leq t) \\
&= P\big(X \leq g^{-1}(t)\big) \\
&= F_X\big(g^{-1}(t)\big)
\end{aligned}
$$

if $g(x)$ is monotonic increasing. This shows how we can get the distribution function for Y from the distribution function for X. If $y = g(x)$ is monotonic decreasing for $x \in R_X$, the event $\{Y \leq t\}$ is equivalent to $\{X \geq g^{-1}(t)\}$ (see Figure 3.5.6) so

$$
\begin{aligned}
F_Y(t) &= P(Y \leq t) \\
&= P\big(X \geq g^{-1}(t)\big) \\
&= 1 - P\big(X \leq g^{-1}(t)\big) \\
&= 1 - F_X\big(g^{-1}(t)\big)
\end{aligned}
$$

when $g(x)$ is monotonic decreasing.

Recall that the density function for a continuous random variable is given by the derivative of its distribution function. Thus if $Y = g(X)$ and

$g(x)$ is monotonic increasing

$$f_Y(t) = \frac{d}{dt} F_Y(t)$$

$$= \frac{d}{dt} F_X(g^{-1}(t))$$

$$= f_X(g^{-1}(t)) \frac{dg^{-1}(t)}{dt},$$

whereas if $g(x)$ is monotonic decreasing

$$f_Y(t) = \frac{d}{dt} F_Y(t)$$

$$= \frac{d}{dt} [1 - F_X(g^{-1}(t))]$$

$$= -f_X(g^{-1}(t)) \frac{dg^{-1}(t)}{dt}$$

$$= f_X(g^{-1}(t)) \left| \frac{dg^{-1}(t)}{dt} \right|$$

since $dg^{-1}(t)/dt$ is negative in this case. Thus if $Y = g(X)$ and $g(x)$ is either monotonic increasing or decreasing, we have

$$f_Y(t) = f_X(g^{-1}(t)) \left| \frac{dg^{-1}(t)}{dt} \right|$$

and we have proved the following theorem.

Theorem 3.5.1. Let X be a continuous random variable, and let $g(x)$ be a continuous differentiable, monotonic function. Then if $Y = g(X)$, the distribution function for Y is

$$F_Y(t) = F_X(g^{-1}(t)) \qquad \text{if } g(x) \text{ is increasing}$$
$$= 1 - F_X(g^{-1}(t)) \qquad \text{if } g(x) \text{ is decreasing};$$

the density function for Y is

$$f_Y(t) = f_X(g^{-1}(t)) \left| \frac{dg^{-1}(t)}{dt} \right|.$$

EXAMPLE 3.5.4.   The natural logarithm is a frequently used transformation in many scientific fields. Suppose a random variable X is uniform on $(0, 1)$; that is,

$$f_X(x) = 1, \quad 0 < x < 1$$
$$= 0, \quad \text{otherwise}$$

and let $Y = -\ln X$. (Note that $-\ln x$ is positive for all $0 < x < 1$ and the range $R_X = \{x: 0 < x < 1\}$ generates the range for Y, $R_Y = \{y: y = -\ln x,$ $x \in R_X\} = \{y: y > 0\}$.) The inverse transformation is $X = e^{-Y}$, that is, $g^{-1}(t) = e^{-t}$ so $dg^{-1}(t)/dt = -e^{-t}$. Then the density for Y is

$$f_Y(t) = f_X\big(g^{-1}(t)\big) \left| \frac{dg^{-1}(t)}{dt} \right|$$

$$= e^{-t}, \quad t > 0.$$

If we reverse this process, that is, let Y be a random variable with density

$$f_Y(y) = e^{-y}, \quad y > 0$$

and define $X = e^{-Y}$, we should arrive back at the density

$$f_X(x) = 1, \quad 0 < x < 1.$$

It is left for the reader to verify that this is in fact the case.   ∎

EXAMPLE 3.5.5.   The transformation in Example 3.5.4 is a special case of the probability integral transform, which has been (and still is) used to transform from the uniform distribution to any desired continuous distribution. Computer simulations are used for many purposes, from determining the "best" locations for soundproofing materials in an automobile or studying the effect of various types of shocks to a building of a new design, to examinations of the impact certain governmental fiscal policies might be expected to have on the U.S. economy. Many of these simulations require observed values of random variables with specified probability laws. Let us take for granted that it is simple for the computer to generate an observed value of a random variable X that is uniform (equally likely observed values) on the interval $(0, 1)$; that is, X has density $f_X(x) = 1$, $0 < x < 1$ and distribution function

$$F_X(t) = \begin{cases} 0, & t \le 0 \\ t, & 0 < t < 1 \\ 1, & t \ge 1. \end{cases}$$

Suppose a computer simulation requires the observed value of a continuous random variable Y with distribution function $F_Y(t)$. As we know from

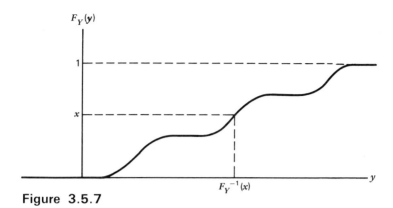

Figure 3.5.7

Section 3.2, the function $F_Y(t)$ must in fact be continuous, nondecreasing and lie between 0 and 1 for all t (see Figure 3.5.7).

Any observed value for X must lie between 0 and 1 since X has density $f_X(x) = 1, 0 < x < 1$ and, for any such x, we could use the inverse function of $F_Y(t)$ to define $y = F_Y^{-1}(x)$ (see Figure 3.5.7). In terms of random variables, then Y is defined by

$$Y = F_Y^{-1}(X), \qquad \text{whose inverse is } X = F_Y(Y).$$

What is the distribution function for Y? Since $F_Y^{-1}(t)$ is nondecreasing we have from Theorem 3.5.1

$$F_Y(t) = F_X(F_Y(t)) = F_Y(t), \qquad 0 < F_Y(t) < 1.$$

That is, given X is uniform, if we define $Y = F_Y^{-1}(X)$ then Y has distribution function $F_Y(t)$ as desired. If you examine this reasoning carefully you might have noticed that $F_Y(t)$ might actually have some flat places (the one pictured in Figure 3.5.7 has such a place) and $F_Y^{-1}(x)$ is not well defined over the interval where $F_Y(t)$ is flat since there is an interval of values corresponding to the same x. This observed value (and any other) occurs with zero probability so it should never be of concern. More practically, we can conveniently define $F_Y^{-1}(x)$ at this point to be the smallest y in the interval of values where $x = F_Y(y)$ and everything goes through all right. ■

We have stressed the derivation of the probability law for $Y = g(X)$ where $g(x)$ is a monotonic function, partly because this case does occur frequently in many areas of application (all linear transformations fall in this category). If we want the probability law for $Y = g(X)$ where the function $g(x)$ is not monotonic, the same general line of reasoning already used is still appropriate, but the details of its application can be more tedious to implement. That is, assume X is a continuous random variable with known distribution function $F_X(t)$ and suppose we want the probability law for $Y = g(X)$ where $g(x)$ is not a monotonic function. Then we still could get the distribution function for Y from equating probabilities of equivalent events.

$$F_Y(t) = P(Y \le t)$$
$$= P[g(X) \le t].$$

The evaluation of $P(g(X) \le t)$, for all t, may be cumbersome when $g(x)$ is not a monotonic function.

EXAMPLE 3.5.6.   One of the most frequently used nonmonotonic functions is the square function $Y = X^2$. Assume X is uniform on $(-1, 1)$ with distribution function

$$F_X(t) = 0, \qquad t \le -1$$
$$= \frac{t + 1}{2}, \qquad -1 < t < 1$$
$$= 1, \qquad t \ge 1.$$

What is the probability law for $Y = X^2$? First, let us recognize that the observed values for X must lie between -1 and 1 and, granted the observed value for Y will be the square of the observed value for X, the observed values for Y must lie between 0 and 1. Thus we know immediately that

$$F_Y(t) = 0, \qquad t < 0$$
$$= 1, \qquad t > 1.$$

Thus we need only find the value for $F_Y(t)$ for $0 \le t \le 1$. Now for any t between 0 and 1 the event $\{Y \le t\}$ is equivalent to the event $\{-\sqrt{t} \le X \le \sqrt{t}\}$ (see Figure 3.5.8). Thus for $0 \le t \le 1$

$$F_Y(t) = P(X^2 \le t)$$
$$= P(-\sqrt{t} \le X \le \sqrt{t})$$
$$= F_X(\sqrt{t}) - F_X(-\sqrt{t})$$
$$= \frac{\sqrt{t} + 1}{2} - \frac{-\sqrt{t} + 1}{2}$$
$$= \sqrt{t}.$$

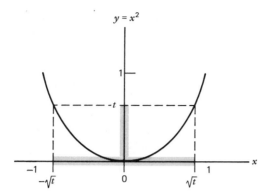

Figure 3.5.8

The density function for Y is

$$f_Y(t) = \frac{d}{dt} F_Y(t)$$

$$= \frac{1}{2\sqrt{t}}, \qquad 0 < t < 1$$

$$= 0, \qquad \text{otherwise.}$$

For nonmonotonic functions the distribution function $F_Y(t)$ is generally given by the sum of probabilities that X lies in any of several intervals.

EXERCISE 3.5

1. Let X be a random variable with distribution function $F_X(t)$ and let $Y = a + bX$ where $b < 0$. Derive the distribution function for Y.
2. Find the optimal number of cakes for the baker to make each day for the situation discussed in Example 3.5.2. That is, he could bake 1, 3, or 4 cakes for special orders; compute his expected marginal daily net income for each of these cases and see which number is best.
3. Suppose that $b = 0$ in Exercise 3.5.1. Derive the distribution function for Y, defined as in that problem.
4. Given

$$F_X(t) = 0, \qquad t < -1$$

$$= \frac{t+1}{2}, \qquad -1 \le t \le 1$$

$$= 1, \qquad t > 1,$$

find the distribution function for $Y = 15 + 2X$ and the density function for Y.

5. Suppose

$$
\begin{aligned}
F_W(t) &= 0, & t &< 0 \\
&= t^3, & 0 &\le t \le 1 \\
&= 1, & t &> 1
\end{aligned}
$$

and let $Z = W - 1$. Find $F_Z(t)$ and $f_Z(t)$.

6. If

$$
\begin{aligned}
F_X(t) &= 0, & t &< -10 \\
&= \tfrac{1}{4}, & -10 &\le t < 0 \\
&= \tfrac{3}{4}, & 0 &\le t < 10 \\
&= 1, & t &\ge 10,
\end{aligned}
$$

find the distribution function for

$$
U = 7X - 50 \qquad \text{and} \qquad p_U(u).
$$

7. If

$$
\begin{aligned}
F_Y(t) &= 1 - e^{-t}, & t &\ge 0 \\
&= 0, & t &< 0,
\end{aligned}
$$

find $F_X(t)$ and $f_X(t)$ where $X = 2Y - 7$.

8. Assume X is a *positive random variable*. (This means $F_X(0) = 0$, that R_X has no negative elements.) Let $Y = \sqrt{X}$ and find $F_Y(t)$ in terms of $F_X(t)$.

9. Assume X, in Exercise 3.5.8, is continuous and find $f_Y(t)$ in terms of $f_X(t)$.

10. Let X have the distribution given in Example 3.5.6; find the distribution function and density function for $Y = |X|$.

*11. Assume X is a continuous random variable with density function

$$
\begin{aligned}
f_X(x) &= \frac{x + 1}{4}, & -1 &\le x \le 1 \\
&= \frac{3 - x}{4}, & 1 &< x \le 3.
\end{aligned}
$$

(a) Find the distribution function for X.
(b) Find the distribution function for $Y = |X|$.
(c) Find the density function for $Y = |X|$.

*12. Let X have the density given in Exercise 3.5.11. Find the distribution function and density function for $Z = X^2$.

13. The random variable Y has density function

$$f_Y(y) = e^{-y} \qquad y > 0.$$

Find the probability function for $X = [Y]$, the integer part of Y. (That is, if $2 \le y < 3$, then $x = [y] = 2$.)

3.6 Summary

Random variable: Real-valued function of the elements of a sample space.
Range of a random variable: Set of possible observed values.
Discrete random variable: A random variable whose range is discrete.
Probability function: $p_X(x) = P(X = x)$, where $p_X(x) \ge 0$ for all x, $\sum p_X(x) = 1$.
Distribution function for a random variable: $F_X(t) = P(X \le t)$ for all real t.

$$P(a < X \le b) = F_X(b) - F_X(a)$$

Continuous random variable X: A random variable whose distribution function $F_X(t)$ is continuous.
Density function for continuous random variable:

$$f_X(t) = \frac{d}{dt} F_X(t).$$

Expected value:

$$E[G(X)] = \sum_{x \in R_X} G(x) p_X(x), \qquad X \text{ discrete}$$

$$= \int_{-\infty}^{\infty} G(x) f_X(x) \, dx, \qquad X \text{ continuous}$$

$E[c] = c$, where c is any constant.
$E[cG(X)] = cE[G(X)]$, where c is any constant.
$E[G(X) + H(X)] = E[G(X)] + E[H(X)]$
Mean value: $\mu_X = E[X]$
Variance: $\sigma_X^2 = E[(X - \mu)^2] = E[X^2] - \mu^2$
Standard deviation: $\sigma_X = \sqrt{\sigma_X^2}$.
If $Y = aX + b$, $\mu_Y = a\mu_X + b$, $\sigma_Y = |a| \sigma_X$, where a and b are any constants.
Quantile of continuous random variable: t_k is ($100k$th) quantile if $F_X(t_k) = k$, $0 < k < 1$
Median: $t_{.5}$
Quartiles: $t_{.25}, t_{.50}, t_{.75}$
Interquartile range: $t_{.75} - t_{.25}$

kth moment of X: $m_k = E[X^k]$

Moment generating function: $m_X(t) = E[e^{tX}]$,

$$m_X^{(k)}(t)|_{t=0} = m_k$$

If $Y = aX + b$, $m_Y(t) = e^{tb}m_X(at)$

Cumulant generating function: $c_X(t) = \ln m_X(t)$

Factorial moment generating function (probability generating function):

$$\psi_X(t) = E[t^X] = m_X(\ln t)$$

If X is discrete and $Y = g(X)$,

$$p_Y(y) = P(Y = y) = \sum_{x \in B_Y} p_X(x), \qquad B_Y = \{x : g(x) = y\}$$

If X is continuous and $Y = g(X)$, where $g(X)$ is monotonic,

$$f_Y(y) = f_X(g^{-1}(y)) \left| \frac{dg^{-1}(y)}{dy} \right|$$

4

Some Standard Probability Laws

There are a few probability laws that have found application in diverse areas. In this chapter we will study some of these frequently occurring probability laws and examine the assumed chance mechanisms that lead to their usage. Section 4.7 presents two tables, summarizing many of the important features of these probability laws.

4.1 The Bernoulli and Binomial Probability Laws

The simplest possible experiment is one that may result in either of two possible outcomes. (If there were only one possible outcome, we really would not have much need for probability in modeling the experiment.) We can think of many examples of such experiments: a flip of a coin (head or tail), performance of a student in a course (pass or fail), the sex of a yet-to-be-born child (male or female), placing a satellite in orbit around the earth (successful or not). We will call an experiment with two possible outcomes a *Bernoulli* trial and will label the two outcomes success (s) and failure (f):

DEFINITION 4.1.1. A *Bernoulli* trial is an experiment with two possible outcomes, success or failure. The sample space for a Bernoulli trial is $S = \{s, f\}$. ∎

Actually, any experiment can be used to define a Bernoulli trial simply by labeling some event A as success and calling its complement $\bar{A}$ a failure.

155

The probability distribution for a Bernoulli trial is easily specified and depends only on the single parameter $p = P(\{s\})$ and then $P(\{f\}) = 1 - p = q$. In order that it be possible for either outcome to occur, the parameter p lies between 0 and 1, noninclusive. A Bernoulli random variable can be defined on a Bernoulli trial or on any more complicated sample space:

DEFINITION 4.1.2.   Let S be the sample space for an experiment, let $A \subset S$ be any event with $p = P(A)$, $0 < p < 1$, and define

$$X(\omega) = 1 \quad \text{if} \quad \omega \in A$$
$$ = 0 \quad \text{if} \quad \omega \in \bar{A}.$$

Then X is called the *Bernoulli random variable* with parameter p.* If the experiment is actually a Bernoulli trial, we simply take $A = \{s\}$.   ∎

The probability law for a Bernoulli random variable follows directly from the probability distribution for S. Since $X = 1$ if and only if event A occurs we have $P(X = 1) = P(A) = p$, and since $X = 0$ if and only if event $\bar{A}$ occurs, it also follows that $P(X = 0) = P(\bar{A}) = 1 - p = q$. No other value can occur for X. Thus the probability law for a Bernoulli random variable X is defined by its mass function

$$p_X(1) = p, \qquad p_X(0) = q, \qquad q + p = 1$$

which can also be written as

$$p_X(k) = p^k q^{1-k}, \qquad k = 0, 1, \qquad p + q = 1.$$

The expected value for a Bernoulli random variable is

$$E[X] = 0 \cdot q + 1 \cdot p = p$$

Indeed, since $0^k = 0$, $1^k = 1$ for any $k = 1, 2, 3, \ldots$, the kth moment for a a Bernoulli random variable is

$$E[X^k] = 0^k \cdot q + 1^k \cdot p = p, \qquad k = 1, 2, 3, \ldots$$

from which it follows that the variance of X is

$$\sigma^2 = E[X^2] - (E[X])^2 = p - p^2 = pq.$$

The moment generating function for X is

$$m_X(t) = E[e^{tX}] = e^0 \cdot q + e^t \cdot p = q + pe^t$$

---

* A random variable defined as in Definition 4.1.2 is also called an *indicator variable* (of the set A) or a *binary* variable.

and the factorial moment generating function is

$$\psi_X(t) = E[t^X] = t^0 \cdot q + t^1 \cdot p = q + pt.$$

Since the Bernoulli probability law depends only on p, we will say that it has one *parameter*, the quantity p. The mean, variance, moments, and generating functions are all simple functions of this one parameter.

Bernoulli random variables are extremely simple and, because of this simplicity, not terribly interesting in themselves. They do, however, provide convenient building blocks for defining many other discrete random variables. The first one we will study is called the *binomial* random variable. To begin, let us discuss a binomial experiment.

DEFINITION 4.1.3. An experiment that consists of n (fixed) repeated independent Bernoulli trials, each with probability of success p, is called a *binomial* experiment with n trials and parameter p. ∎

The phrase "independent trials" means that the trials are "independent events," what occurs on one trial has no effect on the outcome to be observed for any other trial. More exactly, if A is any event whose occurrence or nonoccurrence depends on some subset of the trials and B is any event whose occurrence or nonoccurrence depends on a disjoint subset of the trials, then A and B are independent and $P(A \cap B) = P(A)P(B)$.

The natural sample space for a binomial experiment is the Cartesian product of the Bernoulli trial sample space with itself n times. That is, the binomial sample space is

$$S = S_1 \times S_2 \times \cdots \times S_n,$$

where $S_j = \{s, f\}, j = 1, 2, \ldots, n$. Each element of S is an n-tuple $(x_1, x_2, \ldots, x_n)$ where $x_j = s$ or f, representing success or failure on trial $j, j = 1, 2, \ldots, n$. For each Bernoulli trial

$$P_j(\{s\}) = p, \qquad P_j(\{f\}) = q = 1 - p$$

and because the trials are independent, we compute the probabilities of occurrence of single-element events (single n-tuple element) by multiplying the probabilities of occurrence of the appropriate outcomes for the given trials. Thus the single-element event

$$A = \{(s, s, \ldots, s)\}$$

whose single element represents success on every trial is assigned probability $p \cdot p \cdots p = p^n$, whereas the single-element event

$$B = \{(f, f, \ldots, f)\}$$

whose single element represents failure on every trial is assigned probability
$q \cdot q \cdots q = q^n$; the single-element event, whose single element represents
success on the first trial and failure on the rest, is assigned probability
$p \cdot q \cdot q \cdots q = pq^{n-1}$, and so on. Table 4.1.1 lists all the elements for a
binomial experiment with $n = 3$ and their probabilities of occurrence. You
can verify that the sum of the probabilities assigned to the single-element
events does in fact add to 1, as it must.

Table 4.1.1

Element	Probability of Single-Element Event
(s, s, s)	$ppp = p^3$
(s, s, f)	$ppq = p^2q$
(s, f, s)	$pqp = p^2q$
(s, f, f)	$pqq = pq^2$
(f, s, s)	$qpp = p^2q$
(f, s, f)	$qpq = pq^2$
(f, f, s)	$qqp = pq^2$
(f, f, f)	$qqq = q^3$

DEFINITION 4.1.4. Let X be the total number of successes in a binomial
experiment with n trials and parameter p. Then X is called the *binomial
random variable* with parameters n and p. ∎

Since the binomial random variable X counts the number of successes
observed in a binomial experiment with n trials, it is discrete with range $R_X = \{0, 1, 2, \ldots, n\}$. There is only one element of S for which X equals 0 (failure
on every trial) and its probability of occurrence is q^n; thus $P(X = 0) = q^n$.
Similarly, there is only one element of S for which $X = n$ (success on
every trial) and its probability of occurrence is p^n so $P(X = n) = p^n$. The
other possible observed values for X (the integers 1 through $n - 1$) are
assigned to more than one element of S, thus the probabilities for X equaling these values will be the sum of probabilities of single-element events.
How many elements of S are assigned the value $X = k$? This number is
simply $\binom{n}{k}$, the number of ways we could choose k positions for the s's
in the n-tuple (the remaining $n - k$ positions are all filled with f). Because
any single-element event whose single n-tuple contains k s's and $n - k$ f's is
assigned probability $p^k q^{n-k}$, we immediately have

$$P(X = k) = \binom{n}{k} p^k q^{n-k}, \quad k = 1, 2, \ldots, n - 1,$$

and since $\binom{n}{0} = \binom{n}{n} = 1$, we actually have the following theorem.

Theorem 4.1.1. If X is a binomial random variable with parameters n and p, its probability function is

$$p_X(k) = \binom{n}{k} p^k q^{n-k}, \qquad k = 0, 1, \dots, n. \qquad \blacksquare$$

You can easily verify that $\sum_{k=0}^{n} p_X(k) = 1$ by using the binomial theorem.

EXAMPLE 4.1.1. Assume a student is given a pop quiz with 10 true-false questions. Also assume the student is totally unprepared for the quiz and guesses at the answer to every question. If X is the number of questions answered correctly (by this student), what is the probability law for X?

If the student is really guessing at the right answer for every question, we assume each question is a Bernoulli trial with success = right answer and then $p = \frac{1}{2}$, because either answer should be chosen with equal probability. Furthermore, in such total ignorance, the trials should be independent because the answer chosen for any single question should have no affect on which answer will be selected for any subsequent question. Thus the quiz represents a binomial experiment with $n = 10$, $p = \frac{1}{2}$, and X, the number answered correctly, is then the binomial random variable with $n = 10$, $p = \frac{1}{2}$. The probability that the student answers all questions correctly is

$$P(X = 10) = \binom{10}{10} \left(\frac{1}{2}\right)^{10} \left(\frac{1}{2}\right)^{0} = \left(\frac{1}{2}\right)^{10} = .00098,$$

whereas the probability that (exactly) 5 questions are answered correctly is

$$P(X = 5) = \binom{10}{5} \left(\frac{1}{2}\right)^{5} \left(\frac{1}{2}\right)^{5} = \frac{252}{1024} = .24609.$$

If the student passes the quiz by getting 7 or more correct, the probability that she passes is

$$P(X \geq 7) = \sum_{k=7}^{10} \binom{10}{k} \left(\frac{1}{2}\right)^{10} = .17188. \qquad \blacksquare$$

EXAMPLE 4.1.2. The science of genetics has many models that lead to the binomial probability law. One such model says that if two particular strains of the same type of flower are crossed (pollen from one, ovum from the other), then $\frac{3}{4}$ of the resulting flowers will have ruffled edges and the remainder will have smooth edges. Furthermore, the individual seeds produced by the cross behave like independent Bernoulli trials. Thus if, say, 150 seeds are produced by such a cross, the number of these seeds that will produce ruffled edges is a binomial random variable X with parameters $n = 150$, $p = \frac{3}{4}$. What is the probability law for $Y = 150 - X$, the number of these seeds that will produce smooth-edged flowers?   ▪

The expected value of a binomial random variable is

$$E[X] = \sum_{k=0}^{n} k \binom{n}{k} p^k q^{n-k}$$

$$= \sum_{k=1}^{n} k \frac{n!}{k!(n-k)!} p^k q^{n-k}$$

$$= np \sum_{k=1}^{n} \frac{(n-1)!}{(k-1)!(n-k)!} p^{k-1} q^{n-k}$$

$$= np \sum_{j=0}^{n-1} \binom{n-1}{j} p^j q^{n-1-j}, \qquad j = k - 1$$

$$= np(p + q)^{n-1} = np,$$

the product of the two parameters. It is easy to see that the binomial moment generating function is

$$m_X(t) = E[e^{tX}] = \sum_{k=0}^{n} e^{tk} \binom{n}{k} p^k q^{n-k}$$

$$= \sum_{k=0}^{n} \binom{n}{k} (pe^t)^k q^{n-k}$$

$$= (pe^t + q)^n$$

and that the factorial moment generating function is

$$\psi_X(t) = E[t^X] = E[e^{X \ln t}]$$
$$= m_X(\ln t) = (pe^{\ln t} + q)^n$$
$$= (pt + q)^n.$$

Then

$$\psi_X''(t) = n(n-1) p^2 (pt + q)^{n-2}$$

and since

$$\psi_X''(1) = n(n-1) p^2 = E[X(X-1)] = E[X^2] - E[X]$$

we also have

$$E[X^2] = n(n-1)p^2 + E[X]$$
$$= n^2p^2 - np^2 + np$$

giving

$$\text{Var}[X] = \sigma_X^2 = n^2p^2 - np^2 + np - (np)^2$$
$$= np - np^2 = npq.$$

In Example 4.1.1 the guessing student could expect to get $10\left(\frac{1}{2}\right) = 5$ questions correct; the variance of the number correct is $10\left(\frac{1}{2}\right)\left(\frac{1}{2}\right) = 2.5$. The expected number of ruffle-edged flowers produced by the cross discussed in Example 4.1.2 is $(150)\left(\frac{3}{4}\right) = 112.5$ and its variance is $(150)\left(\frac{3}{4}\right)\left(\frac{1}{4}\right) = 28.125$. The expected value and variance of the number of smooth-edged flowers is $(150)\left(\frac{1}{4}\right) = 37.5$ and 28.125.

The distribution function for a binomial random variable is a step function with jumps at the integers $0, 1, 2, \ldots, n$. The size of the jump at each of these integers is given by the probability function at that integer. Thus if k is any integer between 0 and n,

$$F_X(k) = \sum_{j=0}^{k} \binom{n}{j} p^j q^{n-j}.$$

It is quite straightforward to evaluate the probability function, $p_X(k)$, on a modern hand-held calculator, especially ones with $n!$ and y^x keys, and also to find $F_X(k)$, $k = 0, 1, 2, \ldots, n$ by accumulating these values.

Histograms, or bar charts, are useful in graphically representing discrete probability laws. These are constructed by centering a bar on the real line above each value k from the range of the random variable; the area of the bar is equal to $p_X(k)$, the probability that the random variable equals that observed value and thus areas represent probabilities as they do for continuous random variables. When the discrete random variable is integer-valued (like the binomial), the width of each bar is 1, so the height of the bar is equal to the probability $p_X(k)$. Figure 4.1.1 presents histograms for the binomial probability laws with $n = 10, 20$, $p = .5, .9$. With $p = .5$, the binomial histogram is symmetric about $\mu_X = np$ for any n, whereas as p shifts toward 1 (or zero) the bulk of the probability is shifted toward higher (or lower) values, with the maximum at an integer close to $\mu_X = np$.

It is useful for some purposes to visualize the outcomes of repeated independent Bernoulli trials as defining the positions of a particle performing a random walk. Let the horizontal axis be labeled the trial number (n) and let the vertical axis record the accumulated number of successes. The particle starts at $(0, 0)$ and, at each succeeding step moves one unit to the

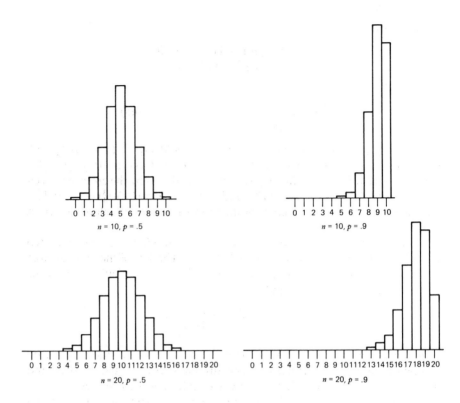

Figure 4.1.1

right horizontally (if a failure was observed) or along the 45° line (if a success was observed). Figure 4.1.2 pictures the results of two sequences of 20 Bernoulli trials. In the solid line sequence, successes were observed on trials 4, 5, 8, 10, 13, 14, 15, and 19, whereas in the dashed sequence, successes were observed on trials 1, 2, 5, 6, 8, 9, 10, 12, 13, 14, 15, 18, 19, 20. The binomial probability law with $n = 10$ gives the probabilities that the trace of the sequence passes through the various possible values at $n = 10$, marked by the circles at trial number 10. The binomial probability law with $n = 20$ gives the probabilities that the trace of the sequence passes through the various possible values at trial number 20, again marked by circles at $n = 20$. Thus the binomial probability law for any fixed n gives the distribution for the location of the particle after a fixed number (n) of steps. As n increases, the number of possible locations increases, so the total probability of 1 is spread over a larger span of values. The *expected* position of the particle is np, described by a straight line through the origin with slope p.

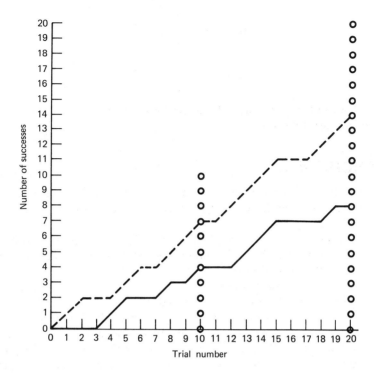

Figure 4.1.2

It is also useful to realize that a binomial random variable with parameters n and p is in fact the sum of n Bernoulli random variables. Although
we will defer discussion of the probability laws for several random variables
until Chapter 5, let us note this binomial-Bernoulli connection now. Suppose
n independent Bernoulli trials, each with probability p, are to be performed.
Define

$$Y_i = 1 \quad \text{if we observe a success on trial } i$$
$$= 0, \quad \text{otherwise}$$

for $i = 1, 2, 3, \ldots, n$. Then $\sum_{i=1}^{n} Y_i$ gives the total number of successes in
the n trials; $X = \sum_{i=1}^{n} Y_i$ is the binomial random variable with parameters
n and p. Note as well that

$$E[Y_i] = p, \qquad \text{Var}[Y_i] = pq$$

and

$$\mu_X = np = \sum_{i=1}^{n} E[Y_i]$$

$$\sigma_X^2 = npq = \sum_{i=1}^{n} \text{Var}[Y_i].$$

We will see in Chapter 5 why these relations between means and between variances must hold true.

EXERCISE 4.1

1. Five fair dice are rolled one time. Let X be the number of 1's that occur. Compute the mean of X, the variance of X, $P(1 \le X < 4)$, and $P(X \ge 2)$.

2. An urn contains 8 red and 2 black balls. Twenty balls are drawn with replacement. Let Y be the number of red balls that occur; compute μ_Y, σ_Y, $P(Y = 16)$, $P(Y < 14)$, and $P(Y > 18)$.

3. Ten coins are tossed onto a table. Let Z be the number of coins that land head up. Compute $P(Z = 5)$ and μ_Z.

4. Suppose that it is known that 10 percent of the glasses made by a certain glass-blowing machine will be defective in some way. If we randomly select 10 glasses made by this machine, what is the probability that none of them are defective? How many would we expect to be defective?

5. A commuter drives to work each morning. The route she takes each day includes seven stoplights. Assume the probability each stoplight is red when she gets to it is .2 and that these trials (stoplights) are independent. What is the distribution for X, the number of times she must stop for a red light on her way to work? Evaluate $P(X = 0)$ and $P(X \le 5)$.

6. Y is known to be a binomial random variable with mean $\mu_Y = 6$ and variance $\sigma_Y^2 = 4$. Find the probability distribution for Y (that is, evaluate n and p).

7. The probability that an individual seed of a certain type will germinate is known to be .9. A nursery man wants to sell flats of this type of plant and will claim that each flat contains 100 plants. If he plants 110 seeds in a flat (which we assume will sprout independently), how many plants should we expect an "average" flat to contain? Is there any number of seeds he can plant in a flat in order to be certain that the flat will contain 100 plants?

8. At a county fair a ring toss game may be played for 25¢. You are given three rings and then attempt to toss them individually onto a peg. If

you successfully get one ring on a peg you win a prize worth 50¢. If you get two on, you get a prize worth $1 and if you get all three on, you win a prize worth $5. Assuming the probability that you ring the peg is .1 each try, what is your expected gain if you play this game one time? Ten times?

9. Five fair dice are rolled one time. What is the probability of getting at least one 3? At least two 3's?

10. An investor buys five single-residence dwellings as an investment. He assumes that the probability he will make a profit on each is .9. Assuming independence,
    (a) What is the probability he makes a profit on every one?
    (b) What is the probability he takes a loss on each one?

11. A manufacturer of small parts ships his parts in lots of size 20 to his customers. Assume each part either is or is not defective and that the probability an individual part is defective is 0.05.
    (a) What is the expected number of defectives per lot?
    (b) What is the probability that any particular lot contains no defectives?

12. Assume you are a customer of the manufacturer mentioned in Exercise 4.1.11 and you receive 10 lots from him.
    (a) What is the expected number of lots that have no defectives?
    (b) What is the probability that you have received no defectives in the 10 lots?

13. A rat maze consists of a straight corridor, at the end of which is a branch; at the branching point the rat must either turn right or left. Assume 10 rats are placed in the maze, one at a time. If each is choosing one of the two branches at random, what is the distribution of the number that turn right? That turn left? What is the probability that at least 9 will turn the same way?

14. A radio-television dealer extends credit to people buying his sets. Assume in the past 10 percent of all those to whom he extended credit did not pay and he took a loss on each sale. The other 90 percent paid in full and he made a profit on those sales. Assume this dealer has 10 identical television sets and that he will sell them individually and independently on credit to 10 people. If the buyer does not pay he takes a loss of $200; if the buyer does pay in full he makes a profit of $100.
    (a) What is the distribution of the amount of profit he will make on these 10 sales?
    (b) What is his expected profit on these 10 sales?

15. It is, of course, possible to model an experiment with independent trials, but with unequal probabilities of success from trial to trial.

Assume $n = 2$ independent trials are to be performed; the probability
of success for the first trial is .4 and the probability of success for the
second is .8. Let X be the total number of successes observed and
evaluate its probability function, mean, and variance. Is X binomial?

16. Reconsider the commuter situation discussed in Exercise 4.1.5. Let us
assume the probability she must stop at the first light is .2 and that
the lights are synchronized and as long as she drives at 30 mph, each
succeeding light will be green upon her arrival. Define X as in Exercise
4.1.5 and answer the questions posed there, assuming she drives at
30 mph.

4.2 Geometric and Negative Binomial Probability Laws

In the last section we discussed independent Bernoulli trials and saw
how the binomial probability law occurs. Both the geometric and negative
binomial probability laws can also be defined on sequences of independent
Bernoulli trials.

Suppose independent Bernoulli trials are performed, each with probability p of success, and we let X be the number of trials necessary to get
the first success. Then

$$P(X = 1) = p,$$

since the probability of a success is p for each trial. We will observe $X = 2$
if and only if we have a failure on the first trial and then a success on the
second, so $P(X = 2) = pq$. Similarly, for any integer $k \geq 3$, we will observe
$X = k$ if and only if we have failures on the first $k - 1$ trials, followed by a
success on the kth trial, so $P(X = k) = pq^{k-1}$. This random number of
trials needed to get the first success is called the *geometric* random variable
with parameter p, because the values of its probability function form a
geometric progression. We have then established the following theorem.

Theorem 4.2.1. Independent Bernoulli trials, each with probability p of
success, are performed. If X is the number of trials needed to get the first
success, then X is called the geometric random variable with parameter p;
its probability function is

$$p_X(k) = pq^{k-1}, \qquad k = 1, 2, 3, \ldots,$$

where $q = 1 - p$. ∎

The expected value of a geometric random variable is

$$E[X] = \sum_{k=1}^{\infty} kpq^{k-1}$$

$$= p \sum_{k=1}^{\infty} kq^{k-1} = p\frac{1}{(1-q)^2} = \frac{1}{p}$$

(see the geometric series results in Section 2.8). We also have

$$E[X(X-1)] = \sum_{k=1}^{\infty} k(k-1)pq^{k-1}$$

$$= 2pq \sum_{k=2}^{\infty} \binom{k}{2} q^{k-2}$$

$$= \frac{2pq}{(1-q)^3} = \frac{2q}{p^2}$$

so

$$E[X^2] = E[X(X-1)] + E(X)$$

$$= \frac{2q}{p^2} + \frac{1}{p} = \frac{2}{p^2} - \frac{1}{p}$$

and we have

$$\text{Var}[X] = E[X^2] - (E[X])^2$$

$$= \frac{2}{p^2} - \frac{1}{p} - \frac{1}{p^2} = \frac{q}{p^2}.$$

The factorial moment generating function for the geometric probability law is

$$\psi_X(t) = E[t^X] = \sum_{k=1}^{\infty} t^k pq^{k-1}$$

$$= pt \sum_{k=1}^{\infty} (qt)^{k-1}$$

$$= \frac{pt}{1-qt}$$

EXAMPLE 4.2.1.   To investigate the effects of a certain drug on a given condition, a medical research team must first locate a person with the specified condition. Suppose 10 percent of the population has the given condition and the research team will interview people until they find a person with the condition; they assume each person interviewed is a Bernoulli trial (success = person has the condition) with $p = .1$ and that the trials are independent. Thus if X is the number of people they must interview to locate one person with the condition, X is geometric with parameter $p = .1$ and

$$p_X(k) = (.1)(.9)^{k-1}, \qquad k = 1, 2, 3, \dots.$$

The expected value for X is

$$E[X] = \frac{1}{.1} = 10$$

and the variance for X is

$$\text{Var}[X] = \frac{.9}{(.1)^2} = 90;$$

the standard deviation for X is

$$\sigma_X = \sqrt{90} = 9.49.$$

Since σ_X is quite large, there would be a large variation in the number of people interviewed to locate one with the condition, if the interviewing process were repeated several times. ∎

The histogram for a geometric probability law has a very simple form. The area of the bar over $k = 1$ is p and the area of each succeeding bar is q times the preceding bar's area. Figure 4.2.1 gives the histogram for the geometric probability law with $p = \frac{1}{2}$. The area of the first bar is $\frac{1}{2}$ and each succeeding bar has half the area of the one preceding.

The distribution function for the geometric random variable is easily evaluated. We have

$$P(X > t) = \sum_{k=t+1}^{\infty} pq^{k-1}$$

$$= pq^t \frac{1}{1-q} = q^t, \qquad t = 0, 1, 2, 3, \dots$$

and thus

$$F_X(t) = 1 - P(X > t)$$
$$= 1 - q^t, \qquad t = 0, 1, 2, \dots.$$

The geometric probability law has a "memoryless" property, which is shared by no other discrete distribution. This property derives from the

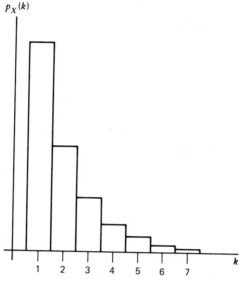

Figure 4.2.1

value of $P(X > a + b \mid X > a)$, the conditional probability that more than $a + b$ trials will be needed to get the first success, given that more than a trials are needed; a and b are positive integers. To use our earlier definitions for conditional probability, we let A be the event that $X > a$ and let B be the event that $X > a + b$; we seek the value for

$$P(B \mid A) = \frac{P(A \cap B)}{P(A)} .$$

A little reflection easily shows that B is a subset of A, since X must necessarily be larger than a if it is larger than $a + b$. But when $B \subset A$, we have $A \cap B = B$ and thus

$$P(B \mid A) = \frac{P(B)}{P(A)}$$

$$= \frac{P(X > a + b)}{P(X > a)}$$

$$= \frac{q^{a+b}}{q^a} = q^b = P(X > b).$$

This says the probability that more than b additional trials are required to observe the first success, granted no successes in the first a trials, is the same as the original probability that more than b trials would be required.

In terms of the situation described in Example 4.2.1, let $a = 5$, $b = 10$, and $X =$ the number of people interviewed to get the first success. Then if the first $5(a)$ people interviewed did not yield a success, the probability it will take at least $10(b)$ more people to find the first success is the same as it was at the start of the process. The probability law (conditionally) has no memory and does not change. This is, of course, a direct consequence of the assumptions that the trials are independent, each with the same non-changing probability of a success. The geometric probability law is the only discrete distribution with this property.

The *negative binomial* (also called the Pascal) probability law gives the distribution for the number of independent Bernoulli trials necessary to observe the rth success, $r = 2, 3, 4, \dots$. Thus again, assume independent Bernoulli trials are performed, each with probability of success p, and now let Y be the number of trials necessary to observe the rth success, $r = 2, 3, \dots$. Then clearly, the range for Y is $R_Y = \{r, r + 1, r + 2, \dots\}$ since at least r trials must be performed to observe r successes. We will observe $Y = r$ if and only if a success occurs on each of the first r trials so $P(Y = r) = p^r$. In order to observe $Y = r + 1$, the rth success must occur on the $(r + 1)$th trial *and* there must have been exactly $r - 1$ successes in the first r trials. Thus

$$P(Y = r + 1) = p \binom{r}{r-1} p^{r-1} q = \binom{r}{r-1} p^r q.$$

Similarly, for any integer $k > r$, we will observe $Y = k$ if and only if the rth success occurs on trial k (and thus there must be exactly $r - 1$ successes in the first $k - 1$ trials) so

$$P(Y = k) = \binom{k-1}{r-1} p^r q^{k-r}.$$

This establishes the following theorem.

Theorem 4.2.2. Independent Bernoulli trials, each with probability p of success, are performed. If Y is the number of trials needed to observe the rth success, $r = 2, 3, \dots$, then Y is called the negative binomial random variable with parameters r and p; the probability function for Y is

$$p_Y(k) = \binom{k-1}{r-1} p^r q^{k-r}, \qquad k = r, r + 1, r + 2, \dots. \qquad \blacksquare$$

The factorial moment generating function for the negative binomial random variable Y is

$$\psi_Y(t) = E[t^Y] = \sum_{k=r}^{\infty} t^k \binom{k-1}{r-1} p^r q^{k-r}$$

$$= (pt)^r \sum_{k=r}^{\infty} \binom{k-1}{r-1} (qt)^{k-r}$$

$$= \left(\frac{pt}{1-qt} \right)^r$$

from which it is easily found that the mean for Y is

$$E[Y] = \frac{r}{p}$$

and the variance for Y is

$$\text{Var}[Y] = \frac{rq}{p^2}.$$

Figure 4.2.2 gives the histograms for a negative binomial probability law with $r = 3, 6$ and $p = .5, .8$.

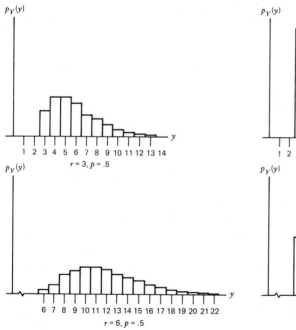

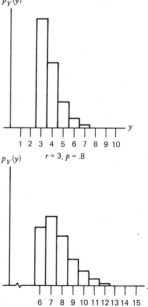

Figure 4.2.2

EXAMPLE 4.2.2.   Suppose the medical research team mentioned in Example 4.2.1 wants to find $r = 10$ persons with the stated condition, rather than just 1, and let Y be the number of people they must interview to accomplish this. Then Y has the negative binomial probability law with $r = 10$, $p = .1$. The expected number of people they would have to interview is $\frac{10}{.1} = 100$ and the variance of Y is $10(.9)/(.1)^2 = 900$; the standard deviation of Y is $\sqrt{900} = 30$.   ∎

EXAMPLE 4.2.3.   The negative binomial probability law has been employed in constructing simple models for failures of various sorts, such as metal fatigue in an aircraft wing. Metal fatigue is evidenced by such things as minute cracks in the surface of the metal. During flight the wing receives various sorts of shocks, especially from turbulent air. Call each such shock a Bernoulli trial and assume there is a constant probability p (hopefully small) of "damage" being caused by each shock; once r damaging shocks have occurred, a hairline crack will appear in the wing's surface. Then if Y is the number of shocks received at the time the hairline crack appears, Y is negative binomial with parameters r and p.   ∎

It may have occurred to you that the negative binomial probability law with $r = 1$ should be the same as the geometric probability law; that this is so is easily verified simply by setting $r = 1$. It is also easy to see that if we let X_1 be the number of Bernoulli trials to get the first success, let X_2 be the number of *additional* trials (beyond X_1) to get the second success, let X_3 be the number of *additional* trials (beyond $X_1 + X_2$) to get the third success, and so on, then $X_1, X_2, X_3, \ldots$ are each geometric random variables with parameter p. Also, $X_1 + X_2 + \cdots + X_r$ then is the number of trials to get the rth success and has the negative binomial distribution with parameters r and p; that is, the negative binomial probability law occurs as the sum of certain geometric random variables, the parameter r being equal to the number of geometric random variables added together. This situation is quite analogous to the relation between the binomial and Bernoulli random variables.

It is also instructive to consider the geometric and negative binomial probability laws in terms of the random walk of a particle (moves one step to the right with a failure, one to the right and up with a success), as we did in Section 4.1.1. The observed value for the geometric random variable then is the random step number at which the number of successes first equals 1, whereas the observed value for the negative binomial random variable, with parameter r, is the step number at which the number of successes first equals r. Figure 4.2.3 pictures the results of two sequences of Bernoulli trials; the circles opposite 1 success indicate the possible observed values for the geometric random variable (trial number at which

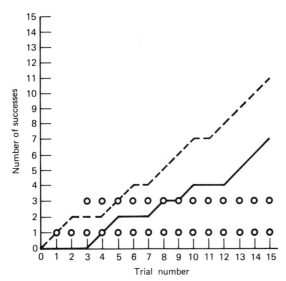

Figure 4.2.3

the random walk first hits level 1) and for the negative binomial random variable with $r = 3$ (trial number at which the walk first hits level 3).

The geometric and negative binomial probability laws occur in situations that are, in a sense, the reverse of those in which the Bernoulli and binomial probability laws occur. For both the Bernoulli and binomial cases the number of trials to be performed is fixed (at 1 or n, respectively) and the observed value for the random variable is the random number of successes to occur. For both the geometric and negative binomial situations the number of successes to be observed is fixed (at 1 or r, respectively) and the observed value for the random variable is the random number of trials required to generate the fixed number of successes. From this point of view it would seem reasonable to call the geometric probability law the "negative Bernoulli."

EXERCISE 4.2

1. A fair coin is flipped until a head occurs. What is the probability that less than 3 flips are required? That less than 4 flips are required?
2. An American roulette wheel commonly has 38 spots on it of which 18 are black, 18 are red, and 2 are green. Let X be the number of spins necessary to get the first red number. Give the probability function for X and the mean for X.

3. Let Y be the number of spins necessary to observe the first green number for the roulette wheel mentioned in Exercise 4.2.2. What is the probability function, mean, and variance for Y?

4. Suppose an urn contains 10 balls of which 1 is black. Let Z be the number of draws, with replacement, necessary to observe the black ball. What is the probability function for Z? The mean of Z?

5. An ice cream company makes chocolate-covered ice cream bars on sticks that sell for 10 cents. Suppose they put a star on every fiftieth stick; anyone who buys a bar with a starred stick gets a free ice cream bar. If you decide to buy ice cream bars until you get a free one, how much would you expect to spend before getting a free bar?

6. Evaluate the moment generating functions for the geometric and negative binomial probability laws.

7. A basketball player makes repeated shots from the free throw line. Assume his shots are independent Bernoulli trials with $p = .7$ (probability that the ball goes through the basket).
   (a) What is the probability that it takes him less than 5 shots to make his first basket?
   (b) What is the probability that it takes him less than 5 shots to make his second basket?
   (c) What is the probability that it takes him an odd number of shots to make his first basket?

8. A small manufactured item is produced on an assembly line. Each item either is or is not defective; defectives occur independently with probability .05. Suppose we start observing items at an arbitrary point on the line and let X be the number of items we inspect until we find the first defective item.
   (a) What is the probability function for X?
   (b) What is the expected number of items we must inspect to find the first defective item?

9. Assume the same situation as in Exercise 4.2.8, except now we will continue inspecting until we find 5 defective items. Answer the questions posed in Exercise 4.2.8.

10. A defective particle counter has probability .7 of counting each particle that enters its aperture, independently from one particle to another. What is the distribution for the number of particles it does not register, before the first one to be counted?

11. Assume every time you drive your car the probability is .001 that you will get a ticket for speeding. Also assume you will lose your license once you have received three such tickets. Let X be the number of times you will drive your car until you get the third such ticket and derive the probability function for X (assume you have the same

probability, .001, of getting a ticket each time you drive and that the occurrences of such tickets are independent).

12. A coin is flipped until the first head occurs. Assume the flips are independent and that the probability of a head occurring each time is p.
    (a) Show that the probability an odd number of flips is required is $p/(1 - q^2)$.
    (b) Find the value of p so that the probability is .6 an odd number of flips is required.
    (c) Can you find a value of p so that the probability is .5 that an odd number of flips is required?

4.3 Sampling and the Hypergeometric Probability Law

Sample surveys are used to a great extent in modern society for diverse purposes. These include the Nielson ratings of television shows, which are quite instrumental in determining which particular programs will continue to be shown, and political polls, which candidates for office use in determining the strategy they will employ in trying to win an election. Market research firms use sample surveys to try to judge the possible market appeal new products might enjoy. In this section we will study some simple problems connected with sampling from a population.

To do so, we will refer to selecting balls from an urn at random. The urn contains m balls, of which w are white and the remaining b are blue; thus $w + b = m$. The balls are meant to represent eligible voters, for example, and the two different colors represent two different possible responses (favors the candidate, does not favor the candidate). A sample of n balls is selected at random from the urn without replacement. By this we mean any possible subset of size n of the m balls in the urn could be the n selected. Each of these possible subsets has the same probability of being selected.

Now let Y be the number of white balls in the n selected for the sample. Provided that $w \geq n$ and $b \geq n$, the range of Y is the set of integers $0, 1, ..., n$ (the balls in the sample selected could vary from all blue to all white). The total number of different samples that can be selected from the urn is $\binom{m}{n}$, the number of subsets of size n that can be constructed from a set with m elements. Because the sample is selected at random from the urn, each of these subsets has the same probability, $1 / \binom{m}{n}$, of being the one selected.

The number of these subsets that contain exactly y white balls (and thus also $n - y$ blue balls) is $\binom{w}{y}\binom{b}{n-y}$, because any y of the white balls could be chosen together with any $n - y$ of the blue balls. Thus the probability function for Y, the number of white balls in the sample, is

$$p_Y(y) = \frac{\binom{w}{y}\binom{b}{n-y}}{\binom{m}{n}}, \qquad y = 0, 1, 2, \ldots, n$$

and we have proved the following theorem.

Theorem 4.3.1. A sample of n balls is selected at random without replacement from an urn that contains w white balls and $b = m - w$ blue balls. The probability function for Y, the number of white balls in the sample, is

$$p_Y(y) = \frac{\binom{w}{y}\binom{b}{n-y}}{\binom{m}{n}}, \qquad y = 0, 1, 2, \ldots, n.$$

Y is called the *hypergeometric* random variable with parameters n, m, and w.

■

In the preceding derivation we assumed that both w and b were at least as large as n (so that the range of Y is the integers 0 through n). This is not really necessary as long as we realize that some of the integers 0 through n may have probability zero of occurring. To illustrate this, take $w = 3$, $b = 2$, $m = 3 + 2 = 5$, $n = 4$. Then in any sample of 4 balls from this urn we must get at least 2 white balls and cannot get any more than 3 white balls; the probability of getting 0, 1, or 4 white balls is zero. Recall we can write

$$\binom{w}{y} = \frac{w(w-1)\cdots(w-y+1)}{y(y-1)\cdots 2 \cdot 1}$$

and the product in the numerator, $w(w-1)\cdots(w-y+1)$, is in fact zero if $y > w$ (because one of the terms in the product is zero); thus it is natural to define $\binom{w}{y} = 0$ for any positive integers with $y > w$. With this convention you can verify that the formula given for the probability function for Y

in Theorem 4.3.1 actually works for all possible cases. Those integers in $R_Y = \{0, 1, 2, ..., n\}$ that cannot occur are assigned probability zero.

EXAMPLE 4.3.1. Assume there are 20 television sets in use in a small town at a given time and that exactly 9 (less than a majority) are tuned to the CBS network; the remaining 11 are split between ABC, NBC, and the local educational station. Let us also assume that a sample of 5 of these 20 sets is randomly selected at this specific time and it is observed, for each of the 5, whether the set is tuned to CBS (white ball) or not (blue ball). Then if Y is the number of sets in the sample that are tuned to CBS, the probability function for Y is

$$p_Y(y) = \frac{\binom{9}{y}\binom{11}{5-y}}{\binom{20}{5}}, \qquad y = 0, 1, ..., 5.$$

The probability that exactly 3 of the sets in the sample are tuned to CBS is

$$P(Y = 3) = \frac{\binom{9}{3}\binom{11}{2}}{\binom{20}{5}} = .2980.$$

The probability that a majority in the sample is tuned to CBS is

$$P(Y \geq 3) = \frac{\binom{9}{3}\binom{11}{2}}{\binom{20}{5}} + \frac{\binom{9}{4}\binom{11}{1}}{\binom{20}{5}} + \frac{\binom{9}{5}\binom{11}{0}}{\binom{20}{5}}$$

$$= .3925. \qquad \blacksquare$$

If the hypergeometric probability function stated in Theorem 4.3.1 is correct, it is necessary that

$$\sum_{y=0}^{n} \binom{w}{y}\binom{b}{n-y} = \binom{m}{n},$$

so that

$$\sum_{R_Y} p_Y(y) = 1.$$

This equality is a special case of the Vandermonde convolution. To see that this result is correct, we know that

$$(1 + x)^w (1 + x)^b = (1 + x)^m,$$

where $m = b + w$. From the binomial theorem, the multiplier of x^n in the expansion of $(1 + x)^m$ is $\binom{m}{n}$. On the left-hand side of this equation the term x^n occurs by taking x^y from $(1 + x)^w$ times x^{n-y} from $(1 + x)^b$, $y = 0, 1, \ldots, n$ (since $x^y x^{n-y} = x^n$). Thus the total coefficient of x^n from the left-hand side is

$$\sum_{y=0}^{n} \binom{w}{y} \binom{b}{n-y}$$

and since the multipliers of the same term from both sides must be equal, we have

$$\sum_{y=0}^{n} \binom{w}{y} \binom{b}{n-y} = \binom{m}{n},$$

as desired.

The expected value for a hypergeometric random variable is

$$E[Y] = \frac{1}{\binom{m}{n}} \sum_{y=0}^{n} y \binom{w}{y} \binom{b}{n-y}$$

$$= \frac{1}{\binom{m}{n}} \sum_{y=1}^{n} y \frac{w!}{y!(w-y)!} \binom{b}{n-y}$$

$$= \frac{w}{\binom{m}{n}} \sum_{y=1}^{n} \frac{(w-1)!}{(y-1)!(w-y)!} \binom{b}{n-y}$$

$$= \frac{w}{\binom{m}{n}} \sum_{x=0}^{n-1} \binom{w-1}{x} \binom{b}{n-1-x}, \quad \text{(with } x = y - 1\text{)}$$

$$= \frac{w}{\binom{m}{n}} \binom{m-1}{n-1} = n\frac{w}{m}.$$

You can also easily verify that

$$E[Y(Y-1)] = \frac{w(w-1)}{\binom{m}{n}}\binom{m-2}{n-2} = \frac{w(w-1)n(n-1)}{m(m-1)}$$

and from this

$$\begin{aligned}\mathrm{Var}[Y] &= E[Y(Y-1)] + E[Y] - (E[Y])^2 \\ &= n\frac{w}{m}\frac{(m-w)}{m}\frac{(m-n)}{(m-1)}.\end{aligned}$$

Neither the factorial moment generating function nor the moment generating function is very useful for the hypergeometric probability law. It is quite straightforward to evaluate any of the moments of Y directly without resorting to these generating functions. Figure 4.3.1 gives the histograms

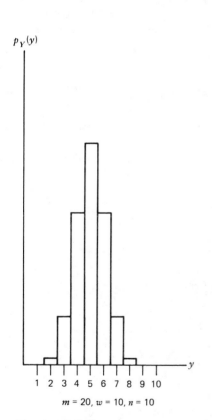

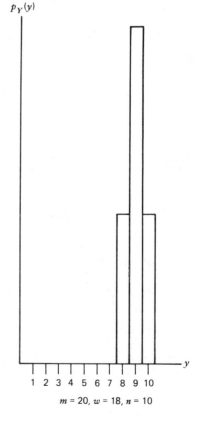

$m = 20, w = 10, n = 10$

$m = 20, w = 18, n = 10$

Figure 4.3.1

of the hypergeometric probability laws with $m = 20$, $w = 10$, $n = 10$, and $m = 20$, $w = 18$, $n = 10$.

EXAMPLE 4.3.2.   Sampling procedures are frequently used by organizations and groups that buy materials in large lots. In such situations the buyer and supplier have agreed to some acceptable level of quality; if the lot is large it may be too time consuming and expensive to inspect each individual item in the lot, so only a random sample of the items in the lot will actually be inspected. The whole lot is accepted as being good or rejected as being unacceptable, based on the results of the inspection of the sample. As a simple example of this type, suppose a lot contains 100 items, each of which is either defective (white ball) or nondefective (blue ball). Let us also assume two items are to be selected at random for inspection. If both items selected are nondefective, the lot is accepted; if either (or both) is defective the lot is rejected. Let Y be the number of defectives in the sample of 2. Then Y is hypergeometric with $n = 2$, $m = 100$, and the lot will be accepted if and only if we observe $Y = 0$; the probability that $Y = 0$ depends critically on the value of w, the number of defectives in the lot. If $w = 0$ (and thus $b = 100$), we have

$$P(\text{lot is accepted}) = P(Y = 0)$$

$$= \frac{\binom{0}{0}\binom{100}{2}}{\binom{100}{2}} = 1,$$

and if $w = 5$

$$P(Y = 0) = \frac{\binom{5}{0}\binom{95}{2}}{\binom{100}{2}} = .902;$$

if the lot contains 10 defectives

$$P(Y = 0) = \frac{\binom{10}{0}\binom{90}{2}}{\binom{100}{2}} = .809$$

and, if $w = 20$,

$$P(Y = 0) = \frac{\binom{20}{0}\binom{80}{2}}{\binom{100}{2}} = .638 .$$

The larger w is, the smaller the chance that the lot will be accepted. By allowing n, the sample size, to vary (and the number of defectives in the sample for an acceptable lot), one has a wide choice of inspection plans and essentially any desired properties (in terms of probability of lot acceptability) can be implemented. ∎

We have described sampling without replacement as though all n items in the sample are simultaneously selected from the urn. Clearly, this is not necessary; we could select a first item and, without replacing it, a second item, and, without replacing either of those two, select a third item, and so on, until we have chosen n items for the sample. Conceiving of sampling in this way makes the phrase "sampling without replacement" perhaps seem more descriptive. Additionally, if the sampling were done one item at a time like this, we could think of each selection as being a Bernoulli trial, because each ball drawn is either white (success) or blue (failure). But now the probability of a success is *not* the same for each trial. The probability that the first ball drawn is white is simply w/m; the probability that the second ball drawn is white is either $(w - 1)/(m - 1)$ or $w/(m - 1)$, depending on whether the first ball was white or blue, respectively. Thus we can also say the trials are *dependent* when sampling without replacement, because the probability of success on any given trial after the first depends on which balls were removed on previous draws.

Suppose we were to select balls from the urn *with* replacement, meaning after we select the first ball and observe its color, we replace it in the urn before drawing the next ball; this procedure is followed for all draws in sampling with replacement and thus there are always m balls in the urn at each draw. Thus if the sample were drawn at random with replacement, the probability of a white ball being drawn is $p = w/m$ for each trial and the trials are independent. Then Y, the number of white balls drawn for the sample, would be binomial with parameters n and $p = w/m$. The expected value for Y (with replacement) is $np = n(w/m)$, the same as $E[Y]$ when the sampling is done without replacement. The variance for Y (with replacement) is $npq = n(w/m)(m - w)/m$, which differs from the hypergeometric variance only by the factor $(m - n)/(m - 1)$ (sometimes called the finite population correction factor). Thus the distribution of the number of white balls in the sample is hypergeometric or binomial, depending on whether the sampling is done without replacement or with replacement.

If both the number of white balls (w) and the number of blue balls (b) in the urn are "large" compared to the sample size n, it would not seem to matter much whether the sampling is done with or without replacement. For example, if $w = 400$, $b = 600$ and we take a sample of, say, $n = 5$ balls without replacement, the exact probability that all the balls are white is

$$\frac{400 \cdot 399 \cdot 398 \cdot 397 \cdot 396}{1000 \cdot 999 \cdot 998 \cdot 997 \cdot 996} = .01009,$$

which differs negligibly from $(.4)^5 = .01024$. It follows that we might expect the binomial probability law to give a good approximation to the hypergeometric probability law in this case. In fact, if n is no larger than the smaller of $.2b$ or $.2w$, the binomial probability law gives good approximations to the exact hypergeometric values.

EXAMPLE 4.3.3.   Inadvertently one burned-out light bulb was placed into a box together with six good bulbs. The bulbs are tested at random one after another, without replacement, to locate the bad bulb. If we let X be the number of bulbs tested to locate the bad bulb, the probability function for X is easily determined. The probability that the first bulb tested is the bad bulb is $P(X = 1) = \frac{1}{7}$. The bad bulb is found on the second test only if the first bulb is good and the second is the bad bulb; thus $(P(X = 2) = \frac{6}{7} \cdot \frac{1}{6} = \frac{1}{7}$. Similarly, the bad bulb is found on the third test only if both the first two tested are good and the third is the bad one so $P(X = 3) = \frac{6}{7} \cdot \frac{5}{6} \cdot \frac{1}{5} = \frac{1}{7}$. By now the pattern should be clear; the probability function for X is

$$p_X(x) = \tfrac{1}{7}, \qquad x = 1, 2, \ldots, 7.$$

This is called the discrete uniform distribution on the integers 1 through 7. You may have noted one particular weak point in the discussion of this problem. Suppose five bulbs have been tested and all are good, leaving only two bulbs in the box, one good and one bad. Then actually only one more bulb needs to be tested. If it is bad then the bad one has been located and, if it is good, the one remaining untested must be the bad bulb. With this point recognized the probability function for X is

$$
\begin{aligned}
p_X(x) &= \tfrac{1}{7}, & x &= 1, 2, \ldots, 5 \\
&= \tfrac{2}{7}, & x &= 6.
\end{aligned}
$$ ∎

EXERCISE 4.3

1. A box contains 5 marbles of which 3 are chipped. Two marbles are chosen randomly without replacement from the box. What is the probability function for the number of chipped marbles in the sample?
2. Thirteen cards are drawn randomly without replacement from a regular 52-card deck. What is the probability function for the number of red cards in the sample? What are the mean and the variance of the number of red cards?
3. A bag contains 10 flashbulbs, 8 of which are good. If 5 flashbulbs are chosen from the bag at random, what is the probability function for the number of good flashbulbs? For the number of bad flashbulbs?

4. A panel of 7 judges is to decide which of 2 final contestants in a beauty contest will be declared the winner; a simple majority of the judges will determine the winner. Assume 4 of the judges will vote for Marie and that the other 3 will vote for Sue. If we randomly select 3 of the judges and ask them who they are going to vote for, what is the probability that a majority of the judges in the sample will favor Marie?

5. An incoming lot of material contains 100 items. A sample of 5 items will be inspected; each item selected will be classified as defective or nondefective. If the sample contains 1 or fewer defectives it will be accepted, otherwise it will be rejected. If the lot contains 5 defectives, what is the probability that it will be accepted? If the lot contains 15 defectives, what is the probability that it will be rejected?

6. An urn contains 10 white, 20 blue, 5 red, and 10 green balls. A sample of 10 balls is selected from the urn at random, without replacement. If X is the number of white balls in the sample, what is the probability function for X? The mean and variance for X?

7. A box contains 7 light bulbs, of which 2 are bad. The bulbs are tested one after another without replacement. Let X be the number of bulbs tested until the first bad bulb is found. What is the probability distribution for X?

8. In Exercise 4.3.7, let Y be the number of bulbs tested to locate the second bad bulb. Find the probability function for Y.

9. A hat contains 100 slips of paper, of which 30 are green and 70 are red. A random sample of 5 slips is selected at random from the hat. Let W be the number of green slips in the sample and compare the exact hypergeometric probability law for W with its binomial approximation. (That is, evalute $P(W = w)$, $w = 0, 1, 2, \ldots, 5$ in two different ways.)

10. In one version of the game of Keeno, you, as the player, may mark any 5 of the first 100 integers. Then the casino randomly selects 20 of the first 100 integers. Let X be the number of the integers you selected that are included in those chosen by the casino. What is the probability function for X?

4.4 The Poisson Probability Law

Many important applications of probability theory are concerned with modeling the time instants at which events occur. For example, in planning the number of telephone lines and the type of equipment that might best serve the needs of a given organization, one must have some idea of the number of incoming and outgoing telephone calls that might be expected during various periods of time. The efficient design of a roadway system depends on the expected number and times of arrival of vehicles that will use the system. A hospital, department store, or any other group that maintains a staff to serve individuals is very interested in the number of, and times at which, demands for service may occur. In this section we will study the simplest and perhaps most frequently used model for the time instants at which events are observed.

We will assume we are going to observe the phenomenon of interest (telephone call arrival times, automobile arrivals, customer arrival times) for a period of time; the time instant at which we begin observing the phenomenon will be labeled 0, the origin for our time scale. First, let us assume we will observe the phenomenon for a fixed time period of length t, where t is any positive number. The number of events to occur in this fixed interval $(0, t]$ is a random variable X. Since the value for X will be the number of events that occur, the range of X is discrete, so X is a discrete random variable. The probability law for X, of course, depends on the manner in which the events occur.

Now let us make the following assumptions about the way in which the events occur:

1. In a sufficiently short length of time, say of length Δt, only 0 or 1 event can occur (two or more simultaneous occurrences are impossible).
2. The probability of exactly 1 event occurring in this short interval of length Δt is equal to $\lambda \Delta t$, proportional to the length of the interval.
3. Any nonoverlapping intervals of length Δt are independent Bernoulli trials.

With these assumptions (called the assumptions for a Poisson process with parameter λ), we can subdivide our interval of length t into $n = t/\Delta t$ nonoverlapping, equal length pieces (see Figure 4.4.1). These small intervals of time then are independent Bernoulli trials, each with probability of success (an event occurs) equal to $p = \lambda \Delta t$. The probability of no event occurring on each trial is $q = 1 - \lambda \Delta t$. Then X, the number of events in the interval of length t, is binomial, n, $p = \lambda \Delta t = \lambda t/n$ and

$$p_X(k) = \binom{n}{k} (\lambda \Delta t)^k (1 - \lambda \Delta t)^{n-k}$$

$$= \binom{n}{k} \left(\frac{\lambda t}{n}\right)^k \left(1 - \frac{\lambda t}{n}\right)^{n-k}$$

Figure 4.4.1

If we now take the limit of this probability function as $\Delta t \to 0$ (and thus $n \to \infty$), we arrive at the *Poisson* probability law, which gives the probability of occurrence for any number k of events in the continuous interval of length t, as long as assumptions 1, 2, and 3 are true regarding the process generating the events. We can write

$$p_X(k) = \binom{n}{k}\left(\frac{\lambda t}{n}\right)^k\left(1 - \frac{\lambda t}{n}\right)^{n-k}$$

$$= \frac{n!}{k!(n-k)!}\frac{(\lambda t)^k}{n^k}\left(1 - \frac{\lambda t}{n}\right)^{-k}\left(1 - \frac{\lambda t}{n}\right)^n$$

$$= \frac{(\lambda t)^k}{k!}\left(1 - \frac{\lambda t}{n}\right)^n\left(1 - \frac{\lambda t}{n}\right)^{-k}\frac{n(n-1)\cdots(n-k+1)}{n^k}.$$

In the limit as $n \to \infty$,

$$\left(1 - \frac{\lambda t}{n}\right)^n \to e^{-\lambda t},$$

$$\left(1 - \frac{\lambda t}{n}\right)^{-k} \to 1,$$

$$\frac{n(n-1)\cdots(n-k+1)}{n^k} = 1\cdot\left(1 - \frac{1}{n}\right)\left(1 - \frac{2}{n}\right)\cdots\left(1 - \frac{k+1}{n}\right) \to 1$$

and thus

$$p_X(k) \to \frac{(\lambda t)^k}{k!}e^{-\lambda t}$$

for each $k = 0, 1, 2, \ldots$ and we have proved the following theorem.

Theorem 4.4.1. Events are generated consistent with assumptions 1, 2, 3. If X is the number of events in an interval of fixed length t, then the probability function for X is

$$p_X(k) = \frac{(\lambda t)^k}{k!}e^{-\lambda t}, \qquad k = 0, 1, 2, \ldots.$$

X is called a Poisson random variable with parameter λt. ∎

You are familiar with the Taylor series expansion for e^x,

$$e^x = 1 + x + \frac{x^2}{2!} + \frac{x^3}{3!} + \cdots = \sum_{j=0}^{\infty} \frac{x^j}{j!},$$

valid for any real x. If we sum the probability function for the Poisson random variable over its range, we have

$$\sum_{k=0}^{\infty} p_X(k) = \sum_{k=0}^{\infty} \frac{(\lambda t)^k}{k!} e^{-\lambda t} = e^{\lambda t} e^{-\lambda t} = 1,$$

as of course we must. The expected value for X is

$$E[X] = \sum_{k=0}^{\infty} k \frac{(\lambda t)^k}{k!} e^{-\lambda t} = \lambda t \sum_{k=1}^{\infty} \frac{(\lambda t)^{k-1}}{(k-1)!} e^{-\lambda t}$$

$$= \lambda t.$$

To evaluate $\mathrm{Var}[X]$, we need to find $E[X^2]$. This can be found directly or, as we know, it can also be evaluated from either the moment generating or factorial moment generating function. We have been using t as the argument for the factorial moment generating function; this argument has no relationship, of course, to the length of time for which we count events in the Poisson process, which we have also labeled t. To avoid confusion between the two, let us use μ to represent the parameter of the Poisson random variable X; as we have just seen, this parameter is in fact equal to $E[X]$. Thus if we count the number of events to occur in a period of length t, the parameter of the distribution for X is $\mu = \lambda t$, where λ is the number expected in a unit length of time and the probability function for X is

$$p_X(k) = \frac{\mu^k}{k!} e^{-\mu}, \qquad k = 0, 1, 2, \ldots .$$

The factorial moment generating function for X then is

$$\psi_X(t) = E[t^X] = \sum_{k=0}^{\infty} t^k \frac{\mu^k}{k!} e^{-\mu}$$

$$= e^{-\mu} \sum_{k=0}^{\infty} \frac{(\mu t)^k}{k!}$$

$$= e^{-\mu} e^{\mu t} = e^{\mu(t-1)}.$$

You can easily verify by differentiation that $E[X(X-1)] = \mu^2 + \mu$ and thus $\mathrm{Var}[X] = \mu$. The mean and the variance are equal for the Poisson probability law.

EXAMPLE 4.4.1. It is frequently assumed phenomena such as injury-causing accidents in a large industrial plant satisfy the three assumptions for a Poisson process. Suppose these assumptions are true for a particular plant, occurring at a rate of, say, $\lambda = \frac{1}{2}$ per week. If X represents the number of such accidents to occur in the next $t = 6$ weeks, then X is a Poisson random variable with parameter $\mu = \frac{1}{2}(6) = 3$ and its probability function is

$$p_X(k) = \frac{3^k}{k!}e^{-3}, \qquad k = 0, 1, 2, \ldots .$$

Both the mean and variance for X equal 3. The probability of exactly 3 accidents in this period is

$$p_X(3) = \frac{3^3}{3!}e^{-3} = .224,$$

and the probability of at least 4 accidents is

$$\sum_{k=4}^{\infty} \frac{3^k}{k!}e^{-3} = .353.$$

The probability there will be no accidents is

$$p_X(0) = \frac{3^0}{0!}e^{-3} = .050.$$

This last probability could easily be computed in a second way. If there are no accidents in the 6-week period, then there must be no accidents in each of three successive 2-week periods. But the number of accidents in a 2-week period is a Poisson random variable Y with parameter $\mu = \frac{1}{2}(2) = 1$ and the probability of zero accidents then is $\frac{1^0}{0!}e^{-1} = e^{-1}$. Non-overlapping periods of time in a Poisson process are independent trials. Thus our three successive 2-week periods are three independent Bernoulli trials with probability of success (no accident) being e^{-1} for each trial. The probability of three successes in a row thus is $(e^{-1})^3 = e^{-3}$, the same answer we got from the probability function for X. ∎

Figure 4.4.2 gives histograms for the Poisson probability laws with $\mu = 1$ and $\mu = 8$. It is quite straightforward to evaluate the individual terms of the Poisson probability function on a hand-held calculator. Since the sequence of probabilities for the integers $0, 1, 2, \ldots$ is $e^{-\mu}, \mu e^{-\mu}, (\mu^2/2!)e^{-\mu}, \ldots$ the first probability (that $X = 0$) is simply $e^{-\mu}$; the second term is μ times this value, the third is $\mu/2$ times the preceding, and so on. In fact,

$$\frac{p_X(k+1)}{p_X(k)} = \frac{\dfrac{\mu^{k+1}}{(k+1)!}e^{-\mu}}{\dfrac{\mu^k}{k!}e^{-\mu}} = \frac{\mu}{k+1}$$

so each succeeding term is simply μ over an integer times the preceding term, making it very easy to evaluate as many individual terms as desired or to evaluate the distribution function for X, $F_X(t) = \sum_{k=0}^{t} p_X(k)$, by simply accumulating the terms as they are computed. Note that if the parameter μ is an integer, say $\mu = k + 1$, then

$$\frac{\mu}{k+1} = \frac{p_X(k+1)}{p_X(k)} = 1$$

and the two probabilities, $p_X(k)$ and $p_X(k+1)$, are equal to each other. For values of k smaller than $\mu - 1$ the probabilities are increasing and for k greater than μ they are decreasing, as illustrated in Figure 4.4.2.

EXAMPLE 4.4.2. Let us assume telephone calls to a toll-free 800 number are made in accord with the Poisson process assumptions at the rate of 120 per hour during the period 9 A.M. to 12 noon. Then the expected number of calls in any single minute during this period is $\mu = 120\left(\frac{1}{60}\right) = 2$

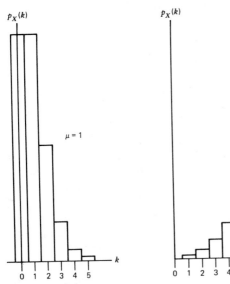
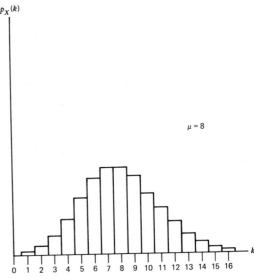

Figure 4.4.2

and the expected number in any single second is $\mu = 120\left(\frac{1}{3600}\right) = \frac{1}{30}$. The probability of at least one call being made in any given minute is $1 - e^{-2} = .865$ and the probability of at least one call being made during any given second is $1 - e^{-1/30} = .0328$. ∎

We have described the Poisson process assumptions in terms of events happening in the continuum called time. The same assumptions can be (and have been) made regarding occurrences in other continuously varying media, such as distance (number of insulation defects along a wire), area (number of boulders of a given size per square kilometer on the moon), volume (number of bacteria per cubic centimeter of liquid). The important thing to realize in using the Poisson assumptions is that whatever event is being modeled, they imply that the events are "uniformly" or evenly distributed, one at a time, at a constant rate throughout the medium and that they occur independently from one small piece of the medium to the next.

EXAMPLE 4.4.3.   Assume molecules of a rare gas occur at an average rate of 3 per cubic foot of air. Then, if it is reasonable to assume that the molecules of this particular gas are distributed independently and randomly in air, the number of molecules we would find in a cubic foot sample of air is a Poisson random variable with parameter 3, and the probability function for X is

$$p_X(k) = \frac{3^k}{k!}e^{-3}, \qquad k = 0, 1, 2, \dots.$$

Thus the probability we would find no molecules of this gas in our sample of air is $p_X(0) = .050$, the probability we would find exactly 1 molecule is $p_X(1) = .149$, and so on. The mean number we would find is $\mu_X = 3$ and the variance of the number we would find is $\sigma_X^2 = 3$.

Suppose we wanted to take a sufficiently large amount of air (say, t cubic feet) such that the probability of our finding at least one molecule of this gas in the sample is at least .99. How large must t be? If we let Y be the number of molecules of this gas in a sample of t cubic feet, then Y is a Poisson random variable with parameter $3t$, and

$$\begin{aligned} P(Y \geq 1) &= 1 - P(Y = 0) \\ &= 1 - e^{-3t}; \end{aligned}$$

then if we are to have at least probability .99 of getting 1 or more molecules, we must have $e^{-3t} \leq .01$; that is,

$$-3t \leq \ln(.01) = -4.605$$

or $t \geq 1.535$. Thus if we select a sample of 1.535 cubic feet of air (or more) the probability is at least .99 that we will find one or more molecules of the gas in the sample. ■

We derived the Poisson probability law by taking the limit of the binomial probability law with parameters n and $p = \lambda t/n$, as $n \to \infty$. Note that the way this limiting process was set up the product of the binomial parameters is $np = n(\lambda t/n) = \lambda t = \mu$, and is held constant as $n \to \infty$. At the same time, the individual parameters are changing; n is getting larger as $p = \lambda t/n$ is getting smaller and the value for the binomial probability function converges to the value for the Poisson probability function with parameter $\mu = np$. Thus if n is large and p is small we might expect the value for $P(X = k)$, computed from the Poisson probability law with $\mu = np$, to provide a good approximation for the same binomial probability. This is illustrated in the following example.

EXAMPLE 4.4.4. A college professor, based on his past experience, feels that there is a probability of .001 that he will be late to any given class and that being late or not for any class has no effect on whether or not he is late for any other class. Then the number of times X that he will be late to his next 100 classes is a binomial random variable with parameters $n = 100$ and $p = .001$, and the exact probabilities that he is late exactly 0 times and exactly 1 time are, respectively,

$$P(X = 0) = (.999)^{100} = .9048$$

$$P(X = 1) = \binom{100}{1}(.999)^{99}(.001) = .0906.$$

Note that in this case n is fairly large and p is rather small so that we would expect the preceding probabilities to be well approximated by those for a Poisson random variable with $\lambda t = np = 100(.001) = .1$. The Poisson values are

$$\frac{(.1)^0}{0!}e^{-.1} = .9048$$

$$\frac{(.1)^1}{1!}e^{-.1} = .0905$$

and the approximation is quite good. ■

A number of rules of thumb have been given that describe when this Poisson approximation is accurate; many texts say that if $n \geq 20$ and $p \leq .05$, it is quite accurate. It is very good if $n \geq 100$ and $np \leq 10$.

EXERCISE 4.4

1. It has been observed that cars pass a certain point on a rural road at the average rate of 3 per hour. Assume that the instants at which the cars pass are independent and let X be the number that pass this point in a 20-minute interval. Compute $P(X = 0)$, $P(X \geq 2)$.

2. It has been observed empirically that deaths per hour, due to traffic accidents, occur at a rate of 8 per hour on long holiday weekends in the United States. Assuming these deaths occur independently, compute the probability that a 1-hour period would pass with no deaths; that a 15-minute period would pass with no deaths; that 4 consecutive, nonoverlapping 15-minute periods would pass with no deaths.

3. It has been observed that packages of Hamm's beer are removed from the shelf of a particular supermarket at a rate of 10 per hour during rush periods. What is the probability that at least 1 package is removed during the first 6 minutes of a rush period? What is the probability that at least 1 is removed from the shelf during each of 3 consecutive, nonoverlapping 6-minute intervals?

4. At a certain manufacturing plant, accidents have been occurring at the rate of 1 every 2 months. Assuming the accidents occur independently, what is the expected number of accidents per year? What is the standard deviation of the number of accidents per year? What is the probability that there will be no accidents in a given month?

5. Assume boulders of diameter 1 meter or more are distributed at random on the surface of the moon at the rate of 1000 per square kilometer (satisfying the Poisson process assumptions).
   (a) What is the probability that no such boulders will be found in a particular square meter of the moon's surface?
   (b) Assuming a particular 25 square meter portion of the surface is targeted for the landing of a spacecraft, what is the probability that there are no boulders of this size in the area selected?

6. With the assumptions stated in Exercise 4.4.5:
   (a) What is the smallest area that you would expect to contain one boulder?
   (b) Assume we can select a portion of the moon of any desired area (say, the area is t square meters). The area selected either will or will not contain at least one boulder. What is the smallest area (value for t) such that the probability that the area selected contains at least one boulder is .9?

7. Derive the moment generating function for the Poisson random variable.

8. Assume 1 baby in 10,000 is born blind. If a large city hospital had 5000

births in 1970, approximate the probability that none of the babies born that year was blind at birth. Also approximate the probabilities that exactly 1 was born blind and that at least 2 were born blind. (Use the Poisson approximation.)

9. Assume 1 new tire in 1000 has a weak spot in its side wall. If you buy 4 new cars, approximate the probability that none of your cars have a tire with a weak spot and compare this with the exact binomial value.

10. A bakery makes chocolate chip cookies; a batch consists of 1000 cookies. A total of 3000 chocolate chips is added to the batter for each batch and the batter is well mixed. If we select 1 cookie at random from a batch, what is the probability that it contains no chocolate chips? That it contains exactly 3 chips? How many cookies with exactly 1 chocolate chip each would you expect in a batch?

11. Assume the sales made by a used car salesman occur like events in a Poisson process with parameter $\lambda = 1$ per week.
    (a) What is the probability he makes (exactly) 3 sales in a 2-week period? At least 3? At most 3?
    (b) What is the probability of three 2-week periods in a row with no sales?

12. Assume a printed page in a book contains 40 lines, and each line contains 75 positions (each of which may be left blank or filled with some symbol). Thus each page has 3000 positions to be set. Assume a particular typesetter makes one error per 6000 positions on the average.
    (a) What is the distribution for X, the number of errors per page?
    (b) Compute the probability that a page contains no errors.
    (c) What is the probability that a 16-page chapter contains no errors?

13. In a given semester, a large university will process 100,000 grades. In the past, .1 percent of all grades have been erroneously reported. Assume you are taking five courses at this university in one semester. What is the probability that all your grades are correctly reported?

4.5 The Uniform, Exponential and Gamma Probability Laws

Perhaps the simplest possible continuous probability law is the uniform distribution, which we have in fact already used several times in our earlier discussions. We will say a continuous random variable X is uniformly distributed over the interval from a to b (written (a, b)) if the probability that the observed value for X falls in any interval of length Δt, in (a, b), is proportional to Δt. The probability that the observed value for X

is smaller than a or greater than b is zero. Thus the density for X is positive only for values between a and b, $f_X(x) > 0$ for $a < x < b$, and if the probability that the observed value for X lies in an interval is proportional to the length of that interval, $f_X(x)$ must be constant for $a < x < b$. That is, $f_X(x) = c, a < x < b$. But the integral of $f_X(x)$ over the whole range for X is

$$\int_a^b c\,dx = c(b - a)$$

and for this to equal 1 it must be true that $c = 1/(b - a)$. Thus we have established the following theorem.

Theorem 4.5.1. If X is a continuous random variable, uniformly distributed over the interval (a, b), the density for X is

$$f_X(x) = \frac{1}{b - a}, \qquad a < x < b. \qquad \blacksquare$$

Recalling that the mean of a continuous probability law locates the point at which its density function balances, it is obvious that the mean of a uniform random variable on (a, b) must be $\mu_X = (b + a)/2$. (This is confirmed by

$$E[X] = \int_a^b \frac{x}{b - a}\,dx = \frac{b^2 - a^2}{2(b - a)} = \frac{b + a}{2}.\Bigg)$$

We also have

$$E[X^2] = \int_a^b \frac{x^2}{b - a}\,dx = \frac{b^3 - a^3}{3(b - a)} = \frac{b^2 + ab + a^2}{3}$$

from which it follows that

$$\sigma_X^2 = E[X^2] - \mu_X^2 = \frac{(b - a)^2}{12}$$

and

$$\sigma_X = \frac{b - a}{\sqrt{12}},$$

so the standard deviation for a uniform random variable is proportional to the length of the interval it is uniformly distributed over. The moment generating function for X is

$$m_X(t) = E[e^{tX}] = \int_a^b \frac{e^{tx}}{b - a}\,dx = \frac{e^{tb} - e^{ta}}{t(b - a)}, \qquad t \neq 0.$$

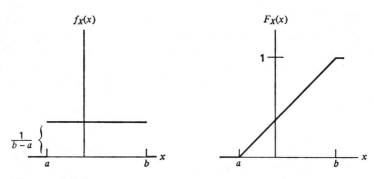

Figure 4.5.1

The distribution function for the uniform random variable is

$$F_X(t) = 0, \qquad t < a$$

$$= \int_a^t \frac{1}{b-a}\,dx = \frac{t-a}{b-a}, \qquad a \le t \le b$$

$$= 1, \qquad t > b.$$

Both the density function and distribution function for a uniform random variable on (a, b) are pictured in Figure 4.5.1.

EXAMPLE 4.5.1.   Suppose a computer simulation is to be used to study the effects of various types and locations of insulating materials on the heat retention of a new building design. Let us also assume this simulation calls for the use of observed values of Bernoulli random variables with parameter p. Such observed values are frequently generated by using the continuous uniform random variable on the interval $(0, 1)$. Let X be the uniform $(0, 1)$ random variable and suppose we desire the observed value for a Bernoulli random variable Y with specified parameter p. Then if we define

$$Y = 1 \qquad \text{if} \qquad X \le p$$
$$= 0 \qquad \text{if} \qquad X > p,$$

we have

$$P(Y = 1) = P(X \le p) = F_X(p) = p,$$

and Y has the desired Bernoulli distribution. To be more specific, if we want to generate the observed value for a Bernoulli random variable with parameter $p = .75$, we can generate an observed value for X chosen uniformly on $(0, 1)$. If the observed value for X is .75 or less we set $Y = 1$, otherwise we set $Y = 0$. Then Y is Bernoulli, $p = .75$.   ∎

In Section 4.4 we studied the assumptions for a Poisson process with parameter λ. The exponential and Erlang random variables are both easily defined on this process. First, we will study the exponential probability law.

Recall that in a Poisson process events are occurring independently at random and at a uniform rate per unit of time. Suppose again we label as time zero the instant at which we begin observing the Poisson process and now let T be the time at which the first event occurs. T then is a continuous random variable and its range is $R_T = \{t : t > 0\}$. Let t be any fixed positive number and consider the event $\{T > t\}$, that the time of the first event is greater than t. This event occurs if and only if there are zero events in the fixed interval $(0, t]$. But we saw in Section 4.4 that the probability of zero events occurring in any interval of fixed length t is $e^{-\lambda t} = P(X = 0)$, where X is the number of events in $(0, t]$. Because these events are equivalent their probabilities must be equal and thus

$$P(T > t) = e^{-\lambda t} = 1 - F_T(t),$$

from which we find the distribution function for T is

$$F_T(t) = 1 - e^{-\lambda t}, \qquad t > 0$$

and its density function is

$$f_T(t) = \frac{d}{dt} F_T(t) = \lambda e^{-\lambda t}, \qquad t > 0.$$

This is called the exponential density function and we have proved the following theorem.

Theorem 4.5.2. Let T be the elapsed time from the beginning of observation until the first event occurs in a Poisson process with parameter λ. Then T has the exponential density function.

$$f_T(t) = \lambda e^{-\lambda t}, \qquad t > 0,$$

and distribution function

$$\begin{aligned} F_T(t) &= 0, & t < 0 \\ &= 1 - e^{-\lambda t}, & t \geq 0. \end{aligned} \qquad \blacksquare$$

The expected value for an exponential random variable is

$$E[T] = \int_0^\infty t\lambda e^{-\lambda t}\, dt$$

$$= -\frac{(\lambda t + 1)}{\lambda} e^{-\lambda t} \bigg|_0^\infty$$

$$= \frac{1}{\lambda}$$

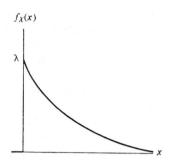
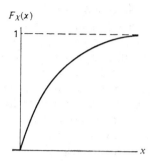

Figure 4.5.2

and the moment generating function is

$$m_T(t) = E[e^{tT}] = \int_0^\infty e^{ts} \lambda e^{-\lambda s}\, ds$$

$$= \int_0^\infty \lambda e^{-s(\lambda-t)}\, ds$$

$$= -\left. \frac{\lambda e^{-s(\lambda-t)}}{\lambda - t} \right|_0^\infty = \frac{\lambda}{\lambda - t}, \qquad t < \lambda$$

from which you can easily verify that the variance of the exponential random variable is $\mathrm{Var}[T] = 1/\lambda^2$ and thus the standard deviation is $\sigma_T = 1/\lambda$, the same as the mean value for T. Figure 4.5.2 plots the density and distribution functions for the exponential random variable with parameter λ.

EXAMPLE 4.5.2. Assume, as in Example 4.4.1, injury-causing accidents occur in a large industrial plant at a rate of $\lambda = \frac{1}{2}$ per week, which, with a 5-day work week, is equivalent to occurrences at the rate of $\frac{1}{10}$ per day. If we begin observing the occurrences of these accidents at the start of work on Monday, for some given week, and let T be the number of days until the first accident occurs, then T is exponential, $\lambda = \frac{1}{10}$. The probability that the first week is accident free is

$$P(T > 5) = e^{-1(5)/10} = .607.$$

The probability that the first accident occurs on Friday of the first week is

$$P(4 < T \le 5) = (1 - e^{-1(5)/10}) - (1 - e^{-1(4)/10})$$
$$= .064,$$

and the probability that it occurs on Wednesday of the second week is

$$P(7 < T \le 8) = (1 - e^{-1(8)/10}) - (1 - e^{-1(7)/10})$$
$$= .047.$$

The expected number of days to the first accident is

$$E[T] = \left(\frac{1}{10}\right)^{-1} = 10$$

and the standard deviation for T is

$$\sigma_T = 10 \text{ days.} \qquad \blacksquare$$

The exponential probability law has the same memoryless property that we found for the geometric probability law. That is, if T is exponential with parameter λ and a and b are any positive constants,

$$P(T > a + b \mid T > a) = \frac{P(T > a + b)}{P(T > a)}$$

$$= \frac{e^{-\lambda(a+b)}}{e^{-\lambda a}} = e^{-\lambda b} = P(T > b).$$

Thus if in the preceding example we have observed $a = 4$ days with no accidents, the probability that it will be at least $b = 2$ days until the first accident is unchanged from the original value for this probability when we began observation. The exponential distribution is the only continuous probability law with this property.

The reason that the geometric and exponential probability laws should both have this property can be seen by remembering that the continuous time Poisson process can be derived as a limit of a sequence of independent Bernoulli trials. A geometric random variable Y is the number of Bernoulli trials until the first success occurs and the exponential random variable T is defined analogously on the Poisson process: It is the time of occurrence of the first event (success). In fact, as seen in Section 4.2, if Y is geometric with parameter p, then

$$P(Y > n) = (1 - p)^n.$$

In deriving the Poisson process in Section 4.4, we set $p = \lambda \Delta t = \lambda t/n$, having subdivided $(0, t]$ into n pieces of length Δt. But then the events $\{Y > n\}$ and $\{T > t\}$ are equivalent and

$$P(T > t) = \lim_{n \to \infty} P(Y > n)$$

$$= \lim_{n \to \infty} \left(1 - \frac{\lambda t}{n}\right)^n$$

$$= e^{-\lambda t}$$

so the exponential distribution function is the limit of the geometric distribution function; the exponential random variable "inherits" the memoryless property from the geometric probability law.

Now let us discuss the Erlang and gamma probability laws. Suppose we again begin observing a Poisson process at time zero and let T_r be the time of occurrence of the rth event, $r \geq 1$. (This random variable is analogous to the negative binomial random variable defined on independent Bernoulli trials.) Again let t be any fixed positive number and consider the event $\{T_r > t\}$, that the time of the rth event is greater than this fixed t. This event $\{T_r > t\}$ is equivalent to the event $\{X \leq r - 1\}$, where X is the number of events that occur in $(0, t]$, because the time of the rth event can exceed t only if there are $r - 1$ or fewer events in $(0, t]$. Since X is a Poisson random variable with parameter $\mu = \lambda t$,

$$P(T_r > t) = P(X \leq r - 1)$$

$$= \sum_{k=0}^{r-1} \frac{(\lambda t)^k}{k!} e^{-\lambda t}$$

and the distribution function for T_r, the time of the rth occurence, is

$$F_{T_r}(t) = P(T_r \leq t) = 1 - P(T_r > t)$$

$$= 1 - \sum_{k=0}^{r-1} \frac{(\lambda t)^k}{k!} e^{-\lambda t}.$$

T_r is called the *Erlang* random variable with parameters r and λ. The density function for T_r is

$$f_{T_r}(t) = \frac{d}{dt} F_{T_r}(t)$$

$$= \frac{d}{dt} \left(1 - e^{-\lambda t} - \lambda t e^{-\lambda t} - \frac{(\lambda t)^2}{2!} e^{-\lambda t} - \cdots - \frac{(\lambda t)^{r-1}}{(r-1)!} e^{-\lambda t} \right)$$

$$= \lambda e^{-\lambda t} - \lambda e^{-\lambda t} + \lambda^2 t e^{-\lambda t} - \lambda^2 t e^{-\lambda t} + \frac{\lambda^3 t^2}{2!} e^{-\lambda t} - \cdots$$

$$- \frac{\lambda^{r-1} t^{r-2}}{(r-2)!} e^{-\lambda t} + \frac{\lambda^r t^{r-1}}{(r-1)!} e^{-\lambda t}$$

$$= \frac{\lambda^r t^{r-1}}{(r-1)!} e^{-\lambda t}, \qquad t > 0.$$

Notice that with the alternating signs the only term that remains is the last one. This Erlang probability law is a particular case of the *gamma* probability law that can be defined in terms of the gamma function. It can be shown that

$$\int_0^\infty u^{n-1} e^{-u} \, du$$

exists (has a finite value) for any $n > 0$ and its value is denoted by $\Gamma(n)$:

$$\Gamma(n) = \int_0^\infty u^{n-1} e^{-u} \, du.$$

$\Gamma(n)$ is called the *gamma function*. Using integration by parts, we can easily show that

$$\Gamma(n) = (n-1)\Gamma(n-1)$$

and thus if r is a positive integer

$$\begin{aligned}
\Gamma(r) &= (r-1)\Gamma(r-1) \\
&= (r-1)(r-2)\Gamma(r-2) \\
&\qquad\cdots \\
&= (r-1)(r-2)\cdots 2 \cdot 1 \Gamma(1) \\
&= (r-1)!
\end{aligned}$$

since $\Gamma(1) = \int_0^\infty e^{-u} \, du = 1$. It can also be shown that $\Gamma(\tfrac{1}{2}) = \sqrt{\pi}$, that is,

$$\Gamma(\tfrac{1}{2}) = \int_0^\infty \frac{1}{\sqrt{u}} e^{-u} \, du = \sqrt{\pi},$$

which proves useful in evaluating certain integrals.

By making the change of variable $u = \lambda x$, $du = \lambda \, dx$, we can also write

$$\Gamma(n) = \int_0^\infty \lambda^n x^{n-1} e^{-\lambda x} \, dx$$

so that

$$1 = \int_0^\infty \frac{\lambda^n x^{n-1}}{\Gamma(n)} e^{-\lambda x} \, dx.$$

Thus

$$f_X(x) = \frac{\lambda^n x^{n-1}}{\Gamma(n)} e^{-\lambda x}, \qquad x > 0$$

is a density function and defines the *gamma* probability law with parameters n and λ. Notice that with $n = r$, a positive integer, this is identical with the Erlang density function previously derived as the density for the time to the rth occurrence in a Poisson process with parameter λ. We will summarize much of this as the following theorem.

Theorem 4.5.3.   If X is a continuous random variable with density

$$f_X(x) = \frac{\lambda^n x^{n-1}}{\Gamma(n)} e^{-\lambda x}, \qquad x > 0,$$

then X is called the *gamma* random variable with parameters $n > 0$ and $\lambda > 0$. If n is a positive integer, X is the time of occurrence of the nth event in a Poisson process with parameter λ.   ∎

The moment generating function for the gamma random variable is

$$m_X(t) = E[e^{tX}] = \int_0^\infty e^{tx} \frac{\lambda^n x^{n-1}}{\Gamma(n)} e^{-\lambda x}\, dx$$

$$= \frac{\lambda^n}{\Gamma(n)} \int_0^\infty x^{n-1} e^{-x(\lambda - t)}\, dx$$

$$= \frac{\lambda^n}{\Gamma(n)} \int_0^\infty \left(\frac{u}{\lambda - t}\right)^{n-1} e^{-u} \frac{du}{\lambda - t}$$

$$= \left(\frac{\lambda}{\lambda - t}\right)^n \frac{1}{\Gamma(n)} \int_0^\infty u^{n-1} e^{-u}\, du$$

$$= \left(\frac{\lambda}{\lambda - t}\right)^n,$$

where we have made the change of variable $u = x(\lambda - t)$. It is easy to verify that $E[X] = n/\lambda$, $\text{Var}[X] = n/\lambda^2$. Figure 4.5.3 graphs the densities for the Erlang probability law with $r = 1, 2, 3$, $\lambda = 1$. The graphs of the gamma densities with $n \geq 1$ are similar to these, having a unique maximum value at $x = (n - 1)/\lambda$; thus the probability that the observed value for a gamma random variable lies in an interval of length Δx is maximized for the interval centered at $x = (n - 1)/\lambda$. With $0 < n < 1$, the gamma density is unbounded as $x \to 0$; the graph of one of these densities is given in Figure 4.5.4.

EXAMPLE 4.5.3.   Assume, as in Example 4.4.2, the instants at which telephone calls are made to a toll-free 800 number form a Poisson process with $\lambda = 120/\text{hour}$, during the hours from 9 A.M. to 12 noon. Let T_{10} be the time at which the tenth call is made, starting from 9 A.M., using minutes as our unit of time. Then T_{10} is Erlang with $r = 10$, $\lambda = 2$ per minute and the expected time of the tenth call is

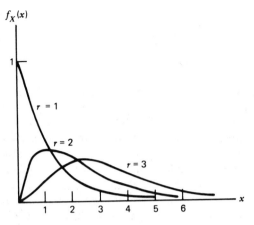

Figure 4.5.3

$$E[T_{10}] = \tfrac{10}{2} = 5,$$

so we would expect the tenth call at 9:05 A.M. The probability that the tenth call occurs before 9:05 A.M. is

$$P(T_{10} < 5) = 1 - \sum_{k=0}^{9} \frac{(10)^k}{k!} e^{-10} = .542,$$

which is easily evaluated by using the Poisson calculations discussed in Section 4.4. The probability that the tenth call is received between 9:05 A.M. and 9:07 A.M. is

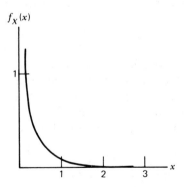

Figure 4.5.4

$$P(5 < T_{10} \le 7) = \left(1 - \sum_{k=0}^{9} \frac{(14)^k}{k!} e^{-14}\right) - \left(1 - \sum_{k=0}^{9} \frac{(10)^k}{k!} e^{-10}\right)$$
$$= .349,$$

again evaluated by using the Poisson calculations. ∎

We will see many uses for the *chi-square* (χ^2) probability law in our study of statistics in the latter half of this volume. The chi-square probability law is also a special case of the gamma probability law. If the parameter $\lambda = \frac{1}{2}$ and the parameter n equals $\frac{1}{2}$, 1, $\frac{3}{2}$, 2, $\frac{5}{2}$, ... in the gamma probability law, then the random variable with this distribution is called a χ^2 random variable with $v = 2n$ degrees of freedom. Thus the single parameter for the χ^2 probability law is called "degrees of freedom" and can equal any positive integer. We will see later why this parameter has such a colorful name.

EXERCISE 4.5

1. A random variable could be uniformly distributed ("equally likely outcomes," probability of observed value in an interval of length Δt is proportional to Δt) over two (or more) disjoint intervals. Assume X is uniform over the two intervals (0, 1) and (2, 3). Evaluate the density for X, $E[X]$, and σ_X.

2. For the random variable X described in Exercise 4.5.1, derive the density for $Y = X^2$.

3. If X is uniformly distributed on the interval (1, 4), derive the density function of $Z = X^{1/2}$.

4. Let X be uniform on $(-2, -1)$ and (1, 2) (see Exercise 4.5.1) and derive the density function for $W = X^2$. Compare this density with that of $U = V^2$, where V is uniform over (1, 2).

5. X is uniformly distributed on (0, 2) and Y is exponential with parameter λ. Find the value of λ such that $P(X < 1) = P(Y < 1)$.

6. Calls arrive at a switchboard according to a Poisson process with parameter $\lambda = 5$ per hour. If we are at the switchboard, what is the probability that it is at least 15 minutes until the next call? That it is no more than 10 minutes? That it is exactly 5 minutes?

7. Let X be uniform on the interval (0, 10) and define $Y = [X]$, where $[x]$ is the greatest integer function. That is, if $0 \le x < 1$, then $[x] = 0$, if $1 \le x < 2$, then $[x] = 1$, if $2 \le x < 3$, then $[x] = 2$, and so on. What is the probability function for Y?

8. X is uniform on $(-1, 3)$ and Y is exponential with parameter λ. Find λ such that $\sigma_X^2 = \sigma_Y^2$.

9. X is geometric with parameter p and Y is exponential with parameter λ. Find λ such that $P(X > 1) = P(Y > 1)$.

10. Assume in a Poisson process with parameter λ exactly one event occurs in the interval $(0, 1]$. Given this is true, the actual time of occurrence of the single event is uniform on $(0, 1]$. Which of the Poisson process assumptions directly implies this conclusion?

11. Assume we are observing a Poisson process with parameter λ starting at time zero. Let X_1 be the time of occurrence of the first event, let X_2 be the *additional* time (beyond X_1) until the occurrence of the second event, let X_3 be the *additional* time (beyond $X_1 + X_2$) until the occurrence of the third event. What is the probability law for $Y = X_1 + X_2 + X_3$?

*12. Let X be an exponential random variable with parameter λ and define $Y = [X]$, the largest integer in X (see Exercise 4.5.7).
    (a) Find the probability function for Y.
    (b) Show that $Y + 1$ is geometric and evaluate p in terms of λ.

13. Automobile accidents occur in the United States over a 72-hour holiday period like events in a Poisson process with parameter $\lambda = 10$ per hour. Let Y be the time until the first accident (from the beginning of the holiday period).
    (a) What is the mean for Y?
    (b) What is σ_Y?
    (c) Evaluate the probability that Y exceeds 15 minutes.

14. Assume the same situation as described in Exercise 4.5.13 and let V be the time until the tenth accident.
    (a) Evaluate $E[V]$ and σ_V (in hours).
    (b) What is the probability that V is larger than 1 hour?

15. Let Y be the time to the second occurrence in a Poisson process with parameter λ. Evaluate $P(Y > a + b \,|\, Y > a)$ and compare this probability with $P(Y > b)$.

16. Assume X is uniform on $(0, 10)$ and let $a > 0, b > 0$ be given constants such that $a + b \leq 10$. Evaluate $P(X > a + b \,|\, X > a)$. Does this uniform distribution have the memoryless property?

4.6 The Beta and Normal Probability Laws

We will study two additional continuous probability laws in this section, the beta and the normal probability laws. The normal probability law occupies a very central position in probability theory; we will sketch a simple argument that is indicative of how it occurs. The beta probability law can occur in several ways, one of which is indicated here.

Beginning with the beta probability law, let us assume $X_1, X_2, ..., X_n$ are each uniform random variables on the interval $(0, 1)$. If t is any fixed point in the interval $(0, 1)$, assume the events $A_i = \{X_i \leq t\}$, $i = 1, 2, ..., n$ are independent (as we will see in Chapter 5, this is necessarily true if X_1, $X_2, ..., X_n$ are what will be called independent random variables). Thus the uniform random variables define independent Bernoulli trials, where success on trial i means event A_i occurs, failure means $\bar{A}_i$ occurs. The probability of success on each trial is

$$P(A_i) = P(X_i \leq t) = t, \qquad i = 1, 2, ..., n.$$

Now let Y_r denote the rth largest of $X_1, X_2, ..., X_n$ and consider the event $\{Y_r > t\}$. The rth largest of $X_1, X_2, ..., X_n$ will exceed the constant t if and only if $r - 1$ or fewer of $X_1, X_2, ..., X_n$ are smaller than t, that is, if and only if we observe $r - 1$ or fewer successes in the n Bernoulli trials. But the number of successes in n independent Bernoulli trials, each with probability of success equal to t, is binomial with parameters n and t. Thus

$$P(Y_r > t) = \sum_{k=0}^{r-1} \binom{n}{k} t^k (1 - t)^{n-k} \qquad 0 < t < 1$$

and the distribution function for Y_r then is

$$F_{Y_r}(t) = 1 - \sum_{k=0}^{r-1} \binom{n}{k} t^k (1 - t)^{n-k}$$

$$= \sum_{k=r}^{n} \binom{n}{k} t^k (1 - t)^{n-k}$$

$$= \binom{n}{r} t^r (1 - t)^{n-r} + \binom{n}{r+1} t^{r+1} (1 - t)^{n-r-1}$$

$$+ \cdots + nt^{n-1}(1 - t) + t^n.$$

The density function for Y_r is

$$f_{Y_r}(t) = \frac{d}{dt} F_{Y_r}(t)$$

$$= \binom{n}{r} rt^{r-1}(1 - t)^{n-r} - \binom{n}{r}(n - r) t^r (1 - t)^{n-r-1}$$

$$+ \binom{n}{r+1}(r + 1) t^r (1 - t)^{n-r-1}$$

$$- \binom{n}{r+1}(n - r - 1) t^r (1 - t)^{n-r-2}$$

$$+ \cdots + n(n-1) t^{n-2}(1-t) - nt^{n-1} + nt^{n-1}$$

$$= \binom{n}{r} rt^{r-1}(1-t)^{n-r}$$

$$= \frac{n!}{(r-1)!(n-r)!} t^{r-1}(1-t)^{n-r}, \qquad 0 < t < 1;$$

because all terms after the first $\binom{n}{k}(n-k) = \binom{n}{k+1}(k+1)$,
cancel, as is easily seen by writing out the factorials involved. Thus we have
established the following theorem.

Theorem 4.6.1.   Let Y_r be the rth largest of n independent uniform $(0, 1)$
random variables, $r = 1, 2, ..., n$. Then the density function for Y_r is

$$f_{Y_r}(t) = \frac{\Gamma(n+1)}{\Gamma(r)\Gamma(n-r+1)} t^{r-1}(1-t)^{n-r}, \qquad 0 < t < 1. \qquad \blacksquare$$

EXAMPLE 4.6.1.   Suppose a computer generates 9 numbers, each uni-
form on $(0, 1)$. Then letting Y_1 be the smallest of the 9, the density for Y_1 is

$$f_{Y_1}(t) = \frac{9!}{0!\,8!} t^0(1-t)^8 = 9(1-t)^8, \qquad 0 < t < 1.$$

The probability that this smallest value is smaller than .2 is

$$P(Y_1 < .2) = \int_0^{.2} 9(1-t)^8 \, dt = -(1-t)^9 \Big|_0^{.2}$$

$$= 1 - .8^9 = .866$$

and the probability that it exceeds .5 is

$$P(Y_1 > .5) = \int_{.5}^1 9(1-t)^8 \, dt = -(1-t)^9 \Big|_{.5}^1$$

$$= .5^9 = .002.$$

The expected value of the smallest of the 9 uniform variables is

$$E[Y_1] = \int_0^1 t\, 9(1-t)^8 \, dt = \frac{1}{10},$$

using integration by parts. The largest of the 9 is Y_9 with density

$$f_{Y_9}(t) = \frac{9!}{8!\,0!} t^8(1-t)^0 = 9t^8, \qquad 0 < t < 1.$$

The probability that this largest value is smaller than .2 is

$$P(Y_9 < .2) = \int_0^{.2} 9t^8 \, dt = t^9 \Big|_0^{.2} = .2^9$$
$$= 5.12 \times 10^{-7},$$

and the probability that this largest value exceeds .5 is

$$P(Y_9 > .5) = \int_{.5}^1 9t^8 \, dt = t^9 \Big|_{.5}^1 = 1 - .5^9$$
$$= .998.$$

The expected value of Y_9, the largest of the 9 uniform random variables, is

$$E[Y_9] = \int_0^1 t 9t^8 \, dt = \frac{9}{10}.$$

The ordered values $Y_1, Y_2, \ldots, Y_9$ are called *order statistics*; they are used in certain problems of statistical inference. As this example shows, the distributions for the ordered values are generally quite different from the distributions of the original random variables. This point was overlooked and proved somewhat confusing to some of the early students of science. ∎

   This distribution of the rth largest among n uniform $(0, 1)$ random variables is a particular example of the *beta* probability law. The beta probability law with parameters $\alpha > 0, \beta > 0$ has density function

$$f(x) = \frac{\Gamma(\alpha + \beta)}{\Gamma(\alpha)\Gamma(\beta)} x^{\alpha-1}(1 - x)^{\beta-1}, \qquad 0 < x < 1.$$

Notice that with $\alpha = r, \beta = n - r + 1$, this is exactly the density for the rth largest of n uniform random variables. It can be shown that

$$\int_0^1 x^{\alpha-1}(1 - x)^{\beta-1} \, dx = \frac{\Gamma(\alpha)\Gamma(\beta)}{\Gamma(\alpha + \beta)}$$

for any $\alpha > 0, \beta > 0$, thus $f(x)$ is a legitimate density function. It is quite straightforward to evaluate the moments of a beta random variable.

$$E[X^k] = \frac{\Gamma(\alpha + \beta)}{\Gamma(\alpha)\Gamma(\beta)} \int_0^1 x^{k+\alpha-1}(1 - x)^{\beta-1} \, dx$$
$$= \frac{\Gamma(\alpha + \beta)}{\Gamma(\alpha)\Gamma(\beta)} \frac{\Gamma(\alpha + k)\Gamma(\beta)}{\Gamma(\alpha + \beta + k)}$$
$$= \frac{(\alpha + k - 1)(\alpha + k - 2)\cdots(\alpha + 1)\alpha}{(\alpha + \beta + k - 1)(\alpha + \beta + k - 2)\cdots(\alpha + \beta + 1)(\alpha + \beta)},$$

from which it is easily seen that the mean for the beta distribution is

$$E[X] = \frac{\alpha}{\alpha + \beta}$$

and the variance is

$$\text{Var}[X] = \frac{\alpha\beta}{(\alpha + \beta)^2 (\alpha + \beta + 1)}.$$

The moment generating function for the beta distribution is neither simple nor useful. The beta density can take on a variety of shapes, including the uniform on $(0, 1)$ with $\alpha = \beta = 1$; Figure 4.6.1 indicates how these shapes change with the values for the parameters.

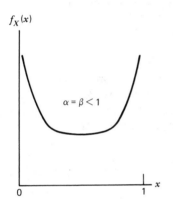

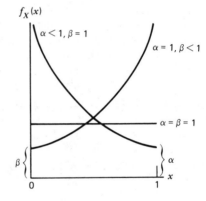

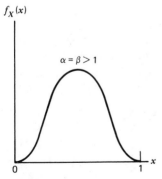

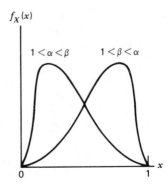

Figure 4.6.1

Now let us discuss the normal probability law, without doubt the single most frequently used probability law. This normal probability law is important both because it seems to provide an adequate model for various observed measurements and, as we will see in Chapter 5, because it provides an accurate approximation to a wide variety of probability laws.

In Section 3.5 we briefly discussed a surveying example and it was mentioned that the error in the final measurement might very well have the normal distribution. When a survey party measures the distance between two (widely separated) points they actually make measurements (as accurately as possible) between pairs of intermediate points; their measure of the distance between the two points (in the simplest case) then is the sum of the distances between the intermediate pairs of points. We will consider a very simple model of this situation and the fact that a certain continuous probability law (the normal distribution) occurs as the description of the final measurement error, if certain assumptions are made.

Let us assume then the final error made in our measurement is a random variable Y and Y itself is the sum of n more fundamental errors $X_1, X_2, ..., X_n$, that is, $Y = \sum_{i=1}^{n} X_i$. Furthermore, let us assume each fundamental error, X_i, can equal only two values, 1 or -1 with equal probability; the X_i's are called *symmetric* Bernoulli random variables. Thus we have $P(X_i = -1) = P(X_i = 1) = \frac{1}{2}$, for $i = 1, 2, ..., n$ and $E[X_i] = 0$, $\text{Var}[X_i] = 1$. Actually, these symmetric Bernoulli random variables are simple linear functions of the standard Bernoulli random variable we studied in Section 4.1. Notice that if V is Bernoulli with $p = \frac{1}{2}$, then $X = 2V - 1$ will equal 1 when $V = 1$ and $X = 2V - 1$ will equal -1 when $V = 0$, so the symmetric Bernoulli X actually equals $2V - 1$, where V is Bernoulli, with $p = \frac{1}{2}$. Thus we can actually write our final measurement error

$$Y = \sum_{i=1}^{n} X_i = \sum_{i=1}^{n} (2V_i - 1) = 2 \sum_{i=1}^{n} V_i - n$$

in terms of Bernoulli random variables $V_1, V_2, ..., V_n$, each with $p = \frac{1}{2}$. Now let us add our final assumption about the fundamental errors: $V_1, V_2, ..., V_n$ are defined on independent Bernoulli trials. With this assumption we know that $X = \sum_{i=1}^{n} V_i$ is binomial with parameters n and $p = \frac{1}{2}$ so our final measurement error $Y = 2X - n$, with X a binomial random variable. This, of course, allows us to write down the probability function for Y.

$$P(Y = k) = P(2X - n = k) = P\left(X = \frac{n+k}{2}\right)$$

$$= \binom{n}{(n+k)/2}\left(\frac{1}{2}\right)^k\left(\frac{1}{2}\right)^{n-k} = \binom{n}{(n+k)/2}\frac{1}{2^n},$$

$$k = -n, -n+2, -n+4, ..., n-2, n.$$

(Remember that Y is the sum of n values, each 1 or -1; thus the smallest possible value for Y is $-n$, the largest is n, and consecutive possible observed values differ by 2, giving the range of Y to be $R_Y = \{-n, -n + 2, ..., n - 2, n\}$.) Since Y is a linear function of the binomial random variable X,

$$E[Y] = E[2X - n] = 2E[X] - n = 2\left(\frac{n}{2}\right) - n = 0,$$

$$\text{Var}[Y] = 4\,\text{Var}[X] = 4n\left(\frac{1}{2}\right)\left(\frac{1}{2}\right) = n.$$

As n, the number of fundamental errors summed to get Y, increases, the distribution for Y remains centered at 0 (since $E[Y] = 0$ for all n), but it also gets more dispersed (since $\text{Var}[Y] = n$ increases with n). If we define $Z = Y/\sqrt{n}$ to be the "standardized" error, the distribution for Z remains stable in a sense, since $E[Z] = 0$, $\text{Var}[Z] = 1$ for all n.

What happens to this distribution for the standardized error Z as n increases without limit? More precisely, what happens to

$$P(z \le Z \le z + \Delta z),$$

where z is any fixed value and $\Delta z > 0$ is a small increment? It can be shown that in the limit as $n \to \infty$

$$P(z \le Z \le z + \Delta z) \doteq \frac{1}{\sqrt{2\pi}}e^{-z^2/2}\Delta z,$$

that is, the distribution for Z converges to a continuous probability law with density function

$$n(z) = \frac{1}{\sqrt{2\pi}}e^{-z^2/2}, \qquad -\infty < z < \infty.$$

This is called the *standard normal density function*. It is a symmetric, bell-shaped curve centered at 0, which asymptotically goes to zero as $z \to \infty$ or $z \to -\infty$; Figure 4.6.2 graphs this density function. Not surprisingly, if Z has this standard normal distribution $E[Z] = 0$, $\text{Var}[Z] = 1$; since we standardized the measurement error to have mean 0 and variance 1 for all n, the limiting distribution retains these values. In Exercise 4.6.9 an argument is sketched that you can use to show that

$$\int_{-\infty}^{\infty} \frac{1}{\sqrt{2\pi}}e^{-z^2/2}\,dz = 1$$

so $n(z)$ is a density function.

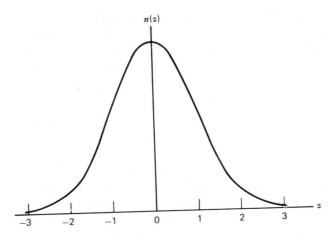

Figure 4.6.2

This simple argument using equal magnitude measurement errors is only one of many different ways in which the normal probability law occurs. In Chapter 5 we will discuss one version of the central limit theorem, which shows that a much wider class of phenomena also lead to the normal probability law.

The standard normal distribution function is

$$P(Z \leq z) = \int_{-\infty}^{z} \frac{1}{\sqrt{2\pi}} e^{-x^2/2} \, dx.$$

This integration cannot be carried out in closed form in terms of elementary functions. Numerical integration can, of course, be used; this results in the values in Table 1 of the Appendix. After formally defining the notation we will use for the standard normal distribution, we will look at some examples of the use of Table 1.

DEFINITION 4.6.1. Z is called the *standard normal* random variable if its density function is

$$n(z) = \frac{1}{\sqrt{2\pi}} e^{-z^2/2}$$

and its distribution function is

$$N(z) = \int_{-\infty}^{z} \frac{1}{\sqrt{2\pi}} e^{-x^2/2} \, dx. \qquad \blacksquare$$

Because of the importance of the standard normal distribution, we

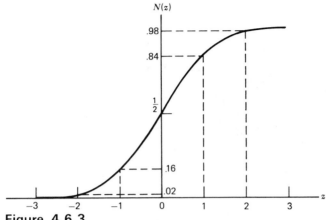

Figure 4.6.3

will reserve $n(z)$ and $N(z)$ as symbols for the density and distribution functions, respectively. Figure 4.6.3 graphs the standard normal distribution function. Notice that because of the symmetry of the normal density about $z = 0$, the area to the left of z must be identical with the area to the right of $-z$ (see Figure 4.6.4). Thus

$$N(z) = 1 - N(-z),$$

a relationship that frequently proves useful. The following example illustrates the use of the standard normal distribution function table (Table 1, Appendix).

EXAMPLE 4.6.2.   From Table 1 we see that

$$N(1.44) = .9251, \quad N(1.85) = .9678, \quad N(-.77) = .2206,$$

and thus if Z is a standard normal random variable

$$P(Z \le 1.44) = .9251, \quad P(Z \le 1.85) = .9678, \quad P(Z \le -.77) = .2206.$$

Also,

$$P(-.77 < Z < 1.44) = N(1.44) - N(-.77) = .7045$$
$$P(Z > -1.85) = 1 - N(1.85) = .0322$$
$$P(|Z| < 1.85) = P(-1.85 < Z < 1.85)$$
$$= N(1.85) - N(-1.85)$$
$$= N(1.85) - (1 - N(1.85))$$
$$= 2N(1.85) - 1 = .9356$$
$$P(-.77 < Z < 0) = \tfrac{1}{2} - .2206 = .2794.$$

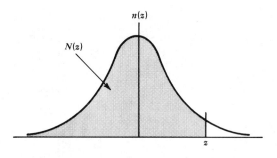

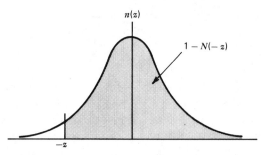

Figure 4.6.4

We can also use Table 1 for inverse table lookups, cases in which the value for $N(z)$ is fixed and we want to find z. For example, if we wanted the value for b such that $N(b) = .75$, we can skim the entries in Table 1 and find $N(.68) = .7517$, $N(.67) = .7486$ so b lies between .67 and .68. For examples worked out in this text we will simply take the closest value rather than interpolate, because the added accuracy is not worth the trouble; thus we would use $b = .67$ as the solution to $N(b) = .75$. Should more accuracy be needed, of course, various methods of integration can be employed, or tables with a finer grid can be consulted. Suppose we wanted to find the value c such that $P(|Z| < c) = .95$, that is, $P(-c < Z < c) = N(c) - N(-c) = .95$. Since $N(-c) = 1 - N(c)$ from the symmetry already mentioned, this is equivalent to finding c such that

$$2N(c) - 1 = .95$$

or

$$N(c) = .975,$$

which gives $c = 1.96$. There are an infinite number of solutions for a and d such that

$$P(a < Z < d) = .95,$$

ranging from $a = -\infty$, $d = 1.64$ to $a = -2.33$, $d = 1.75$ to $a = -1.96$, $d = 1.96$ to $a = -1.75$, $d = 2.33$ to $a = -1.64$, $d = \infty$. The symmetrically placed values, $a = -1.96$, $d = 1.96$, are unique and give the shortest interval that includes probability .95. In working with areas under the standard normal density we frequently find it useful to sketch a picture of the normal density and the area to be evaluated. ∎

We have seen one way in which the standard normal density occurs; it is easy to define a normal probability law with any desired mean value μ and standard deviation σ. To do so, let Z be standard normal and define

$$X = \sigma Z + \mu = g(Z),$$

where $\sigma > 0$ and μ are any desired values. The inverse function is

$$Z = \frac{X - \mu}{\sigma} = g^{-1}(X).$$

Then the mean value for X is

$$E[X] = \sigma E[Z] + \mu = \mu$$

and the variance for X is

$$\text{Var}[X] = \sigma^2 \, \text{Var}[Z] = \sigma^2,$$

so the constant σ is the standard deviation for X. The density function for X is

$$f_X(x) = n\left(\frac{x - \mu}{\sigma}\right) \left|\frac{dg^{-1}(x)}{dx}\right|$$

$$= \frac{1}{\sigma\sqrt{2\pi}} e^{-(x-\mu)^2/2\sigma^2}.$$

The density function for a normal random variable X with arbitrary mean μ and standard deviation σ is a symmetric bell-shaped curve centered at μ; the value for σ controls how relatively peaked or spread out the bell is. This is illustrated in Figure 4.6.5.

The distribution function for X is

$$F_X(x) = N\left(\frac{x - \mu}{\sigma}\right).$$

This shows, in particular, that the standard normal table can be used to evaluate probability statements about any normal random variable X. We formally define the normal probability law as follows.

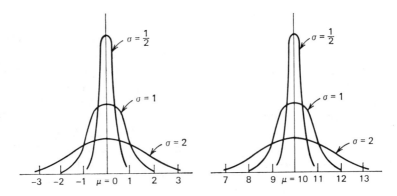

Figure 4.6.5

DEFINITION 4.6.2.   If X is a random variable with density function

$$f_X(x) = \frac{1}{\sigma\sqrt{2\pi}} e^{-(x-\mu)^2/2\sigma^2},$$

where $\sigma > 0$, μ are any given real constants, we will call X the normal random variable with mean μ and variance σ^2.   ■

EXAMPLE 4.6.3.   Normal probability laws are frequently assumed to be good descriptors of human abilities. Assume a college athlete is going to toss the shot (a large steel ball) and assume the distance it will travel is X, a normal random variable with $\mu = 17$ meters, $\sigma = 2$ meters. Then the probability that he throws it no more than 18.5 meters is

$$P(X \leq 18.5) = F_X(18.5) = N\left(\frac{18.5 - 17}{2}\right)$$

$$= N(.75) = .7734$$

and the probability that he throws it at least 15 meters is

$$P(X \geq 15) = 1 - F_X(15)$$

$$= 1 - N\left(\frac{15 - 17}{2}\right)$$

$$= 1 - N(-1) = 1 - .1587 = .8413.$$

The distance d that his toss will exceed with probability .95 can be found by solving

$$P(X > d) = 1 - F_X(d)$$

$$= 1 - N\left(\frac{d - 17}{2}\right)$$

$$= N\left(\frac{17 - d}{2}\right) = .95,$$

which gives

$$\frac{17 - d}{2} = 1.64$$

so

$$d = 13.72.$$ ∎

In Section 4.5 we saw that the χ^2 density, whose parameter was called "degrees of freedom," is a special gamma density. If we let $W = Z^2$, where Z is a standard normal random variable,

$$\begin{aligned}F_W(t) &= P(W \le t) = P(Z^2 \le t)\\ &= P(-\sqrt{t} \le Z \le \sqrt{t}) \quad \text{for} \quad t \ge 0\\ &= 2N(\sqrt{t}) - 1\end{aligned}$$

and the density function for W is

$$f_W(t) = \frac{d}{dt} F_W(t) = \frac{2n(\sqrt{t})}{2\sqrt{t}}$$

$$= \frac{1}{\sqrt{2\pi t}} e^{-t/2}, \quad t > 0.$$

Thus W is a χ^2 random variable with 1 degree of freedom. The degrees of freedom parameter equals 1 because W is the square of a single standard normal random variable; there is only one source of variability in W, the value for Z. Since the χ^2 density with 1 degree of freedom is the same as a gamma density with $n = \frac{1}{2}$, $\lambda = \frac{1}{2}$, the moment generating function for a χ^2 probability law with 1 degree of freedom is

$$m(t) = \left(\frac{\frac{1}{2}}{\frac{1}{2} - t}\right)^{1/2} = \frac{1}{(1 - 2t)^{1/2}}.$$

Its mean is $\mu = 1$ and its variance is $\sigma^2 = 2$.

EXERCISE 4.6

1. Five independent uniform $(0, 1)$ random variables are generated. The ordered values are $Y_1, Y_2, ..., Y_5$.
   (a) What is the value for the mean and variance of each uniform random variable?
   (b) Compute the mean and variance for each of the ordered values $Y_1, Y_2, ..., Y_5$.

2. The middle-ordered value, in a set of n uniform random variables with odd n, is called the sample median. Let $n = 5$ as in Exercise 4.6.1 and evaluate and sketch the density for Y_3, the median of the 5 uniform random variables.

3. Assume the time X required for a distance runner to run a mile is a normal random variable with parameters $\mu = 4$ minutes, 1 second and $\sigma = 2$ seconds. What is the probability that this athlete will run the mile in less than 4 minutes? In more than 3 minutes, 55 seconds?

4. The length X of an adult rock cod caught in Monterey Bay is a normal random variable with parameters $\mu = 30$ inches and $\sigma = 2$ inches. If you catch one of these fish, what is the probability that it will be at least 31 inches long? That it will be no more than 32 inches long? That its length will be between 24 inches and 28 inches?

5. If Z is a standard normal random variable and we define $U = |Z|$, then U is called the *folded* standard normal variable. Express $F_U(t)$ in terms of $N(t)$ and derive the density $f_U(t)$.

6. Suppose we are given a target with a vertical straight line drawn through its center. Let us assume that if we throw a dart at this target and measure the horizontal distance Z between the point we hit and the center line, then Z is a standard normal random variable (if the dart lands right of the centerline the measurement is positive, if it lands to the left of the centerline the measurement is negative). Then the distance from the point we hit to the centerline is $|Z| = U$, the folded normal random variable defined in Exercise 4.6.5. Compute $P(U > 1)$ and $P(U < \frac{1}{2})$.

7. Find the median and the interquartile range for a standard normal random variable Z.

8. Find the median and the interquartile range for a normal variable X with parameters μ and σ.

9. To show that the integral of the standard normal density function, over the whole real line, equals 1, define

$$A = \int_{-\infty}^{\infty} \frac{1}{\sqrt{2\pi}} e^{-x^2/2} \, dx,$$

and then

$$A^2 = \left\{ \int_{-\infty}^{\infty} \frac{1}{\sqrt{2\pi}} e^{-x^2/2} \, dx \right\} \left\{ \int_{-\infty}^{\infty} \frac{1}{\sqrt{2\pi}} e^{-y^2/2} \, dy \right\}$$

$$= \frac{1}{2\pi} \int_{-\infty}^{\infty} \int_{-\infty}^{\infty} e^{-(x^2+y^2)/2} \, dxdy.$$

Transform the double integral A^2 to polar coordinates, $x = r \cos \theta$, $y = r \sin \theta$ to show that

$$A^2 = 1, \quad \text{which implies} \quad A = 1$$

since $A > 0$.

10. (a) Show that $E[Z] = 0$, if Z is standard normal (this function is directly integrable).

(b) To show that the variance of the standard normal random variable is 1, as claimed,

$$E[Z^2] = \int_{-\infty}^{\infty} \frac{z^2}{\sqrt{2\pi}} e^{-z^2/2} \, dz = \frac{2}{\sqrt{2\pi}} \int_0^{\infty} z^2 e^{-z^2/2} \, dz$$

because the integrand is an even function. Make the change of variable $z = \sqrt{2u}$ and use the fact that $\Gamma(\frac{1}{2}) = \sqrt{\pi}$.

11. (a) Show that the moment generating function for a standard normal random variable is $m_Z(t) = e^{t^2/2}$.

(b) Using (a) show that the moment generating function for $X = \sigma Z + \mu$ is

$$m_X(t) = e^{t\mu + t^2\sigma^2/2}.$$

12. Find the median and interquartile range for a χ^2 random variable with 1 degree of freedom.

13. Assume the high price of a commodity tomorrow is a normal random variable with $\mu = 10$, $\sigma = \frac{1}{4}$. What is the shortest interval that has probability .95 of including tomorrow's high price for this commodity?

14. The height a university high-jumper will clear, each time he jumps, is a normal random variable with mean 2 meters and standard deviation 10 centimeters.

(a) What is the greatest height he will jump with probability .95?

(b) What is the height he will clear only 10 percent of the time?

15. Scores on a national educational achievement test are assumed to be normally distributed with $\mu = 500$, $\sigma = 100$.

(a) What is the ninetieth quantile for this distribution?

(b) What are the quartiles of this distribution?

(c) Assume you take this exam and make a score of 585. What proportion of all scores are smaller than yours?

16. Show that the points of inflection of the normal density function occur at $\mu \pm \sigma$.

17. The ninetieth percentile of a normal distribution is known to equal 50 while its fifteenth percentile is 25.
    (a) Find μ and σ.
    (b) What is the value of the fortieth percentile?

18. Express the percentiles of the χ^2, 1 degree of freedom, distribution in terms of the percentiles of a standard normal distribution.

19. Express the percentiles of a folded standard normal distribution in terms of the percentiles of the standard normal distribution.

4.7 Summary of Some Standard Probability Laws

Discrete Probability Laws

Name	Bernoulli	Binomial	Geometric
Parameters	$0 < p < 1$	$n = 1, 2, 3, \ldots$ $0 < p < 1$	$0 < p < 1$
Probability function	$P(X = k) = p^k(1-p)^{1-k}$	$P(X = k) = \binom{n}{k} p^k(1-p)^{n-k}$	$P(X = k) = pq^{k-1}$
Range	$k = 0, 1$	$k = 0, 1, 2, \ldots, n$	$k = 1, 2, 3, \ldots$
Mean	p	np	$1/p$
Variance	$p(1-p)$	$np(1-p)$	$\dfrac{(1-p)}{p^2}$
Factorial moment generating function	$q + pt$ where $q = 1 - p$	$(q + pt)^n$ where $q = 1 - p$	$\dfrac{pt}{1 - qt}$ where $q = 1 - p$

Continuous Probability Laws

Name	Uniform	Exponential	Gamma
Parameters	$-\infty < a < b < \infty$	$\lambda > 0$	$n > 0$ $\lambda > 0$
Density function	$f(x) = \dfrac{1}{b - a}$	$f(x) = \lambda e^{-\lambda x}$	$f(x) = \dfrac{\lambda^n x^{n-1}}{\Gamma(n)} e^{-\lambda x}$
Range	$a < x < b$	$x > 0$	$x > 0$
Mean	$\dfrac{a + b}{2}$	$\dfrac{1}{\lambda}$	$\dfrac{n}{\lambda}$
Variance	$\dfrac{(b - a)^2}{12}$	$\dfrac{1}{\lambda^2}$	$\dfrac{n}{\lambda^2}$
Moment generating function	$\dfrac{e^{tb} - e^{ta}}{t(b - a)}$	$\dfrac{\lambda}{\lambda - t}$	$\left(\dfrac{\lambda}{\lambda - t} \right)^n$

Negative Binomial	Poisson	Hypergeometric
$r = 1, 2, 3, ...$ $0 < p < 1$	$\mu > 0$	$m = 1, 2, 3, ...$ $w = 0, 1, 2, ...$ $n = 0, 1, 2, ..., m$
$P(X = k) = \binom{k-1}{r-1} p^r q^{k-r}$	$P(X = k) = \dfrac{\mu^k}{k!} e^{-\mu}$	$P(X = k) = \dfrac{\binom{w}{k}\binom{m-w}{n-k}}{\binom{m}{n}}$
$k = r, r+1, r+2, ...$	$k = 0, 1, 2, ...$	$k = 0, 1, ..., n$
r/p	μ	$n\dfrac{w}{m}$
$\dfrac{r(1-p)}{p^2}$	μ	$n\dfrac{w}{m}\dfrac{(m-w)}{m}\dfrac{(m-n)}{(m-1)}$
$\left(\dfrac{pt}{1-qt}\right)^r$ where $q = 1 - p$	$e^{\mu(t-1)}$	not useful

Chi-Square	Beta	Normal
$v = 1, 2, 3, ...$	$\alpha > 0, \beta > 0$	$-\infty < \mu < \infty$ $\sigma > 0$
$f(x) = \dfrac{x^{v/2-1}}{2^{v/2}\,\Gamma\left(\dfrac{v}{2}\right)} e^{-x/2}$	$f(x) = \dfrac{\Gamma(\alpha+\beta)}{\Gamma(\alpha)\,\Gamma(\beta)} x^{\alpha-1}(1-x)^{\beta-1}$	$f(x) = \dfrac{1}{\sigma\sqrt{2\pi}} e^{-(x-\mu)^2/2\sigma^2}$
$x > 0$	$0 < x < 1$	$-\infty < x < \infty$
v	$\dfrac{\alpha}{\alpha+\beta}$	μ
$2v$	$\dfrac{\alpha\beta}{(\alpha+\beta)^2(\alpha+\beta+1)}$	σ^2
$\left(\dfrac{1}{1-2t}\right)^{v/2}$	not useful	$e^{t\mu + t^2\sigma^2/2}$

5

Jointly Distributed Random Variables

We have spent some time studying the concept of a random variable and have seen some simple models that lead to several of the most frequently used probability laws. The random variables considered thus far are called one dimensional, because the observed value for a random variable can be thought of as a single point on the real line; we need only a one-dimensional space (the real line) to represent the range for the random variable. In almost all applications random variables do not occur singly. We will have need for the tools necessary to describe or model the behavior of two, three, or in general some integer $n > 1$, random variables simultaneously. For example, if an economist were trying to model the inflation rate in the United States, and perhaps to give predictions about the value for the rate 12 months in the future, he or she would certainly need to include data (observed values for random variables) from all parts of the country as well as from many different segments of industry. This calls for a model incorporating some number $n > 1$ random variables simultaneously. In medical research it is important to investigate the effects of any new drug proposed for the treatment of a certain condition for $n > 1$ individual patients. Again, then, the model describing the effects of the drug must be able to incorporate several random variables at the same time (measuring the effects on the n patients used). In this chapter we will study what is necessary to describe the probability law for several random variables simultaneously.

5.1   Vector Random Variables

We began our study of random variables in Chapter 2 with the idea that a random variable associates a (single) real number with each possible outcome of an experiment. The probability measure for the experiment then was used to define the probability law for the (one-dimensional) random variable. There is no reason, of course, that a single number is necessarily associated with each experimental outcome. We could equally well define rules that would associate two numbers, or three numbers, or indeed, n numbers with experimental outcomes. These rules are actually examples of what we will call two-dimensional, three-dimensional or, for the general case, n-dimensional random variables (or random vectors). Again, the probability law for the n-dimensional random variable (random vector) derives directly from the probability measure for the experiment.

There is a rich literature already available regarding n-dimensional random variables. Thus in this text, for lack of space, we will only touch very lightly on many concepts and results that apply to n-dimensional random variables. We will, in general, discuss what is needed to specify the probability law for an n-dimensional random vector and to derive some simple summarizations of these probability laws.

First, as already mentioned, the random variables we have studied thus far are one dimensional, the range of each is a set of values from the real line. A two-dimensional random vector (X_1, X_2) associates an ordered pair of real numbers, a 2-tuple (x_1, x_2), with each experimental outcome; thus the range of (X_1, X_2) can be thought of as a set of points in a two-dimensional space, the usual (x_1, x_2) Cartesian plane. A three-dimensional random vector (X_1, X_2, X_3) associates a 3-tuple of real numbers, (x_1, x_2, x_3), with every experimental outcome and an n-dimensional random vector $(X_1, X_2, ..., X_n)$ associates an n-tuple of real numbers, $(x_1, x_2, ..., x_n)$, with every outcome. Thus the range of (X_1, X_2, X_3) can be thought of as a set of points in a three-dimensional space and the range of $(X_1, X_2, ..., X_n)$ can be thought of as a set of points in an n-dimensional space.

Events are subsets of the range of the random vector—subsets for which we can derive a unique number called the probability of the event. Because two- and higher-dimensional spaces are in many senses richer than a one-dimensional space, we will see that the collections of possible events for two- and higher-dimensional random vectors are rich indeed; the technical difficulties involved with computing probabilities of certain events for an n-dimensional random vector can be quite nontrivial. The important concept here is that any higher-dimensional random vector is in all important ways like the random variables we have studied. It is simply a rule associating n-tuples (rather than 1-tuples or scalars) with outcomes for an experiment. We will now define an n-dimensional random vector.

DEFINITION 5.1.1. Let $(X_1, X_2, \ldots, X_n)$ be a rule (or equivalently an ordered collection of n rules) associating an n-tuple with each element ω of a sample space S. Then $(X_1, X_2, \ldots, X_n)$ is called an n-*dimensional random vector*. (We will equivalently say $X_1, X_2, \ldots, X_n$ are jointly *distributed random variables*.) ∎

EXAMPLE 5.1.1. Suppose our experiment consists of three repeated independent Bernoulli trials, each with probability of success p. (For example, an unprepared student guesses at the answer to each of three true-false questions, thus $p = \frac{1}{2}$.) Then let

$$X_1 = 1 \quad \text{if we observe a success on trial 1}$$
$$= 0 \quad \text{if we observe a failure on trial 1}$$

$$X_2 = 1 \quad \text{if we observe a success on trial 2}$$
$$= 0 \quad \text{if we observe a failure on trial 2}$$

$$X_3 = 1 \quad \text{if we observe a success on trial 3}$$
$$= 0 \quad \text{if we observe a failure on trial 3}$$

and (X_1, X_2, X_3) associates a 3-tuple with every experimental outcome. (X_1, X_2, X_3) is an example of a three-dimensional random vector; its range is

$$R_{(X_1, X_2, X_3)} = \{(0, 0, 0), (0, 0, 1), (0, 1, 0), (0, 1, 1), (1, 0, 0), (1, 0, 1),$$
$$(1, 1, 0), (1, 1, 1)\},$$

a collection of 3-tuples or, equivalently, points in a three-dimensional space. Each of X_1, X_2, X_3 (singly) is a Bernoulli random variable. ∎

EXAMPLE 5.1.2. An experienced player throws a dart at a bull's-eye target with a 10 centimeter radius and a superimposed (x_1, x_2) coordinate system, whose origin is at the center of the bull's-eye (see Figure 5.1.1). Then each of the different points on the target that his dart might strike can be represented by their Cartesian coordinates (x_1, x_2) and, if we let (X_1, X_2) be the point at which the dart will land, (X_1, X_2) is a two-dimensional random vector; the range for (X_1, X_2) is

$$R_{(X_1, X_2)} = \{(x_1, x_2): x_1^2 + x_2^2 \leq 100\},$$

under the assumption the dart will necessarily hit the target somewhere within its perimeter. ∎

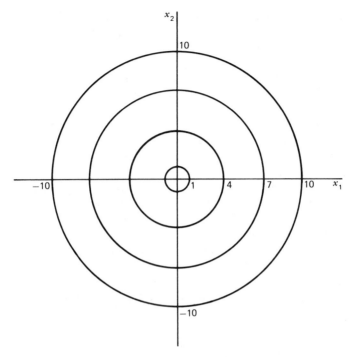

Figure 5.1.1

Random vectors differ from random variables only in the dimensionality, or number of elements, of the -tuples associated with the experimental outcomes. In a general sense the probability law for the random vector is derived from the probability measure of the experiment in exactly the same way we have already discussed for random variables. The fact that the random vector has several dimensions permits a larger variety of possibilities, though. For example, a box might contain 4 numbered light bulbs; we select 1 bulb at random from the 4, place it in service and then observe the number of hours it stays lit until it fails. We could define X_1 to be the number on the bulb selected and X_2 to be the number of hours of service it provides; the pair (X_1, X_2) then is a two-dimensional random vector. These two random variables, X_1 and X_2, differ considerably in their range of possible values. X_1 must equal one of the integers 1, 2, 3, or 4, whereas X_2 can conceivably equal any positive real number; using the terms of Chapter 3, we can see that X_1 is discrete and X_2 is continuous. The probability law for (X_1, X_2) would have to incorporate this basic difference in range for them. We will not delve into this more general topic but will only discuss the probability law for a random vector $(X_1, X_2, ..., X_n)$ in which (1) each of $X_1, X_2, ..., X_n$ is discrete, *or* (2) each of $X_1, X_2, ..., X_n$ is continuous. Thus the random vector (X_1, X_2, X_3) in Example 5.1.1 will be called discrete (since each of X_1, X_2, and X_3 has a discrete range of values, 0 or 1), and the random vector (X_1, X_2) in Example 5.1.2 will be called continuous (since both X_1 and X_2 have continuous ranges, from -10 to 10 centimeters for each).

Granted $(X_1, X_2, ..., X_n)$ is an *n*-dimensional random variable, we can visualize the possible observed values, $(x_1, x_2, ..., x_n)$, as being points in an *n*-dimensional space. Just as with (one-dimensional) random variables, events are collections of possible observed values; thus for an *n*-dimensional random variable an event is a collection of points or some region in *n*-dimensional space. This added feature of higher dimensionality lends a special richness to events for *n*-dimensional random variables, which we will not explore in depth. The two cases we will discuss, as already mentioned, are those in which $(X_1, X_2, ..., X_n)$ is discrete (the range of each of $X_1, X_2, ..., X_n$ is discrete) or in which $(X_1, X_2, ..., X_n)$ is continuous (the range of each of $X_1, X_2, ..., X_n$ is continuous). When $(X_1, X_2, ..., X_n)$ is discrete, its probability law may be specified by its probability function,

$$p_{X_1,X_2,...,X_n}(x_1, x_2, ..., x_n) = P(X_1 = x_1, X_2 = x_2, ..., X_n = x_n),$$

which gives the probability of occurrence of individual points (in the *n*-dimensional space); as in the one-dimensional case, the probability of any event is computed by summing the probabilities of occurrence of individual points that belong to the event of interest. If $(X_1, X_2, ..., X_n)$ is discrete, its probability function must be nonnegative,

$$p_{X_1,X_2,...,X_n}(x_1, x_2, ..., x_n) \geq 0$$

for all $(x_1, x_2, ..., x_n)$ and must sum to 1, if we add over all possible observed vectors.

When $(X_1, X_2, ..., X_n)$ is continuous, its probability law may be specified by its density function $f_{X_1,X_2,...,X_n}(x_1, x_2, ..., x_n)$, which is proportional to the probability that the random vector is equal to the argument $(x_1, x_2, ..., x_n)$; that is,

$$P(x_1 \leq X_1 \leq x_1 + \Delta x_1, x_2 \leq X_2 \leq x_2 + \Delta x_2, ..., x_n \leq X_n \leq x_n + \Delta x_n)$$
$$\doteq f_{X_1,X_2,...,X_n}(x_1, x_2, ..., x_n) \Delta x_1 \Delta x_2 ... \Delta x_n.$$

We evaluate the probability of any event A (a region in *n*-dimensional space) by integrating the density function over the region defined by A. If $(X_1, X_2, ..., X_n)$ is continuous, its density function must be nonnegative,

$$f_{X_1,X_2,...,X_n}(x_1, x_2, ..., x_n) \geq 0$$

for all $(x_1, x_2, \ldots, x_n)$ and must integrate to 1, if the limits for each x_i are $-\infty$ to $+\infty$ (so that the total probability of the whole space is 1). The following examples illustrate these ideas.

EXAMPLE 5.1.3. Consider the situation of Example 5.1.1 in which we have three independent Bernoulli trials, each with probability of success p, and $X_i = 1$ if success occurs on trial i and $X_i = 0$, otherwise, $i = 1, 2, 3$. Then the elements of the sample space, the probabilities of the corresponding single-element events and the observed values for (X_1, X_2, X_3) are as follows (where $q = 1 - p$).

Elements of S	(s, s, s)	(s, s, f)	(s, f, s)	(s, f, f)	(f, s, s)
Probability	p^3	$p^2 q$	pqp	pq^2	qp^2
$(X_1, X_2, X_3) =$	$(1, 1, 1)$	$(1, 1, 0)$	$(1, 0, 1)$	$(1, 0, 0)$	$(0, 1, 1)$

Elements of S	(f, s, f)	(f, f, s)	(f, f, f)
Probability	qpq	$q^2 p$	q^3
$(X_1, X_2, X_3) =$	$(0, 1, 0)$	$(0, 0, 1)$	$(0, 0, 0)$

Just as in the one-dimensional case, the probability function for (X_1, X_2, X_3) derives directly from the probability measure for S. You can easily verify that the probability function for (X_1, X_2, X_3) is

$$
\begin{aligned}
p_{X_1, X_2, X_3}(x_1, x_2, x_3) &= P(X_1 = x_1, X_2 = x_2, X_3 = x_3) \\
&= p^{x_1 + x_2 + x_3} q^{3 - x_1 - x_2 - x_3}, \qquad x_i = 0, 1 \\
& \hspace{5cm} i = 1, 2, 3,
\end{aligned}
$$

where again, $p_{X_1, X_2, X_3}(x_1, x_2, x_3) = 0$ for all 3-tuples not mentioned. You can verify that if we sum $p_{X_1, X_2, X_3}(x_1, x_2, x_3)$ over the eight possible 3-tuples, (x_1, x_2, x_3), we get 1 as required.

Suppose for this same experiment we define Y_1 to be the number of successes in the first two trials and Y_2 to be the number of successes in the last two trials (out of the three performed). Then (Y_1, Y_2) is a discrete two-dimensional random variable whose observed values are associated with the elements of S as follows.

Elements of S	(s, s, s)	(s, s, f)	(s, f, s)	(s, f, f)	(f, s, s)
$(Y_1, Y_2) =$	$(2, 2)$	$(2, 1)$	$(1, 1)$	$(1, 0)$	$(1, 2)$

Elements of S	(f, s, f)	(f, f, s)	(f, f, f)
$(Y_1, Y_2) =$	$(1, 1)$	$(0, 1)$	$(0, 0)$

The probabilities of occurrence of the single-element events are, of course, as previously listed. These again enable us to compute the probabilities of occurrence of the possible observed vectors (y_1, y_2). For example,

$$P(Y_1 = 2, Y_2 = 2) = P(\{(s, s, s)\}) = p^3$$
$$P(Y_1 = 2, Y_2 = 1) = P(\{(s, s, f)\}) = p^2 q$$
$$P(Y_1 = 1, Y_2 = 1) = P(\{(s, f, s), (f, s, f)\}) = p^2 q + pq^2 = pq,$$

and so on. The full probability function for (Y_1, Y_2) is given by the entries in the following table.

	$y_1 =$ 0	1	2
0	q^3	pq^2	0
$y_2 = 1$	pq^2	pq	$p^2 q$
2	0	$p^2 q$	p^3

Again, if we sum the probability function for (Y_1, Y_2) over the seven nonzero cells, it totals to one, as required. Once we have the probability function for (Y_1, Y_2), we can use it to evaluate any probabilities of interest. For example,

$$P(Y_1 = Y_2) = P(Y_1 = 0, Y_2 = 0) + P(Y_1 = 1, Y_2 = 1) + P(Y_1 = 2, Y_2 = 2)$$
$$= q^3 + pq + p^3$$
$$P(Y_1 < Y_2) = P(Y_1 = 0, Y_2 = 1) + P(Y_1 = 1, Y_2 = 2)$$
$$= pq^2 + p^2 q = pq. \qquad \blacksquare$$

EXAMPLE 5.1.4.   Suppose in the dart throwing discussion in Example 5.1.2 the dart is "equally likely" to strike any point on the target. We use the phrase "equally likely" in the same sense as with one-dimensional continuous random variables. That is, the probability that the dart will land in any small rectangular area, with dimensions Δx_1 and Δx_2, is the same, no matter where the rectangle is located on the target,

$$P\left(x_1 - \frac{\Delta x_1}{2} \le X_1 \le x_1 + \frac{\Delta x_1}{2}, x_2 - \frac{\Delta x_2}{2} \le X_2 \le x_2 + \frac{\Delta x_2}{2}\right)$$

is $c \Delta x_1 \Delta x_2$ for all such rectangles on the target. But this assumption implies the joint density function, $f_{X_1, X_2}(x_1, x_2)$ must be constant for all points on the target.

$$f_{X_1, X_2}(x_1, x_2) = c, \qquad \text{for} \qquad x_1^2 + x_2^2 \le 10^2$$
$$= 0, \qquad \text{otherwise.}$$

The requirement that the density integrate to 1 then says

$$\int_{-\infty}^{\infty} \int_{-\infty}^{\infty} f_{X_1, X_2}(x_1, x_2) \, dx_1 \, dx_2 = 1;$$

that is,

$$\int_{x_1^2 + x_2^2 \leq 10^2} \int c \, dx_1 \, dx_2 = 1.$$

Integrating the constant c over the circle $x_1^2 + x_2^2 \leq 10^2$ gives c times the area of the circle, so we require $c(100\pi) = 1$ and the joint density for the coordinates of the point the dart hits is

$$f_{X_1, X_2}(x_1, x_2) = \frac{1}{100\pi}, \qquad x_1^2 + x_2^2 \leq 100$$

$$= 0, \qquad \text{otherwise.}$$

This is called the *uniform* density over the circle with radius 10. We can use this density to evaluate the probability of occurrence of any event (region of the target) by integrating this constant density over the region defined by the event; this, of course, just gives the area of the region divided by the area of the target. Suppose the radii of the circles on the target are 1, 4, and 7 centimeters, as indicated in Figure 5.1.1. Then the probability that the dart would land somewhere in the smallest circle is $\pi/100\pi = \frac{1}{100}$ and the probability it would land somewhere inside the circle labeled 4 is $\pi(4)^2/100\pi = .16$. Remember this density was derived by assuming each small rectangle has the same probability of being hit, no matter where it is located on the target, hardly a reasonable assumption if the dart thrower is experienced. We should more realistically assume that a rectangle of area $\Delta x_1 \Delta x_2$ has higher probability of being hit, the closer it is to the center of the target. ■

Granted $(X_1, X_2, \ldots, X_n)$ is a random vector associating an n-tuple of real numbers with every element of a sample space S, the first coordinate X_1 (or any other single coordinate) by itself (ignoring all the others) then associates a real number with each element of S; that is, X_1 is a (one-dimensional) random variable. Its probability law could, of course, be derived from the probability measure defined for S as we discussed in Chapter 2. Alternatively, we can also derive the probability law for X_1 from the probability law of the vector $(X_1, X_2, \ldots, X_n)$.

First, assume $(X_1, X_2, \ldots, X_n)$ is a discrete vector with probability function $p_{X_1, X_2, \ldots, X_n}(x_1, x_2, \ldots, x_n)$ and consider the event $\{X_1 = k\}$ where k is one of the possible observed values for X_1. Each of the other random variables $X_2, X_3, \ldots, X_n$ could equal any of their possible values, in conjunction with X_1 equaling k, and the event $\{X_1 = k\}$ then occurs. Thus we must hold x_1 constant at k and sum the probability function for $(X_1, X_2, \ldots, X_n)$ over all the possible observed values for $X_2, X_3, \ldots, X_n$

to get the total probability of the event $\{X_1 = k\}$. That is,

$$P(X_1 = k) = \sum \sum \cdots \sum_{\substack{\text{all } (x_1, x_2, \ldots, x_n) \\ \text{with } x_1 = k}} p_{X_1, X_2, \ldots, X_n}(k, x_2, x_3, \ldots, x_n).$$

But $P(X_1 = k) = p_{X_1}(k)$, so considering in turn all the possible observed values k, we can derive the probability function for X_1 from the probability function for $(X_1, X_2, \ldots, X_n)$. In a sense the probability law for X_1 is inherited from the probability law for any vector that has X_1 as a coordinate. This probability function for X_1 is called a *marginal* probability function because its values occur as marginal totals in a tabular presentation of the joint probability law for $(X_1, X_2, \ldots, X_n)$, as illustrated in Example 5.1.5. We can also derive the marginal probability law for X_2 or X_3 or ... or X_n from the joint probability function for $(X_1, X_2, \ldots, X_n)$ by summing it over all the possible values for the other variables.

EXAMPLE 5.1.5. Consider the joint probability function for (X_1, X_2, X_3), discussed in Example 5.1.3,

$$p_{X_1, X_2, X_3}(x_1, x_2, x_3) = p^{x_1 + x_2 + x_3}(1 - p)^{3 - x_1 - x_2 - x_3}, \qquad \begin{aligned} x_i &= 0, 1 \\ i &= 1, 2, 3. \end{aligned}$$

The marginal probability function for X_1 is given by summing over x_2 and x_3.

$$\begin{aligned}
p_{X_1}(x_1) &= p^{x_1 + 0 + 0}(1 - p)^{3 - x_1 - 0 - 0} + p^{x_1 + 0 + 1}(1 - p)^{3 - x_1 - 0 - 1} \\
&\quad + p^{x_1 + 1 + 0}(1 - p)^{3 - x_1 - 1 - 0} + p^{x_1 + 1 + 1}(1 - p)^{3 - x_1 - 1 - 1} \\
&= p^{x_1}(1 - p)^{1 - x_1}\left((1 - p)^2 + p(1 - p) + p(1 - p) + p^2\right) \\
&= p^{x_1}(1 - p)^{1 - x_1}\left((1 - p) + p\right)^2 \\
&= p^{x_1}(1 - p)^{1 - x_1}, \qquad x_1 = 0, 1,
\end{aligned}$$

which is hardly surprising since we already noted each X_i alone is a Bernoulli random variable. The marginal probability laws for X_2 and X_3 are identical with this marginal law for X_1, as is easily verified by summing $p_{X_1, X_2, X_3}(x_1, x_2, x_3)$. We can also derive the marginal probability function for, say, X_1 and X_3 by summing over X_2

$$\begin{aligned}
p_{X_1, X_3}(x_1, x_3) &= p^{x_1 + 0 + x_3}(1 - p)^{3 - x_1 - 0 - x_3} + p^{x_1 + 1 + x_3}(1 - p)^{3 - x_1 - 1 - x_3} \\
&= p^{x_1 + x_3}(1 - p)^{2 - x_1 - x_3}(1 - p + p) \\
&= p^{x_1 + x_3}(1 - p)^{2 - x_1 - x_3}, \qquad (x_1, x_3) = (0, 0), (0, 1), \\
&\hphantom{= p^{x_1 + x_3}(1 - p)^{2 - x_1 - x_3}, \qquad} (1, 0), (1, 1).
\end{aligned}$$

We also discussed the probability law for Y_1 and Y_2 in Example 5.1.3.

	$y_1 =$ 0	1	2	Total
$y_2 = $ 0	q^3	pq^2	0	q^2
1	pq^2	pq	p^2q	$2pq$
2	0	p^2q	p^3	p^2
Total	q^2	$2pq$	p^2	1

The marginal row totals are

$$\sum_{y_1=0}^{2} p_{Y_1,Y_2}(y_1,y_2) = p_{Y_2}(y_2), \qquad y_2 = 0,1,2$$

and the marginal column totals are

$$\sum_{y_2=0}^{2} p_{Y_1,Y_2}(y_1,y_2) = p_{Y_1}(y_1), \qquad y_1 = 0,1,2.$$

Thus as already mentioned, these (marginal) probability laws occur as marginal totals of the two-dimensional table. ∎

The continuous case follows the same line of reasoning. Assume $(X_1, X_2, \ldots, X_n)$ is continuous with density function $f_{X_1,X_2,\ldots,X_n}(x_1, x_2, \ldots, x_n)$ and consider the event $A = \{x \le X_1 \le x + \Delta x\}$, where x is any possible observed value for X_1 and Δx is small. Then the event A occurs if the observed value for X_1 falls in the interval $[x, x + \Delta x]$, regardless of the values for $X_2, X_3, \ldots, X_n$; thus we have

$$P(A) = P(x \le X_1 \le x + \Delta x)$$
$$= \int_{x_1=x}^{x+\Delta x} \int_{x_2=-\infty}^{\infty} \cdots \int_{x_n=-\infty}^{\infty} f_{X_1,X_2,\ldots,X_n}(x_1, x_2, \ldots, x_n) \prod_1^n dx_i$$
$$\doteq \int_{x_2=-\infty}^{\infty} \cdots \int_{x_n=-\infty}^{\infty} f_{X_1,X_2,\ldots,X_n}(x, x_2, \ldots, x_n) \prod_2^n dx_i \, \Delta x.$$

Since $P(x \le X_1 \le x + \Delta x) \doteq f_{X_1}(x)\,\Delta x$, where $f_{X_1}(x)$ is the density for X_1, we have

$$f_{X_1}(x)\,\Delta x = \int_{x_2=-\infty}^{\infty} \cdots \int_{x_n=-\infty}^{\infty} f_{X_1,X_2,\ldots,X_n}(x, x_2, \ldots, x_n) \prod_2^n dx_i \, \Delta x \,;$$

that is, the marginal density for X_1 is

$$f_{X_1}(x) = \int_{x_2 = -\infty}^{\infty} \cdots \int_{x_n = -\infty}^{\infty} f_{X_1, X_2, \ldots, X_n}(x, x_2, \ldots, x_n) \prod_2^n dx_i.$$

We get the marginal density for X_1 by integrating the joint density over all the other variables, $x_2, x_3, \ldots, x_n$. The marginal density for X_3, say, is obtained by integrating the joint density over all its other coordinates, $x_1, x_2, x_4, \ldots, x_n$.

EXAMPLE 5.1.6. In Example 5.1.4 we made an "equally likely" assumption about the point on the target that the dart will hit and found the density for (X_1, X_2) to be

$$f_{X_1, X_2}(x_1, x_2) = \frac{1}{100\pi}, \qquad x_1^2 + x_2^2 \le 100.$$

To find the marginal density for X_1, we must integrate over x_2 (for the given value of x_1, say, $x_1 = x$, see Figure 5.1.2), so we have

$$f_{X_1}(x) = \int_{-\sqrt{100 - x^2}}^{\sqrt{100 - x^2}} \frac{1}{100\pi} dx_2 = \frac{1}{50\pi} \sqrt{100 - x^2}, \qquad -10 < x < 10.$$

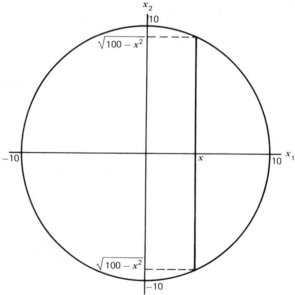

Figure 5.1.2

The marginal density for X_2 is the same, since the limits of integration for x_1 are also from $-\sqrt{100 - x^2}$ to $\sqrt{100 - x^2}$ for X_2 equal to any fixed value x. Thus even though the two-dimensional vector (X_1, X_2) is uniform over the circle, neither of the two marginal densities is uniform; this is caused by the fact that the limits of integration for x_2 change with the given value x for X_1. This common density function for X_1 and X_2 is graphed in Figure 5.1.3. ∎

The joint probability law for $(X_1, X_2, ..., X_n)$ also gives the marginal probability laws for each of the individual random variables $X_1, X_2, ..., X_n$; indeed, it gives the probability law for any subcollection of random variables from $(X_1, X_2, ..., X_n)$. We cannot in general go the other way though. That is, if we know the marginal probability laws for each of $X_1, X_2, ..., X_n$, this is not usually sufficient to identify the joint probability law for $(X_1, X_2, ..., X_n)$. Put another way, there are many different probability laws for $(X_1, X_2, ..., X_n)$ that produce the same marginals. The following example illustrates one such case.

EXAMPLE 5.1.7.   Suppose you are told that a coin is to be flipped two times and that the probability of a head is .5 for each flip. This information actually defines two marginal distributions. Let $X_1 = 1$ if a head occurs on the first flip (and $X_1 = 0$ if not) and let $X_2 = 1$ if a head occurs on the second flip ($X_2 = 0$ if not). Then the given information states that X_1 and X_2 are each Bernoulli, $p = \frac{1}{2}$; their common probability function is

$$p_X(x) = \tfrac{1}{2}, \qquad x = 0, 1.$$

We still do not know the joint probability law for (X_1, X_2). One possibility is that the coin is fair and the two flips are independent Bernoulli trials,

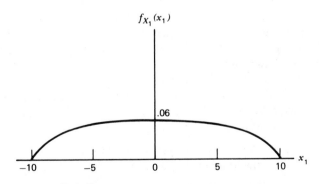

Figure 5.1.3

in which case the joint probability function is

$$p_{X_1,X_2}(x_1, x_2) = \tfrac{1}{4}, \quad \text{for} \quad (x_1, x_2) = (0, 0), (0, 1), (1, 0), (1, 1).$$

This probability law does give these two marginals for X_1 and X_2, as you can verify. Another possibility is that the person tossing the coin has two coins available, one with heads on both sides and the other with tails on both sides. He chooses one of the coins at random and flips it two times. Thus the only possible outcomes are two heads or two tails, depending on which coin was chosen, and the equally likely sample space is

$$S = \{HH, TT\}.$$

If X_1 and X_2 are as previously defined, their joint probability law is given in the following table.

	$x_1 =$ 0	1	Total
$x_2 = 0$	$\frac{1}{2}$	0	$\frac{1}{2}$
1	0	$\frac{1}{2}$	$\frac{1}{2}$
Total	$\frac{1}{2}$	$\frac{1}{2}$	1

This probability law also has the required two marginal distributions. Thus knowing

$$p_{X_1}(x) = p_{X_2}(x) = \tfrac{1}{2}, \quad x = 0, 1$$

does not uniquely specify the probability law for the vector (X_1, X_2). ∎

EXERCISE 5.1

1. A student takes a true-false exam that has 4 questions; assume she is guessing at the answer to each question. Define $X_1 =$ the number she gets right of the first 2 questions, $X_2 =$ the number she gets right of the last 2 questions.
   (a) Derive the probability law for (X_1, X_2).
   (b) Repeat this exercise assuming each exam question is a multiple choice with 4 possible responses.
2. For the case discussed in Exercise 5.1.1 define

   $Y_1 =$ the number she gets right of the first 3 questions,
   $Y_2 =$ the number she gets right of the last 3 questions.

   Answer (a) and (b) for (Y_1, Y_2).

3. What are the marginal probability laws for X_1, X_2, Y_1, Y_2 defined in Exercises 5.1.1 and 5.1.2?

4. What must A equal if

$$f_{X,Y}(x, y) = A\frac{x}{y}, \qquad 0 < x < 1, \qquad 1 < y < 2$$

$$= 0, \qquad \text{otherwise}$$

is to be a density function?

5. A family has two young boys. Let X be the adult height of the older boy and let Y be the adult height of the younger boy. Suppose (X, Y) is equally likely to fall in the rectangle with corners $(66, 68)$, $(66, 72)$, $(71, 68)$, $(71, 72)$. Compute the probability that the older boy will be taller than the younger as adults.

6. Construct a third joint probability law for (X_1, X_2) that has the same marginal probability functions as discussed in Example 5.1.7.

7. Assume (X, Y) is a two-dimensional continuous random variable with density

$$f_{X,Y}(x, y) = \frac{1}{x}, \qquad 0 < y < x, \qquad 0 < x < 1$$

$$= 0, \qquad \text{otherwise.}$$

Find the marginal densities for X and Y.

8. Assume (X, Y) has density

$$f_{X,Y}(x, y) = \tfrac{1}{2}$$

for (x, y) inside the square with corners $(a, a), (a, -a), (-a, a), (-a, -a)$ and that $f_{X,Y}(x, y)$ is zero, otherwise.
(a) Find a.
(b) Find the marginal densities for X and Y.

9. Assume (X, Y) has density

$$f_{X,Y}(x, y) = \tfrac{1}{2}$$

over the square with corners $(a, 0)$, $(-a, 0)$, $(0, a)$, $(0, -a)$ and that $f_{X,Y}(x, y)$ is zero, otherwise.
(a) Find a.
(b) Find the marginal densities for X and Y and compare them with the marginals found in Exercise 5.1.8.

10. Assume two people are waiting in the same line for a bank teller and let X_1 be the time at which person 1 completes his business and let X_2 be the time at which person 2 completes his business. Since person

1 is in front of person 2, he will be finished first and certainly $X_1 < X_2$. The joint density for (X_1, X_2) is

$$f_{X_1, X_2}(x_1, x_2) = \lambda^2 e^{-\lambda x_2}, \qquad 0 < x_1 < x_2$$

Find the marginal densities for X_1 and X_2.

*11. As a more realistic model for the dart thrower discussed in Examples 5.1.2 and 5.1.4, assume the density for the impact point (X_1, X_2) is constant (equals a, say) for (X_1, X_2) in the bull's-eye ($x_1^2 + x_2^2 \leq 1$) and for all other (x_1, x_2) on the target the density decreases linearly with the radius $r = \sqrt{x_1^2 + x_2^2}$ and equals zero for $r^2 \geq 100$ (density equals $c - br$). Thus $f_{X_1, X_2}(x_1, x_2)$ has a flat top and decreases for $x_1^2 + x_2^2 > 1$. Derive the density function if he has probability $\frac{1}{2}$ of hitting the bull's-eye.

*12. For the density described in Exercise 5.1.11, evaluate the probabilities his dart lands in the 3 annuli pictured in Figure 5.1.1.

*13. Repeated independent Bernoulli trials are performed, each with probability p of success. Let X_1 be the trial number of the first success and let X_2 be the trial number of the second success. What is the joint probability function for (X_1, X_2)? Derive the marginal probability laws for X_1 and X_2.

14. For the density given in Exercise 5.1.10, evaluate

(a) $P\left(X_1 < \frac{1}{\lambda}, X_2 < \frac{1}{\lambda} \right)$

(b) $P\left(X_1 < \frac{1}{\lambda}, X_2 > \frac{1}{\lambda} \right)$

5.2 Conditional Distributions and Independence

In Section 2.6 we studied conditional probability and saw how to evaluate the conditional probability of an event A, given that another event B has already occurred. When working with vector random variables, we may have already observed the value of one of the coordinates, say, we have observed $X_1 = x$, and then we want to evaluate the probability that the remaining random variables $X_2, \ldots, X_n$ lie in a given region. The conditional probability law for $X_2, X_3, \ldots, X_n$, given $X_1 = x$, is used in such cases. We will start this section by considering the conditional probability law for several random variables, given that one (or more) other random variables have known values.

The case in which $(X_1, X_2, \ldots, X_n)$ is discrete is quite straightforward and follows directly from the rules for conditional probability, discussed in Section 2.6. Suppose we know the probability function for $(X_1, X_2, \ldots, X_n)$ and want to evaluate

$$P(X_2 = x_2, X_3 = x_3, \ldots, X_n = x_n | X_1 = x),$$

where we again are using the bar within the parentheses to indicate conditional probability. We can simply let A be the event $\{X_2 = x_2, X_3 = x_3, \ldots, X_n = x_n\}$ and let B be the event $\{X_1 = x\}$ and then we are asking for the value $P(A|B)$. But from Section 2.6,

$$
\begin{aligned}
P(A|B) &= \frac{P(A \cap B)}{P(B)} \\
&= \frac{P(X_2 = x_2, X_3 = x_3, \ldots, X_n = x_n \text{ and } X_1 = x)}{P(X_1 = x)} \\
&= \frac{p_{X_1, X_2, \ldots, X_n}(x, x_2, \ldots, x_n)}{p_{X_1}(x)},
\end{aligned}
$$

granted $p_{X_1}(x) > 0$ (we could not be given that $X_1 = x$ if this were not true). This ratio of the joint probability function over the marginal probability function for X_1 is called the conditional probability function for $X_2, X_3, \ldots, X_n$, given $X_1 = x$, as stated in the following definition.

DEFINITION 5.2.1. Let $(X_1, X_2, \ldots, X_n)$ be a discrete random vector with probability function $p_{X_1, X_2, \ldots, X_n}(x_1, x_2, \ldots, x_n)$. The conditional probability function for $X_2, X_3, \ldots, X_n$ given $X_1 = x$ is

$$p_{X_2, X_3, \ldots, X_n | X_1}(x_2, x_3, \ldots, x_n | x) = \frac{p_{X_1, X_2, \ldots, X_n}(x, x_2, \ldots, x_n)}{p_{X_1}(x)},$$

where $p_{X_1}(x)$ is the marginal probability function for X_1 evaluated at x; the conditional probability function for any subcollection of size $n - 1$ of $(X_1, X_2, \ldots, X_n)$, given the value for the one random variable that is not included, is defined analogously. More generally, if $X_{j_1}, X_{j_2}, \ldots, X_{j_k}$ are any k of $(X_1, X_2, \ldots, X_n)$ and $X_{m_1}, X_{m_2}, \ldots, X_{m_{n-k}}$ are the remaining random variables out of the n, the conditional probability function for $X_{j_1}, X_{j_2}, \ldots, X_{j_k}$, given $X_{m_1} = y_1, \ldots, X_{m_{n-k}} = y_{n-k}$, is the ratio of the joint probability function for all n random variables divided by the marginal probability function for $X_{m_1}, X_{m_2}, \ldots, X_{m_{n-k}}$. ∎

EXAMPLE 5.2.1. In Example 5.1.3 we discussed three independent Bernoulli trials, each with probability p of success. Suppose we assume this model represents the sexes of the three children born to a family where success represents a female birth, failure a male birth; also assume for this situation $p = \frac{1}{2}$, which in a crude sense serves as a fairly realistic model. As in the earlier example, let Y_1 be the number of successes (females) in the first two births and let Y_2 be the number of successes in the last two births. The joint probability law for (Y_1, Y_2) then is

	$y_1 =$ 0	1	2	Marginal for Y_2
$y_2 = $ 0	$\frac{1}{8}$	$\frac{1}{8}$	0	$\frac{1}{4}$
1	$\frac{1}{8}$	$\frac{1}{4}$	$\frac{1}{8}$	$\frac{1}{2}$
2	0	$\frac{1}{8}$	$\frac{1}{8}$	$\frac{1}{4}$
Marginal for Y_1	$\frac{1}{4}$	$\frac{1}{2}$	$\frac{1}{4}$	

The conditional probability laws for Y_2, given the different possible values for Y_1, are simply the ratios of the entries in the table divided by the column totals (marginal probability function for Y_1).

Conditional Probability Law for Y_2	Given $Y_1 = $ 0	1	2
$y_2 = $ 0	$\frac{1}{2}$	$\frac{1}{4}$	0
1	$\frac{1}{2}$	$\frac{1}{2}$	$\frac{1}{2}$
2	0	$\frac{1}{4}$	$\frac{1}{2}$
Total	1	1	1

Each of these conditional probability laws totals to 1, of course, as it must. We see from this table that $P(Y_2 = 2 \mid Y_1 = 1) = \frac{1}{4}$. Suppose the first two children born to this family are males, followed by a female, so $Y_1 = 1$. The only way in which we could also have $Y_2 = 2$ is for the last child to be female. Is it true that the probability is $\frac{1}{4}$ for the last child to be a girl? ∎

EXAMPLE 5.2.2. Conditional probability functions are frequently used in constructing joint probability laws. If (X_1, X_2) are jointly distributed discrete random variables and, for example, we know the marginal probability function for X_1, $p_{X_1}(x_1)$, and the conditional probability functions for X_2, given $X_1 = x_1$, $p_{X_2|X_1}(x_2|x_1)$, for each possible x_1, then we can construct the joint probability function for (X_1, X_2) from the fact that

$$p_{X_1,X_2}(x_1, x_2) = p_{X_2|X_1}(x_2|x_1) p_{X_1}(x_1).$$

As an example, assume a small appliance is constructed on an assembly line and that the probability is q that the appliance is defective in some way (it is not defective with probability $p = 1 - q$); we also assume successive appliances produced form independent Bernoulli trials. A large department store orders 1000 of these appliances in one lot. The number of these which are defective then is a binomial random variable, X_1, with parameters $n = 1000$ and q. Before accepting the shipment of 1000 appliances, the store's buyer selects five of the appliances at random from the 1000 and observes X_2, the number of defectives in the five selected. Granted there are x_1 (observed value for X_1) defectives in the 1000, X_2 is a hypergeometric random variable with $m = 1000$, $w = x_1$, $n = 5$ so

$$p_{X_2|X_1}(x_2|x_1) = \frac{\binom{x_1}{x_2}\binom{1000 - x_1}{5 - x_2}}{\binom{1000}{5}}, \qquad \begin{array}{l} x_2 = 0, 1, \ldots, x_1, \\ \\ x_1 = 0, 1, 2, \ldots, 1000. \end{array}$$

The joint probability function for (X_1, X_2) then is

$$p_{X_1,X_2}(x_1, x_2) = p_{X_2|X_1}(x_2|x_1) p_{X_1}(x_1)$$

$$= \frac{\binom{x_1}{x_2}\binom{1000 - x_1}{5 - x_2}}{\binom{1000}{5}} \binom{1000}{x_1} q^{x_1} p^{1000 - x_1}$$

so

$$p_{X_1,X_2}(x_1, x_2) = \binom{5}{x_2}\binom{995}{x_1 - x_2} q^{x_1} p^{1000 - x_1}, \quad \text{for} \quad \begin{array}{l} x_1 = 0, 1, \ldots, 1000 \\ \\ x_2 = 0, 1, \ldots, x_1 \end{array}$$

after canceling some common factors from the combinatorial coefficients. From this joint probability function we can also find the marginal distribution for X_2, the number of defectives in the sample of five inspected:

$$
\begin{aligned}
p_{X_2}(x_2) &= \sum_{x_1 = x_2}^{1000} \binom{5}{x_2}\binom{995}{x_1 - x_2} q^{x_1} p^{1000 - x_1} \\
&= \binom{5}{x_2} q^{x_2} p^{5 - x_2} \sum_{x_1 - x_2 = 0}^{995} \binom{995}{x_1 - x_2} q^{x_1 - x_2} p^{995 - (x_1 - x_2)} \\
&= \binom{5}{x_2} q^{x_2} p^{5 - x_2} (q + p)^{995} \\
&= \binom{5}{x_2} q^{x_2} p^{5 - x_2}, \qquad x_2 = 0, 1, \ldots, 5.
\end{aligned}
$$

Not too surprisingly, the marginal probability law for the number of defectives among the five selected from the lot is binomial with parameters 5 and q. ∎

If $(X_1, X_2, \ldots, X_n)$ is continuous, very similar reasoning is appropriate to define the conditional probability law for $(X_2, X_3, \ldots, X_n)$, given $X_1 = x_1$. One difficulty, at first glance, is that the event $\{X_1 = x_1\}$ has probability zero when X_1 is continuous; thus we cannot directly use our conditional probability definition from Section 2.6 because the denominator would be zero. To keep our discussion simple, let us take $n = 2$ and thus (X_1, X_2) is continuous with density $f_{X_1, X_2}(x_1, x_2)$. This joint density defines a surface over the (X_1, X_2) plane and is positive over those points that can be observed. The volume under this surface, over any region A, is the probability that the observed point lies in A. Now suppose we have already observed $X_1 = x$, say, and want to describe the probability law for X_2, whose value has not yet been observed. Granted $X_1 = x$, we are then certain that the observed point must lie somewhere on the line (x, x_2) in the (x_1, x_2) plane; we still do not know the particular x_2 value that will be observed.

The trace of the joint density over the line (x, x_2) is $f_{X_1, X_2}(x, x_2)$, the height of the density over the line; note that x is fixed, only x_2 is free to vary. This trace of the joint density gives the relative probability that the observed point will equal the different possible values on the line (x, x_2), in the sense that

$$
\begin{aligned}
P(x \le X_1 \le x + \Delta x, 1 \le X_2 \le 1 + \Delta x_2) &\doteq f_{X_1, X_2}(x, 1)\, \Delta x \Delta x_2 \\
P(x \le X_1 \le x + \Delta x, 2 \le X_2 \le 2 + \Delta x_2) &\doteq f_{X_1, X_2}(x, 2)\, \Delta x \Delta x_2,
\end{aligned}
$$

so

$$
\frac{f_{X_1, X_2}(x, 1)}{f_{X_1, X_2}(x, 2)}
$$

is essentially the ratio of the probabilities that X_2 lies in the interval $(1, 1 + \Delta x_2)$ to the probability that X_2 lies in the interval $(2, 2 + \Delta x_2)$, when X_1 is in the interval $(x, x + \Delta x)$. It then follows that the relative probability that the observed value for X_2 lies in the interval (a_1, b_1) versus in the interval (a_2, b_2), granted $X_1 = x$, should be given by the ratio

$$\frac{\int_{a_1}^{b_1} f_{X_1, X_2}(x, x_2)\, dx_2}{\int_{a_2}^{b_2} f_{X_1, X_2}(x, x_2)\, dx_2}$$

that is, the value of

$$\int_{a_1}^{b_1} f_{X_1, X_2}(x, x_2)\, dx_2$$

is a relative measure of the probability of the event $\{a_1 \le X_2 \le b_1\}$ when $X_1 = x$. Since for any possible observed x

$$\int_{-\infty}^{\infty} f_{X_1, X_2}(x, x_2)\, dx_2 = f_{X_1}(x) > 0,$$

the marginal density function for X_1 at x, it must be true that

$$\int_{-\infty}^{\infty} \frac{f_{X_1, X_2}(x, x_2)}{f_{X_1}(x)}\, dx_2 = 1;$$

that is

$$f_{X_2 | X_1}(x_2 | x) = \frac{f_{X_1, X_2}(x, x_2)}{f_{X_1}(x)}$$

is a density function. $f_{X_2 | X_1}(x_2 | x)$ is the *conditional density function* for X_2, given $X_1 = x$, and is used to evaluate probability statements about X_2, given $X_1 = x$. Our formal definition for conditional density functions follows.

Definition 5.2.2. Let $(X_1, X_2, \ldots, X_n)$ be a continuous random vector with density function $f_{X_1, X_2, \ldots, X_n}(x_1, x_2, \ldots, x_n)$. The conditional density function for $(X_2, X_3, \ldots, X_n)$, given $X_1 = x$, is

$$f_{X_2, \ldots, X_n | X_1}(x_2, \ldots, x_n | x) = \frac{f_{X_1, X_2, \ldots, X_n}(x, x_2, \ldots, x_n)}{f_{X_1}(x)}.$$

If $X_{j_1}, X_{j_2}, \ldots, X_{j_k}$ are any k of $(X_1, X_2, \ldots, X_n)$ and $X_{m_1}, X_{m_2}, \ldots, X_{m_{n-k}}$ are those remaining, the conditional density function for $X_{j_1}, X_{j_2}, \ldots, X_{j_k}$, given $X_{m_1} = x_{m_1}, \ldots, X_{m_{n-k}} = x_{m_{n-k}}$, is the ratio

$$\frac{f_{X_1, X_2, \ldots X_n}(x_1, \ldots, x_n)}{f_{X_{m_1}, \ldots, X_{m_{n-k}}}(x_{m_1}, \ldots, x_{m_{n-k}})}. \qquad \blacksquare$$

EXAMPLE 5.2.3.   In Example 5.1.6 we discussed the density

$$f_{X_1, X_2}(x_1, x_2) = \frac{1}{100\pi}, \qquad x_1^2 + x_2^2 \le 100$$

and found the marginal densities for X_1 and X_2 to be

$$f_{X_1}(x) = f_{X_2}(x) = \frac{1}{50\pi}\sqrt{100 - x^2}, \qquad -10 < x < 10.$$

Then if we are given $X_1 = 5$, say, the conditional density for X_2 is

$$f_{X_2|X_1}(x_2|5) = \frac{1/100\pi}{(1/50\pi)\sqrt{100 - 25}} = \frac{1}{2\sqrt{75}}, \qquad -\sqrt{75} < x_2 < \sqrt{75},$$

a uniform density (no surprise since $f_{X_1, X_2}(x_1, x_2)$ is uniform). The conditional density for X_1, given $X_2 = 2$, is

$$f_{X_1|X_2}(x_1|2) = \frac{1/100\pi}{(1/50\pi)\sqrt{100 - 4}} = \frac{1}{2\sqrt{96}}, \qquad -\sqrt{96} < x_1 < \sqrt{96}.$$

also a uniform density function. $\blacksquare$

EXAMPLE 5.2.4.   Suppose the continuous vector (X_1, X_2) has density function

$$f_{X_1, X_2}(x_1, x_2) = \lambda^2 e^{-\lambda x_2}, \qquad 0 < x_1 < x_2.$$

Then the marginal density for X_1 is

$$f_{X_1}(x_1) = \int_{x_1}^{\infty} \lambda^2 e^{-\lambda x_2}\, dx_2 = \lambda e^{-\lambda x_1}$$

and the marginal density for X_2 is

$$f_{X_2}(x_2) = \int_{0}^{x_2} \lambda^2 e^{-\lambda x_2}\, dx_1 = \lambda^2 x_2 e^{-\lambda x_2};$$

thus X_1 is exponential with parameter λ and X_2 is gamma with parameters $r = 2$ and λ. (Actually, it can be shown that X_1 and X_2 are the times of occurrence of the first and second events, respectively, in a Poisson process with parameter λ.) The conditional density for X_2, given $X_1 = x$, is

$$f_{X_2|X_1}(x_2|x) = \frac{\lambda^2 e^{-\lambda x_2}}{\lambda e^{-\lambda x}} = \lambda e^{-\lambda(x_2 - x)}, \qquad 0 < x < x_2$$

an exponential density with shifted origin. The conditional density for X_1 given $X_2 = x$, is

$$f_{X_1|X_2}(x_1|x) = \frac{\lambda^2 e^{-\lambda x}}{\lambda^2 x e^{-\lambda x}} = \frac{1}{x}, \qquad 0 < x_1 < x,$$

uniform over the interval $(0, x)$. Granted that this density does describe the times of occurrence of the first two events in a Poisson process, this says that the time of occurrence of the first, given the second at time x_2, is "equally likely" to be anywhere in the interval $(0, x_2)$. This is a direct consequence of the Poisson process assumptions. ∎

It is possible, of course, that the conditional probability law for a random variable is identical with its marginal probability law. That is, suppose (X_1, X_2) is continuous and

$$f_{X_2|X_1}(x_2|x) = f_{X_2}(x_2),$$

for any x_2 and x. This would say then that the probability of any event $\{a \le X_2 \le b\}$ does not depend on the given value x for X_1; it has value

$$P(a \le X_2 \le b | X_1 = x) = \int_a^b f_{X_2}(x_2)\, dx_2$$

$$= P(a \le X_2 \le b)$$

no matter what the given value x is. In this case then

$$f_{X_2|X_1}(x_2|x) = \frac{f_{X_1,X_2}(x, x_2)}{f_{X_1}(x)} = f_{X_2}(x_2)$$

for any x_2 and x, which would imply that

$$f_{X_1,X_2}(x, x_2) = f_{X_1}(x) f_{X_2}(x_2)$$

and thus, in turn,

$$f_{X_1|X_2}(x|x_2) = \frac{f_{X_1,X_2}(x, x_2)}{f_{X_2}(x_2)}$$

$$= \frac{f_{X_1}(x) f_{X_2}(x_2)}{f_{X_2}(x_2)} = f_{X_1}(x)$$

for any x and x_2. It would also be true that the probability of any event $\{c \leq X_1 \leq d\}$ does not depend on the given value x_2 for X_2 and, in fact, the two events $\{a \leq X_2 \leq b\}$ and $\{c \leq X_1 \leq d\}$ are independent, since

$$
\begin{aligned}
P(c \leq X_1 \leq d, a \leq X_2 \leq b) &= \int_c^d \int_a^b f_{X_1}(x_1) f_{X_2}(x_2) \, dx_1 \, dx_2 \\
&= \int_c^d f_{X_1}(x_1) \, dx_1 \int_a^b f_{X_2}(x_2) \, dx_2 \\
&= P(c \leq X_1 \leq d) P(a \leq X_2 \leq b).
\end{aligned}
$$

The random variables X_1 and X_2 for which

$$
f_{X_1, X_2}(x_1, x_2) = f_{X_1}(x_1) f_{X_2}(x_2), \qquad \text{for all } x_1, x_2
$$

will be called independent. Our formal definition of independent random variables follows.

DEFINITION 5.2.3.
(a) The random variables in the discrete vector $(X_1, X_2, \ldots, X_n)$ are *independent* if and only if

$$
p_{X_1, X_2, \ldots, X_n}(x_1, x_2, \ldots, x_n) = p_{X_1}(x_1) p_{X_2}(x_2) \cdots p_{X_n}(x_n),
$$
for all $x_1, x_2, \ldots, x_n$.

(b) The random variables in the continuous vector $(X_1, X_2, \ldots, X_n)$ are independent if and only if

$$
f_{X_1, X_2, \ldots, X_n}(x_1, x_2, \ldots, x_n) = f_{X_1}(x_1) f_{X_2}(x_2) \cdots f_{X_n}(x_n),
$$
for all $x_1, x_2, \ldots, x_n$. ∎

The concept of independence is very important and will be used extensively in our study of statistical methods. Notice that if $X_1, X_2, \ldots, X_n$ are independent, then their joint probability law is totally specified by the individual marginal probability laws. This tremendously simplifies the specification of the probability law for a vector. Checking the independence of $X_1, X_2, \ldots, X_n$ requires knowledge of their joint probability law and then simply seeing if it is equal to the product of the marginal probability laws at *all* observed values; if this is not true for any possible observed value, then $X_1, X_2, \ldots, X_n$ are not independent. The major usage of the concept of independence is in the construction of joint probability laws from the marginal laws, under the assumption of independence.

EXAMPLE 5.2.5. If (X_1, X_2) has density function

$$f_{X_1,X_2}(x_1, x_2) = \frac{1}{100\pi}, \qquad x_1^2 + x_2^2 \le 100$$

with marginals

$$f_{X_1}(x) = f_{X_2}(x) = \frac{1}{50\pi}\sqrt{100 - x^2}, \qquad -10 \le x \le 10,$$

then X_1 and X_2 are not independent, since

$$f_{X_1,X_2}(9,9) = 0$$

for example, and

$$f_{X_1}(9)\, f_{X_2}(9) = \left(\frac{1}{50\pi}\sqrt{100 - 81}\right)^2$$

$$= \frac{19}{2500\pi^2},$$

which are not equal. ∎

If (X_1, X_2) is discrete with probability function $p_{X_1,X_2}(x_1, x_2)$, then X_1 and X_2 are independent only if

$$p_{X_1,X_2}(x_1, x_2) = p_{X_1}(x_1)\, p_{X_2}(x_2) \qquad \text{for all} \qquad x_1, x_2.$$

Suppose, for example, $p_{X_1}(x_1)$ is positive only for $x_1 = 1, 2, 3, 4$ and $p_{X_2}(x_2)$ is positive only for $x_2 = 0, 1, 2$. Then if X_1 and X_2 are independent the requirement that $p_{X_1,X_2}(x_1, x_2)$ equals the product of the marginal probability functions implies that $p_{X_1,X_2}(x_1, x_2)$ is positive only for the 12 points pictured in Figure 5.2.1, a rectangular lattice. If $p_{X_1,X_2}(x_1, x_2)$ is positive

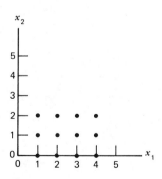

Figure 5.2.1

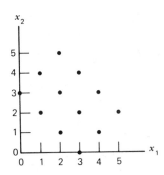

Figure 5.2.2

only for the 12 points shown in Figure 5.2.2, it is not possible for X_1 and X_2 to be independent. This follows from the fact that, for example, $p_{X_1}(0) > 0$ and $p_{X_2}(0) > 0$, for the case in Figure 5.2.2, so $p_{X_1}(0) p_{X_2}(0) > 0$, but $p_{X_1, X_2}(0, 0) = 0$. The same reasoning is appropriate for the continuous case. If $f_{X_1, X_2}(x_1, x_2)$ is positive only for points in a bounded region A, it is necessary (but not sufficient) that A be rectangular with boundaries parallel to the x_1 and x_2 axes before X_1 and X_2 could be independent. Since $f_{X_1, X_2}(x_1, x_2) > 0$ only for (x_1, x_2) in a circle, in Example 5.2.5, this shows immediately that X_1 and X_2 are not independent.

EXAMPLE 5.2.6. Frequently, critical electronic devices are built redundantly, providing an automatic replacement for a failed component. Suppose two identical components are in the device in parallel, as diagramed in Figure 5.2.3. This diagram is meant to show that a current can flow from a to b as long as either component 1 or component 2 is working; the device fails only if *both* components have failed. Let T_1 be the time to failure for component 1 and let T_2 be the time to failure for component 2; also assume T_1 and T_2 are both independent and exponential with parameter λ. The

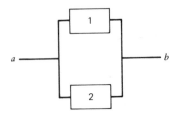

Figure 5.2.3

joint density for (T_1, T_2) then is

$$f_{T_1,T_2}(t_1, t_2) = \lambda e^{-\lambda t_1} \lambda e^{-\lambda t_2}$$
$$= \lambda^2 e^{-\lambda(t_1 + t_2)}, \qquad t_1 > 0, t_2 > 0.$$

Now let Y be the time to failure for the simple device pictured in Figure 5.2.3; that is, $Y = \max(T_1, T_2)$, since the device will continue to operate as long as *either* component is still operating. For any fixed $t > 0$ the event $\{Y \le t\}$ is equivalent to $\{X_1 \le t \text{ and } X_2 \le t\}$, since if either X_1 or X_2 exceeded t, then the maximum, or larger, of the two must as well. Thus the distribution function for Y is

$$F_Y(t) = P(Y \le t)$$
$$= P(X_1 \le t, X_2 \le t)$$
$$= \int_0^t \lambda e^{-\lambda x_1} \, dx_1 \int_0^t \lambda e^{-\lambda x_2} \, dx_2$$
$$= (1 - e^{-\lambda t})^2, \qquad t > 0$$

and the density function for Y is

$$f_Y(t) = \frac{d}{dt} F_Y(t) = 2\lambda e^{-\lambda t}(1 - e^{-\lambda t}), \qquad t > 0.$$

This can be used to evaluate the probabilities of any events of interest for Y, the time to failure for the device, in terms of λ, the parameter of the original exponential marginal densities. Thus for any fixed constant $a > 0$

$$P(Y > a) = 1 - F_Y(a)$$
$$= e^{-\lambda a}(2 - e^{-\lambda a}).$$ ∎

EXERCISE 5.2

1. Given X and Y are jointly discrete random variables with

$$p_{X,Y}(x, y) = \frac{1}{n^2}, \qquad x = 1, 2, \dots, n, \qquad y = 1, 2, \dots, n$$
$$= 0 \qquad \text{otherwise,}$$

verify that X and Y are independent.

2. Given X and Y are jointly discrete random variables with

$$p_{X,Y}(x, y) = \frac{2}{n(n + 1)}, \qquad x = 1, 2, \ldots, n, \qquad y = 1, 2, \ldots, x$$

$$= 0, \qquad\qquad \text{otherwise,}$$

show that X and Y are not independent.

3. Evaluate the conditional probability functions for Y_2, given $Y_1 = y$, for the probability function discussed in Exercise 5.1.2.

4. Evaluate the conditional densities for Y, given X, and for X, given Y, for the joint density provided in Exercise 5.1.4.

5. X and Y are jointly continuous with density function

$$f_{X,Y}(x, y) = 4, \qquad 0 < x < 1, \qquad 0 < y < \tfrac{1}{4}$$

$$= 0, \qquad \text{otherwise.}$$

Verify that X and Y are independent.

6. Suppose that X and Y have joint density

$$f_{X,Y}(x, y) = \tfrac{3}{2}, \qquad 0 < x < 1, \qquad -(x - 1)^2 < y < (x - 1)^2$$

$$= 0, \qquad \text{otherwise.}$$

Show that X and Y are not independent.

7. An assembly line is used to make pens with nylon tips. Assume there is one chance in 100 that the nylon tip (and ink) will not be inserted properly; these errors occur independently from pen to pen. The pens are then placed in boxes that hold 12 pens. These boxes are then placed in cartons, 100 boxes to a carton. A sample of 4 boxes is selected from a carton and 2 pens are examined from each box selected (8 pens). Granted there are 100 defective pens in the carton, what is the conditional probability law for X_2, the number of defective pens in the sample of 8 selected? What is the joint distribution for $X_1 = $ the number of defective pens in the carton and X_2? What is the unconditional probability law for X_2?

*8. A liquor store warehouse contains 1000 bottles of the same label wine, of which 20 are spoiled (which is not evident until the bottle is opened). One hundred of these bottles are selected at random from the 1000 and are placed on sale in the store. You, as a customer, buy a case of 12 of these bottles, selected at random from the 100. What is the probability law for X_2, the number of your bottles that are spoiled? (Consider $X_1 = $ the number in the 100 that are spoiled and the conditional distribution for X_2, given $X_1 = x$.)

9. X and Y have the joint density function

$$f_{X,Y}(x, y) = \frac{3}{2\sqrt{x}}, \qquad 0 < y < x < 1.$$

Find the conditional density for Y, given $X = \frac{1}{2}$, and the conditional density for X, given $Y = \frac{1}{2}$.

10. The conditional density for Y, given $X = x > 0$, is

$$f_{Y|X}(y|x) = \frac{3x + y}{3x + 1} e^{-y}, \qquad y > 0$$

and the marginal density for X is

$$f_X(x) = \frac{3x + 1}{4} e^{-x}, \qquad x > 0.$$

Find the joint density for X and Y and the conditional density for X, given $Y = y$.

11. A code word contains five digits, each either 0 or 1; to be valid the word must contain exactly three 1's and two 0's. One word is selected at random from the valid code words. Define X_1 to be the first (left most) digit and let X_2 be the second digit in the word selected. Derive the probability law for (X_1, X_2). (*Hint.* Consider the conditional probability functions for X_2, given $X_1 = 1$ and given $X_1 = 0$.)

12. Let X_3 be the third digit in the code word described in Exercise 5.2.11 and derive the probability law for (X_1, X_2, X_3).

*13. Assume the number of automobile collisions that occur on a given stretch of highway per year is a Poisson random variable X with $\mu = 20$. The probability is $p = .05$ that there will be one or more fatalities in each accident; occurrences of fatalities are independent from one collision to the next. If Y is the number of collisions with one or more fatalities on this stretch of road in one year, find the probability law for Y. (*Hint.* Consider the conditional probability law for Y, given $X = x$.)

14. A random digit generator ideally produces digits $0, 1, 2, \ldots, 9$, each with probability $\frac{1}{10}$. Furthermore, if the generator produces digits sequentially, they are independent; thus if X_1 is the first digit produced and X_2 is the second, X_3 is the third, and so on, then $X_1, X_2, X_3, \ldots$ are independent random variables.

(a) Suppose this ideal generator produces two digits, X_1 and X_2. What is the probability law for (X_1, X_2)?

(b) The two digits produced are used to define a two-place decimal,

$.X_1 X_2$, which can vary from .00 to .99. What is the probability that the generated two-place decimal is smaller than $\frac{1}{3}$?

15. Assume the random digit generator of Exercise 5.2.14 produces three digits, X_1, X_2, X_3, which are used to define a three-place decimal, $.X_1 X_2 X_3$. Compute the probability that the three-place decimal exceeds .75.

5.3 Expected Values and Moments

We have seen how expected values and moments can be used to summarize or describe various aspects of one-dimensional probability laws. These same concepts prove useful as well with vector random variables.

You will recall from Section 3.3 that an expected value is a special kind of average value; the averaging is done with respect to a probability law. This carries over directly to the vector case. Thus suppose (X_1, X_2) is a discrete random vector with the probability function given in Table 5.3.1. Just as in the one-dimensional case, the values for the probability function can be thought of as the relative frequency of occurrence for the possible observed vectors (x_1, x_2); 45 percent of the time we should observe $(x_1, x_2) = (2, 3)$, 9 percent of the time we should observe $(x_1, x_2) = (7, 3)$, 5 percent of the time we should observe $(x_1, x_2) = (4, 1)$, and so on. If we define $G(X_1, X_2) = X_1 + X_2$, the expected or average value for $G(X_1, X_2)$ is

$$(2 + 1)(.05) + (2 + 2)(.2) + (2 + 3)(.45) + (4 + 1)(.05) + (4 + 2)(0)$$
$$+ (4 + 3)(.1) + (7 + 1)(.01) + (7 + 2)(.05)$$
$$+ (7 + 3)(.09) = 5.58;$$

if we define $H(X_1, X_2) = X_1^2 / X_2$, the average or expected value for

Table 5.3.1

		$x_2 =$		
		1	2	3
	2	.05	.2	.45
$x_1 = $	4	.05	0	.1
	7	.01	.05	.09

$H(X_1, X_2)$ is

$$\frac{(2)^2}{1}(.05) + \frac{(2)^2}{2}(.2) + \frac{(2)^2}{3}(.45) + \frac{(4)^2}{1}(.05) + \frac{(4)^2}{2}(0) + \frac{(4)^2}{3}(.1) + \frac{(7)^2}{1}(.01)$$

$$+ \frac{(7)^2}{2}(.05) + \frac{(7)^2}{3}(.09) = 5.718.$$

We simply multiply all the possible observed values for the function by their probabilities of occurrence and sum them together, just as in the one-dimensional case. If (X_1, X_2) is continuous with density $f_{X_1,X_2}(x_1, x_2)$, the same concept is involved in defining the expected value for a function $G(X_1, X_2)$. The probability that the observed vector will be in a neighborhood of any point (x_1, x_2) is approximately $f_{X_1,X_2}(x_1, x_2) \Delta x_1 \Delta x_2$. Thus the expected value for $G(X_1, X_2)$ is approximately given by

$$\sum_{\text{all } (x_1, x_2)} \sum G(x_1, x_2) f_{X_1,X_2}(x_1, x_2) \Delta x_1 \Delta x_2.$$

The limit of this double sum as $\Delta x_1 \to 0$, $\Delta x_2 \to 0$ is the double integral

$$\int_{-\infty}^{\infty} \int_{-\infty}^{\infty} G(x_1, x_2) f_{X_1,X_2}(x_1, x_2) \, dx_1 \, dx_2$$

if it exists. Thus to find the expected value for functions of continuous random variables, we employ (multiple) integration, just as integration was called for in the one-dimensional case. The formal definition of the expected value for a function of several random variables follows.

DEFINITION 5.3.1. Let $(X_1, X_2, \ldots, X_n)$ be a random vector and let $G(X_1, X_2, \ldots, X_n)$ be a function of $(X_1, X_2, \ldots, X_n)$. The expected value of $G(X_1, X_2, \ldots, X_n)$ is

$$E[G(X_1, X_2, \ldots, X_n)] = \sum_{\text{all } (x_1, x_2, \ldots, x_n)} \sum \cdots \sum G(x_1, x_2, \ldots, x_n)$$

$$\cdot p_{X_1,X_2,\ldots,X_n}(x_1, x_2, \ldots, x_n)$$

$$\text{if } (X_1, X_2, \ldots, X_n) \text{ is discrete,}$$

$$= \int_{-\infty}^{\infty} \int_{-\infty}^{\infty} \cdots \int_{-\infty}^{\infty} G(x_1, x_2, \ldots, x_n)$$

$$\cdot f_{X_1,X_2,\ldots,X_n}(x_1, x_2, \ldots, x_n) \prod_{i=1}^{n} dx_i$$

$$\text{if } (X_1, X_2, \ldots, X_n) \text{ is continuous,}$$

so long as the sum or integral converges absolutely. ∎

EXAMPLE 5.3.1. Let (Y_1, Y_2) be discrete with the probability function given in Table 5.3.2.

Table 5.3.2

		$y_1 =$	
	0	1	2
0	q^3	pq^2	0
$y_2 = 1$	pq^2	pq	p^2q
2	0	p^2q	p^3

The expected value for $Y_1 - Y_2$ is

$$E[Y_1 - Y_2] = (0 - 0)\, q^3 + (0 - 1)\, pq^2 + (1 - 0)\, pq^2 + (1 - 1)\, pq$$
$$+ (1 - 2)\, p^2q + (2 - 1)\, p^2q + (2 - 2)\, p^3 = 0$$

and the expected value for $Y_1 + Y_2$ is

$$E[Y_1 + Y_2] = (0 + 0)\, q^3 + (0 + 1)pq^2 + (1 + 0)\, pq^2 + (1 + 1)\, pq$$
$$+ (1 + 2)\, p^2q + (2 + 1)\, p^2q + (2 + 2)\, p^3$$
$$= 2pq^2 + 2pq + 6p^2q + 4p^3$$
$$= 2p(q^2 + 2pq + p^2 + q + pq + p^2)$$
$$= 4p.$$

If (X_1, X_2) is continuous with density

$$f_{X_1,X_2}(x_1, x_2) = \lambda^2 e^{-\lambda x_2}, \qquad 0 < x_1 < x_2$$

the expected value for $X_1 X_2$ is

$$E[X_1 X_2] = \int_{x_2=0}^{\infty} \int_{x_1=0}^{x_2} x_1 x_2 \lambda^2 e^{-\lambda x_2}\, dx_1\, dx_2$$

$$= \int_0^{\infty} \frac{\lambda^2}{2} x_2^3 e^{-\lambda x_2}\, dx_2$$

$$= \frac{1}{2\lambda^2} \int_0^{\infty} u^3 e^{-u}\, du \qquad \text{(substituting } u = \lambda x_2)$$

$$= \frac{3!}{2\lambda^2} = \frac{3}{\lambda^2}. \qquad \blacksquare$$

It is easy to see that the rules for expectations discussed in Section 3.3 also carry over to expectations taken with respect to joint probability laws. Let $(X_1, X_2, ..., X_n)$ be discrete (the same results hold true for the continuous case, replacing the probability function by the density, integrate instead of sum). If $G(X_1, X_2, ..., X_n) = c$, a constant, then

$$E[c] = \sum_{\text{all } (x_1, x_2, ..., x_n)} \sum \cdots \sum c p_{X_1, X_2, ..., X_n}(x_1, x_2, ..., x_n)$$

$$= c \sum_{\text{all } (x_1, x_2, ..., x_n)} \sum \cdots \sum p_{X_1, X_2, ..., X_n}(x_1, x_2, ..., x_n)$$

$$= c \cdot 1 = c,$$

since the probability function must sum to 1; thus the expected value of any constant is itself. Next, suppose $G(X_1, X_2, ..., X_n) = cH(X_1, X_2, ..., X_n)$, where again c is any constant; you can easily verify that

$$E[cH(X_1, X_2, ..., X_n)] = cE[H(X_1, X_2, ..., X_n)];$$

that is, constants can be factored through the expectation. Finally, suppose $G(X_1, X_2, ..., X_n)$ is an additive function, say,

$$G(X_1, X_2, ..., X_n) = H_1(X_1, X_2, ..., X_n) + H_2(X_1, X_2, ..., X_n).$$

Then

$$E[H_1(X_1, X_2, ..., X_n) + H_2(X_1, X_2, ..., X_n)]$$

$$= \sum \sum_{\text{all } (x_1, x_2, ..., x_n)} \cdots \sum (H_1(x_1, x_2, ..., x_n) + H_2(x_1, x_2, ..., x_n))$$

$$\cdot p_{X_1, X_2 \quad X_n}(x_1, x_2, ..., x_n)$$

$$= \sum \sum_{\text{all } (x_1, x_2, ..., x_n)} \cdots \sum H_1(x_1, x_2, ..., x_n) p_{X_1, X_2, ..., X_n}(x_1, x_2, ..., x_n)$$

$$+ \sum \sum_{\text{all } (x_1, x_2, ..., x_n)} \cdots \sum H_2(x_1, x_2, ..., x_n) p_{X_1, X_2, ..., X_n}(x_1, x_2, ..., x_n)$$

$$= E[H_1(X_1, X_2, ..., X_n)] + E[H_2(X_1, X_2, ..., X_n)],$$

so the expectation of a sum is the sum of the expectations. To summarize, just as in the scalar case,

1. The expected value of a constant is itself.
2. Constants factor through the expectation.
3. The expected value for a sum is the sum of the expected values.

Definition 5.3.1 includes such functions as $G(X_1, X_2, ..., X_n) = X_1$, $G(X_1, X_2, ..., X_n) = X_1^2$, $G(X_1, X_2, ..., X_n) = X_2$, and so on. That is, if

$(X_1, X_2, \ldots, X_n)$ is discrete, definition 5.3.1 says

$$E[X_1] = \sum \sum_{\text{all } (x_1, x_2, \ldots, x_n)} \cdots \sum x_1 p_{X_1, X_2, \ldots, X_n}(x_1, x_2, \ldots, x_n).$$

Using our discussion from Section 3.3, we would also have

$$E[X_1] = \sum_{x_1} x_1 p_{X_1}(x_1)$$

where $p_{X_1}(x_1)$ is the marginal probability function for X_1. Must these two quantities be equal, as our notation would imply? The fact that, for each x_1,

$$\sum_{\text{all } (x_2, x_3, \ldots, x_n)} \sum \cdots \sum p_{X_1, X_2, \ldots, X_n}(x_1, x_2, \ldots, x_n) = p_{X_1}(x_1)$$

shows that these two expected values are indeed necessarily always equal and, not surprisingly, all the moments for each of $X_1, X_2, \ldots, X_n$ can be evaluated from the probability law for the vector $(X_1, X_2, \ldots, X_n)$. Thus the descriptors of the marginal probability laws, called their moments, can all be evaluated from the joint probability law. But we can do more than that; we can also evaluate $E[X_1^{j_1} X_2^{j_2} \cdots X_n^{j_n}]$, where each j_i is a nonnegative integer. These are called the joint moments of the probability law for $(X_1, X_2, \ldots, X_n)$ and can be used to describe various aspects of the joint probability law (as long as they exist).

DEFINITION 5.3.2. The $(j_1, j_2, \ldots, j_n)$th moment of the probability law for $(X_1, X_2, \ldots, X_n)$ is $E[X_1^{j_1} X_2^{j_2} \cdots X_n^{j_n}]$, where each j_i is a nonnegative integer, so long as the expectation exists. ∎

Thus with $j_1 = 1, j_2 = j_3 = \cdots = j_n = 0$, the $(1, 0, 0, \ldots, 0)$ joint moment is

$$E[X_1^1 X_2^0 \cdots X_n^0] = E[X_1],$$

with $j_1 = 2, j_2 = j_3 = \cdots = j_n = 0$, the $(2, 0, 0, \ldots, 0)$ moment is

$$E[X_1^2 X_2^0 \cdots X_n^0] = E[X_1^2],$$

and so on. All the moments of the marginal probability laws are included among the moments of the joint probability law. It is easy to define the moment generating function for $(X_1, X_2, \ldots, X_n)$. Since

$$\frac{\partial^{j_1 + j_2 + \cdots + j_n}}{\partial t_1^{j_1} \partial t_2^{j_2} \cdots \partial t_n^{j_n}} \cdot \exp(t_1 x_1 + t_2 x_2 + \cdots + t_n x_n)$$

$$= x_1^{j_1} x_2^{j_2} \cdots x_n^{j_n} \cdot \exp(t_1 x_1 + t_2 x_2 + \cdots + t_n x_n)$$

we would have

$$\frac{\partial^{j_1 + j_2 + \cdots + j_n}}{\partial t_1^{j_1} \partial t_2^{j_2} \cdots \partial t_n^{j_n}} E[\exp(t_1 X_1 + t_2 X_2 + \cdots + t_n X_n)]$$

$$= E[X_1^{j_1} X_2^{j_2} \cdots X_n^{j_n} \cdot \exp(t_1 X_1 + t_2 X_2 + \cdots + t_n X_n)]$$

(assuming the expectations exist and we can interchange differentiation and expectation as written); thus setting $t_1 = t_2 = \cdots = t_n = 0$ after the differentiation would give us $E[X_1^{j_1} X_2^{j_2} \cdots X_n^{j_n}]$, the $(j_1, j_2, \ldots, j_n)$th moment. Therefore $E[\exp(t_1 X_1 + t_2 X_2 + \cdots + t_n X_n)]$ can be used to generate the moments of the joint probability law and is called the moment generating function.

DEFINITION 5.3.3.   The moment generating function for $(X_1, X_2, \ldots, X_n)$ is

$$m_{X_1, X_2, \ldots, X_n}(t_1, t_2, \ldots, t_n) = E[\exp(t_1 X_1 + t_2 X_2 + \cdots + t_n X_n)],$$

as long as the expectation exists. ∎

EXAMPLE 5.3.2.   Let (X_1, X_2) be continuous with density

$$f_{X_1, X_2}(x_1, x_2) = \lambda^2 e^{-\lambda x_2}, \qquad 0 < x_1 < x_2.$$

The moment generating function for (X_1, X_2) is

$$m_{X_1, X_2}(t_1, t_2) = E[\exp(t_1 X_1 + t_2 X_2)]$$

$$= \int_{x_2=0}^{\infty} \int_{x_1=0}^{x_2} \exp(t_1 x_1 + t_2 x_2) \lambda^2 e^{-\lambda x_2} \, dx_1 \, dx_2$$

$$= \lambda^2 \int_0^{\infty} \frac{1}{t_1} (\exp(t_1 x_2) - 1) \exp[-x_2(\lambda - t_2)] \, dx_2$$

$$= \frac{\lambda^2}{t_1} \left(\frac{1}{\lambda - t_1 - t_2} - \frac{1}{\lambda - t_2} \right)$$

$$= \frac{\lambda^2}{(\lambda - t_1 - t_2)(\lambda - t_2)}, \qquad \text{for} \quad t_2 < \lambda, t_1 < \lambda - t_2.$$

From this we have

$$\frac{\partial}{\partial t_1} m_{X_1, X_2}(t_1, t_2) = \frac{\lambda^2}{(\lambda - t_1 - t_2)^2 (\lambda - t_2)}$$

$$\frac{\partial^2}{\partial t_1 \partial t_2} m_{X_1, X_2}(t_1, t_2) = \frac{3\lambda^2(\lambda - t_2) - \lambda^2 t_1}{(\lambda - t_1 - t_2)^3 (\lambda - t_2)^2}$$

$$\frac{\partial^3}{\partial t_1^2 \partial t_2} m_{X_1, X_2}(t_1, t_2) = \frac{8\lambda^2(\lambda - t_2) - 3\lambda^2 t_1}{(\lambda - t_1 - t_2)^4 (\lambda - t_2)^2}$$

and thus

$$E[X_1] = \frac{\lambda^2}{(\lambda - 0 - 0)^2 (\lambda - 0)} = \frac{1}{\lambda}$$

$$E[X_1 X_2] = \frac{3\lambda^2(\lambda - 0) - \lambda^2(0)}{(\lambda - 0 - 0)^3 (\lambda - 0)^2} = \frac{3}{\lambda^2} \text{ (as seen earlier)}$$

$$E[X_1^2 X_2] = \frac{8\lambda^2(\lambda - 0) - 3\lambda^2(0)}{(\lambda - 0 - 0)^4 (\lambda - 0)^2} = \frac{8}{\lambda^3}, \text{ and so on.}$$

The moments for (X_1, X_2) can be generated from the partial derivatives of $m_{X_1, X_2}(t_1, t_2)$ with $t_1 = t_2 = 0$. ∎

The *covariance* of two random variables is frequently used as a crude measure of how they vary together. It is defined as follows.

DEFINITION 5.3.4. The covariance of X_1 and X_2 is $\text{Cov}[X_1, X_2] = E[(X_1 - \mu_1)(X_2 - \mu_2)]$, where $\mu_1 = E[X_1]$, $\mu_2 = E[X_2]$, so long as the expectations exist. ∎

Recalling that expectations are averages with respect to the probability law for (X_1, X_2), we can see that the covariance is the average of the product of the deviation of X_1 from its mean, times the deviation of X_2 from its mean. This average value could be negative, zero, or positive, depending on the way in which X_1 and X_2 vary together. If X_1 is "small" ($< \mu_1$) when X_2 is "large" ($> \mu_2$), and vice versa, with high probability, then the covariance will be negative. If X_1 is "large" ($> \mu_1$) when X_2 is "large" ($> \mu_2$), with high probability, the covariance will be positive. If $X_1 - \mu_1$ is equally probable to be positive or negative, when $X_2 - \mu_2$ is either positive or negative, the covariance will be zero. Realizing that

$$\begin{aligned}
\text{Cov}[X_1, X_2] &= E[(X_1 - \mu_1)(X_2 - \mu_2)] \\
&= E[X_1 X_2 - \mu_1 X_2 - X_1 \mu_2 + \mu_1 \mu_2] \\
&= E[X_1 X_2] - \mu_1 E[X_2] - \mu_2 E[X_1] + \mu_1 \mu_2 \\
&= E[X_1 X_2] - \mu_1 \mu_2
\end{aligned}$$

we can see that the covariance is simply the difference between the average of the product $X_1 X_2$ and the product of the two averages μ_1 and μ_2. It is negative, zero, or positive, depending on whether $E[X_1 X_2]$ is smaller than, equal to, or larger than $\mu_1 \mu_2$.

The covariance of two random variables is a relative measure of "covariability"; its magnitude depends on $\mathrm{Var}[X_1]$ and $\mathrm{Var}[X_2]$. Note that if we define $Y_1 = aX_1$, $Y_2 = bX_2$, where a and b are constants, then

$$E[Y_1] = aE[X_1] = a\mu_1, \qquad E[Y_2] = bE[X_2] = b\mu_2$$

and the covariance of Y_1 and Y_2 is

$$\begin{aligned}
\mathrm{Cov}[Y_1, Y_2] &= E[(Y_1 - E[Y_1])(Y_2 - E[Y_2])] \\
&= E[(aX_1 - a\mu_1)(bX_2 - b\mu_2)] \\
&= abE[(X_1 - \mu_1)(X_2 - \mu_2)] \\
&= ab\,\mathrm{Cov}[X_1, X_2].
\end{aligned}$$

If the covariance between X_1 and X_2 is nonzero, we can choose a and b to make $\mathrm{Cov}[Y_1, Y_2]$ equal to any real number, positive or negative. Thus the covariance is an unbounded measure of how two random variables vary together. The *correlation* between X_1 and X_2 provides a bounded measure of association between two random variables. This is now defined.

DEFINITION 5.3.5. The *correlation* between X_1 and X_2 is

$$\rho = \frac{\mathrm{Cov}[X_1, X_2]}{\sigma_1 \sigma_2},$$

where σ_1 and σ_2 are the standard deviations for X_1 and X_2, respectively. ∎

We can see then that $\rho = 0$ if and only if the covariance is zero and, since $\sigma_1 > 0$, $\sigma_2 > 0$, the sign of ρ is determined by the sign of the covariance.

EXAMPLE 5.3.3. For the random variable (X_1, X_2) discussed in Example 5.3.1 we found

$$E[X_1 X_2] = \frac{3}{\lambda^2}$$

and, since X_1 is exponential with parameter λ, whereas X_2 is gamma with parameters $r = 2$ and λ, as mentioned in Example 5.2.4, we also have $\mu_1 = 1/\lambda$, $\mu_2 = 2/\lambda$, $\sigma_1 = 1/\lambda$, $\sigma_2 = \sqrt{2}/\lambda$. Thus the covariance between

X_1 and X_2 is

$$\text{Cov}[X_1, X_2] = E[X_1 X_2] - \mu_1 \mu_2 = \frac{3}{\lambda^2} - \frac{1}{\lambda}\left(\frac{2}{\lambda}\right) = \frac{1}{\lambda^2}$$

whereas their correlation is

$$\rho = \frac{1/\lambda^2}{(1/\lambda)(\sqrt{2}/\lambda)} = \frac{1}{\sqrt{2}},$$

regardless of the value for λ. Because the covariance (and correlation) is positive, X_2 tends also to be large if X_1 is large; this is hardly surprising if one remembers X_1 is the time of the first occurrence and X_2 is the time of the second occurrence for events in a Poisson process. Because the second occurrence must come after the first, any increase in the time to the first occurrence will also increase the time to the second. ∎

Let X and Y be random variables such that $E[X^2] = E[Y^2] = 1$. Then $(X - Y)^2$ is nonnegative and its expected value must be nonnegative,

$$0 \le E[(X - Y)^2] = E[X^2] + E[Y^2] - 2E[XY]$$
$$= 2(1 - E[XY]),$$

which implies that

$$E[XY] \le 1.$$

Similarly, $(X + Y)^2$ must have a nonnegative expectation

$$0 \le E[(X + Y)^2] = 2(1 + E[XY]),$$

which implies that

$$E[XY] \ge -1$$

and thus we have

$$|E[XY]| \le 1$$

as long as $E[X^2] = E[Y^2] = 1$. Now let X_1 be a random variable with mean μ_1 and variance σ_1^2 and let X_2 be a random variable with mean μ_2 and variance σ_2^2. Then if we define

$$X = \frac{X_1 - \mu_1}{\sigma_1}, \qquad Y = \frac{X_2 - \mu_2}{\sigma_2},$$

it is easily verified that $E[X^2] = E[Y^2] = 1$, so the preceding inequality applies. That is, we have

$$|E[XY]| = \left| E\left[\frac{(X_1 - \mu_1)}{\sigma_1} \frac{(X_2 - \mu_2)}{\sigma_2}\right] \right| = |\rho| \le 1$$

and

$$|E[(X_1 - \mu_1)(X_2 - \mu_2)]| \le \sigma_1\sigma_2.$$

This establishes the following theorem.

Theorem 5.3.1. The correlation, ρ, between any two random variables X_1 and X_2 must lie between -1 and 1, inclusive. ■

The two extreme values for ρ, 1 and -1, only occur in "degenerate" cases, in a sense. You can see from the preceding argument that $\rho = 1$ only if $E[(X - Y)^2] = 0$; but if the nonnegative random variable $(X - Y)^2$ has mean 0, it must equal 0 with probability 1, $P(X - Y = 0) = 1$ or $P(X = Y) = 1$. But this, in turn, says

$$\frac{X_1 - \mu_1}{\sigma_1} = \frac{X_2 - \mu_2}{\sigma_2} \quad \text{or} \quad X_1 = \frac{\sigma_1}{\sigma_2} X_2 - \frac{\sigma_1}{\sigma_2}\mu_2 + \mu_1$$

so X_1 is a linear function of X_2, with slope σ_1/σ_2, with probability 1. This means every possible observed pair (x_1, x_2) must lie on a straight line with positive slope and the value for X_1 is completely determined by the value for X_2 (or vice versa). Similarly, $\rho = -1$ only if $(X + Y)^2$ has mean zero, implying $P(X = -Y) = 1$, which again says that every possible observed pair (x_1, x_2) must lie on a straight line, now with a negative slope $-\sigma_1/\sigma_2$ in the (x_1, x_2) plane. Whenever $|\rho| < 1$, the observed pairs are not restricted to lying on a straight line.

When $|\rho| = 1$, the conditional distribution for X_1, given $X_2 = x_2$, is degenerate, in the sense that the value for X_1 then is fixed at

$$\mu_1 \pm \frac{\sigma_1}{\sigma_2}(x_2 - \mu_2),$$

depending on the sign of ρ. The conditional expected value for X_1, given $X_2 = x_2$, is this same value and its conditional variance is zero. If the conditional probability law for X_1, given $X_2 = x_2$, is not degenerate, meaning that there are two or more possible observed values for X_1, then the observed value for X_1 is not specified by the fact that $X_2 = x_2$. This conditional probability law can then be summarized like any other. The conditional mean for X_1, given $X_2 = x_2$, is called the *regression* of X_1 on X_2 and may or may not be a function of the given value x_2. It is denoted by $E[X_1 | X_2 = x_2]$ and is defined by

$$E[X_1 | X_2 = x_2] = \sum_{x_1} x_1 p_{X_1|X_2}(x_1|x_2) \qquad \text{if } X_1 \text{ is discrete}$$

$$= \int_{-\infty}^{\infty} x_1 f_{X_1|X_2}(x_1|x_2)\, dx_1 \qquad \text{if } X_1 \text{ is continuous.}$$

Define the function $G(x_2)$ by

$$G(x_2) = E[X_1 | X_2 = x_2],$$

which, as already noted, may not depend on x_2 and then is a constant. Then, if (X_1, X_2) is discrete,

$$E[G(X_2)] = E[E[X_1 | X_2]] = \sum_{x_2} G(x_2)\, p_{X_2}(x_2)$$

$$= \sum_{x_2} \left(\sum_{x_1} x_1 p_{X_1|X_2}(x_1|x_2) \right) p_{X_2}(x_2)$$

$$= \sum_{x_2} \left(\sum_{x_1} x_1 \frac{p_{X_1,X_2}(x_1,x_2)}{p_{X_2}(x_2)} \right) p_{X_2}(x_2)$$

$$= \sum_{x_2} \sum_{x_1} x_1 p_{X_1,X_2}(x_1,x_2) = E[X_1].$$

That is, the expected value of the conditional mean of X_1, given $X_2 = x_2$, is the unconditional mean for X_1; the same result holds for the continuous case. If the regression function is averaged over the possible observed values x_2, the result is the unconditional mean for X_1.

EXAMPLE 5.3.4.   Assume a radioactive source emits particles according to a Poisson process with parameter λ; emissions of particles are the events of the process. Also, assume a defective particle counter is used to count the particles emitted in a unit length of time ($t = 1$). The counter is defective in the sense that each particle may or may not be registered by the counter; we assume the probability each particle is registered is p and the registrations form independent Bernoulli trials. Now let X_1 be the number of particles emitted during the unit length period and let X_2 be the number of these that are registered by the counter. Then, clearly, we must have $X_2 \leq X_1$. Given $X_1 = x_1$ particles were in fact emitted during the period, the number registered, X_2, then is binomial with $n = x_1$ and parameter p. Thus the conditional probability law for X_2, given $X_1 = x_1$, is

$$p_{X_2|X_1}(x_2|x_1) = \binom{x_1}{x_2} p^{x_2}(1-p)^{x_1-x_2}, \qquad x_2 = 0, 1, \ldots, x_1,$$

and $E[X_2 | X_1 = x_1] = x_1 p$ is the regression of X_2 on x_1. We are also given that the particle emissions form a Poisson process and thus the marginal probability law for X_1 is Poisson with parameter λ

$$p_{X_1}(x_1) = \frac{\lambda^{x_1}}{x_1!} e^{-\lambda}, \qquad x_1 = 0, 1, 2, \ldots.$$

The expected value of the regression function $g(X_1) = pX_1 = E[X_2 | X_1]$ then is simply

$$E[g(X_1)] = pE[X_1] = p\lambda.$$

This must be equal to $E[X_2]$, the unconditional average number of particles counted in the unit time interval. To verify this, we need the joint probability law for (X_1, X_2) and/or the marginal probability function for X_2. From the definition of the conditional probability function for X_2, given $X_1 = x_1$,

$$p_{X_2|X_1}(x_2 | x_1) = \frac{p_{X_1, X_2}(x_1, x_2)}{p_{X_1}(x_1)}$$

from which it follows that

$$p_{X_1, X_2}(x_1, x_2) = p_{X_1}(x_1) \, p_{X_2|X_1}(x_2 | x_1).$$

Thus

$$p_{X_1, X_2}(x_1, x_2) = \frac{\lambda^{x_1}}{x_1!} e^{-\lambda} \binom{x_1}{x_2} p^{x_2}(1 - p)^{x_1 - x_2}$$

$$= e^{-\lambda} \frac{p^{x_2} \lambda^{x_2} \left(\lambda(1 - p) \right)^{x_1 - x_2}}{x_2! \ (x_1 - x_2)!}$$

for $x_2 = 0, 1, 2, \ldots, x_1 = x_2, x_2 + 1, x_2 + 2, \ldots$. The marginal probability function for X_2 then is

$$p_{X_2}(x_2) = e^{-\lambda} \frac{p^{x_2} \lambda^{x_2}}{x_2!} \sum_{x_1 = x_2}^{\infty} \frac{\left(\lambda(1 - p) \right)^{x_1 - x_2}}{(x_1 - x_2)!}$$

$$= e^{-\lambda} \frac{(p\lambda)^{x_2}}{x_2!} \sum_{j=0}^{\infty} \frac{\left(\lambda(1 - p) \right)^{j}}{j!} \qquad \text{(letting } j = x_1 - x_2\text{)}$$

$$= e^{-\lambda} \frac{(p\lambda)^{x_2}}{x_2!} e^{\lambda(1 - p)}$$

$$= \frac{(p\lambda)^{x_2}}{x_2!} e^{-p\lambda}, \qquad x_2 = 0, 1, 2, \ldots.$$

That is, X_2, the number of particles counted, is itself a Poisson random variable with parameter $p\lambda$; its expectation is the value of its parameter, $p\lambda$, the same as $E[E[X_2|X_1]]$.   ∎

Now we will discuss some special results that hold true for independent random variables. Thus assume $X_1, X_2, \ldots, X_n$ are independent continuous random variables with joint density function

$$f_{X_1, X_2, \ldots, X_n}(x_1, x_2, \ldots, x_n) = f_{X_1}(x_1) f_{X_2}(x_2) \cdots f_{X_n}(x_n)$$

and let $G(X_1, X_2, \ldots, X_n) = H_1(X_1) H_2(X_2) \cdots H_n(X_n)$ be a general multiplicative function of $X_1, X_2, \ldots, X_n$. Then

$$E[G(X_1, X_2, \ldots, X_n)] = E[H_1(X_1) H_2(X_2) \cdots H_n(X_n)]$$

$$= \int_{-\infty}^{\infty} \cdots \int_{-\infty}^{\infty} H_1(x_1) H_2(x_2) \cdots H_n(x_n) f_{X_1}(x_1) f_{X_2}(x_2) \cdots f_{X_n}(x_n) \pi \, dx_i$$

$$= \int_{-\infty}^{\infty} H_1(x_1) f_{X_1}(x_1) \, dx_1 \int_{-\infty}^{\infty} H_2(x_2) f_{X_2}(x_2) \, dx_2 \cdots$$

$$\cdots \int_{-\infty}^{\infty} H_n(x_n) f_{X_n}(x_n) \, dx_n$$

$$= E[H_1(X_1)] E[H_2(X_2)] \cdots E[H_n(X_n)],$$

since the n-fold integral is the product of n one-dimensional integrals when the joint density is the product of the marginal densities (the same result holds when $(X_1, X_2, \ldots, X_n)$ is discrete). That is, the expected value of a product of independent random variables is given by the product of the expectations. This establishes the following theorem.

Theorem 5.3.2. If $X_1, X_2, \ldots, X_n$ are independent random variables, the expected value of any multiplicative function of $X_1, X_2, \ldots, X_n$ is equal to the product of the individual expectations.   ∎

A number of important facts follow directly from this result. First, if $X_1, X_2, \ldots, X_n$ are independent

$$E[X_1^{j_1} X_2^{j_2} \cdots X_n^{j_n}] = E[X_1^{j_1}] E[X_2^{j_2}] \cdots E[X_n^{j_n}],$$

which says that the $(j_1, j_2, \ldots, j_n)$th moment of $(X_1, X_2, \ldots, X_n)$ is simply the product of the individual marginal moments; the joint moments are all specified by the values for the marginal moments. Second, since

$$\exp(t_1 X_1 + t_2 X_2 + \cdots + t_n X_n) = \exp(t_1 X_1) \exp(t_2 X_2) \cdots \exp(t_n X_n)$$

the joint moment generating function is

$$m_{X_1,X_2,\ldots,X_n}(t_1, t_2, \ldots, t_n) = E[\exp(t_1 X_1)\exp(t_2 X_2) \cdots \exp(t_n X_n)]$$
$$= E[\exp(t_1 X_1)] E[\exp(t_2 X_2)] \cdots E[\exp(t_n X_n)]$$
$$= m_{X_1}(t_1) m_{X_2}(t_2) \cdots m_{X_n}(t_n),$$

the product of the marginal moment generating functions. This could in fact be taken as the definition of independence: $X_1, X_2, \ldots, X_n$ are independent if and only if the joint moment generating function is the product of the marginal moment generating functions. Finally, if X_1 and X_2 are independent

$$E[X_1 X_2] = E[X_1] E[X_2] = \mu_1 \mu_2$$

and thus the covariance between X_1 and X_2 is zero:

$$E[(X_1 - \mu_1)(X_2 - \mu_2)] = E[X_1 X_2] - \mu_1 \mu_2 = \mu_1 \mu_2 - \mu_1 \mu_2 = 0.$$

This also implies that the correlation between X_1 and X_2 must be zero when the two are independent.

It is important to realize that if X_1 and X_2 are independent, their covariance (and correlation) must be zero, but the converse does not hold in general. That is, the covariance (and correlation) between X_1 and X_2 can be zero when X_1 and X_2 are not independent. This is illustrated in the following example.

EXAMPLE 5.3.5. Let X_1 be a discrete random variable with probability function

$$p_{X_1}(x_1) = \tfrac{1}{3}, \qquad x_1 = -1, 0, 1$$

and define $X_2 = X_1^2$. Then X_2 is certainly not independent of X_1 and the possible observed values for (X_1, X_2) are $(x_1, x_2) = (-1, 1), (0, 0), (1, 1)$; the probability function for (X_1, X_2) is

$$p_{X_1,X_2}(x_1, x_2) = \tfrac{1}{3} \qquad \text{at} \qquad (x_1, x_2) = (-1, 1), (0, 0), (1, 1).$$

Thus

$$E[X_1] = (-1)(\tfrac{1}{3}) + (0)(\tfrac{1}{3}) + (1)(\tfrac{1}{3}) = 0$$
$$E[X_2] = (1)(\tfrac{1}{3}) + (0)(\tfrac{1}{3}) + (1)(\tfrac{1}{3}) = \tfrac{2}{3}$$
$$E[X_1 X_2] = (-1)(1)(\tfrac{1}{3}) + (0)(0)(\tfrac{1}{3}) + (1)(1)(\tfrac{1}{3})$$
$$= 0$$

and the covariance between X_1 and X_2 is zero (as is their correlation). Thus two dependent random variables may have zero covariance and correlation. ∎

EXERCISE 5.3

1. X and Y have the joint density

    $$f_{X,Y}(x, y) = \tfrac{3}{2}, \qquad 0 < x < 1, \qquad -(x - 1)^2 < y < (x - 1)^2.$$

    Evaluate $E[X + Y]$, $E[X \mid Y = 0]$, $E[Y \mid X = \tfrac{1}{2}]$.
2. (X, Y) have probability function

    $$p_{X,Y}(x, y) = \frac{2}{n(n + 1)}, \qquad x = 1, 2, \ldots, n, \ y = 1, 2, \ldots, x.$$

    Evaluate the regression of X on Y and the regression of Y on X.
3. Find $\text{Cov}[X, Y]$
    (a) If (X, Y) has the density given in Exercise 5.3.1.
    (b) If (X, Y) has the probability function given in Exercise 5.3.2.
4. Assume (X, Y) has the density described in Exercise 5.2.10 and evaluate the correlation between X and Y.
5. Assume (X, Y) has the density described in Exercise 5.2.9 and evaluate $\text{Cov}[X, Y]$ and ρ.
6. A lighthouse has 10 bulbs on hand for the main light. The expected lifetime for each bulb is 6 months. What is the expected length of time the lighthouse can stay lit with these 10 bulbs, when each bulb burned out is replaced instantaneously with a new bulb?
7. Assume X and Y are independent random variables. Is it true that

    $$E\left[\frac{X}{Y}\right] = \frac{E[X]}{E[Y]}?$$

8. What is the regression of X_1 on X_2 if they have the joint probability function defined in Example 5.3.5?
9. Show that the covariance of two random variables is invariant with shifts of the origin. That is, if a and b are any constants, $Y_1 = X_1 + a$, $Y_2 = X_2 + b$, then

    $$\text{Cov}[Y_1, Y_2] = \text{Cov}[X_1, X_2].$$

*10. Construct an example to show that it is possible to have the regression of Y on X constant (does not depend on x), but the regression of X on Y is not constant (does depend on y).

5.4 Sums of Random Variables

Many statistical applications make use of sums of random variables. We will study a number of results applicable to sums of random variables in this section.

First assume $(X_1, X_2, \ldots, X_n)$ is a random vector and let $a_1, a_2, \ldots, a_n$ be any constants. Then

$$Y = a_1 X_1 + a_2 X_2 + \cdots + a_n X_n = \sum_{i=1}^{n} a_i X_i$$

is a random variable; granted $(X_1, X_2, \ldots, X_n)$ associates a vector of values with every outcome of an experiment, $Y = \sum a_i X_i$ associates a real number with each experimental outcome. We could in fact derive the probability law for Y from the probability law for $(X_1, X_2, \ldots, X_n)$ and then use this probability law to find μ_Y and σ_Y^2. Actually, this is not necessary; we can evaluate these quantities directly from the probability law for $(X_1, X_2, \ldots, X_n)$. Although the proof of this statement is beyond the level of this text, it is nonetheless true as long as the expected values exist. In fact,

$$\mu_Y = E[Y] = E\left[\sum_{i=1}^{n} a_i X_i\right]$$

$$= \sum_{i=1}^{n} a_i E[X_i]$$

$$= \sum_{i=1}^{n} a_i \mu_i$$

where $E[X_i] = \mu_i$. Similarly,

$$\sigma_Y^2 = E[(Y - \mu_Y)^2]$$

$$= E\left[\left(\sum_{i=1}^{n} a_i X_i - \sum_{i=1}^{n} a_i \mu_i\right)^2\right]$$

$$= E\left[\left(\sum_{i=1}^{n} a_i (X_i - \mu_i)\right)^2\right]$$

$$= E\left[\sum_{i=1}^{n} a_i^2 (X_i - \mu_i)^2 + 2 \sum \sum_{i<j} a_i a_j (X_i - \mu_i)(X_j - \mu_j)\right]$$

$$= \sum_{i=1}^{n} a_i^2 E[(X_i - \mu_i)^2] + 2 \sum \sum_{i<j} a_i a_j E[(X_i - \mu_i)(X_j - \mu_j)]$$

$$= \sum_{i=1}^{n} a_i^2 \sigma_i^2 + 2 \sum \sum_{i<j} a_i a_j \operatorname{Cov}(X_i, X_j).$$

This establishes the following theorem.

Theorem 5.4.1. If $(X_1, X_2, \ldots, X_n)$ is a random vector, $a_1, a_2, \ldots, a_n$ is any set of constants and $Y = \sum_{i=1}^{n} a_i X_i$, then the mean and variance of Y are:

$$\mu_Y = \sum_{i=1}^{n} a_i \mu_i, \qquad \sigma_Y^2 = \sum_{i=1}^{n} a_i^2 \sigma_i^2 + 2 \sum \sum_{i<j} a_i a_j \operatorname{Cov}[X_i, X_j]$$

where

$$\mu_i = E[X_i], \qquad \sigma_i^2 = \operatorname{Var}[X_i],$$
$$\operatorname{Cov}[X_i, X_j] = E[(X_i - \mu_i)(X_j - \mu_j)]. \qquad \blacksquare$$

EXAMPLE 5.4.1. Suppose (X_1, X_2) is a random vector with $\mu_1 = \mu_2 = \mu$, $\sigma_1^2 = \sigma_2^2 = \sigma^2$ and $\operatorname{Cov}(X_1, X_2) = \rho\sigma^2$, so the correlation between X_1 and X_2 is ρ, and thus $|\rho| \le 1$. If we define $Y = X_1 + X_2$ $(a_1 = a_2 = 1)$, for example, then

$$\mu_Y = 2\mu, \qquad \sigma_Y^2 = 2\sigma^2 + 2\rho\sigma^2 = 2\sigma^2(\rho + 1).$$

If we define $Y = X_1 - X_2$ $(a_1 = 1, a_2 = -1)$, then

$$\mu_Y = \mu - \mu = 0, \qquad \sigma_Y^2 = 2\sigma^2 - 2\rho\sigma^2 = 2\sigma^2(1 - \rho).$$

If we wanted $Y = a_1 X_1 + a_2 X_2$ to have the same expectation as each of X_1 and X_2, we would need

$$\mu_Y = a_1\mu + a_2\mu = \mu(a_1 + a_2) = \mu,$$

which requires $a_1 + a_2 = 1$, so let us take $a_2 = 1 - a_1$. For this choice we also have

$$\sigma_Y^2 = a_1^2\sigma^2 + (1 - a_1)^2\sigma^2 + 2a_1(1 - a_1)\rho\sigma^2$$
$$= \sigma^2[a_1^2 + (1 - a_1)^2 + 2a_1(1 - a_1)\rho].$$

Furthermore, if we wanted to choose a_1 to minimize σ_Y^2 this is accomplished by minimizing

$$q(a_1) = a_1^2 + (1 - a_1)^2 + 2a_1(1 - a_1)\rho.$$

Since

$$\frac{dq(a_1)}{da_1} = 2a_1 - 2(1 - a_1) + 2\rho - 4a_1\rho$$
$$= 4a_1(1 - \rho) - 2(1 - \rho)$$

we find the minimizing value for $q(a_1)$ by setting

$$0 = \frac{dq(a_1)}{da_1} = 4a_1(1 - \rho) - 2(1 - \rho),$$

which implies $a_1 = \frac{1}{2}$, no matter what the value for ρ. Thus $Y = \frac{1}{2}(X_1 + X_2)$ is the linear function of X_1 and X_2, satisfying $\mu_Y = \mu$, which has the smallest variance for any value of ρ, the correlation between X_1 and X_2. If, for example, we were to select some qualified expert at random on, say, March 1st of some year and get her guess for the U.S. inflation rate for the following year (call this number X_1), and we also contact this same expert on September 1st and get her then current guess for the inflation rate of the following year (call this number X_2), it might be reasonable to assume $E[X_1] = E[X_2] = \mu$, where μ is the actual (unknown) inflation rate for the following year. Since X_1 and X_2 are responses from the same individual (separated by 6 months in time) it also seems that X_1 and X_2 may be correlated (if she is too high in the estimate in March, she might also tend to be too high in September). Then the preceding derivation shows that if we want to find the linear function $Y = a_1 X_1 + a_2 X_2$ of her two guesses, which also has expectation $\mu_Y = \mu$, we get the least variable (about μ) function by taking $Y = \frac{1}{2}(X_1 + X_2)$, no matter what the value for ρ. If the expert guesses in March that the new inflation rate will be $X_1 = 10$ (percent) and then in September changes her guess to $X_2 = 6$ (percent), the linear function that has the smallest variance about μ, the assumed "true" rate, is $Y = \frac{1}{2}(X_1 + X_2)$, whose observed value is $\frac{1}{2}(10 + 6) = 8$ (percent) for the given numbers. ■

An important immediate corollary to Theorem 5.4.1 describes this result when $X_1, X_2, ..., X_n$ are all uncorrelated. If the n random variables are in fact independent, then the covariance (and correlation) between any pair of them is zero; this is the form in which the result is frequently used, but actually we need only assume $X_1, X_2, ..., X_n$ are uncorrelated. We will state the corollary with the weaker condition.

Corollary 5.4.1. If $X_1, X_2, ..., X_n$ are uncorrelated random variables, $a_1, a_2, ..., a_n$ are any constants and $Y = \sum_{i=1}^{n} a_i X_i$, then

$$\mu_Y = \sum_{i=1}^{n} a_i \mu_i, \qquad \sigma_Y^2 = \sum_{i=1}^{n} a_i^2 \sigma_i^2,$$

where

$$\mu_i = E[X_i], \qquad \sigma_i^2 = \text{Var}[X_i]. \qquad ■$$

EXAMPLE 5.4.2. Assume X_1 is exponential with parameter $\lambda = .01$, X_2 is exponential with parameter $\lambda = .05$, and X_1 and X_2 are independent. Then the means for X_1 and X_2 are, respectively, $\mu_1 = 100$, $\mu_2 = 20$, and their standard deviations are $\sigma_1 = 100$, $\sigma_2 = 20$. If we define $Y_1 = X_1 + X_2$, then the mean and standard deviation for Y are:

$$\mu_Y = 100 + 20 = 120 \quad \text{and} \quad \sigma_Y = \sqrt{100^2 + 20^2} = \sqrt{10,400}$$
$$= 101.98.$$

If we let $W = X_1 - X_2$, the mean for W is $\mu_W = 100 - 20 = 80$ and the standard deviation for W is $\sigma_W = \sqrt{100^2 + 20^2} = 101.98$, the same value as σ_Y. ∎

EXAMPLE 5.4.3.   Assume $X_1, X_2, \ldots, X_n$ are uncorrelated, each with the same mean μ and the same variance σ^2. If we let $Y = \sum_{i=1}^{n} a_i X_i$, then Corollary 5.4.1 shows that

$$\mu_Y = \mu \sum a_i \quad \text{and} \quad \sigma_Y^2 = \sigma^2 \sum a_i^2.$$

To have $\mu_Y = \mu$, then, requires $\sum a_i = 1$, which we can satisfy by taking $a_n = 1 - a_1 - a_2 - \cdots - a_{n-1}$, leaving $a_1, a_2, \ldots, a_{n-1}$ yet to be determined. For this choice we would also have

$$\sigma_Y^2 = \sigma^2 \big(a_1^2 + a_2^2 + \cdots + a_{n-1}^2 + (1 - a_1 - a_2 - \cdots - a_{n-1})^2\big).$$

What choice of $a_1, a_2, \ldots, a_{n-1}$ gives the smallest value for σ_Y^2? We can answer this by minimizing

$$Q(a_1, a_2, \ldots, a_{n-1}) = a_1^2 + a_2^2 + \cdots + a_{n-1}^2 + (1 - a_1 - a_2 - \cdots - a_{n-1})^2.$$

It is easy to see that

$$\frac{\partial Q}{\partial a_i} = 2a_i + 2(1 - a_1 - a_2 - \cdots - a_{n-1})(-1), \quad i = 1, 2, \ldots, n-1,$$

and thus if we want $\partial Q / \partial a_i = 0$ for each i, we require

$$a_1 = 1 - a_1 - a_2 - \cdots - a_{n-1}$$
$$a_2 = 1 - a_1 - a_2 - \cdots - a_{n-1}$$
$$\vdots \qquad \vdots \qquad \qquad \vdots$$
$$a_{n-1} = 1 - a_1 - a_2 - \cdots - a_{n-1}.$$

Adding these equations gives

$$\sum_{i=1}^{n-1} a_i = (n-1)\left(1 - \sum_{i=1}^{n-1} a_i\right)$$

or

$$\sum_{i=1}^{n-1} a_i (1 + (n-1)) = n - 1$$

so

$$\sum_{i=1}^{n-1} a_i = \frac{n-1}{n},$$

which, substituted in the preceding expression gives, in turn,

$$a_1 = 1 - \frac{n-1}{n} \cdot \frac{1}{n} = \frac{1}{n}, \qquad a_2 = 1 - \frac{n-1}{n} \cdot \frac{1}{n} = \frac{1}{n}, \dots,$$

$$a_{n-1} = 1 - \frac{n-1}{n} = \frac{1}{n}$$

and, of course,

$$a_n = 1 - \sum_{i=1}^{n-1} a_i = \frac{1}{n}$$

as well. That is,

$$Y = \sum_{i=1}^{n} \frac{1}{n} X_i = \frac{1}{n} \sum_{i=1}^{n} X_i$$

is the linear function of $X_1, \dots, X_n$, which satisfies $\mu_Y = \mu$ and has minimum variance. This function will be denoted by $\bar{X}$ and is called the mean or average value of $X_1, X_2, \dots, X_n$. It is used in many statistical techniques as we will see. ∎

Many statistical methods make use of two or more linear functions of the same random variables. Let $X_1, X_2, \dots, X_n$ be uncorrelated (certainly true if they are independent) random variables with means $\mu_1, \mu_2, \dots, \mu_n$ and variances $\sigma_1^2, \sigma_2^2, \dots, \sigma_n^2$ and let $a_1, a_2, \dots, a_n$ and $b_1, b_2, \dots, b_n$ be any two sets of constants. Now define the two linear functions

$$U = \sum_{i=1}^{n} a_i X_i, \qquad V = \sum_{i=1}^{n} b_i X_i.$$

Immediately, from Corollary 5.4.1,

$$\mu_U = \sum a_i \mu_i, \qquad \sigma_U^2 = \sum a_i^2 \sigma_i^2, \qquad \mu_V = \sum b_i \mu_i, \qquad \sigma_V^2 = \sum b_i^2 \sigma_i^2.$$

What is the covariance between the two linear functions U and V? We have

$$\begin{aligned}
\mathrm{Cov}[U, V] &= E[(U - \mu_U)(V - \mu_V)] \\
&= E[(\sum a_i X_i - \sum a_i \mu_i)(\sum b_i X_i - \sum b_i \mu_i)] \\
&= E[(\sum a_i (X_i - \mu_i))(\sum b_i (X_i - \mu_i))] \\
&= \sum a_i b_i E[(X_i - \mu_i)^2] + 2 \sum_{i<j} \sum a_i b_j E[(X_i - \mu_i)(X_j - \mu_j)] \\
&= \sum a_i b_i \sigma_i^2
\end{aligned}$$

since the covariance between any two of the X_i's is zero. This establishes the following theorem.

Theorem 5.4.2. If $X_1, X_2, ..., X_n$ are uncorrelated random variables, and $U = \sum a_i X_i$, $V = \sum b_i X_i$, then the covariance between U and V is

$$\text{Cov}[U, V] = \sum_{i=1}^{n} a_i b_i \sigma_i^2. \quad ■$$

EXAMPLE 5.4.4.   By the appropriate choice of the constants $a_1, a_2, ..., a_n, b_1, b_2, ..., b_n$ it is possible to make the covariance (and the correlation) equal essentially any desired value, including zero when they are chosen to satisfy $\sum_{i=1}^{n} a_i b_i \sigma_i^2 = 0$. Let X_1 and X_2 be uncorrelated random variables, each with the same mean μ and the same variance σ^2. If we let

$$U = X_1 + X_2$$
$$V = X_1 - X_2,$$

then $\mu_U = 2\mu$, $\mu_V = 0$, $\sigma_U^2 = \sigma_V^2 = 2\sigma^2$, and $\text{Cov}(U, V) = 0$, so U and V are themselves uncorrelated random variables. In a sense this says the value of the sum of two random variables does not tell us anything about the difference of the two. The sum could be large and the difference small or large since the two are uncorrelated.   ■

EXAMPLE 5.4.5.   Let $X_1, X_2, ..., X_n$ be uncorrelated, each with mean μ and variance σ^2. Let

$$\bar{X} = \frac{1}{n} \sum_{i=1}^{n} X_i,$$

as in Example 5.4.3, and let

$$U = X_1 - \bar{X} = \frac{n-1}{n} X_1 - \frac{1}{n} X_2 - \frac{1}{n} X_3 - \cdots - \frac{1}{n} X_n.$$

Then from Theorem 5.4.2 the covariance between $\bar{X}$ and U is

$$\text{Cov}(\bar{X}, U) = \sum_{i=1}^{n} a_i b_i \sigma_i^2$$

$$= \sigma^2 \left(\frac{1}{n} \frac{(n-1)}{n} + \left(\frac{1}{n} \right) \left(-\frac{1}{n} \right) + \cdots + \left(\frac{1}{n} \right) \left(-\frac{1}{n} \right) \right)$$

$$= \sigma^2 \left(\frac{n-1}{n^2} - \frac{n-1}{n^2} \right) = 0$$

so $\overline{X}$ and U are uncorrelated. The same result shows that $\overline{X}$ and $X_2 - \overline{X}$, $\overline{X}$ and $X_3 - \overline{X}, \ldots, \overline{X}$ and $X_n - \overline{X}$ are uncorrelated. That is, $\overline{X}$ is uncorrelated with each of $X_1 - \overline{X}, X_2 - \overline{X}, \ldots, X_n - \overline{X}$. This fact also proves of great use in many statistical techniques. ∎

We will study many statistical methods that are used with independent random variables $X_1, X_2, \ldots, X_n$. The moment generating function for $Y = \sum_{i=1}^n a_i X_i$ is easily expressed in terms of the moment generating functions of $X_1, X_2, \ldots, X_n$. In fact,

$$
\begin{aligned}
m_Y(t) = E[e^{tY}] &= E\left[\exp\left(t \sum a_i X_i\right)\right] \\
&= E[\exp(ta_1 X_1)\exp(ta_2 X_2)\cdots\exp(ta_n X_n)] \\
&= E[\exp(ta_1 X_1)]\,E[\exp(ta_2 X_2)]\cdots E[\exp(ta_n X_n)] \\
&= m_{X_1}(ta_1)\,m_{X_2}(ta_2)\cdots m_{X_n}(ta_n),
\end{aligned}
$$

the product of the marginal moment generating functions with appropriate arguments. This establishes the following theorem.

Theorem 5.4.3. Let $X_1, X_2, \ldots, X_n$ be independent and define $Y = \sum a_i X_i$. Then the moment generating function for Y is

$$
m_Y(t) = \prod_{i=1}^n m_{X_i}(ta_i). \quad ∎
$$

EXAMPLE 5.4.6. Let X_1 and X_2 be independent exponential random variables, each with parameter λ; thus the moment generating function for each X_i is $m(t) = \lambda/(\lambda - t)$, for $t < \lambda$. If we let $U = X_1 + X_2$ the moment generating function for U is

$$
m_U(t) = m_{X_1}(t)\,m_{X_2}(t) = \frac{\lambda}{\lambda - t}\frac{\lambda}{\lambda - t} = \left(\frac{\lambda}{\lambda - t}\right)^2.
$$

The moment generating function for $V = X_1 - X_2$ is

$$
m_V(t) = m_{X_1}(t)\,m_{X_2}(-t) = \frac{\lambda}{\lambda - t}\frac{\lambda}{\lambda + t} = \frac{\lambda^2}{\lambda^2 - t^2}. \quad ∎
$$

The result given in Theorem 5.4.3 is useful in establishing the probability law for linear functions of independent random variables. Although it is beyond the level of this text, it can be shown that if X and Y are random variables with moment generating functions, $m_X(t)$ and $m_Y(t)$, and $m_X(t) = m_Y(t)$ for all t in some neighborhood of 0, then X and Y must have the same probability law. Thus if we use Theorem 5.4.3 to evaluate the moment generating function for $\sum a_i X_i$, and then recognize this moment generating function as being from a normal probability law, the probability law for $\sum a_i X_i$ is normal.

EXAMPLE 5.4.7. Assume $X_1, X_2, \ldots, X_n$ are independent, normal, with means $\mu_1, \mu_2, \ldots, \mu_n$ and variances $\sigma_1^2, \sigma_2^2, \ldots, \sigma_n^2$, respectively. Then the marginal moment generating function for X_i is

$$m_{X_i}(t) = \exp\left(t\mu_i + \frac{t^2\sigma_i^2}{2} \right).$$

Now let $Y = \sum a_i X_i$, where $a_1, a_2, \ldots, a_n$ are any constants. The moment generating function for Y is

$$m_Y(t) = \prod_{i=1}^{n} m_{X_i}(ta_i)$$

$$= \prod_{i=1}^{n} \exp\left(ta_i\mu_i + \frac{t^2 a_i^2 \sigma_i^2}{2} \right)$$

$$= \exp\left(t\sum a_i\mu_i + \frac{t^2 \sum a_i^2 \sigma_i^2}{2} \right),$$

which again is a normal moment generating function whose mean is the multiplier of t, so $\mu_Y = \sum a_i\mu_i$, and whose variance is the multiplier of $t^2/2$, so $\sigma_Y^2 = \sum a_i^2 \sigma_i^2$. Thus any linear function of independent normal random variables again has a normal probability law. Actually, this result may be strengthened; any linear combination of any normal random variables, independent or not, has a normal probability law. ∎

The result given in Example 5.4.7, that linear functions of normal random variables again have the normal distribution, is called the *reproductive property* for the normal probability law. The normal probability law reproduces itself through linear functions. Several other probability laws reproduce themselves through summation. You should be able to use Theorem 5.4.3, and the subsequent discussion, to verify that

1. If $X_1, X_2, \ldots, X_k$ are independent binomial, where X_i has parameters n_i and p, then $\sum X_i$ has the binomial probability law with parameters $\sum n_i$ and p.
2. If $X_1, X_2, \ldots, X_k$ are independent negative binomial, where X_i has parameters r_i and p, then $\sum X_i$ again has the negative binomial probability law with parameters $\sum r_i$ and p.
3. If $X_1, X_2, \ldots, X_k$ are independent Poisson random variables, where X_i has parameter λ_i, then $\sum X_i$ has the Poisson probability law with parameter $\sum \lambda_i$.
4. If $X_1, X_2, \ldots, X_k$ are independent gamma random variables, where X_i has parameters r_i and λ, then $\sum X_i$ has the gamma probability law with parameters $\sum r_i$ and λ.

In addition to these results, it is also easy to use Theorem 5.4.3 to verify (as was mentioned in Chapter 3) that the sum of n independent Bernoulli random variables, each with parameter p, is binomial n, p and the sum of r independent exponential random variables, each with parameter λ, is a gamma (actually, Erlang) random variable with parameters r and λ.

We will close this section by mentioning one more result. Suppose $X_1, X_2, \ldots, X_n$ are independent random variables and let $Y_1 = G_1(X_1)$, $Y_2 = G_2(X_2), \ldots, Y_n = G_n(X_n)$, that is, Y_j is a function only of X_j, $j = 1, 2, \ldots, n$. Then the moment generating function for $Y_1, Y_2, \ldots, Y_n$ is

$$
\begin{aligned}
m_{Y_1, Y_2, \ldots, Y_n}(t_1, t_2, \ldots, t_n) &= E\big[\exp(t_1 Y_1 + t_2 Y_2 + \cdots + t_n Y_n)\big] \\
&= E\big[\exp\big(t_1 G_1(X_1) + t_2 G_2(X_2) + \cdots + t_n G_n(X_n)\big)\big] \\
&= E\big[\exp\big(t_1 G_1(X_1)\big)\big] E\big[\exp\big(t_2 G_2(X_2)\big) \cdots \\
&\quad E\big[\exp\big(t_n G_n(X_n)\big)\big] \\
&= m_{Y_1}(t_1) m_{Y_2}(t_2) \cdots m_{Y_n}(t_n),
\end{aligned}
$$

provided the expectations exist; the factorization occurs because $X_1, X_2, \ldots, X_n$ are independent. But as remarked in Section 5.3, the fact that the joint moment generating function for $Y_1, Y_2, \ldots, Y_n$ equals the product of the marginal moment generating functions implies $Y_1, Y_2, \ldots, Y_n$ are independent random variables, and we have the following theorem.

Theorem 5.4.4. If $X_1, X_2, \ldots, X_n$ are independent random variables and $Y_1 = G_1(X_1)$, $Y_2 = G_2(X_2), \ldots, Y_n = G_n(X_n)$, then $Y_1, Y_2, \ldots, Y_n$ are independent random variables. ∎

As a corollary to this result, let $X_1, X_2, \ldots, X_n$ be independent random variables and let $Y_1 = G_1(X_1, \ldots, X_{j_1})$, $Y_2 = G_2(X_{j_1+1}, \ldots, X_{j_2}), \ldots, Y_k = G_k(X_{j_{k-1}+1}, \ldots, X_{j_k})$, where the Y_j's are any functions of mutually exclusive subsets of $X_1, X_2, \ldots, X_n$; that is, no two Y_j's are functions of the same X_r. Then it can be shown that $Y_1, Y_2, \ldots, Y_k$ are also independent random variables. Although the proof of this result is beyond the level of this text, intuitively Theorem 5.4.4 should help one believe that it must be true. We will see several uses of this result in Section 5.7.

EXAMPLE 5.4.8. As mentioned earlier, the χ^2 probability law is used in many statistical methods; let us illustrate Theorem 5.4.4 and discuss this probability law a little more. Recall that we saw in Section 4.6 that if $W = Z^2$, where Z is standard normal, then W has the χ^2 probability law with

1 degree of freedom (df). The moment generating function for W is

$$m_W(t) = \frac{1}{(1 - 2t)^{1/2}},$$

as was also mentioned in Section 5.4.6. Now assume $Z_1, Z_2, \ldots, Z_v$ are independent standard normal random variables and let $V = Z_1^2 + Z_2^2 + \cdots + Z_v^2$, where $v = 2, 3, 4, \ldots$. Each Z_i^2 is a χ^2 random variable with 1 degree of freedom and since $Z_1, Z_2, \ldots, Z_v$ are independent, so are $Z_1^2, Z_2^2, \ldots, Z_v^2$ from Theorem 5.4.4. Thus V is a sum of v independent random variables and its moment generating function is the product of the marginal moment generating functions for $Z_1^2, Z_2^2, \ldots, Z_v^2$ (from Theorem 5.4.3), so

$$m_V(t) = \prod_{i=1}^{v} m_{Z_i^2}(t)$$

$$= \prod_{i=1}^{v} \frac{1}{(1 - 2t)^{1/2}} = \frac{1}{(1 - 2t)^{v/2}}.$$

This is easily seen to be the moment generating function of a gamma random variable with $n = v/2$ and $\lambda = \frac{1}{2}$ and, as mentioned in Section 5.4.5, this random variable is called a χ^2 random variable with v degrees of freedom, its only parameter. Perhaps now we can see more clearly why the parameter of the χ^2 probability law is called degrees of freedom: V is the sum of squares of v independent standard normal random variables. If the value of any one of the standard normal random variables is changed, with the others held constant, the value for V is also changed; thus there are v independent ways in which the value for V could be changed or perturbed. In mechanics, independently varying quantities are called "degrees of freedom," hence (?) the parameter in the probability law for V is also called degrees of freedom. You can also easily see that the χ^2 probability law reproduces itself. That is, if $V_1, V_2, \ldots, V_m$ are independent χ^2 random variables with degrees of freedom $v_1, v_2, \ldots, v_m$, respectively, then

$$U = \sum_{i=1}^{m} V_i$$

is a χ^2 random variable with $v = \sum_{i=1}^{m} v_i$ degrees of freedom. ■

From the preceding corollary, notice as well that if $Z_1, Z_2, \ldots, Z_{20}$, say, are independent standard normal random variables, and we define

$$Y_1 = Z_1^2 + Z_2^2 + \cdots + Z_7^2$$
$$Y_2 = Z_8^2 + Z_9^2 + \cdots + Z_{12}^2$$
$$Y_3 = Z_{13}^2 + Z_{14}^2 + \cdots + Z_{20}^2,$$

then Y_1, Y_2, Y_3 are also independent random variables. What are the probability laws for Y_1, Y_2, and Y_3?

EXERCISE 5.4

1. X_1 and X_2 are independent random variables with $\mu_1 = 5$, $\sigma_1 = 10$, $\mu_2 = 2$, $\sigma_2 = 4$. Compute the mean and standard deviation for
   (a) $U_1 = X_1 + X_2$
   (b) $U_2 = X_1 - X_2$
   (c) $U_3 = X_1 + 4X_2$
   (d) $U_4 = 2X_1 - 5X_2$.

2. The weight of one "large" hen's egg is a random variable X with mean $\mu = 90$ grams, $\sigma = 5$ grams. You buy one dozen of these eggs; assuming their weights $X_1, X_2, ..., X_{12}$ are independent with $\mu = 90$, $\sigma = 5$, what is the expected value and standard deviation of $Y = \sum X_i$, the net weight of the dozen eggs?

3. Y_1 and Y_2 are random variables with $\mu_1 = 15$, $\mu_2 = 10$, $\sigma_1 = 5$, $\sigma_2 = 10$, $\rho = -.6$. Evaluate the mean and standard deviation for
   (a) $V_1 = Y_1 + Y_2$
   (b) $V_2 = Y_1 - Y_2$
   (c) $V_3 = 2Y_1 - 3Y_2$.

4. For the random variables U_1, U_2, U_3, U_4 defined in Exercise 5.4.1 evaluate the covariance between
   (a) U_1 and U_2
   (b) U_1 and U_3
   (c) U_2 and U_3
   (d) U_2 and U_4.

*5. Evaluate the covariance between (a) V_1 and V_2, (b) V_1 and V_3, (c) V_2 and V_3, for the random variables defined in Exercise 5.4.3.

6. On a holiday weekend it is assumed that fatal highway accidents, beginning at 6:00 P.M. on Friday, occur in California like events in a Poisson process with $\lambda = 1$ per hour. Let X_1 be the time (after 6:00 P.M. Friday) of the first fatal accident and let X_2 be the subsequent time (after X_1) to the second. Then X_1 and X_2 are independent, exponential, $\lambda = 1$ (with 1 hour as the unit on the time scale), and $Y = X_1 + X_2$ is the time of the second fatal accident, in hours after 6:00 P.M. Evaluate
   (a) $P(Y \le 1.5)$
   (b) $P(Y > 2.5)$.

7. In Exercise 5.4.2 add the assumption that the weight of each egg is a normal random variable. Evaluate
   (a) $P(Y > 1100)$

(b) $P(1050 < Y < 1150)$
where the units used are grams.

8. A businessman is in the habit of playing backgammon during his lunch hour at his club. Each day he plays three games; assume his probability of winning each game is .45 and that the outcomes of the games are independent. Let X_1, X_2, X_3, X_4, X_5 be the number of games he wins on Monday through Friday, respectively, of 1 week and define $Y = \sum X_i$, the number he wins in the week. Evaluate the probability
   (a) He wins more than half the games ($Y \geq 8$).
   (b) He wins no more than $\frac{1}{3}$ of the games ($Y \leq 5$).

9. Let Y be defined as in Exercise 5.4.8 and define $W = 15 - Y$, the number of games he loses.
   (a) What is the probability law for W?
   (b) What is the joint probability law for Y and W?
   (c) Evaluate ρ, the correlation between Y and W.

10. The number of cars to pass a farmhouse on a remote country road, per day, is a Poisson random variable with parameter $\mu = 2/\text{day}$. The numbers of cars passing from one day to another are independent random variables, with the same parameter. Let $X_1, X_2, ..., X_7$ be the numbers to pass this house during a 7-day week; $Y = \sum X_i$ then is the total number of cars to pass the house in the week. Evaluate
    (a) μ_Y and σ_Y
    (b) $P(Y \leq 12)$.

*11. A dart is tossed at a target, with a Cartesian (X_1, X_2) coordinate system centered in the middle of the bull's-eye. Assume X_1, the horizontal miss distance, and X_2, the vertical miss distance, are independent, normal random variables, each with $\mu = 0$, $\sigma^2 = 2$. The radial miss distance then is $R = \sqrt{X_1^2 + X_2^2}$; evaluate
    (a) $P(R > 2)$
    (b) The radius, r, of the circle centered at $(0, 0)$, which has probability .5 of containing (X_1, X_2).

12. Packages that are labeled "Contents: 100 bolts" are in fact filled by weight. Assume the weight of any individual bolt is normal, $\mu = 5$ grams, $\sigma = .5$ gram, and the weights of individual bolts are independent.
    (a) What is the probability a package with $n = 100$ bolts weighs at least 495 grams?
    (b) What is the probability a package with $n = 99$ bolts weighs at least 495 grams?

13. A plaque beside an elevator reads "Capacity: 950 pounds, 6 persons." Assume the weights of people entering the elevator are independent, normal, $\mu = 150$, $\sigma = 30$ (both in pounds).

(a) If there are six people in the elevator, what is the probability that the weight limit is exceeded?

(b) Answer (a) if seven persons crowd in.

*(c) Can you answer (b) if you are one of the seven and your weight is $w_7 = 140$ pounds?

14. X_1 is a random variable with $\mu_1 = 5, \sigma_1 = 5$ and X_2 is a random variable with $\mu_2 = 2, \sigma_2 = 1$. The correlation between X_1 and X_2 is ρ. Define $Y_1 = a_1 X_1 + a_2 X_2$, $Y_2 = b_1 X_1 + b_2 X_2$.

(a) What values for a_1, a_2 make $\mu_{Y_1} = 0$?

(b) What values for a_1, a_2, b_1, b_2 make $\text{Cov}[Y_1, Y_2] = 0$?

15. If V is a χ^2 random variable with v degrees of freedom, evaluate the mean and variance for V.

5.5 The Chebyshev Inequality and the Law of Large Numbers

We have used a basically "frequentist" approach to probability in all our discussions. By frequentist approach we mean that the probability of an event A, say, should "match" or be equal to the long-run relative frequency of occurrence of A, over a very large number of independent repetitions of the experiment. One way of justifying this type of interpretation is given by the law of large numbers, which we will discuss in this section. Before doing so, however, let us first establish the Chebyshev inequality, derived by the Russian mathematician for whom it is named, in 1867. Assume X is a random variable and ε is any positive constant. Define the random variable Y_ε by

$$Y_\varepsilon = 1 \quad \text{if} \quad |X| \geq \varepsilon$$
$$\quad\, = 0 \quad \text{if} \quad |X| < \varepsilon.$$

Thus

$$E[Y_\varepsilon] = 1 \cdot P(|X| \geq \varepsilon) + 0 \cdot P(|X| < \varepsilon) = P(|X| \geq \varepsilon)$$

and it certainly must be true that

$$X^2 \geq X^2 Y_\varepsilon \geq \varepsilon^2 Y_\varepsilon.$$

From this it also follows that

$$E[X^2] \geq E[\varepsilon^2 Y_\varepsilon] = \varepsilon^2 E[Y_\varepsilon] = \varepsilon^2 P(|X| \geq \varepsilon)$$

and thus

$$P(|X| \geq \varepsilon) \leq \frac{1}{\varepsilon^2} E[X^2].$$

This gives the *Chebyshev inequality*, which we state as a theorem.

Theorem 5.5.1. Assume X is any random variable such that $E[X^2]$ exists. Then

$$P(|X| \geq \varepsilon) \leq \frac{1}{\varepsilon^2} E[X^2]$$

where ε is any positive constant.   ∎

$$\Rightarrow \quad P(|X - \mu| \geq \varepsilon) \leq \frac{\sigma^2}{\varepsilon^2}$$

It is rather remarkable that this simple inequality must hold for any random variable X, either discrete or continuous. If X is uniform on $(-1, 1)$, then

$$f_X(x) = \tfrac{1}{2}, \qquad -1 < x < 1$$

$$E[X^2] = \int_{-1}^{1} \frac{x^2}{2}\, dx = \tfrac{1}{3}$$

$$P(|X| \geq \varepsilon) = 0, \qquad \varepsilon \geq 1$$
$$= 1 - \varepsilon, \qquad \varepsilon < 1,$$

and, as claimed,

$$1 - \varepsilon \leq \frac{1}{3\varepsilon^2};$$

both $P(|X| \geq \varepsilon)$ and $E[X^2]/\varepsilon^2$ are pictured in Figure 5.5.1. Similarly, if X is standard normal, then

$$E[X^2] = 1,$$
$$P(|X| \geq \varepsilon) = 1 - P(-\varepsilon < X < \varepsilon)$$
$$= 1 - (2N(\varepsilon) - 1)$$
$$= 2(1 - N(\varepsilon));$$

both $1/\varepsilon^2$ and $2[1 - N(\varepsilon)]$ are pictured in Figure 5.5.2. This inequality cannot be improved on, as shown in the following example.

EXAMPLE 5.5.1. Let $\varepsilon > 0$ be any fixed constant and let X be a discrete random variable with probability function

$$p_X(x) = \tfrac{1}{2}, \qquad x = -\varepsilon, \varepsilon.$$

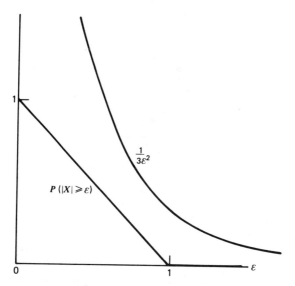

Figure 5.5.1

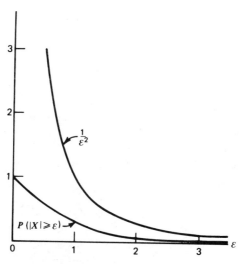

Figure 5.5.2

Then
$$E[X^2] = \tfrac{1}{2}(-\varepsilon)^2 + \tfrac{1}{2}(\varepsilon)^2 = \varepsilon^2$$

so
$$\frac{1}{\varepsilon^2} E[X^2] = \frac{\varepsilon^2}{\varepsilon^2} = 1.$$

Furthermore, for this discrete probability law
$$P(|X| \geq \varepsilon) = P(|X| = \varepsilon) = 1 = \frac{E[X^2]}{\varepsilon^2}$$

so the Chebyshev bound is in fact attained and could not be made any smaller. That is, for any $\varepsilon > 0$ we can define a discrete probability law as previously mentioned for which the bound is actually attained; it follows that the bound cannot be made smaller for any ε. ∎

We have been using the standard deviation as a measure of how "compact" a probability law is around its mean value μ. One justification for this usage can be seen from the Chebyshev inequality. Let Y be a random variable with mean μ and variance σ^2 and define $X = (Y - \mu)/\sigma$. Then

$$E[X^2] = \frac{1}{\sigma^2} E[(Y - \mu)^2] = 1,$$

and the Chebyshev inequality states that

$$P\left(\left|\frac{Y - \mu}{\sigma}\right| \geq \varepsilon\right) \leq \frac{1}{\varepsilon^2};$$

that is,

$$P(|Y - \mu| \geq \varepsilon\sigma) \leq \frac{1}{\varepsilon^2}$$

for any random variable Y whose variance σ^2 exists. Thus, for example, the probability that the random variable Y differs from μ by at least 2σ ($\varepsilon = 2$) can be no larger than $\tfrac{1}{4}$, the probability that it differs from μ by at least 3σ ($\varepsilon = 3$) can be no larger than $\tfrac{1}{9}$, and so on. The standard deviation σ provides a natural scale for the Chebyshev inequality in bounding the amount of probability in intervals centered at μ, for any probability law with finite variance.

This form of the Chebyshev inequality is useful in establishing the (weak) law of large numbers. Before discussing this result, let us briefly mention a line of reasoning that is quite intuitive for most people, and which we relied on earlier in trying to explain certain concepts. Suppose an experiment can be performed (and reperformed) as frequently as we want. Assume these performances are independent and that the same probability measure describes each performance. Let A be any desired event that may or may not occur on each performance of the experiment. The probability of A occurring is $P(A)$, the same for each repetition. For any fixed number n of repetitions of the experiment, let $n(A)$ be the number of times that event A occurs; thus $n(A)/n$ is the proportion of repetitions of the experiment for which event A occurred. The intuitive result referred to regards the value of the proportion, $n(A)/n$, as n, the number of repetitions of the experiment, increases without limit. It seems reasonable that this proportion should in fact converge to $P(A)$, the probability that A occurs for each performance of the experiment. One way of trying to show that this convergence should occur is given by the law of large numbers, as follows.

Let $X_1, X_2, \ldots, X_n, \ldots$ be a sequence of random variables that are uncorrelated and all of which have the same mean μ and the same variance σ^2. If we define

$$\overline{X}_n = \frac{1}{n} \sum_{j=1}^{n} X_j$$

to be the average of the first n of these random variables, $n = 1, 2, 3, \ldots$, we know

$$E[\overline{X}_n] = \mu, \qquad \text{Var}[\overline{X}_n] = \frac{\sigma^2}{n}$$

from Example 5.4.3. From the Chebyshev inequality

$$P(|\overline{X}_n - \mu| \geq \varepsilon \sigma_{\overline{X}_n}) = P\left(|\overline{X}_n - \mu| \geq \varepsilon \frac{\sigma}{\sqrt{n}}\right) \leq \frac{1}{\varepsilon^2}$$

for any $\varepsilon > 0$. Let us choose $\varepsilon = (\delta \sqrt{n})/\sigma > 0$ and we have

$$P(|\overline{X}_n - \mu| \geq \delta) \leq \frac{\sigma^2}{\delta^2 n}.$$

Then as n increases without bound

$$P(|\overline{X}_n - \mu| \geq \delta) \leq \frac{\sigma^2}{\delta^2 n} \to 0$$

for any $\delta > 0$; that is, the probability that $|\overline{X}_n - \mu|$ is δ or larger converges to 0, no matter how small δ is. This result is called the *law of large numbers*, which we state as the following theorem.

Theorem 5.5.2.   Let $X_1, X_2, \ldots, X_n, \ldots$ be a sequence of uncorrelated random variables, each with mean μ and variance σ^2. Then if

$$\overline{X}_n = \frac{1}{n} \sum_{j=1}^{n} X_j,$$

$$\lim_{n \to \infty} P(|\overline{X}_n - \mu| \geq \delta) = 0$$

for any $\delta > 0$.   ∎

It is important to realize that this result describes the limiting value for a sequence of probabilities, as distinguished from the limiting value for $\overline{X}_n$ (if such exists). That is, if we were to compute the difference $|\overline{X}_n - \mu|$ for successively larger values for n, the *probability* that this value exceeds any fixed δ gets smaller and smaller (and converges to 0). This does not necessarily imply that for any given realized sequence $\overline{X}_1, \overline{X}_2, \overline{X}_3, \ldots$ the actual observed difference $|\overline{X}_n - \mu|$ must converge to 0. A stronger notion of convergence is needed for this to be true, which is beyond the level of this text.

EXAMPLE 5.5.2.   Suppose, as discussed earlier, an independent sequence of experiments is performed; on each performance either event A does or does not occur, so we can look at the sequence of experiments as being independent Bernoulli trials with probability of success $p = P(A)$. Now let $X_1, X_2, \ldots, X_n, \ldots$ be the corresponding independent Bernoulli random variables, $X_j = 1$ if A occurs on experiment j and $X_j = 0$, otherwise, $j = 1, 2, 3, \ldots$. $(X_1, X_2, \ldots, X_n$ then are independent, which implies they are uncorrelated.) Then for any fixed n, $\sum_{j=1}^{n} X_j$ gives the number of successes observed and is binomial with parameters n and p;

$$\overline{X}_n = \frac{1}{n} \sum_{j=1}^{n} X_j$$

gives the proportion of successes so

$$E[\overline{X}_n] = \frac{1}{n} E\left[\sum_{j=1}^{n} X_j \right] = \frac{1}{n}(np) = p$$

and

$$\operatorname{Var}[\overline{X}_n] = \frac{1}{n^2} \operatorname{Var}\left[\sum_{j=1}^{n} X_j \right] = \frac{1}{n^2}(np(1-p)) = \frac{p(1-p)}{n}.$$

From the law of large numbers

$$\lim_{n \to \infty} P(|\overline{X}_n - p| \geq \delta) = 0$$

for any $\delta > 0$. This says the probability that the difference between the observed proportion of successes $(\overline{X}_n)$ and the assumed probability of a success (p) exceeds δ converges to 0 as n increases. This suggests that if p is unknown and we know $\bar{x}_n$, the observed proportion of successes in n trials, it is likely that $\bar{x}_n$ does not differ much from p; it is less likely to differ by any fixed amount, the larger that n becomes. In some senses this says that $\bar{x}_n$ is a good guess for the unknown value for p, which seems quite rational. We will see other criteria for deriving estimates, or guesses, for the unknown value of p, based on the observed values for random variables, when we study problems of statistical inference. ∎

EXAMPLE 5.5.3. The line of reasoning used in Example 5.5.2 extends easily to using observed random variables to estimate the unknown value for other parameters beside $p = P(\text{success})$. Suppose we assume the time to failure T, for any silicon diode made in a specified way is an exponential random variable with parameter λ (unknown). Also, assume n of these diodes are independently placed on test until they fail; we will observe $T_1, T_2, \ldots, T_n$, the times to failure for the n diodes. Then $E[T_i] = 1/\lambda$, $i = 1, 2, \ldots, n$ and $T_1, T_2, \ldots, T_n$ are uncorrelated because they are independent; with

$$\overline{T}_n = \frac{1}{n} \sum_{i=1}^{n} T_i$$

$$P\left(\left| \overline{T}_n - \frac{1}{\lambda} \right| \geq \delta \right) \to 0$$

as $n \to \infty$, which suggests that for n sufficiently large the observed value $\bar{t}_n$ should be a good guess or estimate of $1/\lambda$ and $1/\bar{t}_n$ should then provide a reasonable estimate for the unknown value of λ. Again, we will see other criteria for generating estimates of unknown parameters as we proceed. ∎

The Chebyshev inequality can be used to get a crude bound on the number of experiments required so that the observed proportion of the time an event A occurs is within a known amount of $P(A)$, with at least probability γ. This type of reasoning is discussed in the following example.

EXAMPLE 5.5.4. Suppose an experiment can be repeated independently any desired number, n, of times. On each repetition event A either does or does not occur; the probability A occurs is $p = P(A)$ for each repetition. As in Example 5.5.2, let $X_j = 1$ if A occurs on repetition j and $X_j = 0$ if not; thus $X_1, X_2, \ldots, X_n$ are independent, Bernoulli random variables, each with parameter $p = P(A)$. The observed relative frequency of event A over the n trials then is

$$\overline{X}_n = \frac{1}{n} \sum_{j=1}^{n} X_j,$$

and as seen earlier,

$$E[\overline{X}_n] = p, \qquad \mathrm{Var}[\overline{X}_n] = \sigma_{\overline{X}_n}^2 = \frac{p(1-p)}{n}$$

for any fixed n. From the Chebyshev inequality

$$P(|\overline{X}_n - p| \geq \varepsilon\sigma_{\overline{X}n}) \leq \frac{1}{\varepsilon^2}$$

that is,

$$P(|\overline{X}_n - p| \geq \varepsilon\sqrt{p(1-p)/n}) \leq \frac{1}{\varepsilon^2}$$

and thus

$$P(|\overline{X}_n - p| < \varepsilon\sqrt{p(1-p)/n}) > 1 - \frac{1}{\varepsilon^2}$$

for any $\varepsilon > 0$. The value for p must lie between 0 and 1, and clearly, $p(1-p) \leq \frac{1}{4}$ for all $0 \leq p \leq 1$. Thus

$$\sqrt{\frac{p(1-p)}{n}} \leq \sqrt{\frac{1}{4n}} = \frac{1}{2\sqrt{n}}$$

and it must be true that

$$P\left(|\overline{X}_n - p| < \frac{\varepsilon}{2\sqrt{n}}\right) > 1 - \frac{1}{\varepsilon^2}$$

for any $\varepsilon > 0$. Now suppose we want the observed proportion, $\overline{X}_n$, to differ from the true probability p by no more than δ, say; thus set $\delta = \varepsilon/2\sqrt{n}$, giving $\varepsilon = 2\delta\sqrt{n}$, and we have

$$P(|\overline{X}_n - p| < \delta) > 1 - \frac{1}{4n\delta^2}.$$

If we want this probability to equal at least γ, we must then set $\gamma = 1 - 1/4n\delta^2$ from which we can solve for n, giving

$$n = \frac{1}{4\delta^2(1-\gamma)}.$$

Thus if we perform the experiment at least $n = 1/4\delta^2(1 - \gamma)$ times, the probability is at least γ that the observed proportion $\overline{X}_n$ differs from the true probability p by no more than δ. Not surprisingly, the smaller δ gets, or the larger γ gets, the larger n must be. For example, suppose a new process is available for doping silicon chips, used in electronic devices. The probability is p (unknown) that each chip produced in this way is defective, and the chips are defective or not independently. How many chips, n, must we produce and test so that the proportion of defectives found ($\overline{X}_n$) does not differ from p by more than .01, with probability at least .9? That is, we want n such that

$$P(|\overline{X}_n - p| < .01) > .9,$$

so with the preceding notation, $\delta = .01$, $\gamma = .9$, and the number required is

$$n = \frac{1}{4\delta^2(1 - \gamma)} = \frac{1}{4(.01)^2(.1)} = 25{,}000.$$

Assume we do in fact produce $n = 25{,}000$ chips, inspect them all, and find 500 of them are defective; thus the observed proportion of defectives is $\frac{500}{25{,}000} = .02$. From the Chebyshev inequality the probability is at least .9 that this number does not differ from p by more than .01. In a sense then we can be at least 90 percent sure the unknown value for p must lie between .01 and .03, because these are the numbers within distance .01 of the observed proportion. Actually, this probability is a great deal greater than .9, as we will see in Section 5.6. There are two reasons that the exact probability may be a lot bigger than .9: (1) Our reasoning came from the Chebyshev inequality that itself just gives a bound on the probability. (2) We derived the value for n by replacing $p(1 - p)$ by $\frac{1}{4}$, the largest possible value this product might have. If in fact $p = .03$, say, then $p(1 - p) = .0291$, which is much smaller than .25, the value used. After studying the central limit theorem, we will see a much more accurate way of solving for n, the sample size necessary to give the desired probability statement. ■

EXERCISE 5.5

1. Assume X is uniform on $(0, 1)$. Plot $P(|X| \geq \varepsilon)$ and $(1/\varepsilon^2) E[X^2]$ as functions of ε.
2. Let X be a Poisson with parameter μ. For $\varepsilon < 1$ plot $P(|X| \geq \varepsilon)$ and $E[X^2]/\varepsilon^2$.
3. Plot $P(|X| \geq \varepsilon)$ versus $E[X^2]/\varepsilon^2$, for $\varepsilon < 1$, granted X is binomial, n, p.

4. Assume Y is an exponential random variable with parameter λ. Plot $P(|Y| \geq \varepsilon)$ versus $E[Y^2]/\varepsilon^2$.

5. In Example 5.5.4, how large a value for n is required if we want $\bar{X}_n$ to differ from p by no more than .05 with probability .95? with probability .90?

6. Take the problem as stated in Example 5.5.4, $\delta = .01$, $\gamma = .9$, but use $p(1-p) = .0291$ instead of $p(1-p) = \frac{1}{4}$. What is the required value for n?

7. Vegetable seeds germinate or not independently, each with probability p, under the same conditions of light, moisture, nutrients, and so on. How many seeds must be planted so that $\bar{X}_n$, the observed proportion to germinate, will differ from p by no more than .1 with probability .9? $\left(Hint.\ \text{Use } p(1-p) = \frac{1}{4}.\right)$

*8. For the case in Exercise 5.5.7, what is the required value for n if we assume the true value for p is at least .9?

9. It was shown in Example 5.4.3 that if $X_1, X_2, \ldots, X_n$ are uncorrelated, each with mean μ and variance σ^2, then $\bar{X}_n = (1/n)\sum X_i$ has mean μ and variance $\sigma^2_{\bar{X}_n} = \sigma^2/n$. If $\sigma^2 = 1$, how large must n be so that

$$P(|\bar{X}_n - \mu| < 1) > .9?$$

10. Sylvania's 40-watt light bulbs will burn a random time X before failing. Assume X is a random variable with mean μ and standard deviation $\sigma = 100$ hours. If n of these bulbs are placed on test until they burn out, resulting in observed values for $X_1, X_2, \ldots, X_n$, how large should n be so that the probability the average of these observed times differs from μ by less than 50 hours is at least .95?

5.6 The Central Limit Theorem and Probability Law Approximations

We saw one particular way in which the normal probability law occurs in Section 4.6. The situation we used there for deriving the normal density actually is a particular example of what is called the *central limit theorem*, concerned with the probability law of sums of random variables as n, the number of terms in the sum, increases without bound. In Section 4.6 we were concerned with a sum of Bernoulli variables and saw that its probability law was described, in the limit, by a density function—the normal density function. We are now going to generalize this reasoning to consider sums of random variables whose individual probability laws are not restricted to being Bernoulli.

Assume $X_1, X_2, X_3, \ldots, X_n \ldots$ is a sequence of independent, identically distributed random variables, each of which has mean μ, variance σ^2, and moment generating function $m(t)$. Now let us define the sequence of standardized random variables

$$Y_1 = \frac{X_1 - \mu}{\sigma}, \qquad Y_2 = \frac{X_2 - \mu}{\sigma}, \qquad \cdots, \qquad Y_n = \frac{X_n - \mu}{\sigma}, \qquad \cdots ;$$

each Y_i has mean 0, variance 1, and moment generating function

$$m_Y(t) = E[e^{tY}] = E[e^{t(X - \mu)/\sigma}]$$

$$= 1 + \frac{t^2}{2!} + \frac{t^3}{3!} m_3 + \frac{t^4}{4!} m_4 + \cdots,$$

where $m_3, m_4, \ldots$ are the third, fourth, $\ldots$ moments for Y. The linear term is multiplied by zero because $\mu_Y = 0$ and the quadratic term is just $t^2/2!$, since $\sigma_Y^2 = 1$. Furthermore, $Y_1, Y_2, Y_3, \ldots, Y_n, \ldots$ are independent because the original X_j's are independent, from Theorem 5.4.4.

Now let us define a sequence of partial sums by $S_1 = Y_1, S_2 = Y_1 + Y_2, S_3 = Y_1 + Y_2 + Y_3, \ldots, S_n = Y_1 + Y_2 + \cdots + Y_n, \ldots$. The mean of each S_j is 0 and the variance of S_j is j, increasing with j; this means that the probability law for each S_j is "centered" at 0 and the probability laws are more spread out as j increases. In fact, $\lim_{j \to \infty} \text{Var}[S_j]$ diverges to ∞ so in the limit the variance of these partial sums blows up. To correct for this, define the new standardized sequence $Z_1 = S_1, Z_2 = S_2/\sqrt{2}, \ldots, Z_n = S_n/\sqrt{n}, \ldots$. We now have a sequence of random variables $Z_1, Z_2, \ldots, Z_n, \ldots$, whose probability laws are stable in the sense that $E[Z_j] = 0$, $\text{Var}[Z_j] = 1$ for all j. What happens to the probability law for Z_j as $j \to \infty$? We will now show that

$$\lim_{n \to \infty} m_{Z_n}(t) \to e^{t^2/2}$$

the moment generating function for the standard normal probability law, which shows that the probability law for Z_n converges to the standard normal.

For any fixed n

$$m_{Z_n}(t) = E[e^{tZ_n}]$$

$$= E\left[\exp\left(t \sum_{i=1}^{n} Y_i/\sqrt{n} \right) \right]$$

$$= \prod_{i=1}^{n} E[\exp(t Y_i/\sqrt{n})]$$

$$= \prod_{i=1}^{n} m_{Y_i}\left(\frac{t}{\sqrt{n}}\right)$$

$$= \left[m_Y\left(\frac{t}{\sqrt{n}}\right)\right]^n.$$

As already shown,

$$m_Y\left(\frac{t}{\sqrt{n}}\right) = 1 + \frac{t^2}{2n} + \frac{t^3 m_3}{3!n^{3/2}} + \cdots,$$

and thus

$$m_{Z_n}(t) = \left[1 + \frac{t^2}{2n} + \frac{t^3 m_3}{3!n^{3/2}} + \cdots\right]^n.$$

A result from calculus, called Euler's limit, states that if $\lim_{n\to\infty} w_n = w$, then

$$\lim_{n\to\infty}\left(1 + \frac{w_n}{n}\right)^n = e^w.$$

With

$$w_n = \frac{t^2}{2} + \frac{t^3 m_3}{3!n^{1/2}} + \cdots,$$

we have

$$\lim_{n\to\infty} w_n = \frac{t^2}{2}$$

and thus

$$\lim_{n\to\infty} m_{Z_n}(t) = e^{t^2/2}.$$

This establishes a special version of the *central limit theorem*.

Theorem 5.6.1. Let $X_1, X_2, \ldots, X_n, \ldots$ be a sequence of independent identically distributed random variables, each with mean μ, variance σ^2, and moment generating function $m(t)$. If a new sequence of standardized partial sums is defined by

$$Z_n = \frac{\sum_{i=1}^{n} X_i - n\mu}{\sigma\sqrt{n}}, \qquad n = 1, 2, 3, \ldots$$

then the probability law for Z_n converges to the standard normal as $n \to \infty$.

■

The central limit theorem is frequently relied on to justify the assumption of a normal probability law for any random variable whose value can be thought of as the accumulation of a large number of independent quantities; the survey error discussion of Section 4.6 is of this kind. It is also frequently used to approximate exact probability laws for sums of independent random variables. This type of approximation rests on the following reasoning. Suppose S_n is the sum of n independent identically distributed random variables, each with mean μ and variance σ^2. Then Theorem 5.6.1 says

$$\lim_{n \to \infty} P\left(\frac{S_n - n\mu}{\sigma \sqrt{n}} \le z\right) = N(z),$$

the standard normal distribution function at z. If n is finite, but large, we would expect

$$P\left(\frac{S_n - n\mu}{\sigma \sqrt{n}} \le z\right)$$

to be well approximated by $N(z)$. An equivalent statement is that the sum

$$S_n = \sum_{i=1}^{n} X_i$$

is approximately normal with mean $n\mu$ and variance $n\sigma^2$, so that

$$F_{S_n}(s) = P(S_n \le s)$$

$$= P\left(\frac{S_n - n\mu}{\sigma \sqrt{n}} \le \frac{s - n\mu}{\sigma \sqrt{n}}\right)$$

$$\doteq N\left(\frac{s - n\mu}{\sigma \sqrt{n}}\right).$$

EXAMPLE 5.6.1. As mentioned earlier, computer simulations of various phenomena frequently call for the use of observed values for random variables, in particular, normal observed values are often required. One of the earliest ways in which normal observed values were generated from uniform random variables is based on the central limit theorem. Let $X_1, X_2, ..., X_{12}$ be 12 independent uniform random variables on $(0, 1)$ (these are easily generated by the computer). Recall then that

$$\mu = E[X_i] = \frac{0 + 1}{2} = \frac{1}{2}, \qquad \sigma^2 = \text{Var}[X_i] = \frac{(1 - 0)^2}{12} = \frac{1}{12}$$

and thus

$$S_{12} = \sum_{i=1}^{12} X_i$$

has mean $\frac{1}{2}(12) = 6$ and variance $12(\frac{1}{12}) = 1$. Assuming $n = 12$ is "large," Theorem 5.6.1 says that $S_{12} - 6$ is approximately standard normal and, for example,

$$P(S_{12} \leq 7) = P(S_{12} - 6 \leq 1)$$
$$\doteq N(1) = .8413$$
$$P(4 \leq S_{12} \leq 8) = P(-2 \leq S_{12} - 6 \leq 2)$$
$$\doteq N(2) - N(-2) = .9546.$$

The specific value $n = 12$ was used mainly because no division was then required to get a single observed normal value; the division operation was much more time consuming than either addition or generation of a single uniform $(0, 1)$ value. The normal approximation to the exact distribution for S_{12} is quite good, especially in the middle. As long as one does not require good approximations to the "tail" probabilities, the normal assumption for S_{12} is quite tenable. ∎

EXAMPLE 5.6.2.   It is frequently assumed the time to failure for various electronic components is an exponential random variable X with parameter λ, so $\mu = \sigma = 1/\lambda$. Assume a short-wave radio receiver is of modular construction; when a module fails, it can easily (and quickly) be removed and replaced. Suppose one of these receivers has a single critical module; it is the only module that can fail, and if it does fail, the receiver will not operate. Suppose further this receiver is located at a remote arctic site and is on continuously; the station where it is located has 19 spare modules (in addition to the one in use). Assume the time to failure for module i is X_i, an exponential random variable with $\lambda = .002$ (per hour), $i = 1, 2, ..., 20$; the X_i's are independent, and once a given module fails, it is replaced instantly with a new module. The total time that the receiver can be kept working then is $T = \sum_{i=1}^{20} X_i$, the sum of the 20 times to failure for the modules on hand at the station, without bringing in extra spares. From the central limit theorem T is approximately normal with

$$\mu_T = \frac{20}{.002} = 10,000 \text{ hours},$$

$$\sigma_T = \sqrt{20/(.002)^2} = 2236 \text{ hours}.$$

The probability that the receiver can be kept working for at least one year ($24 \times 365 = 8760$ hours) can be approximated by the normal values; that is,

$$P(T > 8760) = 1 - P(T \leq 8760)$$

$$\doteq 1 - N\left(\frac{8760 - 10,000}{2236}\right)$$

$$= 1 - N(-.55) = .71.$$

Thus if this remote station is resupplied only once a year, there is roughly a 71 percent chance that its short-wave receiver will still be operating when supplies next arrive. If the station had 20 spares (plus the one module in operation), the total time it could remain in operation is $T = \sum_{i=1}^{21} X_i$, with

$$\mu_T = 10,500, \qquad \sigma_T = \sqrt{21/(.002)^2} = 2291.$$

You can verify that the approximate probability it then could operate at least a year is increased to .78. ∎

Table 2 in the Appendix presents selected values of the distribution function for a χ^2 random variable with degrees of freedom $v = 1, 2, \ldots, 30$. The columns correspond to different values for the distribution function and the different rows are for the various degrees of freedom. Thus, for example, if V has the χ^2 distribution with $v = 5$ degrees of freedom,

$$P(V \leq 1.15) = .05, \qquad P(V \leq 11.1) = .95,$$

and if V has $v = 20$ degrees of freedom,

$$P(V \leq 12.4) = .1, \qquad P(V \leq 34.2) = .975.$$

Thus Table 2 allows us to evaluate selected values for the distribution functions for χ^2 random variables with degrees of freedom $v = 1, 2, \ldots, 30$. How might we evaluate the distribution function for a χ^2 random variable whose degrees of freedom exceeds 30? Recalling that if V has the χ^2 distribution with v degrees of freedom, then V can be thought of as the sum of squares of v independent standard normal random variables, we can see that the distribution function for V might be approximated by the standard normal (since it is the sum of v independent random variables). The following example discusses the approximation of the χ^2 distribution by the normal.

EXAMPLE 5.6.3. If V is a χ^2 random variable with v degrees of freedom, the moment generating function for V is

$$m(t) = \frac{1}{(1 - 2t)^{v/2}},$$

from which it is straightforward to verify that the mean and variance for V are $\mu = v$, $\sigma^2 = 2v$. Because the sum of v independent random variables, each the square of a standard normal random variable, has this probability law, we can use the central limit theorem to approximate the exact probability law for V for large values for v (certainly, for $v \geq 31$). The straightforward use of the central limit theorem says, for any $v > 0$,

$$P(V \leq v) \doteq N\left(\frac{v - v}{\sqrt{2v}}\right),$$

and thus if we wanted the kth quantile v_k, say, the value such that

$$P(V \leq v_k) = k, \qquad 0 < k < 1$$

the preceding equation gives

$$P(V \leq v_k) \doteq N\left(\frac{v_k - v}{\sqrt{2v}}\right) = k,$$

which in turn says

$$\frac{v_k - v}{\sqrt{2v}} = z_k,$$

where z_k is the kth quantile for the standard normal distribution; thus this approach gives $v_k = v + z_k\sqrt{2v}$. For example, with $v = 35$ degrees of freedom, this approximation gives

$$v_{.95} \doteq 35 + 1.64\sqrt{2(35)} = 48.72$$
$$v_{.7} \doteq 35 + .52\sqrt{2(35)} = 39.35,$$

and

$$v_{.1} \doteq 35 - 1.28\sqrt{2(35)} = 24.29. \qquad \blacksquare$$

More than 50 years ago Sir R. A. Fisher, the originator of many statistical techniques that we will study, showed that a slight adaptation to this approach gives a more accurate quantile approximation for any value for v. The central limit theorem gives a good approximation when v, the number of random variables summed together, is large. The preceding approach gave the equation

$$\frac{v_k - v}{\sqrt{2v}} = z_k.$$

For v large it is obvious that $\sqrt{2v} \doteq \sqrt{2v - 1}$ (which we will use subsequently) and

$$z_k \doteq z_k + \frac{z_k^2 - 1}{2\sqrt{2v}},$$

especially when we remember that $|z_k| \leq 3$, for any reasonable k. Thus instead of our previous solution, we would have

$$\frac{v_k - v}{\sqrt{2v}} = z_k + \frac{z_k^2 - 1}{2\sqrt{2v}}$$

which gives

$$v_k = v + z_k \sqrt{2v} + \frac{z_k^2 - 1}{2}$$

$$= \frac{z_k^2 + 2z_k \sqrt{2v} + 2v - 1}{2}$$

$$\doteq \frac{z_k^2 + 2z_k \sqrt{2v - 1} + 2v - 1}{2}$$

$$= \frac{(z_k + \sqrt{2v - 1})^2}{2}.$$

As already mentioned, Fisher showed that this approximation is in fact more accurate than the straightforward central limit theorem approximation we previously used. Thus for the same cases already discussed, with $v = 35$, we have

$$v_{.95} = \frac{(1.64 + \sqrt{69})^2}{2} = 49.47$$

$$v_{.7} = \frac{(.52 + \sqrt{69})^2}{2} = 38.95$$

$$v_{.1} = \frac{(-1.28 + \sqrt{69})^2}{2} = 24.69,$$

which Fisher showed to be closer to the exact values. Whenever we have need for χ^2 quantiles for $v > 30$ degrees of freedom, we will use this approximate formula; for $v \leq 30$ we have the values given in Table 2.

We have seen the use of the central limit theorem in approximating the probability law for sums of continuous random variables. And actually, our derivation of the normal density function in Section 4.6 could be used to derive a continuous normal approximation to a discrete probability law (sum of independent Bernoulli random variables). We will now examine in a little more detail the use of the central limit theorem in approximating discrete probability laws.

We have seen that if $X_1, X_2, \ldots, X_n$ are independent Bernoulli random variables, each with parameter p, then $Y = \sum_{i=1}^{n} X_i$ is binomial with parameters n and p. Furthermore, the expected value for each X_i is $\mu = p$, and the variance for each X_i is $\sigma^2 = p(1 - p)$; as we know from both Section 5.4 and our study of the binomial probability law, the mean of $Y = \sum X_i$ then is $n\mu = np$ and the variance for Y is $n\sigma^2 = np(1 - p)$. From the central limit theorem we can also see that for large n the probability law for $Y = \sum_{i=1}^{n} X_i$ should be well approximated by

$$P(Y \leq y) \doteq N\left(\frac{y - np}{\sqrt{np(1 - p)}}\right),$$

which corresponds more or less directly to the derivation of the normal probability law we discussed in Section 4.6. To get a better feel for what this approximation says in the discrete case, and to see a modification that generally improves the accuracy of the normal approximation for any discrete probability law, let us discuss the approximation to the binomial with $n = 15$, $p = .6$. If Y is binomial, $n = 15$, $p = .6$, the probability function for Y is

$$p_Y(y) = \binom{15}{y}(.6)^y(.4)^{15-y}, \qquad y = 0, 1, 2, \ldots, 15.$$

The values for this probability function (which are about .01 or more) are given in Table 5.6.1. (The other possible integer values have $p_Y(y) < .005$.) Figure 5.6.1 plots a histogram of this exact probablity law, as well as a superimposed normal density with mean $np = 9$ and variance $np(1 - p) = 3.6$; in drawing the histogram, we have used the convention mentioned earlier: the bar centered over any integer (observed value) is of unit width and has

Table 5.6.1

y	4	5	6	7	8	9	10	11	12	13
$p_Y(y)$	.01	.02	.06	.12	.18	.21	.19	.13	.06	.02

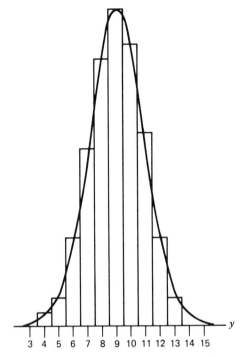

Figure 5.6.1

height equal to the value of the probability function for that value. Thus the exact value for

$$P(6 \leq Y \leq 10) = P(Y = 6) + P(Y = 7) + \cdots + P(Y = 10),$$

for example, is the sum of the areas of the bars centered at 6, 7, 8, 9 and 10, as pictured in Figure 5.6.2; we can also compute $P(6 \leq Y \leq 10) = .76$ from Table 5.6.1, which is the sum of the areas of these five bars. If we apply the central limit theorem directly, the normal approximation for this probability is

$$P(6 \leq Y \leq 10) = P(5 < Y \leq 10) = F_Y(10) - F_Y(5)$$

$$\doteq N\left(\frac{10 - 9}{\sqrt{3.6}}\right) - N\left(\frac{5 - 9}{\sqrt{3.6}}\right) = N(.53) - N(-2.11)$$

$$= .7019 - .0174 = .6845.$$

Graphically, this area is pictured in Figure 5.6.3. It is easy to see that this

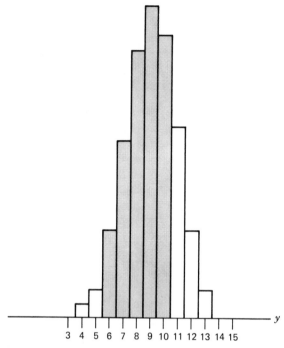

Figure 5.6.2

continuous approximation would be improved if we were to use the area under the normal curve between 5.5 and 10.5 as pictured in Figure 5.6.4, rather than between 5 and 10, because this would then cover the same interval on the y axis as the bars centered at 6, 7, 8, 9, 10 do. This adjustment of $\frac{1}{2}$ unit to make the normal interval match the discrete binomial interval is called a *continuity correction*. For our case the approximation with continuity correction gives

$$P(6 \leq Y \leq 10) = P(5 < Y \leq 10)$$

$$= F_Y(10) - F_Y(5)$$

$$\doteq N\left(\frac{10.5 - 9}{\sqrt{3.6}} \right) - N\left(\frac{5.5 - 9}{\sqrt{3.6}} \right)$$

$$= N(.79) - N(-1.84)$$

$$= .7523,$$

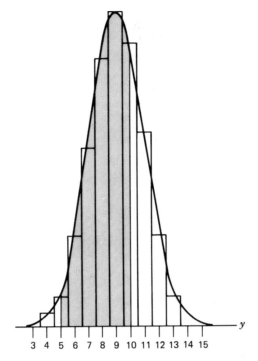

Figure 5.6.3

considerably better than the approximation without the continuity correction.

In general, if Y is binomial with parameters n and p, the normal approximation with continuity correction is

$$F_Y(b) = P(Y \leq b) \doteq N\left(\frac{b + \frac{1}{2} - np}{\sqrt{npq}}\right),$$

$$P(Y \geq a) = 1 - P(Y \leq a - 1)$$

$$\doteq 1 - N\left(\frac{a - \frac{1}{2} - np}{\sqrt{npq}}\right),$$

$$P(a \leq Y \leq b) = F_Y(b) - F_Y(a - 1)$$

$$= N\left(\frac{b + \frac{1}{2} - np}{\sqrt{npq}}\right) - N\left(\frac{a - \frac{1}{2} - np}{\sqrt{npq}}\right).$$

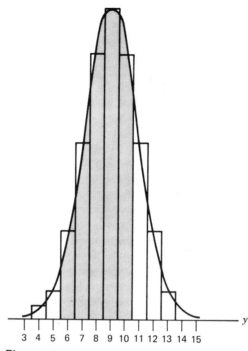

Figure 5.6.4

You can see that the correction has a negligible effect for n very large, since $\frac{1}{2}/\sqrt{npq} \to 0$ as $n \to \infty$. For moderate n, though, the continuity correction gives a noticeably better approximation. The continuity correction is also useful in approximating exact probabilities for other discrete laws as well. The rationale is the same and, if you think of the exact discrete probability as being a sum of areas of bars in histograms, it is not difficult to see how the approximation should be applied for any discrete case.

EXAMPLE 5.6.4.   We saw in Section 5.4 that the sum of independent Poisson random variables again has a Poisson probability law. Thus if $X_1, X_2, \ldots, X_n$ are independent Poisson random variables, each with parameter $\mu = 1$, then $Y = \sum_{i=1}^{n} X_i$ has a Poisson probability law with parameter $\mu = n$; if n is "large," the central limit theorem implies we should be able to approximate $F_Y(y)$ by the normal distribution function, since Y could be regarded as the sum of independent random variables. Thus if Y is Poisson with parameter μ, which is "large" (and not necessarily integer), we approximate the exact distribution for Y by the normal probability law with mean μ and variance μ. Thus using the continuity correction, we have

$$F_Y(y) \doteq N\left(\frac{y + \frac{1}{2} - \mu}{\sqrt{\mu}}\right)$$

and again the continuity correction "disappears" as μ gets larger.

Suppose home burglaries occur in a town of 30,000 people like events in a Poisson process with parameter $\lambda = \frac{1}{2}$ per day. If Y is the number of such burglaries reported in a 30-day month, then Y is Poisson with parameter $\mu = 15$. The probability that no more than 10 burglaries will be reported in the month is

$$P(Y \le 10) = F_Y(10) \doteq N\left(\frac{10.5 - 15}{\sqrt{15}}\right)$$

$$= N(-1.16) = .1230,$$

and the probability that at least 17 burglaries will be reported is

$$P(Y \ge 17) = 1 - P(Y \le 16)$$

$$= 1 - N\left(\frac{16.5 - 15}{\sqrt{15}}\right)$$

$$= 1 - N(.39) = .3483.$$

You can easily verify with a hand-held calculator that the exact values are

$$P(Y \le 10) = e^{-15} \sum_{i=0}^{10} \frac{15^i}{i!} = .118$$

$$P(Y \ge 17) = 1 - e^{-15} \sum_{i=0}^{16} \frac{15^i}{i!} = .336,$$

so the approximations are not bad. Again, the approximation without the continuity correction is not as good, as you can verify by recomputing the preceding probabilities. ∎

In Example 5.5.4 we discussed the number of trials necessary, n, so that the probability is at least γ that the observed proportion of successes, $\bar{X}_n$, does not differ from the probability of success, p, by more than δ. The value for n that we arrived at was derived from the Chebyshev inequality; it was mentioned that a more accurate computation could be made, without relying on the Chebyshev inequality. This more accurate computation makes use of the central limit theorem and is discussed in the following example.

EXAMPLE 5.6.5.   Assume as we did earlier that n independent Bernoulli trials are to be performed, each with probability of success p. The observed proportion of successes, $\overline{X}_n = (1/n)\sum_{i=1}^{n} X_i$, then is approximately normal with $\mu = p$, $\sigma = \sqrt{p(1-p)/n}$; again, for any real value for p, $p(1-p) \leq \frac{1}{4}$ and thus $\sigma \leq 1/2\sqrt{n}$. If we use $\sigma = 1/2\sqrt{n}$, we will in fact have a conservative approach because our approximating normal density uses the largest possible value for σ, which in turn leads to the *smallest* possible (approximate) probability for the random variable taking on a value in any fixed length interval centered at $\mu = p$. To restate the sample size question discussed earlier, we want to find the smallest value for n such that

$$P(|\overline{X}_n - p| < \delta) > \gamma.$$

Anticipating that the required value for n will be "large" (so the normal approximation will be good) we find that the continuity correction will have very little effect and will not use it. Furthermore, because we are using an approximation (a continuous one), it doesn't matter whether or not end points are included, in terms of the continuous probability measure. We will find n such that

$$P(|\overline{X}_n - p| \leq \delta) = \gamma.$$

Clearly,

$$P(|\overline{X}_n - p| \leq \delta) = P(-\delta \leq (\overline{X}_n - p) \leq \delta)$$

$$= P\left(\frac{-\delta}{1/2\sqrt{n}} \leq \frac{(\overline{X}_n - p)}{1/2\sqrt{n}} \leq \frac{\delta}{1/2\sqrt{n}}\right)$$

$$\doteq N(2\delta\sqrt{n}) - N(-2\delta\sqrt{n})$$

$$= 2N(2\delta\sqrt{n}) - 1,$$

and for this probability to equal γ we need

$$N(2\delta\sqrt{n}) = \tfrac{1}{2}(1 + \gamma),$$

which is satisfied by

$$2\delta\sqrt{n} = z_{(1+\gamma)/2}.$$

The required value for n then is

$$n = \frac{z_{(1+\gamma)/2}^2}{4\delta^2}. \qquad \blacksquare$$

Thus if we want

$$P(|\overline{X}_n - p| < .01) > .9,$$

$\delta = .01$, $\gamma = .9$, $z_{.95} = 1.64$, we have

$$n = \frac{(1.64)^2}{4(.01)^2} = 6724,$$

considerably smaller than the value $n = 25,000$ we found from Chebyshev's inequality. ∎

EXERCISE 5.6

1. A pair of dice is rolled 180 times an hour (approximately) at a craps table in Las Vegas. What is the (approximate) probability that 25 or more rolls have a sum of 7 during the first hour? What is the (approximate) probability that between 700 and 750 rolls have a sum of 7 during 24 hours?

2. Define the partial sums $S_1 = Y_1$, $S_2 = Y_1 + Y_2$, $S_3 = Y_1 + Y_2 + Y_3$, ... as we did at the start of this section, where the Y_i's are independent, $\mu_Y = 0$, $\sigma_Y^2 = 1$. What is the value of $\text{Cov}[S_1, S_2]$? of $\text{Cov}[S_2, S_3]$? of $\text{Cov}[S_n, S_{n+1}]$? of $\text{Cov}[S_1, S_n]$?

3. A nursery man plants 115 cuttings of ivy in every flat he prepares. Assume the probability that an individual cutting will develop roots is .9 and approximate the probability that the average number of rooted cuttings (per flat) in 50 flats is less than 100.

4. Experience has shown that the number of accidents to occur along a particular 10-mile stretch of two-lane highway is a Poisson random variable with a mean of 2 per week. What is the (approximate) probability that there will be less than 100 accidents on this stretch of road in a year?

5. The length of a continuous nylon filament that can be drawn without a break occurring is an exponential random variable with mean 5000 feet. What is the (approximate) probability that the average length of 100 filaments lies between 4750 and 5550 feet?

6. The probability that a person survives an attack of cholera (with good medical help) is assumed to be .4. What is the probability that at least half of 100 patients with cholera will survive?

7. How many modules would have to be stocked at the arctic station mentioned in Example 5.6.2 in order that the receiver will operate for at least one year with approximate probability .9?

8. Assume a basketball player has a constant (unknown) probability p

of sinking a free-throw and that his attempts are independent. How many shots should he attempt, if the proportion of shots made is to lie within .1 of p, with probability at least .8?

9. Approximate the 70, 80, 90, and 95th quantiles for a χ^2 random variable with $v = 60$ degrees of freedom. Do the same for a χ^2 random variable with $v = 100$ degrees of freedom.

10. Compare the Fisher approximate 75, 90, 95th χ^2 quantiles, $v = 30$, with the exact value given in Table 2.

11. A big-city used car dealer is open 365 days per year; the number of sales he makes per day is a Poisson random variable with parameter $\mu = 2$, independently from one day to another. Let Y be the number of sales he makes in a year and approximate

    (a) $P(Y \geq 700)$
    (b) $P(Y \leq 800)$
    (c) $P(700 \leq Y \leq 800)$.

12. Areas under the normal density can also be used to approximate the probability a discrete random variable is equal to any value from its range, by realizing

    $$P(Y = y) = F_Y(y) - F_Y(y - 1)$$

    $$\doteq N\left(\frac{y + \frac{1}{2} - \mu}{\sigma}\right) - N\left(\frac{y - \frac{1}{2} - \mu}{\sigma}\right);$$

    we have used the continuity correction to center the normal area over the bar. Use this approach to approximate $P(Y = 729)$, $P(Y = 730)$, $P(Y = 735)$ for the situation described in Exercise 5.6.11.

*13. The normal approximation discussed in Example 5.6.5 again used the biggest possible value for $p(1 - p)$, which is $\frac{1}{4}$. If you are sure $p \leq p_0$, describe how the formula for n, derived in that example, would change.

14. The negative binomial distribution can also be well approximated by a normal distribution for large r, since it is the sum of r independent geometric random variables with parameter p. Assume 10 percent of the U.S. population have a malady and, for research purposes, it is desired to find 50 people with the malady. If people are examined at random until 50 are located, and X is the number examined then X is (at least approximately) a negative binomial random variable with parameters $r = 50$, $p = .1$. What is the (approximate) probability that at least 400 people must be examined to find 50 with the malady?

15. Assume in Exercise 5.6.14 only half of those with the malady would be willing to take part in a medical experiment.

    (a) What is the distribution of Y, the number that must be examined to find 50 who would be willing to take part in an experiment?

(b) What is the (approximate) probability that at least 800 must be examined to find 50 willing to participate in the experiment?

5.7 Some Special Distributions

In this section we will study two new probability laws, the multinomial and bivariate normal probability laws, and some special distributions that occur as the probability laws for certain functions of independent normal random variables. We will derive the multinomial probability law in much the same way we derived the distributions discussed in Chapter 4. The other distributions would require some mathematical techniques that we have not described, so we will not justify them in the same way.

In Section 4.1 we discussed Bernoulli trials, simple experiments that had two different possible results, called success or failure. The natural extension of this idea is called a *multinomial trial*, a simple experiment that can produce $k \geq 3$ different results. Examples of multinomial trials are:

1. A single roll of a single die has $k = 6$ different possible results.
2. The response of a person asked which candidate she will vote for in an election (say, there are two candidates) could result in $k = 4$ different responses:
   (a) Vote for candidate A, (b) vote for candidate B, (c) not certain which candidate she'll vote for, and (d) she will not vote.
3. A manufactured item might have any one of, say, four different types of defects or be nondefective. Inspection of such an item produces one of $k = 5$ different possible results.

A single multinomial trial has k different possible results; if k were 2 it would be a Bernoulli trial.

In the commonly used notation, k parameters are used to describe a multinomial trial; these are defined by

$$p_i = P(\text{Result } i \text{ occurs}), \qquad i = 1, 2, \ldots, k,$$

one of which is actually redundant, since we must have $p_1 + p_2 + \cdots + p_k = 1$, so $p_k = 1 - p_1 - p_2 - \cdots - p_{k-1}$. In the same notation a Bernoulli trial would have two associated parameters, $p_1 = p = P(\text{success})$ and $p_2 = 1 - p = P(\text{failure})$, which is clearly redundant. For a fair die, the $k = 6$ different faces are equally likely to occur so $p_1 = p_2 = \cdots = p_6 = \frac{1}{6}$. If in the election with two candidates we assume 60 percent of the voters will participate, 20 percent have not yet made up their minds, 25 percent will vote for A, and the remainder for B, then

$p_1 = P(\text{selected person says she'll vote for } A) = .25$
$p_2 = P(\text{selected person says she'll vote for } B) = .15$
$p_3 = P(\text{selected person says she's undecided}) = .20$
$p_4 = P(\text{selected person will note vote}) = .40.$

Now suppose an experiment consists of n repeated independent multi-nomial trials, each with parameters $p_1, p_2, \ldots, p_k$. Each outcome for the experiment (element of S) then could be represented by an n-tuple $(x_1, x_2, \ldots, x_n)$, where x_i records which result occurred on trial i; thus we could use $x_i = 1, 2, \ldots, k$ for $i = 1, 2, \ldots, n$. Because the trials are assumed independent, the probability of the single-element event $\{(x_1, x_2, \ldots, x_n)\}$ is a product of n terms, each term being one of the probabilities $p_1, p_2, \ldots, p_k$.

Suppose, for example, crossing two particular hybrid petunias leads to plants with one of four different types of flowers. (1) Red-serrated edge, (2) red-smooth edge, (3) white-serrated edge, (4) white-smooth edge; that is, each plant from this cross is a multinomial trial with four results. Also, assume the separate plants are independent and $p_1 = .12$, $p_2 = .28$, $p_3 = .18$, $p_4 = .42$ for each and three plants are produced from crossing the two hybrid petunias. The sample space then would have $4^3 = 64$ different elements and the probabilities of the single-element events are products of three terms because of the assumed independence. For example,

$$P(\{(1, 1, 1)\}) = (.12)(.12)(.12)$$
$$P(\{(1, 3, 4)\}) = (.12)(.18)(.42)$$
$$P(\{(2, 3, 2)\}) = (.28)(.18)(.28)$$

and so on. Now let Y_j = number of plants produced of type $j, j = 1, 2, 3, 4$; then necessarily $Y_1 + Y_2 + Y_3 + Y_4 = 3$, because each plant must be of one of the four types. To evaluate $P(Y_1 = 2, Y_2 = 1, Y_3 = Y_4 = 0)$, for example, we would sum the probabilities of all the possible single-element events whose 3-tuples contain two 1's and one 2. The probability of each such single-element event is $(.12)^2 (.28)$ and there are $3!/2!1!$ such single-element events (because this is the number of ordered ways we can put three symbols in a row, two of which are alike and the remaining one is different). We thus have

$$P(Y_1 = 2, Y_2 = 1, Y_3 = Y_4 = 0) = \frac{3!}{2!1!}(.12)^2 (.28).$$

More generally, if (y_1, y_2, y_3, y_4) is such that $y_j = 0, 1, 2, 3, j = 1, 2, 3, 4,$

$y_1 + y_2 + y_3 + y_4 = 3$, then

$$P(Y_1 = y_1, Y_2 = y_2, Y_3 = y_3, Y_4 = y_4) = \frac{3!}{y_1!y_2!y_3!y_4!}(.12)^{y_1}(.28)^{y_2}$$
$$\cdot(.18)^{y_3}(.42)^{y_4}.$$

(Y_1, Y_2, Y_3, Y_4) is an example of a multinomial random vector with parameters $n = 3$, $p_1 = .12$, $p_2 = .28$, $p_3 = .18$, $p_4 = .42$. For example,

$$P(Y_1 = 0, Y_2 = 2, Y_3 = 0, Y_4 = 1) = \frac{3!}{0!2!0!1!}(.12)^0(.28)^2(.18)^0(.42)^1$$

$$P(Y_1 = Y_2 = 0, Y_3 = 3, Y_4 = 0) = \frac{3!}{0!0!3!0!}(.12)^0(.28)^0(.18)^3(.42)^0.$$

If $n = 20$ plants were produced from the cross and we define (Y_1, Y_2, Y_3, Y_4) in the same way, then (Y_1, Y_2, Y_3, Y_4) is multinomial, $n = 20$, $p_1 = .12$, $p_2 = .28$, $p_3 = .18$, $p_4 = .42$, and

$$P(Y_1 = y_1, Y_2 = y_2, Y_3 = y_3, Y_4 = y_4) = \frac{20!}{y_1!y_2!y_3!y_4!}(.12)^{y_1}(.28)^{y_2}$$
$$\cdot(.18)^{y_3}(.42)^{y_4}$$

for any (y_1, y_2, y_3, y_4) such that $y_j = 0, 1, 2, ..., 20$, $j = 1, 2, 3, 4$, $y_1 + y_2 + y_3 + y_4 = 20$.

For the general case, n independent multinomial trials are performed, each with probabilities $p_1, p_2, ..., p_k$ of the different results occurring, and the multinomial random vector is $(Y_1, Y_2, ..., Y_k)$, where

Y_j = the number of times result j occurs in the n trials, $j = 1, 2, ..., k$;

the probability function for $(Y_1, Y_2, ..., Y_k)$ is

$$P(Y_1 = y_1, Y_2 = y_2, ..., Y_k = y_k) = \frac{n!}{y_1!y_2!...y_k!}p_1^{y_1}p_2^{y_2}\cdots p_k^{y_k},$$

where $n = 1, 2, 3, ...$, $0 < p_j < 1$, $\sum p_j = 1$, $y_j = 0, 1, ..., n$, $\sum y_j = n$.

The marginal probability function for any single Y_j must be binomial with parameters n and p_j, as is most easily seen by referring to the original experiment. We can simply observe on each of the n independent trials whether or not result j occurs; thus each trial is then Bernoulli with success defined as the occurrence of result j and failure, the occurrence of any other result. Thus Y_j is the number of successes in n trials and must be binomial with parameters n and p_j. We have immediately then that

$$E[Y_j] = np_j, \qquad \text{Var}[Y_j] = np_j(1 - p_j), \qquad j = 1, 2, ..., k.$$

Similarly, let

Y_1 = the number of times result 1 occurs
Y_2 = the number of times result 2 occurs
X_3 = the number of times neither result 1 nor result 2 occurs

and it is easy to see that (Y_1, Y_2, X_3) is multinomial with parameters n, p_1, p_2, and $1 - p_1 - p_2$. The multinomial probability law occurs as the joint distribution for any subset of $(Y_1, Y_2, \ldots, Y_k)$ (augmented by the random variable that counts the number of times none of the results occur).

To evaluate the covariance and correlation between two components of a multinomial random vector, we can use a conditional probability argument. Suppose n independent multinomial trials are performed and we are given that $Y_1 = y_1$, result 1 occurred on y_1 of the n trials. Then result 1 did *not* occur on the other $n - y_1$ trials; on each of these result 2, 3, ... or k must have occurred.

What are the probabilities of results 2, 3, ..., k on each of these $n - y_1$ trials, given result 1 did not occur? Let B be the event "result 2 occurs" and let A be the event "result 1 does not occur." From our basic definition of conditional probability

$$P(B|A) = \frac{P(B \cap A)}{P(A)} = \frac{P(B)}{P(A)} = \frac{p_2}{1 - p_1}$$

since $B \cap A = B$. Similarly, the conditional probabilities of results 3, 4, ..., k occurring, given result 1 did not, are $p_3/(1 - p_1), p_4/(1 - p_1), \ldots, p_k/(1 - p_1)$, for each of these $n - y_1$ trials. Thus the $n - y_1$ trials on which result 1 did not occur are independent multinomial with probabilities $p_2/(1 - p_1)$, $p_3/(1 - p_1), \ldots, p_k/(1 - p_1)$. But then the conditional distribution for Y_2, say, given $Y_1 = y_1$, is binomial with parameters $n - y_1, p_2/(1 - p_1)$ and we have

$$E[Y_2|Y_1 = y_1] = \frac{(n - y_1)p_2}{1 - p_1}.$$

Conditioned on $Y_1 = y_1$, the given value y_1 is a constant, and thus we can also write

$$E[y_1 Y_2 | Y_1 = y_1] = y_1 E[Y_2 | Y_1 = y_1]$$
$$= y_1 \frac{(n - y_1)p_2}{1 - p_1}.$$

Now recall that the expectation of the conditional expectation is the unconditional expectation. That is, we have

$$
\begin{aligned}
E[Y_1 Y_2] &= E[E[Y_1 Y_2 | Y_1]] \\
&= E[Y_1 E[Y_2 | Y_1]] \\
&= E\left[Y_1 \frac{(n - Y_1) p_2}{1 - p_1} \right] \\
&= \frac{p_2}{1 - p_1} \{E[nY_1] - E[Y_1^2]\} \\
&= \frac{p_2}{1 - p_1} \{n^2 p_1 - (np_1(1 - p_1) + n^2 p_1^2)\} \\
&= p_2 (n^2 p_1 - np_1).
\end{aligned}
$$

Thus the covariance of Y_1 and Y_2 is

$$
\begin{aligned}
\text{Cov}[Y_1, Y_2] &= p_2(n^2 p_1 - np_1) - (np_1)(np_2) \\
&= -np_1 p_2
\end{aligned}
$$

and their correlation is

$$
\rho = \frac{-np_1 p_2}{\sqrt{np_1(1 - p_1)}\sqrt{np_2(1 - p_2)}} = -\sqrt{p_1 p_2 / (1 - p_1)(1 - p_2)}.
$$

You will recall from Section 5.3 that a negative correlation (or covariance) means that as one of the random variables increases the other tends to decrease (and vice versa). That is certainly the situation for the multinomial random vector; the higher the observed value for Y_1 is, the more restricted is the possible observed value for Y_2, because it cannot be larger than $n - y_1$.

Let us summarize this discussion in the following theorem.

Theorem 5.7.1. Let $(Y_1, Y_2, \ldots, Y_k)$ be a multinomial random vector with parameters $n, p_1, p_2, \ldots, p_k$. Then
(a) The marginal probability law for Y_j is binomial, $n, p_j, j = 1, 2, \ldots, k$.
(b) The (marginal) probability function for any subset of $(Y_1, Y_2, \ldots, Y_k)$ is also multinomial.
(c) The covariance between Y_q and Y_r is

$$
\text{Cov}(Y_q, Y_r) = -np_q p_r, \qquad
\begin{array}{l}
q = 1, 2, \ldots, k \\
r = 1, 2, \ldots, k,
\end{array}
\quad q \neq r. \qquad \blacksquare
$$

EXAMPLE 5.7.1. Assume the use of a drug by any patient with a specific ailment will lead to one of four possible results: total cure, improved condition, no change, or worsened condition, with probabilities .6, .2, .15, and .05, respectively. If $n = 100$ patients with this condition are given the drug and their reactions are independent and we let Y_1, Y_2, Y_3, Y_4 be the number of the 100 who are cured, have an improved condition, remain unchanged, suffer worsened condition, respectively, then (Y_1, Y_2, Y_3, Y_4) is multinomial with parameters 100, .6, .2, .15, and .05. Thus, for example, Y_1 is binomial $n = 100$, $p = .6$, Y_3 is binomial $n = 100$, $p = .15$, and so on, and the expected numbers of the various results are 60, 20, 15, and 5. The covariance between Y_1 and Y_2 is $-100(.6)(.2) = -12$ and the covariance between Y_3 and Y_4 is $-100(.15)(.05) = -.75$. The correlation between Y_1 and Y_3 is $-\sqrt{(.6)(.15)/(.4)(.85)} = -.51$ and the correlation between Y_2 and Y_4 is $-\sqrt{(.2)(.05)/(.8)(.95)} = -.013$. ∎

We have studied the normal probability law and seen how it can occur through the central limit theorem. The normal density function is symmetric and bell shaped, as already seen. The two-dimensional extension of this probability law is called the *bivariate normal* probability law. The vector (X, Y) has a bivariate normal probability law if its density function is

$$f_{X,Y}(x, y) = \frac{1}{2\pi\sigma_X\sigma_Y\sqrt{1 - \rho^2}}e^{-a/2},$$

where

$$a = \frac{1}{1 - \rho^2}\left(\frac{(x - \mu_X)^2}{\sigma_X^2} + \frac{(y - \mu_Y)^2}{\sigma_Y^2} - 2\rho\frac{(x - \mu_X)(y - \mu_Y)}{\sigma_X\sigma_Y}\right).$$

There are five parameters in this density, $\mu_X, \mu_Y, \sigma_X > 0, \sigma_Y > 0, -1 < \rho < 1$, and as you might expect from the symbols used, μ_X and μ_Y are the expected values for X and Y, σ_X and σ_Y are the standard deviations for X and Y, ρ is the correlation between X and Y. This density function defines a surface over the (x, y) plane that has a unique maximum value at $(x, y) = (\mu_X, \mu_Y)$. The density function is constant for all (x, y) for which a is constant; these points describe an ellipse in the (x, y) plane. The bivariate density function resembles a rounded hill with its highest point at $(x, y) = (\mu_X, \mu_Y)$.

By completing the square in the quadratic function a we can write

$$a = \frac{\left(x - \mu_X - \rho\dfrac{\sigma_X}{\sigma_Y}(y - \mu_Y)\right)^2}{\sigma_X^2(1 - \rho^2)} + \frac{(y - \mu_Y)^2}{\sigma_Y^2}$$

Using this result, you can verify that

$$\int_{-\infty}^{\infty} \int_{-\infty}^{\infty} f_{X,Y}(x, y) \, dx \, dy = 1$$

so $f_{X,Y}(x, y)$ is a density function. This form for a also makes it easy to see that

$$\int_{x=-\infty}^{\infty} f_{X,Y}(x, y) \, dx = \frac{1}{\sigma_Y \sqrt{2\pi}} e^{-(y-\mu_Y)^2/2\sigma_Y^2} = f_Y(y),$$

because the integration with respect to x is that of a normal density whose mean is

$$\mu_X + \rho \frac{\sigma_X}{\sigma_Y}(y - \mu_Y)$$

and whose variance is $\sigma_X^2(1 - \rho^2)$. This just gives the constant $\sigma_X \sqrt{(1 - \rho^2) 2\pi}$ and thus the marginal density for Y, as previously noted, is normal with mean μ_Y and variance σ_Y^2. Reversing the roles of x and y in completing the square in a easily shows that

$$f_X(x) = \frac{1}{\sigma_X \sqrt{2\pi}} e^{-(x-\mu_X)^2/2\sigma_X^2}$$

so the marginal density for X is normal with mean μ_X and variance σ_X^2. This indicates that if (X, Y) is bivariate normal, each of X and Y have marginal normal distributions, not terribly surprising in light of the terminology used.

We have seen that if (X, Y) is bivariate normal with parameters μ_X, μ_Y, σ_X, σ_Y and ρ, then the marginal densities for X and Y are both normal. If the parameter $\rho = 0$, you can easily see that the joint density becomes

$$f_{X,Y}(x, y) = \frac{1}{2\pi\sigma_X\sigma_Y} e^{-((x-\mu_X)^2/2\sigma_X^2 + (y-\mu_Y)^2/2\sigma_Y^2)}$$

$$= \frac{1}{\sigma_X \sqrt{2\pi}} e^{-(x-\mu_X)^2/2\sigma_X^2} \frac{1}{\sigma_Y \sqrt{2\pi}} e^{-(y-\mu_Y)^2/2\sigma_Y^2}$$

$$= f_X(x) f_Y(y).$$

That is, X and Y then are independent random variables. Recall that this result is not true in general; it is quite possible for X and Y not to be independent, even if $\rho = 0$, for other probability laws.

The conditional density for Y, given $X = x$, is

$$f_{Y|X}(y|x) = \frac{f_{X,Y}(x, y)}{f_X(x)}$$

$$= \frac{1}{\sigma_Y\sqrt{2\pi(1 - \rho^2)}} \exp\left[-\left(y - \mu_Y - \rho\frac{\sigma_Y}{\sigma_X}(x - \mu_X)\right)^2 \div 2\sigma_Y^2(1 - \rho^2)\right]$$

as is easily seen with the exponent *a* written in the completed square form; the term $(x - \mu_X)^2/\sigma_X^2$ in the exponent cancels in forming the ratio of the two densities. But this is a normal density with mean $\mu_Y + \rho(\sigma_Y/\sigma_X)(x - \mu_X)$ and variance $\sigma_Y^2(1 - \rho^2)$. The preceding is summarized in the following theorem.

Theorem 5.7.2. If (X, Y) is a bivariate normal random vector with parameters $\mu_X, \mu_Y, \sigma_X, \sigma_Y$, and ρ, then
(a) X is normal with parameters μ_X, σ_X, and Y is normal with parameters μ_Y, σ_Y.
(b) X and Y are independent if and only if $\rho = 0$.
(c) The conditional distribution for Y, given $X = x$, is normal with mean $\mu_Y + \rho(\sigma_Y/\sigma_X)(x - \mu_X)$ and variance $\sigma_Y^2(1 - \rho^2)$. ∎

EXAMPLE 5.7.2.   Many types of measurements have been assumed to follow a bivariate normal distribution. Examples include:

1. Height X and weight Y of a healthy "typical" adult male.
2. The Cartesian coordinates (X, Y) on a flat plane of the impact point of an object (a ball, missile, reentry vehicle).
3. The score made on exam 1 (X) and the score made on exam 2 (Y) by the same individual.
4. The adjusted gross income (X) and medical deduction claimed (Y) on a randomly selected income tax form.

The bivariate normal probability law has been used in many different contexts. Let us assume an individual applying for college takes a mathematics achievement test and makes score X. This same individual takes a second mathematics achievement test four years later, after college, and scores Y. Now suppose (X, Y) has a bivariate normal probability law with $\mu_X = 100$, $\mu_Y = 150$, $\sigma_X = 20$, $\sigma_Y = 30$, $\rho = .5$ (chosen quite arbitrarily). Then the probability is .6 that his initial score exceeds $100 - z_{.6}(20) = 95.0$ and the probability is .9 that his initial score will not exceed $100 + z_{.9}(20) =$ 125.6 (because X is normal, $\mu = 100$, $\sigma = 20$, with our assumptions). Similarly, he has a 60% chance that his second score (Y) will exceed $150 - z_{.6}(30) = 142.5$ and the probability is .9 his second score will not exceed $150 + z_{.9}(30) = 188.4$ (because Y is normal, $\mu = 150$, $\sigma = 30$).

If this student scores $x = 115$ on the first test, the conditional distribution for Y is normal with

$$\mu = 150 + (.5)\frac{(30)}{(20)}(115 - 100) = 161.25,$$

$$\sigma = 30\sqrt{1 - (.5)^2} = 25.98.$$

Thus the conditional probability is .6 that his second score exceeds $161.25 - z_{.6}(25.98) = 154.76$ and the probability is .9 it will not exceed $161.25 + z_{.9}(25.98) = 194.50$. These conditional probabilities for Y are considerably different from the marginal values because $\rho = .5$. If we had assumed $\rho = 0$, the conditional and marginal probability statements would have been identical since X and Y are independent in that case.

It may strike you as strange to assume the two scores to be made by this individual, (X, Y), are a random vector. There are actually several different types of assumptions that make this type of model seem reasonable; let us briefly discuss one of these, which, although rather statistical in nature (and we have not discussed such notions yet), may make the model seem more appealing. Each year a large college receives applications from many students. Each applicant is required to be a high school graduate and to take the mathematics aptitude test (which determines the observed value for X for that student). One can conceive of each of these applicants also taking the mathematics aptitude test four years later, giving the student's Y score (at least conceptually). It is very commonly assumed a bivariate normal density for (X, Y), over all possible applicants, provides an adequate probability model for the proportion of actual observed scores that will fall in any region in the (x, y) plane. The college administrators can then use the applicant's known first score (value for x) to determine the conditional distribution for Y, the future score the applicant will make in the later test. If the college has some absolute minimum score, say, $y = 138$, which it feels every one of its graduates must attain to uphold the standards of the institution, this conditional distribution for Y allows them to evaluate $P(Y \geq 138 | X = x)$ for each applicant. The value for this conditional probability gives one possible way of deciding which applicants should, and which should not, be admitted to the college. Whether or not one really likes this type of reasoning, it is rather commonly employed in many circumstances. ■

EXAMPLE 5.7.3. The circular probable error (CEP) is a frequently used measure of accuracy for various systems that throw or fire a projectile at a point in an (x, y) plane. Suppose a dart thrower tosses a dart at a target with an (x, y) coordinate system centered in the bull's-eye (see Figure 5.1.1, for example). The actual impact point of the dart is (X, Y), assumed bivariate normal with $\mu_X = \mu_Y = 0$, $\sigma_X = \sigma_Y = \sigma$, $\rho = 0$. Then the CEP for this dart thrower is defined to be the radius of the circle, centered at $(0, 0)$, which has probability .5 of containing (X, Y), the impact point. That is, if we let $r = $ CEP, it is defined through

$$P(\sqrt{X^2 + Y^2} \leq r) = .5.$$

We can evaluate r in terms of σ by realizing that X and Y are independent (since $\rho = 0$) and in fact X/σ and Y/σ are independent standard normal with the preceding assumptions. Thus

$$\frac{X^2}{\sigma^2} + \frac{Y^2}{\sigma^2} = \frac{X^2 + Y^2}{\sigma^2}$$

is a χ^2 random variable with 2 degrees of freedom. But a χ^2 random variable with 2 degrees of freedom has the exponential density with $\lambda = \frac{1}{2}$. Thus

$$P(\sqrt{X^2 + Y^2} \leq r) = P\left(\frac{X^2 + Y^2}{\sigma^2} \leq \frac{r^2}{\sigma^2}\right)$$

$$= 1 - e^{-r^2/2\sigma^2}$$

and for this probability to equal $\frac{1}{2}$ we have

$$1 - e^{-r^2/2\sigma^2} = \tfrac{1}{2},$$

which gives

$$r = \sigma\sqrt{2\ln 2}$$

as the value for the CEP for this dart thrower. The smaller that σ becomes, the smaller will be the value for the CEP (and the more you would like this thrower on your team). ∎

Now let us study two probability laws that will be used considerably in certain problems of statistical inference, called the T distribution and the F distribution. Both these probability laws occur for certain functions of normal random variables, as we will see. Because their actual derivation involves mathematical techniques that we have not assumed, we will be content simply to describe how they occur. These probability laws are defined in the following two theorems.

Theorem 5.7.3. Let Z be a standard normal random variable and let V be an independent χ^2 random variable with d degrees of freedom. Then the random variable

$$T = Z \bigg/ \sqrt{\frac{V}{d}}$$

has density function

$$f(t) = \frac{\Gamma((d+1)/2)}{\sqrt{\pi d}\,\Gamma(d/2)} \frac{1}{\left(1 + (t^2/d)\right)^{(d+1)/2}}.$$

This is called the T density function with d degrees of freedom (defines the probability law for a T random variable with d degrees of freedom). ∎

If you examine this T density function, it is easy to see that it has a maximum at $t = 0$ and declines symmetrically for $|t| > 0$; that is, it is a symmetric bell-shaped curve, centered at zero, very similar to the standard normal density function. Figure 5.7.1 plots the T density for 5 degrees of freedom and the standard normal density. In fact as d, the degrees of freedom, increases indefinitely, the T density function converges to the value of the standard normal density for any real number. Also like the normal, there is no simple function (for all d) whose derivative is $f(t)$; thus Table 3 in the Appendix presents selected quantiles from the T distributions, for $d \le 30$; the line labeled $d = \infty$ actually gives values for the standard nor-

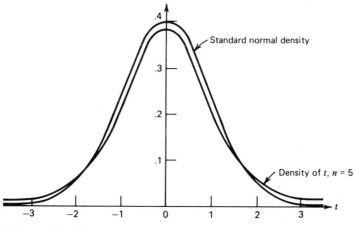

Figure 5.7.1

mal quantiles. We will employ the same type of notation already used earlier to denote quantiles, letting t_k represent the 100kth quantile (which actually depends on the degrees of freedom d but is suppressed for simplicity of notation). Thus from Table 3 in the Appendix, with $d = 5$, we have $t_{.90} = 1.476, t_{.99} = 3.365$, so if a T random variable has 5 degrees of freedom

$$P(T \leq 1.476) = .9, \qquad P(T \leq 3.365) = .99;$$

with $d = 15$, we find $t_{.90} = 1.341, t_{.99} = 2.602$ so for a T random variable with 15 degrees of freedom

$$P(T \leq 1.341) = .9, \qquad P(T \leq 2.602) = .99.$$

Because the T density function is symmetric about 0, we also have, for $d = 5$,

$$P(T \leq -1.476) = .1, \qquad P(T \leq -3.365) = .01,$$
$$P(-1.476 \leq T \leq 1.476) = .80, \qquad P(-3.365 \leq T \leq 3.365) = .98,$$

whereas with $d = 15$

$$P(T \leq -1.341) = .1, \qquad P(T \geq -2.602) = .99$$
$$P(-1.341 \leq T \leq 1.341) = .8, \qquad P(-2.602 \leq T \leq 2.602) = .98.$$

EXAMPLE 5.7.4.   Suppose Z_1 and Z_2 are independent standard normal random variables. Define $Y_1 = Z_1 - Z_2$ and $Y_2 = Z_1 + Z_2$ and we know, from Example 5.4.7, that both Y_1 and Y_2 are normal random variables because each of them is a linear function of normal random variables. Furthermore,

$$E[Y_1] = E[Y_2] = 0 \quad \text{and} \quad \text{Var}[Y_1] = \text{Var}[Y_2] = 2$$

using Theorem 5.4.1, Corollary 5.4.1, and

$$\text{Cov}[Y_1, Y_2] = \text{Cov}[Z_1 - Z_2, Z_1 + Z_2]$$
$$= ((1)(1) + (1)(-1))(1) = 0,$$

from Theorem 5.4.2. But then, since Y_1 and Y_2 are normal with $\rho = 0$, they are also independent from Theorem 5.7.2. Going one step further, $Y_1/\sqrt{2}$ and $Y_2/\sqrt{2}$ are independent standard normal, $(Y_2/\sqrt{2})^2 = Y_2^2/2$ is χ^2 with 1 degree of freedom (and still independent of $Y_1/\sqrt{2}$), so the ratio

$$T = \frac{Y_1}{\sqrt{2}} \bigg/ \sqrt{\frac{Y_2^2}{2}} = \frac{Y_1}{|Y_2|} = \frac{Z_1 - Z_2}{|Z_1 + Z_2|}$$

has the T distribution with 1 degree of freedom (the degrees of freedom of the χ^2 random variable in the denominator). By referring to Table 3 you can see that this T density with 1 degree of freedom is widely dispersed (or has thick tails, as is sometimes said). Since $t_{.95} = 6.314$ an interval 12.628 units long (from -6.314 to 6.314) is needed to have a 90% chance of containing the observed T value. In contrast, the interval from -1.645 to 1.645, length 3.29 units, has the same probability of containing the observed value of a standard normal random variable. ∎

Another probability law that can be derived from normal random variables is called the F distribution. It is described in the following theorem.

Theorem 5.7.4.   Let U be a χ^2 random variable with d_1 degrees of freedom and let V be an independent χ^2 random variable with d_2 degrees of freedom. Then the random variable

$$F = \frac{U}{d_1} \bigg/ \frac{V}{d_2}$$

has density function

$$f(x) = \frac{\Gamma((d_1 + d_2)/2)}{\Gamma(d_1/2)\,\Gamma(d_2/2)} \frac{(d_1/d_2)^{d_1/2}\, x^{(d_1/2)-1}}{(1 + (d_1 x/d_2))^{(d_1 + d_2)/2}}, \qquad x > 0.$$

This is called the F density function with d_1 and d_2 degrees of freedom (or defines the F probability law with d_1 and d_2 degrees of freedom). ∎

The F distribution has two parameters, d_1 and d_2, corresponding to (defined by) the degrees of freedom of the two χ^2 random variables in the ratio; the degrees of freedom for the numerator random variable is listed first and the ordering makes a difference, of course, if $d_1 \neq d_2$. The reciprocal of an F random variable $(1/F = (V/d_2)/(U/d_1))$ again is the ratio of two independent χ^2 random variables, each divided by its degrees of freedom, so it again has the F distribution, now with d_2 and d_1 degrees of freedom.

The F density functions resemble the gamma densities, but actually come from a distinct family. Figure 5.7.2 plots the F density functions for $d_1 = d_2 = 1$ $(f(0) = \infty)$, $d_1 = d_2 = 2$ $(f(0) = 1)$, and $d_1 = 5$, $d_2 = 10$ $(f(0) = 0)$. As long as $d_1 \geq 2$, the density has a maximum value at $x = (d_1 - 2)d_2/(d_1)(d_2 + 2)$. The F density also occurs as the density of

$$F = \frac{\beta X}{\alpha(1 - X)},$$

where X is a beta random variable with parameters α and β; the parameters of the F distribution are $d_1 = \alpha/2$, $d_2 = \beta/2$.

Table 4 in the Appendix gives five different quantiles, $F_{.5}$, $F_{.75}$, $F_{.9}$, $F_{.95}$, $F_{.99}$, for 18 different values for d_1 in conjunction with 27 different

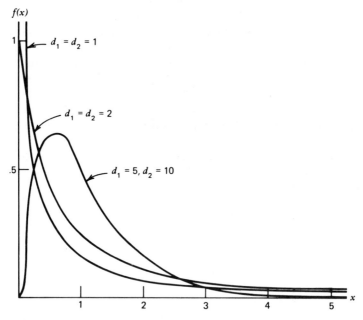

Figure 5.7.2

values for d_2. Each page of the table presents a single quantile for all the (d_1, d_2) combinations. Thus, for example, if F has the F distribution with $d_1 = 7$, $d_2 = 10$, we find $P(F \leq 1.569) = .75$ and $P(F \leq 2.414) = .90$, whereas with $d_1 = 12$, $d_2 = 12$, $P(F \leq 1.490) = .75$ and $P(F \leq 2.147) = .90$. We can make use of the fact that $1/F$ also has an F distribution to evaluate quantiles with $k < .5$. In showing how this is done, let $F(d_1, d_2)$ represent an F random variable with d_1 and d_2 degrees of freedom and let $F_k(d_1, d_2)$ be the $100k$th quantile for this distribution; thus

$$P\bigl(F(d_1, d_2) \leq F_k(d_1, d_2)\bigr) = k.$$

The event $\{F(d_1, d_2) \leq F_k(d_1, d_2)\}$ is equivalent to the event $\{1/F(d_1, d_2) \geq 1/F_k(d_1, d_2)\}$ so

$$P\left(\frac{1}{F(d_1, d_2)} \geq \frac{1}{F_k(d_1, d_2)} \right) = k,$$

their probabilities must be equal. As we know, $1/F(d_1, d_2)$ has an $F(d_2, d_1)$ distribution, and thus

$$P\left(F(d_2, d_1) \geq \frac{1}{F_k(d_1, d_2)} \right) = k$$

$$P\left(F(d_2, d_1) < \frac{1}{F_k(d_1, d_2)} \right) = 1 - k;$$

that is, the $100(1 - k)$th quantile of $F(d_2, d_1)$ is $1/F_k(d_1, d_2)$, $F_{1-k}(d_2, d_1) = 1/F_k(d_1, d_2)$. This relation enables us also to use Table 4 to evaluate the 1st, 5th, 10th, 25th, and 50th quantiles for the degrees of freedom tabulated (interchanging d_1 and d_2). Thus, for example, as already seen, $F_{.75}(7, 10) = 1.569$, $F_{.90}(7, 10) = 2.414$, $F_{.75}(12, 12) = 1.490$, $F_{.90}(12, 12) = 2.147$, so

$$F_{.25}(10, 7) = \frac{1}{1.569} = .637$$

$$F_{.10}(10, 7) = \frac{1}{2.414} = .414$$

$$F_{.25}(12, 12) = \frac{1}{1.490} = .671$$

$$F_{.10}(12, 12) = \frac{1}{2.147} = .466.$$

For our purposes, linear interpolation on the degrees of freedom, holding the quantile fraction constant, should be sufficiently accurate.

EXAMPLE 5.7.4.   Let Z_1, Z_2, Z_3, Z_4, and Z_5 be independent standard normal random variables and define $Y_1 = Z_1^2 + Z_2^2$, $Y_2 = Z_3^2 + Z_4^2 + Z_5^2$; Y_1 and Y_2 are then independent χ^2 random variables with 2 and 3 degrees of freedom, respectively. The ratio

$$F = \frac{Y_1}{2} \bigg/ \frac{Y_2}{3}$$

has the F distribution with $d_1 = 2$ and $d_2 = 3$. From Table 4

$$P\left(\frac{1}{19.16} \le F(2, 3) \le 9.552 \right) = .90$$

since $F_{.95}(2, 3) = 9.552$, $F_{.05}(2, 3) = 1/F_{.95}(3, 2) = 1/19.16 = .052$; thus the probability is .90 that we would find the observed value for

$$\frac{Z_1^2 + Z_2^2}{2} \bigg/ \frac{Z_3^2 + Z_4^2 + Z_5^2}{3}$$

in the interval $(.052, 9.552)$.   ∎

We saw in Theorem 5.7.3 that the ratio of a standard normal random variable to the square root of an independent χ^2 random variable with d degrees of freedom, over d, has the T distribution with d degrees of freedom. If we square this T random variable, it becomes the ratio of a χ^2 random variable, 1 degree of freedom, over an independent χ^2 random variable divided by d, its degrees of freedom. But this ratio must have the $F(1, d)$ distribution from Theorem 5.7.4; that is, if $F = T^2$, where T has the T distribution, d degrees of freedom, then F has the F distribution with 1 and d degrees of freedom.

Let us close this section with a short discussion of ordered (in magnitude) random variables. Assume $X_1, X_2, ..., X_n$ are independent continuous random variables, each with density function $f(x)$ and distribution function $F(x)$. Now let $X_{(1)}, X_{(2)}, ..., X_{(n)}$ denote their ordered values, from smallest $(X_{(1)})$ to largest $(X_{(n)})$; these are called *order statistics* and are useful in some statistical procedures, as we will see. What is the probability law for $X_{(n)}$, the largest or maximum value?

The event $\{X_{(n)} \leq x\}$ occurs if and only if (or is equivalent to) the event $\{X_1 \leq x, X_2 \leq x, ..., X_n \leq x\}$ occurs, since if the largest is smaller than x, all n of the random variables must be smaller than x, where x is any fixed real number. Because the events are equivalent the distribution function for $X_{(n)}$ can be evaluated from

$$
\begin{aligned}
F_{X_{(n)}}(x) &= P(X_{(n)} \leq x) \\
&= P(X_1 \leq x, X_2 \leq x, ..., X_n \leq x) \\
&= P(X_1 \leq x) P(X_2 \leq x) \cdots P(X_n \leq x)
\end{aligned}
$$

since $X_1, X_2, ..., X_n$ are independent. But each of $X_1, X_2, ..., X_n$ has the same distribution function $F(x)$ so

$$ F_{X_{(n)}}(x) = [F(x)]^n. $$

The density function for $X_{(n)}$ then is

$$
\begin{aligned}
f_{X_{(n)}}(x) &= \frac{d}{dx} F_{X_{(n)}}(x) \\
&= \frac{d}{dx} [F(x)]^n \\
&= n[F(x)]^{n-1} f(x)
\end{aligned}
$$

since $dF(x)/dx = f(x)$. Thus both the distribution function and density function for $X_{(n)}$, the maximum of $X_1, X_2, ..., X_n$, are simply related to the common density and distribution functions for the individual X_i's.

A very similar approach works for $X_{(1)}$, the smallest of $X_1, X_2, ..., X_n$. For any fixed x we will observe $X_{(1)} > x$ if and only if all the X_i's exceed x; that is, the events $\{X_{(1)} > x\}$ and $\{X_1 > x, X_2 > x, ..., X_n > x\}$ are equivalent so their probabilities must be equal:

$$P(X_{(1)} > x) = P(X_1 > x, X_2 > x, ..., X_n > x)$$
$$= P(X_1 > x) P(X_2 > x) \cdots P(X_n > x).$$

Again, $X_1, X_2, ..., X_n$ have the same distribution function and $P(X_i > x) = 1 - F(x)$ for each i, so

$$P(X_{(1)} > x) = [1 - F(x)]^n,$$

and the distribution function for $X_{(1)}$ is

$$F_{X_{(1)}}(x) = 1 - P(X_{(1)} > x)$$
$$= 1 - [1 - F(x)]^n.$$

The density function for $X_{(1)}$ is

$$f_{X_{(1)}}(x) = \frac{d}{dx} F_{X_{(1)}}(x)$$

$$= \frac{d}{dx} \{1 - [1 - F(x)]^n\}$$

$$= n[1 - F(x)]^{n-1} f(x),$$

linking the density for the minimum value with the density and distribution function for the individual X_i's. At least conceptually, it is straightforward to evaluate the probability law for either the smallest or the largest of n independent random variables, each with the same distribution.

EXAMPLE 5.7.5. Let $X_1, X_2, ..., X_n$ be independent exponential random variables, each with

$$f(x) = \lambda e^{-\lambda x}, \qquad x > 0,$$
$$F(x) = 1 - e^{-\lambda x}, \qquad x > 0.$$

Then if $X_{(1)}$ is the smallest or minimum of these n exponential random variables

$$F_{X_{(1)}}(x) = 1 - [1 - F(x)]^n$$
$$= 1 - [1 - (1 - e^{-\lambda x})]^n$$
$$= 1 - e^{-n\lambda x}, \qquad x > 0$$

and the density for $X_{(1)}$ is

$$f_{X_{(1)}}(x) = n\lambda e^{-n\lambda x}, \qquad x > 0,$$

so $X_{(1)}$ is again exponential with parameter $n\lambda$. The maximum, or largest value, $X_{(n)}$, has distribution function

$$F_{X_{(n)}}(x) = [1 - e^{-\lambda x}]^n, \qquad x > 0$$

and density function

$$f_{X_{(1)}}(x) = n\lambda e^{-\lambda x}[1 - e^{-\lambda x}]^{n-1}, \qquad x > 0.$$

This maximum value does not have an exponential distribution. ∎

EXERCISE 5.7

1. Let (X_1, X_2) be the number of successes and failures, respectively, in n repeated independent Bernoulli trials, each with parameter p.
   (a) What is the probability law for (X_1, X_2)?
   (b) What is the correlation between X_1 and X_2?

2. A doctor's office has four telephone lines, each of which can be independently dialed by any patient. Assume, on any given business day, the probabilities are $.4, .3, .2, .1$, respectively, the *first* call is received on line 1, 2, 3, or 4, and the first calls are independent from one business day to another.
   (a) For n business days let Y_i be the number of days on which the first call arrives on line i, $i = 1, 2, 3, 4$. What is the distribution for (Y_1, Y_2, Y_3, Y_4)?
   (b) With $n = 10$, evaluate

$$P(Y_1 = Y_2 = 5, Y_3 = Y_4 = 0).$$

   (c) With $n = 12$, evaluate

$$P(Y_1 = Y_2 = Y_3 = Y_4).$$

3. Five fair dice are rolled at one time.
   (a) What is the probability that all five show the same number?
   (b) What is the probability that the five numbers are in sequence? (either 1, 2, 3, 4, 5 or 2, 3, 4, 5, 6).

*4. Just as the Bernoulli trial (sampling with replacement) situation can be generalized to the multinomial, the sampling without replacement model can also be generalized. Assume a bowl contains m slips of paper, of which k_1 are color 1, k_2 are color 2, ..., k_c are color c, so $\sum_{j=1}^c k_j = m$. A random sample (without replacement) of n ($<m$) slips is removed from the bowl at random. Let X_j = number of slips in the sample of color j, $j = 1, 2, ..., c$, and find the probability law for $X_1, X_2, ..., X_c$. This is called the multihypergeometric probability law.

*5. If you have answered Exercise 5.7.4, assume a small state has 100,000 registered voters (for a given party) and there are four candidates on the ballot during a primary election. Of the 100,000 registered voters, assume 45,000 will vote for A; 35,000, for B; 15,000, for C; and the remainder, for D. A sample of 1000 voters is selected at random from those registered, without replacement. Let X_A, X_B, X_C, X_D be the numbers in the sample who will vote for the respective candidates. What is the probability law for (X_A, X_B, X_C, X_D)?

6. Let X be the time necessary for a college mile runner to run his first quarter-mile and let Y be the time necessary for him to run his second quarter-mile (in the same race). (X, Y) is a bivariate normal random vector with $\mu_X = 59$, $\mu_Y = 60$, $\sigma_X = \sigma_Y = 1$, $\rho = -.5$.
(a) Evaluate $P(X \le 60)$ and $P(Y \le 59)$.
(b) Assume the first quarter-mile took $x = 60$ seconds and compute the probability that $Y \le 59$.

7. The amount of rainfall recorded at a U.S. weather station in January is a random variable X and the amount of rainfall recorded at the same station in February of the same year is a random variable Y. Granted (X, Y) is bivariate normal with $\mu_X = 6$ inches, $\mu_Y = 4$, $\sigma_X = 1$, $\sigma_Y = .5$, $\rho = .1$, evaluate
(a) $P(X \le 5)$
(b) $P(Y \le 5)$
(c) $P(Y \le 5 | X = 5)$.

8. Let X be the adjusted gross income and let Y be the medical deduction claimed on a federal income tax form selected at random from those filing joint returns who itemize deductions. Assume X and Y are bivariate normal with $\mu_X = 15,000$, $\mu_Y = 1000$, $\sigma_X = 800$, $\sigma_Y = 100$, $\rho = .6$.
(a) On a form with adjusted gross income of $x = \$16,000$, what is the probability that the medical deduction Y is less than $\$900$?
(b) Find the constant a such that

$$P(Y \le a | X = 16,000) = .95.$$

9. For the case discussed in Example 5.7.3 find the side length a of the square, centered at $(0, 0)$, that has probability .5 of containing the observed point (x, y).

10. Let X be standard normal and let Y be an independent χ^2 random variable with $v = 4$ degrees of freedom. What is the value for a such that

$$P(|X| \le a\sqrt{Y}) = .9?$$

11. Let (X, Y) be bivariate normal with $\mu_X = \mu_Y = 0$, $\sigma_X = \sigma_Y = 1$, $\rho = 0$

and evaluate

$$P\left(\sqrt{\left|\frac{X}{Y}\right|} \leq 8\right)$$

*12. Assume the same dart thrower tosses two darts at the same target. For each throw the impact point (X, Y) is bivariate normal with $\mu_X = \mu_Y = 0$, $\sigma_X = \sigma_Y = 1$, $\rho = 0$ and the two impact points are independent. How can you use the F distribution to evaluate probability statements about the radial miss distances of the two throws? What is the probability that the first impact point is closer to $(0, 0)$ than the second is?

13. Assume $U_1, U_2, \ldots, U_n$ are independent uniform random variables on $(0, 1)$ and let $U_{(1)}, U_{(2)}, \ldots, U_{(n)}$ be the ordered values.
   (a) Find the density for $U_{(1)}$, the smallest value.
   (b) Find the density for $U_{(n)}$, the largest value.
   (c) Evaluate $P(U_{(1)} \leq \frac{1}{2})$ and $P(U_{(n)} \leq \frac{1}{2})$.

*14. Let $U_1, U_2, \ldots, U_n$ be defined as in Exercise 5.7.13 and evaluate the density for $U_{(k)}$, the kth smallest value. Evaluate $P(U_{(k)} \leq \frac{1}{2})$. (*Hint.* The event $\{U_{(k)} \leq t\}$ is equivalent to the event $\{at\ least\ k$ of the U_i's are each smaller than $t\}$. Do you recognize the probability law for $U_{(k)}$?)

15. Let X_1 and X_2 be independent exponential random variables, each with parameter λ, and let $X_{(1)}$ be their minimum value.
   (a) What is the distribution for $2\lambda (X_1 + X_2)$?
   *(b) Compare $P(|2X_{(1)} - 1/\lambda| < \varepsilon)$ and

$$P\left(\left|\frac{X_1 + X_2}{2} - \frac{1}{\lambda}\right| < \varepsilon\right) = P\left(\left|X_1 + X_2 - \frac{2}{\lambda}\right| < 2\varepsilon\right)$$

16. Assume X and Y have probability function

$$P(X = Y = 0) = P(X = Y = 1) = \frac{1 + \rho}{4}$$

$$P(X = 0, Y = 1) = P(X = 1, Y = 0) = \frac{1 - \rho}{4}$$

and show that the correlation between X and Y is ρ, where $-1 \leq \rho \leq 1$. Note then that X and Y are independent if and only if $\rho = 0$, a property shared by the bivariate normal density.

5.8 Summary

Random vector $(X_1, X_2, ..., X_n)$: associates an n-tuple with every element of a sample space.
Probability function for discrete $(X_1, X_2, ..., X_n)$:

$$p_{X_1,X_2,...,X_n}(x_1, x_2, ..., x_n) = P(X_1 = x_1, X_2 = x_2, ..., X_n = x_n)$$

Density function for continuous $(X_1, X_2, ..., X_n)$:

$$P(x_1 \le X_1 \le x_1 + \Delta x_1, ..., x_n \le X_n \le x_n + \Delta x_n)$$
$$\doteq f_{X_1,X_2,...,X_n}(x_1, x_2, ..., x_n)\, \Delta x_1 \, ... \, \Delta x_n$$

Marginal probability function for X_1:

$$p_{X_1}(x) = \sum_{x_2} \sum_{x_3} \cdots \sum_{x_n} p_{X_1,X_2,...,X_n}(x, x_2, ..., x_n)$$

Marginal density function for X_1:

$$f_{X_1}(x) = \int_{x_2 = -\infty}^{\infty} \cdots \int_{x_n = -\infty}^{\infty} f_{X_1,X_2,...,X_n}(x, x_2, ..., x_n) \cdot dx_2 \cdots dx_n$$

Conditional probability function:

$$p_{X_1|X_2,...,X_n}(x \,|\, X_2 = x_2, ..., X_n = x_n) = \frac{p_{X_1,X_2,...,X_n}(x, x_2, ..., x_n)}{p_{X_2,...,X_n}(x_2, ..., x_n)}$$

Conditional density function:

$$f_{X_1|X_2,...,X_n}(x \,|\, X_2 = x_2, ..., X_n = x_n) = \frac{f_{X_1,X_2,...,X_n}(x, x_2, ..., x_n)}{f_{X_2,...,X_n}(x_2, ..., x_n)}$$

$X_1, X_2, ..., X_n$ are independent discrete random variables, if and only if

$$p_{X_1,X_2,...,X_n}(x_1, x_2, ..., x_n) = p_{X_1}(x_1)\, p_{X_2}(x_2) \cdots p_{X_n}(x_n),$$
for all $(x_1, x_2, ..., x_n)$.

$X_1, X_2, ..., X_n$ are independent continuous random variables if and only if

$$f_{X_1,X_2,...,X_n}(x_1, x_2, ..., x_n) = f_{X_1}(x_1)\, f_{X_2}(x_2) \cdots f_{X_n}(x_n),$$
for all $(x_1, x_2, ..., x_n)$.

$$E[G(X_1, X_2, ..., X_n)] = \sum_{x_1} \cdots \sum_{x_n} G(x_1, ..., x_n) \cdot p_{X_1,...,X_n}(x_1, ..., x_n),$$
$$\text{if } (X_1, X_2, ..., X_n) \text{ is discrete}$$

$$= \int_{-\infty}^{\infty} \cdots \int_{-\infty}^{\infty} G(x_1, \ldots, x_n) \cdot f_{X_1, \ldots, X_n}(x_1, \ldots, x_n) \, dx_1, \ldots, dx_n,$$

if $(X_1, X_2, \ldots, X_n)$ is continuous.

Joint moments: $E[X_1^{j_1} \cdots X_n^{j_n}]$

Moment generating function:

$$m_{X_1, \ldots, X_n}(t_1, \ldots, t_n) = E[\exp(t_1 X_1 + \cdots + t_n X_n)]$$

Covariance:

$$\begin{aligned} \mathrm{Cov}[X_1, X_2] &= E[(X_1 - \mu_1)(X_2 - \mu_2)] \\ &= E[X_1 X_2] - \mu_1 \mu_2 \end{aligned}$$

Correlation:

$$\rho = \frac{\mathrm{Cov}[X_1, X_2]}{\sigma_1 \sigma_2}, \quad |\rho| \le 1$$

$$E[E[X_2 | X_1]] = E[X_2]$$

If $X_1, X_2, \ldots, X_n$ are independent, then

$$E[H_1(X_1) H_2(X_2) \cdots H_n(X_n)] = E[H_1(X_1)] \cdots E[H_n(X_n)].$$

If X_1 and X_2 are independent, then

$$\mathrm{Cov}[X_1, X_2] = \rho = 0.$$

If $Y = a_1 X_1 + \cdots + a_n X_n$, then $\mu_Y = a_1 \mu_1 + \cdots + a_n \mu_n$,

$$\sigma_Y^2 = a_1^2 \sigma_1^2 + \cdots + a_n^2 \sigma_n^2 + 2 \sum_{i < j} \sum a_i a_j \, \mathrm{Cov}[X_i, X_j].$$

If $Y = a_1 X_1 + \cdots + a_n X_n$ and $X_1, X_2, \ldots, X_n$ are independent, then $\mu_Y = a_1 \mu_1 + \cdots + a_n \mu_n$, $\sigma_Y^2 = a_1^2 \sigma_1^2 + \cdots + a_n^2 \sigma_n^2$.

If $X_1, X_2, \ldots, X_n$ are uncorrelated and

$$U = \sum a_i X_i, \quad V = \sum b_i X_i, \quad \text{then} \quad \mathrm{Cov}[U, V] = \sum a_i b_i \sigma_i^2$$

If $X_1, X_2, \ldots, X_n$ are independent and $Y = \sum a_i X_i$, then

$$m_Y(t) = \prod m_{X_i}(a_i t)$$

Chebyshev inequality:

$$P(|X| \ge \varepsilon) \le \frac{1}{\varepsilon^2} E[X^2]$$

$$P(|Y - \mu| \ge \varepsilon \sigma) \le \frac{1}{\varepsilon^2}$$

Law of large numbers: If $X_1, X_2, ..., X_n, ...$ is a sequence of uncorrelated
random variables each with mean μ, variance σ^2, $\overline{X}_n = (1/n) \sum_{i=1}^{n} X_i$,
then

$$\lim_{n \to \infty} P(|\overline{X}_n - \mu| \geq \delta) = 0 \qquad \text{for any} \qquad \delta > 0$$

Central limit theorem: If $X_1, X_2, ..., X_n, ...$ is a sequence of independent
random variables each with mean μ and variance σ^2, then the probability law for $Z_n = (\sum_{i=1}^{n} X_i - n\mu)/\sigma \sqrt{n}$ converges to the standard
normal as $n \to \infty$.

Approximations: If n is large, $S_n = \sum_{i=1}^{n} X_i$, $\overline{X}_n = S_n/n$,

for total $F_{S_n}(t) \doteq N \left(\dfrac{t - n\mu}{\sigma \sqrt{n}} \right)$

for mean $F_{\overline{X}_n}(t) \doteq N \left(\dfrac{(t - \mu) \sqrt{n}}{\sigma} \right)$

Multinomial probability law:

Parameters $n = 1, 2, 3, ...,$ $\qquad 0 < p_i < 1,$ $\qquad i = 1, 2, ..., k,$

$$\sum_{1}^{k} p_i = 1,$$

$$P(X_1 = x_1, X_2 = x_2, ..., X_k = x_k) = \binom{n}{x_1 x_2 \cdots x_k} p_1^{x_1} \cdots p_k^{x_k},$$

where $x_i = 0, 1, ..., n,$ $\qquad i = 1, 2, ..., k$ $\qquad \sum x_i = n$

Marginal for X_i is binomial n, p_i

$\text{Cov}[X_i, X_j] = -np_i p_j$

Bivariate normal parameters $\mu_X, \mu_Y, \sigma_X, \sigma_Y, \rho$:
$\quad$ X and Y are independent if $\rho = 0$.
$\quad$ X and Y are each normal.
$\quad$ Conditional probability law for Y, given $X = x$, is normal,

$$\mu = \mu_Y + \rho \frac{\sigma_Y}{\sigma_X}(x - \mu_X), \qquad \sigma^2 = \sigma_Y^2(1 - \rho^2)$$

T distribution: Probability law for the ratio of a standard normal random
variable divided by the square root of an independent χ^2 random
variable over its degrees of freedom.

F distribution: Probability law for the ratio of two independent χ^2 random variables, each divided by its degrees of freedom.

If $X_1, X_2, \ldots, X_n$ are independent, continuous, each with distribution function $F(x)$,

$$F_{X_{(1)}}(x) = 1 - [1 - F(x)]^n$$
$$F_{X_{(n)}}(x) = [F(x)]^n$$

where $X_{(1)}$ is the minimum and $X_{(n)}$ the maximum of $X_1, X_2, \ldots, X_n$

6

Descriptive and Inferential Statistics

We have studied a number of results from probability theory. Although this theory itself is a very interesting subject, its applications in many areas can be even more interesting and useful. One major use for the ideas we have discussed is in the field of statistical inference, which we will introduce in this chapter and expand in the remainder of this book. You should, however, be aware that there are many other modern applications of probability theory, which for lack of space we do not introduce. These include communication theory, which develops the theory of stochastic processes, essentially "random functions" of a time parameter, and which is very widely used in modern communications systems of all sorts, from satellites to telephones to computers. Stochastic processes provide a rich variety of models that have also been used to model population growth, the spread of epidemics, occurrences of earthquakes or lightning strikes, and many other phenomena.

6.1 Descriptive Statistics

The term "descriptive statistics" itself is meant to be descriptive. It is used to denote a smaller number of descriptors (statistics), which in some sense describe certain aspects of larger bodies of numbers. All the descriptive statistics we will examine in this section have been used for many years; because our main interest will be in "inferential" statistics we will examine only the most basic descriptive statistics.

The word "statistics" itself appears to have been coined in the seventeenth century as a label for numbers presented to the sovereign to describe his subjects and various aspects of his realm. People who gathered and summarized these numbers were called statisticians. Hopefully, as we proceed, it will become evident that modern usage of the term *statistician* includes people who do a very broad range of things, one of the most primitive (but still very useful and important) being the summarization or description of a body of data. Also, among the modern tools of a statistician are many methods for making inferences beyond the observed set of data, generalizations which prove of interest in a wide variety of situations from surveys of public opinion (for whom would you vote if the election were held tomorrow) to questions of which of several possible designs of a nuclear reactor appears to be the safest or most reliable in a given location. We will begin our study of statistical inference in Section 6.2, after we examine some descriptive statistics in this section.

First, let us discuss the notation to be used. Granted we have some data available, a collection of numbers, and want to express various quantities that describe the collection, we first need a notation to represent the numbers themselves. We will employ a subscripting system for this purpose; thus if we have a collection of n numbers, we can distinguish among them (identify individual numbers) by denoting them $x_1, x_2, ..., x_n$ (or any other letter with subscripts ranging from 1 to n). For example, Table 6.1.1 presents the results of using a spectrometer 60 times to measure the iron contamination of used engine oil. Each spectrometer reading (in parts per million, denoted ppm) is given by burning a small portion of the oil; the light emitted is analyzed to measure the iron contamination. This procedure was repeated 60 times, each time burning a tiny portion of oil selected from the same reservoir. For various reasons the reading produced is not the same number each time, as is obvious from inspecting Table 6.1.1.

Table 6.1.1

99	100	100	103	102	106	106	100	103	102
104	104	100	101	104	107	107	109	100	110
100	102	99	103	95	102	99	99	108	102
103	100	98	96	96	99	98	96	99	97
103	97	97	108	100	97	99	102	98	98
100	100	100	100	100	100	105	101	102	102

With the subscripting scheme just mentioned, we will represent these numbers by $x_1, x_2, ..., x_{60}$, reading across the rows and thus $x_1 = 99$, $x_{10} = 102$, $x_{51} = 100$, and so on. Presented as in this table, in the order in which the measurements were taken, it is not easy to make much sense from these data, beyond the fact that the repeated iron readings were not all equal. By reading through the numbers in the table, you will find the smallest value is $x_{25} = 95$ and the largest value is $x_{20} = 110$; thus the total range spanned by these data is the difference $110 - 95 = 15$, meaning all 60 numbers fell within an interval of length 15.

A frequency count (or table) of how many readings fell in various intervals is often used to summarize data such as the readings in Table 6.1.1. To construct such a table, the first decision one must make is how many intervals (also called *classes*) should be employed. There is, of course, no hard and fast rule for determining the number of intervals; most frequency tables employ at least five and no more than 15 intervals, the actual number used depending heavily on n, the size of the data set, and the range spanned by the data. The data recorded in Table 6.1.1 contain $n = 60$ numbers, which range from 95 to 110; we will arbitrarily use five classes in constructing our frequency count. Once the number of classes to be employed has been chosen (five in our case), the next decision is to define these classes exactly, in terms of location and boundaries. The boundaries must, of course, be unambiguous so that there is no possible question which class any possible observed value falls into. It is not necessary, but many people tend to use classes of equal length in constructing frequency tables; equal length intervals have certain advantages in terms of constructing visual displays of frequency tables. We will arbitrarily employ five classes to construct a frequency table for the data in Table 6.1.1 and count the readings that fall in the intervals 95 to 97, 98 to 100, 101 to 103, 104 to 106, and 107 to 110, respectively. Each of these classes (intervals) contains three different possible ppm readings (except the last one, which contains 4). These first four classes are called of equal length and the final class has greater length.

If you now return to Table 6.1.1 and count the number of the ppm readings that fall between 95 and 97, inclusive, and the number of those ppm readings that fall between 98 and 100, inclusive, and so on, the result will be the values presented in Table 6.1.2. The total of all the counts gives $n = 60$, the number of readings summarized, as it must. Such a frequency

Table 6.1.2

	ppm Value				
	95 to 97	98 to 100	101 to 103	104 to 106	107 to 110
Count	8	25	15	6	6

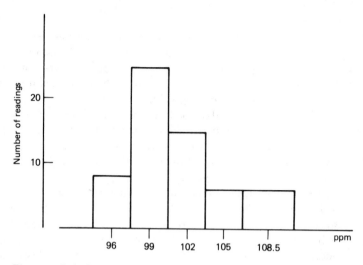

Figure 6.1.1

table makes it quite easy for a reader to see the range of the data, which class (or interval) contained most of the data, roughly how spread out the data are, and so on. A frequency table also lends itself to a natural graphical presentation using a histogram. Figures 6.1.1 and 6.1.2 present histograms of this frequency table. In both histograms the bars used are centered over the classes and are labeled (on the horizontal axis) with the class midpoints. In Figure 6.1.1 the height of each bar is equal to the number of readings found in each class, whereas in Figure 6.1.2 the *areas* of the bars are made

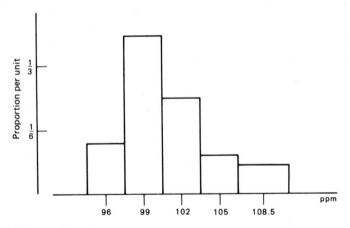

Figure 6.1.2

proportional to the number of measurements falling in the class. As long as all the classes are of equal length, both ways of drawing the histogram give the same picture. If, however, one or more classes has a different length from the others, the height used must be adjusted if the area is to be proportional to the number counted in the class. Thus the final bar in Figure 6.1.2 must be reduced in height (by $\frac{1}{4}$), since the class is four units long rather than three, to keep the area proportional to the number counted in the class. Both methods of construction are used; many people feel the second procedure (area proportional to count) gives a more accurate impression to the casual observer. Frequency tables or histograms provide descriptors of data that display certain features of the data that are certainly not immediately evident from just a tabulation of the raw data.

In addition to frequency tables and histograms, numerical descriptors are frequently useful. We are acquainted with the concepts of the mean, variance, and standard deviation of a random variable. Recall that these give measures of the middle and the variability of a probability law. The same ideas prove useful in describing a set of data. That is, the mean of the data gives a measure of its "middle," a more or less typical value, whereas the variance and standard deviation are useful in describing how variable the set of data is (again the variance will be measured in the square of the units and the standard deviation in the same units as the data themselves). These quantities are defined as follows.

DEFINITION 6.1.1.   Given a set of data $x_1, x_2, \ldots, x_n$, the *mean* (or average value) of the data is $\bar{x} = \frac{1}{n}\sum_{i=1}^{n} x_i$. The *variance* of the data is defined to be

$$s^2 = \frac{1}{n-1} \sum_{i=1}^{n} (x_i - \bar{x})^2$$

and its positive square root, $s = \sqrt{s^2}$, is called the *standard deviation* of the data.   ■

The mean, $\bar{x}$, is simply the total divided by n. It measures the middle of the data in a center of gravity sense and is essentially the point at which the histogram in Figure 6.1.2 would balance. Its descriptive role is quite analogous to that of the mean of a random variable, used in describing its probability law.

The variance, σ^2, of a random variable is the expected or average value of the square of the difference between the random variable and its mean. In defining the variance, s^2, of a set of data, we have used the sum of the squares of the distances between the data values x_i and their mean, $\bar{x}$, divided by $n - 1$, rather than by n. Although this may strike you as curious, there is a good reason for doing so. Since

$$\sum_{i=1}^{n} (x_i - \bar{x}) = \sum_{i=1}^{n} x_i - \sum_{i=1}^{n} \bar{x}$$

$$= \sum_{i=1}^{n} x_i - n\bar{x} = \sum_{i=1}^{n} x_i - \sum_{i=1}^{n} x_i$$

$$= 0,$$

note that the differences $(x_1 - \bar{x}), (x_2 - \bar{x}), ..., (x_n - \bar{x})$, although n in number, always add up to zero; thus knowledge of only the first $n - 1$ of their values also specifies

$$(x_n - \bar{x}) = -(x_1 - \bar{x}) - (x_2 - \bar{x}) - \cdots - (x_{n-1} - \bar{x}),$$

so there really are only $n - 1$ differences that are free to vary. Thus division by $n - 1$ gives the average of the (freely varying) squared differences between the x_i's and $\bar{x}$. We will see a second (perhaps more compelling) reason for dividing by $n - 1$ in Section 6.2. Of course, if n is of any size at all there is very little difference between dividing by $n - 1$ versus n.

As was the case with random variables, direct use of the definition of s^2 is rather cumbersome; the following computational formulas generally are much simpler to employ.

$$\sum_{i=1}^{n} (x_i - \bar{x})^2 = \sum_{i=1}^{n} (x_i^2 - 2x_i\bar{x} + \bar{x}^2)$$

$$= \sum_{i=1}^{n} x_i^2 - 2\bar{x} \sum_{i=1}^{n} x_i + \sum_{i=1}^{n} \bar{x}^2$$

$$= \sum_{i=1}^{n} x_i^2 - 2\bar{x}(n\bar{x}) + n\bar{x}^2$$

$$= \sum_{i=1}^{n} x_i^2 - n\bar{x}^2 = \sum_{i=1}^{n} x_i^2 - n\left(\frac{\sum_{i=1}^{n} x_i}{n}\right)^2$$

$$= \sum_{i=1}^{n} x_i^2 - \frac{\left(\sum_{i=1}^{n} x_i\right)^2}{n}$$

Thus we can compute s^2 from either

$$s^2 = \frac{1}{n-1}\left(\sum_{i=1}^{n} x_i^2 - n\bar{x}^2\right)$$

or

$$s^2 = \frac{1}{n-1} \left(\sum_{i=1}^{n} x_i^2 - \frac{\left(\sum_{i=1}^{n} x_i \right)^2}{n} \right),$$

either of which is more easily evaluated than the original definition. It is easily seen that

$$\sum_{i=1}^{n} (x_i - \bar{x})^2 \geq 0$$

for any set of (real) data, because squares of real numbers must be nonnegative. Thus we also have, using the preceding computational formulas,

$$\sum_{i=1}^{n} x_i^2 - n\bar{x}^2 \geq 0$$

or

$$\frac{1}{n} \sum_{i=1}^{n} x_i^2 \geq \bar{x}^2.$$

The average of the squares of any set of numbers is necessarily greater than or equal to the square of their average.

For the iron ppm readings given in Table 6.1.1 it is easily verified that

$$\sum_{i=1}^{60} x_i = 6067, \qquad \sum_{i=1}^{60} x_i^2 = 614,163,$$

from which we find

$$\bar{x} = \frac{6067}{60} = 101.12$$

$$s^2 = \frac{1}{59} \left(614,163 - \frac{(6067)^2}{60} \right) = 11.664$$

$$s = \sqrt{11.664} = 3.415.$$

Thus the average or "typical" value of the 60 iron readings in Table 6.1.1 is $\bar{x} = 101.12$: some idea of how the data values are spread out around $\bar{x}$ is given by $s = 3.415$, the square root of the average squared distance from the individual x_i's to $\bar{x}$. By referring back to either the data in Table 6.1.1 or to Figure 6.1.2, you can see that $\bar{x}$ does give a fairly nice measure of the middle of the data (35 values are less than $\bar{x}$ and 25 are greater): you can also see that 43 of the 60 numbers fall within one s unit of $\bar{x}$. These data are fairly "well behaved" and $\bar{x}$ and s are good descriptors.

Many of the hand-held programmable calculators have a special built-in function that, after entering the data values once, gives n, $\sum x_i$ and $\sum x_i^2$: indeed, many also automatically compute the mean and standard deviation. If your calculator does automatically compute a variance or standard deviation, you should carefully note their definition: some have routines that divide by n instead of $n - 1$, which can equally well be used to measure the variability of a set of data, but which may cause problems in reproducing numerical values that we will quote. It is, of course, easy to convert from one of these divisors to the other: indeed, with

$$s'^2 = \frac{1}{n}\sum(x_i - \bar{x})^2$$

we have

$$s^2 = \frac{ns'^2}{n-1}$$

and

$$s'^2 = \frac{n-1}{n}s^2.$$

Now let us turn our attention to the data in Table 6.1.3. Given there are the adjusted gross income values (in thousands of dollars) claimed on the 1977 income tax returns for 42 married couples, all living in the same town, and all filing joint returns. If you look through these values you can see immediately that they are much more erratic in some sense than the data in Table 6.1.1. A frequency table for these data is given in Table 6.1.4 and a histogram of these frequencies (with bar areas proportional to frequency) is presented in Figure 6.1.3. Notice that this histogram differs a great deal

Table 6.1.3

1.2	29.3	11.6	14.5	26.8	28.1
17.0	8.2	39.4	151.2	8.2	17.8
23.2	20.6	157.4	10.1	25.8	26.8
36.0	20.1	10.3	92.3	8.0	17.8
74.7	8.8	16.2	7.7	19.4	19.3
125.2	10.7	100.2	47.6	21.2	37.2
19.6	26.0	37.7	29.0	150.1	13.4

Table 6.1.4

Range (in dollars)	Frequency
Less than 10,000	6
More than 10,000, but less than 20,000	13
More than 20,000, but less than 30,000	11
More than 30,000, but less than 50,000	5
More than 50,000, but less than 160,000	7

from being symmetric about some point; it has a "long tail" extending to the right. Data sets that exhibit this type of behavior are said to be positively skewed or skewed to the right. You might find it informative to draw the histogram of the frequency counts given in Table 6.1.3, making the heights of the bars equal to the frequency counts (and keeping the widths as given in Figure 6.1.3); a glance at such a picture gives the mistaken impression that the interval from $50,000 to $160,000 adjusted gross income is more densely occupied than the interval from 0 to $10,000.

For the data given in Table 6.1.3 you will find

$$\sum x_i = 1565.7, \qquad \sum x_i^2 = 128{,}484.93, \qquad \bar{x} = 37.28, \qquad s = 41.35$$

and we find that $\bar{x}$ is not a particularly "typical" value; indeed, 32 of the data values are smaller than $\bar{x}$ and only 10 are larger than $\bar{x}$. The fact that

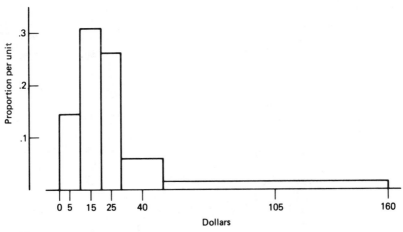

Figure 6.1.3

these data are skewed to the right has rather severely "pulled" $\bar{x}$ to the right, so much so that 75 percent of the data lies to the left of $\bar{x}$. Extremely large and extremely small x_i values exert a great influence on the value for $\bar{x}$ and may cause it to have a value that really is not well centered in the data. Similarly, extremely large and extremely small x_i's tend to inflate the value for the standard deviation s, making it somewhat more difficult to interpret and use.

In short, with data that are skewed heavily either to the right or to the left, $\bar{x}$ and s may not be the most desirable descriptors of the middle of the data set and the variability of the data. Alternative quantities that are used in this case are based on the *ordered* or *ranked* data values $x_{(1)}, x_{(2)}, \ldots, x_{(n)}$. (As before, $x_{(1)} \leq x_{(2)} \leq \cdots \leq x_{(n)}$.) Table 6.1.5 presents the same 42 adjusted gross income values as Table 6.1.3, ranked in order of magnitude. We can easily see that $x_{(1)} = 1.2$, $x_{(42)} = 157.4$, $x_{(21)} = 20.6$, and so on from this table. A second measure of the middle of a set of data is provided by the *median*, m, which is the middle ranked value (we will use $m = x_{((n+1)/2)}$ when n is odd and $m = (x_{(n/2)} + x_{((n+2)/2)})/2$ when n is even). Thus for the income data, we have $m = (x_{(21)} + x_{(22)})/2 = (20.6 + 21.2)/2 = 20.9$. This value is much more representative of the bulk of data than $\bar{x}$ is when the data set is skewed; it has the property that (about) half the data values are below and half above it, so it must be in the "middle" in this sense. A perpendicular line drawn through m in a histogram like Figure 6.1.3 has roughly half the total area of all the bars on either side of it.

The ranked data can also be used to define alternative measures of the variability or spread of the data set. One such measure that is frequently used is called the *interquartile range* q. It is defined to be the length of an interval which includes roughly the middle 50 percent of the data; thus if $x_{.25}$ is a value such that about 25 percent of the data is to its left and $x_{.75}$ is a value such that about 75 percent of the data is to its left (about 25 percent to its right), we have $q = x_{.75} - x_{.25}$. We use the word *roughly*

Table 6.1.5

1.2	7.7	8.0	8.2	8.2	8.8
10.1	10.3	10.7	11.6	13.4	14.5
16.2	17.0	17.8	17.8	19.3	19.4
19.6	20.1	20.6	21.2	23.2	25.8
26.0	26.8	26.8	28.1	29.0	29.3
36.0	37.2	37.7	39.4	47.6	74.7
92.3	100.2	125.2	150.1	151.2	157.4

because not every integer n is divisible by 4. In any particular case we will simply let $x_{.25}$ be the largest number in the data such that the number of data values less than or equal to its value is no larger than $n/4$. Similarly, $x_{.75}$ is the smallest number in the data such that the number of data values greater than or equal to its value is no larger than $n/4$. Although these statements are extremely cumbersome, the basic idea is simple and easy to use. For the income data we have $n = 42$, $n/4 = 10.5$, and thus $x_{.25} = x_{(10)} = 11.6$, $x_{.75} = x_{(33)} = 37.7$, and the interquartile range is $q = 26.1$; an interval of length 26.1 is required to bracket about half the data values (in the center of the data). The two numbers $x_{.25}$ and $x_{.75}$ are called the *quartiles* of the data for obvious reasons. The more spread out the data values are, of course, the larger will be the value for the interquartile range. Notice that for these skewed income data, $\bar{x}$ and m differ considerably, as do s and q; this is because they measure or describe different aspects of the data. Most people feel m and q are the more useful descriptors for skewed data sets.

Table 6.1.6 presents the ordered values for the iron ppm measurements from Table 6.1.1. Recall earlier we found $\bar{x} = 101.12$ and $s = 3.415$ for these data. You can easily see that the median iron reading is

$$m = \frac{(x_{(30)} + x_{(31)})}{2} = \frac{(100 + 100)}{2}$$

$$= 100$$

whereas

$$\frac{n}{4} = \frac{60}{4} = 15$$

so

$$x_{.25} = x_{(15)} = 99$$
$$x_{.75} = x_{(46)} = 103$$

Table 6.1.6

95	96	96	96	97	97	97	97	98	98
98	98	99	99	99	99	99	99	99	100
100	100	100	100	100	100	100	100	100	100
100	100	100	101	101	102	102	102	102	102
102	102	102	103	103	103	103	103	104	104
104	105	106	106	107	107	108	108	109	110

and the interquartile range is

$$q = 103 - 99 = 4.$$

For this (better behaved) set of data either $\bar{x}$ and s or m and q provide useful descriptors.

EXERCISE 6.1

1. The numbers of paid admissions to a swimming pool on 25 successive days are given in the following table.

461	425	415	236	483
218	279	343	554	552
358	593	233	573	350
238	364	570	480	513
107	534	468	599	216

   (a) Compute $\bar{x}$, s, and s^2 for these data.
   (b) Construct the frequency table and histogram for these data using 100 to 199, 200 to 299, 300 to 399, 400 to 499, 500 to 599 as class limits.

2. Twenty identical 75-watt light bulbs were placed in operation and the number of hours of service provided by each (before burning out) was measured. The resulting values are given in the following table.

689	790	723	847	952	805	619	808	558	780
723	803	568	679	835	637	764	734	937	632

   (a) Compute $\bar{x}$, s^2, and s for these data.
   (b) Construct a frequency table and histogram for these data, using four equal length classes.

3. The "tar" content of 25 cigarettes was measured by the FDA, resulting in the following data.

7.6	7.8	7.9	8.6	8.6
8	9.1	7.6	8.8	8
7.7	8	8.7	8.3	8.4
6.9	8.5	8.3	9.6	7.6
7.8	8.1	7.8	7.6	8

(a) Compute $\bar{x}$, s^2, and s for these data.
(b) Construct a frequency table and histogram for these data.

4. A "random number" generator produced the sequence of numbers given in the following table.

28	15	80	53	95	55
62	15	71	9	0	70
23	31	65	68	38	14
84	95	14	40	56	48
96	19	62	65	80	47
20	90	14	41	88	16
36	13	45	45	93	21
90	86	50	81	46	63
82	70	95	28	51	41
87	72	71	70	1	52

(a) Compute $\bar{x}$, s^2, and s for these data.
(b) Using 10 equal length classes, construct the frequency table and draw the histogram for these data.

5. The weights of the first 48 Miss America contest winners are given in the following table.

128	119	125	120	118	121	110	125
135	116	115	124	124	115	118	116
120	114	130	120	116	124	132	118
143	119	106	140	130	123	135	125
130	118	120	120	126	128	120	114
120	112	115	118	138	137	140	108

(a) Compute $\bar{x}$, s^2, and s for these data.
(b) Use 10 equal length classes to construct a frequency table and to draw a histogram for these data.

6. Compute the median and interquartile range for the data given in Exercise 6.1.1.
7. Compute the median and interquartile range for the Miss America weights in Exercise 6.1.5.
8. Compute the median and interquartile range for the random numbers given in Exercise 6.1.4.

9. It is possible for the interquartile range to be zero. Explain how this could happen.

10. Quantiles other than the quartiles can be (and are) used to describe the variability of a set of data. The interquintile range is $x_{.8} - x_{.2}$, where (at most) 20 percent of the data values are no greater than $x_{.2}$ and (at most) 20 percent are no less than $x_{.8}$.

   (a) What proportion (roughly) of the data values are covered by the interquintile range?

   (b) Compute the interquintile range for the income data in Table 6.1.5 and for the iron ppm values in Table 6.1.6.

*11. For a given set of data $x_1, x_2, \ldots, x_n$ show that

$$\sum_{i=1}^{n} (x_i - a)^2 = Q(a)$$

is minimized with $a = \bar{x} = 1/n \sum x_i$. (*Hint.* Set $dQ/da = 0$ and solve.)

*12. For a given set of data $x_1, x_2, \ldots, x_n$ show that

$$M(a) = \sum_{i=1}^{n} |x_i - a| = \sum_{x_i \leq a} (a - x_i) + \sum_{x_i > a} (x_i - a)$$

is minimized with $a = x_{((n+1)/2)}$ if n is odd or any value between $x_{(n/2)}$ and $x_{((n+2)/2)}$ if n is even.

13. Any set of data $x_1, x_2, \ldots, x_n$ can be transformed to "standard scale" by defining the standardized variables

$$z_i = \frac{x_i - \bar{x}}{s}, \qquad i = 1, 2, \ldots, n,$$

where $\bar{x} = \frac{1}{n} \sum x_i$, $\quad s^2 = \frac{1}{n-1} \sum (x_i - \bar{x})^2$.

   (a) Show that $\bar{z} = \frac{1}{n} \sum_{i=1}^{n} z_i = 0$.

   (b) Show that $s_z^2 = \frac{1}{n-1} \sum_{i=1}^{n} z_i^2 = 1$.

*14. The Chebyshev inequality can also be applied to any set of data to bound the number of data values outside a fixed distance of the mean; the natural scale factor for this measure of distance is s (or $\sqrt{(n-1)/(n)} \cdot s$). Let $z_1, z_2, \ldots, z_n$ be a set of data on the "standard scale" (see

Exercise 6.1.13) and define

$$y_i = 1 \quad \text{if} \quad |z_i| > \varepsilon.$$
$$= 0 \quad \text{if} \quad |z_i| \leq \varepsilon.$$

Then $\sum y_i$ = number of z_i's that exceed ε in magnitude; also it is easy to see that $\varepsilon^2 y_i \leq z_i^2 y_i \leq z_i^2$.

(a) Show that the number of z_i's that exceed ε in magnitude $\leq \dfrac{1}{\varepsilon^2} \sum z_i^2$
   $= (n - 1)/\varepsilon^2$.

(b) Use (a) to show that the number of values x_i for which $|x_i - \bar{x}| > s\varepsilon$ can be no larger than $(n - 1)/\varepsilon^2$.

15. Suppose $x_1, x_2, \ldots, x_n$ is a given set of data and $a, b \neq 0$ are any two constants. Then we can define a new set of data values by

$$y_i = a + bx_i, \quad i = 1, 2, \ldots, n.$$

(a) Show that

$$\bar{y} = \frac{1}{n} \sum_{i=1}^{n} y_i = a + b\bar{x}.$$

(b) Show that

$$s_y^2 = \frac{1}{n - 1} \sum (y_i - \bar{y})^2 = b^2 s_x^2.$$

(c) Show that $m_y = a + bm_x$, where m_y, m_x are the medians of the two data sets.

(d) Show that $q_y = b|q_x|$, where q_y and q_x are the interquartile ranges for the two.

The values $y_1, y_2, \ldots, y_n$ are frequently called the "coded" values for $x_1, \ldots, x_n$. If a and b can be chosen so that $\bar{y}$ and s_y are easily evaluated, then computations made with the coded values give a simple way of evaluating $\bar{x}$ and s_x.

16. When a data set has been summarized by a frequency table, it is not possible to exactly evaluate the mean and standard deviation of the original data from just the frequency table (if two or more possible data values occur in the same class). They can, however, be fairly accurately approximated from the frequency table. Let $x_1, x_2, \ldots, x_n$ be the original numbers and assume they have been summarized (counts given) into k classes. Also, let $c_1, c_2, \ldots, c_k$ be the midpoints of the classes (sometimes called the class marks) and let $f_1, f_2, \ldots, f_k$ be the numbers of data values that fall into these classes. Then $f_i c_i$ is approximately the total of the x_j's that fall in class i and $f_i c_i^2$ is approximately

the sum of squares of the x_j's that fall in class i. Using these facts show that

$$\bar{x} \doteq \frac{1}{n} \sum_{i=1}^{k} f_i c_i, \qquad s^2 \doteq \frac{1}{n-1} \sum_{i=1}^{k} f_i (c_i - \bar{c})^2.$$

6.2   Inferential Statistics

No probability theory is used or required in describing sets of data, $x_1, x_2, \ldots, x_n$, as discussed in Section 6.1; the thrust of descriptive techniques is simply to describe, regardless of the origin of the data. In particular, no attempt is made to use the data set to generalize beyond the numbers in hand. Inferential statistics, on the other hand, is primarily concerned with the source of the data itself and with generalizing beyond the numbers in the data set back to this source, attempting to make generalizations beyond the actual data in hand. This is accomplished by constructing a model that describes the origin of the data, a set of assumptions about the data that carry implications beyond the numbers obtained.

Probability theory is very useful in constructing models for data sets. In general, the numbers in the data set are assumed to be observed values for random variables, numbers that have been observed as the result of performing an experiment, for example. This probabilistic model then can be used in making inferences about the phenomena studied in the experiment.

As a simple example of this sort, consider again the 60 numbers in Table 6.1.1, the repeated iron ppm readings obtained from burning small samples of oil from the same oil reservoir. The iron reading produced by the spectrometer is affected by several quantities. If the iron contamination in the reservoir is not uniformly dispersed, it is possible that the actual amount of iron content varies from one to another of these small samples burned on the spectrometer. Even if the true iron content is in fact constant from one sample to another, the spectrometer readings still would vary because of the way the spectrometer works: it analyzes the light emitted when the oil is burned and things such as slight voltage changes in its power supply, variations in room temperature and humidity, purity of the electrodes used, and so on, will cause variability in its readings. We can take these various possible sources of variability into account by assuming the 60 numbers observed are actually observed values of random variables. Thus the variability in these 60 numbers is accounted for by the fact that these are observed values for random variables.

In fact, a useful probabilistic model for these data goes as follows:

We have an oil reservoir whose true iron content is, say, μ parts per million (this number is unknown; one reason the oil is analyzed on the spectrometer is to get information about μ). Let us assume the oil is well mixed and homogeneous and that if we burn a small sample (roughly 10^{-4} cubic centimeters) of the oil on a spectrometer, the reading produced by the spectrometer is a normal random variable with parameters μ and σ. (This assumption of a normal distribution for X seems justified from the way the machine operates, a topic beyond the scope of our discussion, although we will discuss how we might judge whether observed data are consistent with such an assumption.) As long as the spectrometer is well calibrated and good laboratory procedures are followed, the random variables produced by burning repeated samples from the reservoir should in fact be independent, each with the same normal distribution. Thus if 60 samples are burned, we can model the 60 readings to be produced as being $X_1, X_2, ...,$ X_{60}, independent normal random variables, each with mean μ and standard deviation σ. The marginal density for each X_i then is

$$f(x) = \frac{1}{\sigma\sqrt{2\pi}} e^{-(x-\mu)^2/2\sigma^2}$$

and because of the assumed independence of $X_1, X_2, ..., X_{60}$, their joint density is given by the product of their marginal density functions. Note that we have in fact assumed the true iron content μ is the mean of the density for each X_i; the "most probable" value for each reading is μ, the unknown true iron content of the reservoir. After the 60 numbers are known, say, those given in Table 6.1.1, what should be done with them to get information about μ, the iron ppm content of the reservoir? This is a question of statistical inference, how to use observed values for random variables to gather information about their probability law. We will study a number of such questions and procedures for deriving answers to them as we proceed. (The obvious answer, that $\bar{x} = 101.12$ is the "best guess" for the value of μ, is correct from several points of view; we will see several ways for justifying this intuitive result. The procedures we will look at are, of course, also applicable in situations that may not lend themselves to such intuitive solutions.)

As a second simple example of a probabilistic model, assume a university student has an 8 A.M. class every day and there is a constant (unknown) probability p she will be late for this class each day. We can think of her arrival time at this class each day as being a Bernoulli trial; "success" is a late arrival for the class. Let us also assume these Bernoulli trials are independent. For a sequence of n class days, then, we will observe $X_1, X_2,$ $..., X_n$, independent Bernoulli random variables, each with parameter p;

$X_i = 1$ if she is late on day i and $X_i = 0$ if she is not late. How could this sequence of 0's and 1's be used to get information about the value for p, her assumed probability of being late to this 8 A.M. class each day? Again, realizing that $\sum_{i=1}^{n} x_i$ gives the number of days out of the n that she was late, it would seem intuitively clear that $\bar{x} = (1/n) \sum x_i$, the proportion of days she was late, would be a good guess for p.

In both these examples we discussed models in which we observed values of independent random variables $X_1, X_2, ..., X_n$, and each of the X_i's had the same probability law. We will call $X_1, X_2, ..., X_n$ a *random sample* of size n of a random variable X in this case, as given in the following definition.

DEFINITION 6.2.1. If $X_1, X_2, ..., X_n$ are independent, identically distributed, each with the same probability law (the probability law for some random variable X), we will call $X_1, X_2, ..., X_n$ a random sample of the random variable X.   ■

Definition 6.2.1 will be used very frequently in the remainder of this book. If $X_1, X_2, ..., X_n$ is a random sample of X, their joint probability law then is completely specified by their marginal distributions, as we know from Chapter 5; this means we can easily specify their joint probability law. This is very useful in answering questions of statistical inference.

If $X_1, X_2, ..., X_n$ is a random sample of a random variable X, many authors call X the *population random variable*. This terminology comes about in the following way. Consider again the spectrometric oil analysis readings discussed earlier. We could in theory make an arbitrarily large (ideally even infinite) number of repeated burns of oil samples from this same oil reservoir. Each burn gives rise to a number, an iron reading from the spectrometer. Our earlier assumptions, that these iron readings are normally distributed with mean μ and variance σ^2, then can be interpreted to mean that the *proportion* of these readings (numbers) that lie in a small interval of length Δx centered at x is approximately given by

$$f(x)\,\Delta x = \frac{\Delta x}{\sigma\sqrt{2\pi}}\, e^{-(x-\mu)^2/2\sigma^2},$$

the approximate area under the density over the interval, for whatever x we want to consider (for some choice of μ and σ^2). This conceptual population of numbers (iron readings) is described by a normal probability law and the random variable with this probability law is called the population random variable X. We will say the iron readings are normally distributed with mean μ and variance σ^2. The actual readings that we observe,

$x_1, x_2, \ldots, x_{60}$, are called a random sample of values selected from this conceptual population. The probability law for X_1, the first number to be observed (or indeed any of the others), then has the same probability law as X, the population random variable; that is, its probability law is also normal with parameters μ and σ^2. If the values we observe are truly selected "at random" from this conceptual population, $X_1, X_2, \ldots, X_n$ are then independent (actually, this independence is what we mean by selecting at random).

Definition 6.2.1 is not satisfied if n numbers are selected at random without replacement from a finite population. To see this, suppose a small business concern has 10 employees, each paid on an hourly basis. Of the 10 employees, let us also assume 4 are paid \$3.10/hour, 3 are paid \$3.85/hour, 2 are paid \$4.50/hour and 1 is paid \$6.50/hour. This then defines a population of 10 numbers, the hourly wages paid; the population random variable X has mass function

$$
\begin{aligned}
p_X(x) &= \tfrac{4}{10}, & x &= 3.10 \\
&= \tfrac{3}{10}, & x &= 3.85 \\
&= \tfrac{2}{10}, & x &= 4.50 \\
&= \tfrac{1}{10}, & x &= 6.50.
\end{aligned}
$$

Suppose we select 3 employees at random without replacement from the 10 and record their hourly wages; equivalently, we select three numbers at random from the collection 3.10, 3.10, 3.10, 3.10, 3.85, 3.85, 3.85, 4.50, 4.50, 6.50. If X_1, X_2, X_3 represent the three numbers selected, then the marginal (unconditional) probability function for each of the three is $p_X(x)$ as previously defined, so X_1, X_2, X_3 are in fact identically distributed. (Why is this?) They are not independent, though; the conditional probability function for X_2, given $X_1 = 3.10$, is

$$
\begin{aligned}
p_{X_2|X_1}(x_2 | x_1 = 3.10) &= \tfrac{3}{9}, & x_2 &= 3.10 \\
&= \tfrac{3}{9}, & x_2 &= 3.85 \\
&= \tfrac{2}{9}, & x_2 &= 4.50 \\
&= \tfrac{1}{9}, & x_2 &= 6.50,
\end{aligned}
$$

whereas if $X_1 = 6.50$, the conditional probability function for X_2 is

$$
\begin{aligned}
p_{X_2|X_1}(x_2 | x_1 = 6.50) &= \tfrac{4}{9}, & x_2 &= 3.10 \\
&= \tfrac{3}{9}, & x_2 &= 3.85 \\
&= \tfrac{2}{9}, & x_2 &= 4.50.
\end{aligned}
$$

Since $p_{X_2|X_1}(x_2|x_1) \neq p_{X_2}(x_2)$, we know X_1, X_2, and X_3 are not independent and thus X_1, X_2, X_3 is not a random sample of X as previously defined. If a sample of size n is selected at random without replacement from a finite population, and n is "small" compared to the size of the population, then

$X_1, X_2, \ldots, X_n$ approximately satisfies Definition 6.2.1. We will not formally treat the question of sampling from a finite population; the interested reader may consult books on the theory of sample surveys for more information.

We will frequently refer to $X_1, X_2, \ldots, X_n$ as the *elements* of the sample, when $X_1, X_2, \ldots, X_n$ is a random sample of a random variable X. They are jointly distributed random variables and, if we know the population probability law, we can immediately evaluate the probability law for the elements of the sample. Generally, of course, some one or more aspects of the population probability law will be unknown to us, or there would not have been any reason to select the sample in the first place. For example, the true iron content μ of an oil reservoir may be unknown to us, but we are willing to assume $X_1, X_2, \ldots, X_n$ is a random sample of a normal random variable X with parameters μ and σ^2. Or it may be that we are only willing to assume $X_1, X_2, \ldots, X_n$ are independent and want to know whether it seems plausible that X has a normal probability law. Our goal is to use the elements of the sample to answer questions, as best we can, regarding the population probability law.

We will see that certain functions of the elements of a random sample, such as

$$g_1(X_1, X_2, \ldots, X_n) = \frac{1}{n} \sum_{i=1}^{n} X_i = \overline{X}$$

$$g_2(X_1, X_2, \ldots, X_n) = \frac{1}{n-1} \sum (X_i - \overline{X})^2$$

are very useful in answering questions about the population probability law. Any function of the elements of a random sample, which does not depend on any unknown parameters, is called a *statistic*. Thus statistics are themselves random variables whose observed values can be evaluated after the observed values for $X_1, X_2, \ldots, X_n$ are known. Because statistics are random variables we will in general continue to use capital letters to represent them and use lowercase letters to represent their observed values.

As seen in Example 5.4.3, if $X_1, X_2, \ldots, X_n$ are uncorrelated, each with mean μ and variance σ^2, then $\overline{X}$ also has mean μ and variance σ^2/n. Because of the importance of this result, let us restate it here as a theorem, for a random sample of a random variable.

Theorem 6.2.1.   If $X_1, X_2, \ldots, X_n$ is a random sample of a random variable X with mean μ and variance σ^2, $\overline{X} = (1/n) \sum X_i$ also has mean μ and variance σ^2/n. ∎

It is important to realize what this result says (and does not say) about $\overline{X}$, the mean of a random sample. The fact that $E[\overline{X}] = \mu$ does not say that the mean $\bar{x}$ of n observed numbers is equal to the population mean μ. It does say that if we *repeatedly* take samples of size n and compute $\bar{x}$ for each, the (very) long-run average of the $\bar{x}$'s is μ, the same number as the population mean. Similarly, the result $\text{Var}[\bar{X}] = \sigma^2/n$ describes the variability of these observed $\bar{x}$ values over repeated samples of the same size. Thus the probability law for $\bar{X}$ is more concentrated about μ than is the original population probability law. If, for example, the population random variable X has a normal probability law as pictured in Figure 6.2.1, then $\bar{X}$ also has a normal probability law (since it is a linear function of normal random variables), but its standard deviation is $1/\sqrt{n}$ times the population standard deviation. The density for $\bar{X}$ with $n = 25$ is also plotted in Figure 6.2.1; because the density for $\bar{X}$ is much more concentrated about μ, the probability $\bar{X}$ lies close to μ is much greater than the probability an individual sample value selected at random from the population lies within the same distance of μ. This phenomenon is behind the law of large numbers and

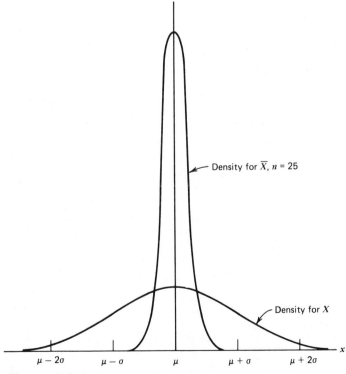

Density for $\bar{X}$, $n = 25$

Density for X

x

$\mu - 2\sigma$ $\mu - \sigma$ μ $\mu + \sigma$ $\mu + 2\sigma$

Figure 6.2.1

provides some impetus for using the observed value, $\bar{x}$, as a guess for μ if μ is unknown. If the population distribution is not normal, of course, then the two densities (for X and $\bar{X}$) are not necessarily so nicely symmetric about the value of μ, but the same relative phenomenon occurs; still the standard deviation for $\bar{X}$ is $\sigma/\sqrt{n}$, so the density for $\bar{X}$ is more concentrated about μ. The central limit theorem (as seen in Section 5.6) shows that the density for $\bar{X}$ tends to the normal as $n \to \infty$ and for large, but finite, n its density will be well approximated by a normal density, even though the population density may be far from normal (so long as σ^2 exists).

When $X_1, X_2, \ldots, X_n$ is a random sample of X, each of $X_1, X_2, \ldots, X_n$ has the same probability law as X. Thus it must be true that

$$E[X_i^2] = E[X^2] = \mu^2 + \sigma^2, \quad i = 1, 2, \ldots, n,$$

where the population mean is μ and its variance is σ^2. Furthermore, since the variance of $\bar{X}$ is σ^2/n and its mean is μ, we have

$$E[\bar{X}^2] = \mu^2 + \frac{\sigma^2}{n},$$

and thus

$$E[X_i^2] - E[\bar{X}^2] = (\mu^2 + \sigma^2) - \left(\mu^2 + \frac{\sigma^2}{n}\right)$$

$$= \frac{(n-1)\sigma^2}{n}, \quad i = 1, 2, \ldots, n;$$

but then

$$E\left[\sum_{i=1}^{n}(X_i - \bar{X})^2\right] = E\left[\sum_{i=1}^{n}X_i^2 - n\bar{X}^2\right]$$

$$= \sum_{i=1}^{n}E[X_i^2] - nE[\bar{X}^2]$$

$$= nE[X^2] - nE[\bar{X}^2]$$

$$= n\left(\frac{(n-1)\sigma^2}{n}\right) = (n-1)\sigma^2.$$

Hence if we define the statistic

$$S^2 = \frac{1}{n-1}\sum_{i=1}^{n}(X_i - \bar{X})^2,$$

called the *sample variance*, it follows that

$$E[S^2] = \frac{1}{n-1} E\left[\sum_{i=1}^{n} (X_i - \bar{X})^2\right] = \frac{(n-1)\sigma^2}{n-1} = \sigma^2.$$

This proves Theorem 6.2.2.

Theorem 6.2.2. If $X_1, X_2, ..., X_n$ is a random sample of a random variable X with mean μ and variance σ^2, the *sample variance*

$$S^2 = \frac{1}{n-1} \sum_{i=1}^{n} (X_i - \bar{X})^2$$

has expected value σ^2. Its positive square root, $S = \sqrt{S^2}$, is called the *sample standard deviation*.   ∎

   Again, it is important to consider carefully what this result says about S^2, the sample variance. If we observe the elements $x_1, x_2, ..., x_n$ of a random sample from a population, it is not true that

$$s^2 = \frac{1}{n-1} \sum_{i=1}^{n} (x_i - \bar{x})^2$$

is equal to the population variance σ^2. If we were repeatedly to take random samples of the same size and compute s^2 for each of them, the long-term average of these s^2 values is σ^2, the same number as the population variance. Because of this result, the observed value s^2 for S^2 seems a reasonable guess for the population variance σ^2; it is also the main reason we used the divisor $n - 1$ instead of n in defining the variance of a set of data (Definition 6.1.1).

   Let us now discuss a very important result that is used frequently when sampling from normal populations. Suppose $Z_1, Z_2, ..., Z_n$ are independent, standard normal random variables and let $\bar{Z} = (1/n) \sum_{i=1}^{n} Z_i$ be their average. Then from Theorem 6.2.1 $\bar{Z}$ is itself normal with mean 0 and variance $1/n$ which in turn says that $(\bar{Z} - 0)/(1/\sqrt{n}) = \sqrt{n}\bar{Z}$ is also standard normal, so $(\sqrt{n}\bar{Z})^2 = n\bar{Z}^2$ is a χ^2 random variable with 1 degree of freedom. In Example 5.4.5 we saw that $X_i - \bar{X}$ and $\bar{X}$ are uncorrelated for each i, whenever $X_1, X_2, ..., X_n$ are uncorrelated. Thus if $Z_1, Z_2, ..., Z_n$ are independent standard normal, we know that $\bar{Z}$ and $Z_i - \bar{Z}$ are uncorrelated for each i, hence actually independent since each $Z_i - \bar{Z}$ is a linear function of normal random variables and is itself normal, and $\bar{Z}$ and $(Z_i - \bar{Z})$ are bivariate normal. But then from Theorem 5.4.4 and its corollary it must be true that $\bar{Z}$ and $\sum_{i=1}^{n} (Z_i - \bar{Z})^2$ are independent random variables (as are $n\bar{Z}^2$ and $\sum_{i=1}^{n} (Z_i - \bar{Z})^2$). From the computational formulas in

Section 6.1,

$$\sum_{i=1}^{n} (Z_i - \bar{Z})^2 = \sum_{i=1}^{n} Z_i^2 - n\bar{Z}^2$$

or

$$\sum_{i=1}^{n} (Z_i - \bar{Z})^2 + n\bar{Z}^2 = \sum_{i=1}^{n} Z_i^2.$$

Since $n\bar{Z}^2$ and $\sum_{i=1}^{n} (Z_i - \bar{Z})^2$ are independent, we have

$$E[\exp(t \sum Z_i^2)] = E[\exp\{t[\sum (Z_i - \bar{Z})^2 + n\bar{Z}^2]\}]$$
$$= E[\exp[t \sum (Z_i - \bar{Z})^2]] E[\exp(tn\bar{Z}^2)].$$

Now $\sum_{i=1}^{n} Z_i^2$ is a χ^2 random variable with n degrees of freedom and (as noted) $n\bar{Z}^2$ is a χ^2 random variable with 1 degree of freedom. But then this equation says

$$\frac{1}{(1 - 2t)^{n/2}} = E[\exp[t \sum (Z_i - \bar{Z})^2]] \frac{1}{(1 - 2t)^{1/2}}$$

from which it follows that

$$E[\exp[t \sum (Z_i - \bar{Z})^2]] = \frac{1}{(1 - 2t)^{(n-1)/2}} ,$$

the moment generating function for a χ^2 random variable with $n - 1$ degrees of freedom; that is, $\sum_{i=1}^{n} (Z_i - \bar{Z})^2$ is a χ^2 random variable with $n - 1$ degrees of freedom and is independent of $\bar{Z}$ when $Z_1, Z_2, ..., Z_n$ are independent standard normal.

What is the significance of these facts for sampling from a normal population? Assume $X_1, X_2, ..., X_n$ is a random sample from a normal population with mean μ and variance σ^2. Then

$$Z_1 = \frac{X_1 - \mu}{\sigma}, \qquad Z_2 = \frac{X_2 - \mu}{\sigma}, \qquad ..., \qquad Z_n = \frac{X_n - \mu}{\sigma}$$

are independent standard normal, where

$$\bar{Z} = \frac{1}{n} \sum_{i=1}^{n} \frac{(X_i - \mu)}{\sigma} = \frac{\bar{X} - \mu}{\sigma}$$

$$Z_i - \bar{Z} = \frac{X_i - \mu}{\sigma} - \frac{\bar{X} - \mu}{\sigma} = \frac{X_i - \bar{X}}{\sigma}$$

and thus we know that

(a) $\bar{Z} = (\bar{X} - \mu)/\sigma$ is independent of

$$\sum_{i=1}^{n} (Z_i - \bar{Z})^2 = \sum_{i=1}^{n} \frac{(X_i - \bar{X})^2}{\sigma^2} = \frac{(n-1)S^2}{\sigma^2},$$

so $\bar{X}$ and S^2 are independent random variables.

(b) $\sum (Z_i - \bar{Z})^2 = (n-1)S^2/\sigma^2$ is a χ^2 random variable with $n-1$ degrees of freedom.

These results are summarized in the following theorem.

Theorem 6.2.3. Assume $X_1, X_2, \ldots, X_n$ is a random sample of a normal random variable with mean μ and variance σ^2. Then

(a) The sample mean, $\bar{X}$, and sample variance, S^2, are independent random variables.

(b) $(n-1)S^2/\sigma^2$ is a χ^2 random variable with $n-1$ degrees of freedom. ∎

From the results in Theorem 6.2.3 it is easy to see what the distribution for $\sqrt{n}(\bar{X} - \mu)/S$ must be, when $X_1, X_2, \ldots, X_n$ is a random sample from a normal population. We know that $(\sqrt{n})(\bar{X} - \mu)/\sigma$ is a standard normal random variable, that $(n-1)S^2/\sigma^2$ is an independent χ^2 random variable with $n-1$ degrees of freedom and thus

$$\frac{\sqrt{n}(\bar{X} - \mu)}{\sigma} \Bigg/ \sqrt{\frac{(n-1)S^2}{\sigma^2(n-1)}} = \frac{\sqrt{n}(\bar{X} - \mu)}{S}$$

is the ratio of a standard normal random variable over the square root of an independent χ^2 random variable, divided by its degrees of freedom. We saw in Section 5.7 that this ratio has a T distribution with $n-1$ degrees of freedom. Thus we have one further theorem.

Theorem 6.2.4. If $X_1, X_2, \ldots, X_n$ is a random sample of a normal random variable with mean μ and variance σ^2, $(\sqrt{n})(\bar{X} - \mu)/S$ has a T distribution with $n-1$ degrees of freedom. ∎

What is remarkable about this result is that the unknown parameter σ cancels in forming the ratio $(\sqrt{n}(\bar{X} - \mu)/\sigma)/\sqrt{S^2/\sigma^2}$ and thus the probability law for $(\sqrt{n})(\bar{X} - \mu)/S$ is the same, no matter what the value for σ. Thus this random variable with a T distribution will prove very useful in making inferences about the mean μ of a normal population whose variance σ^2 is unknown.

One basic idea, which we have in fact already used in other contexts and is frequently used in problems of statistical inference, is that the probability of occurrence of equivalent events must be equal. Two events A and B are called equivalent if the occurrence of A implies the occurrence of B and the occurrence of B implies the occurrence of A; in our earlier language of events this means $A \subset B$ and $B \subset A$, in short $A = B$. We used this idea of equivalent events freely in discussing probability laws for functions of a random variable. For example, if a and $b > 0$ are any two constants and $Y = a + bX$, then $\{Y \leq t\}$ is equivalent to $\{a + bX \leq t\}$, which in turn is equivalent to $\{X \leq (t - a)/b\}$; this reasoning was the basis for the result that

$$F_Y(t) = F_X\left(\frac{t - a}{b}\right)$$

whenever $Y = a + bX$, with $b > 0$. More recently (Section 5.7), we used this equivalent event approach to find $F_{X_{(n)}}(t) = [F_X(t)]^n$ if $X_1, X_2, ..., X_n$ are independent random variables, each with distribution function $F_X(t)$ and $X_{(n)}$ is the maximum of $X_1, X_2, ..., X_n$. The equivalent event concept has many applications in probability theory.

Many interesting questions of statistical inference can also be answered using this equivalent event concept. For example, suppose X is an exponential random variable with density function

$$f(x) = \lambda e^{-\lambda x}, \qquad x > 0,$$

pictured in Figure 6.2.2. It is simple to find the constants a and b such that $P(X \leq a) = .1$, $P(X \geq b) = .1$. In fact

$$P(X \leq a) = 1 - e^{-\lambda a} = .1 \qquad \text{gives} \qquad a = -\frac{1}{\lambda}\ln .9 = \frac{.11}{\lambda}$$

$$P(X \geq b) = e^{-\lambda b} = .1 \qquad \text{gives} \qquad b = -\frac{1}{\lambda}\ln .1 = \frac{2.21}{\lambda};$$

These values are also pictured in Figure 6.2.2. We have then that

$$P\left(\frac{.11}{\lambda} \leq X \leq \frac{2.21}{\lambda}\right) = .8$$

if X is an exponential random variable with parameter λ. The event $\{.11/\lambda \leq X \leq 2.21/\lambda\}$ is equivalent to the event $\{.11/X \leq \lambda \leq 2.21/X\}$, which in turn is equivalent to $\{X/2.21 \leq 1/\lambda \leq X/.11\}$; this latter event says

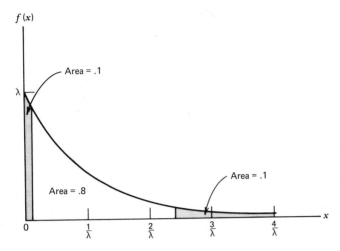

Figure 6.2.2

$1/\lambda = \mu$ (the mean for X) lies between $X/2.21$ and $X/.11$; then since

$$P\left(\frac{.11}{\lambda} \leq X \leq \frac{2.21}{\lambda}\right) = .8,$$

it must also be true that

$$P\left(\frac{X}{2.21} \leq \mu \leq \frac{X}{.11}\right) = .8$$

since the events are equivalent. This latter statement is of some interest in problems of inference. Suppose we assume the times to failure for an electronic component made in a certain way have an exponential probability law with parameter λ (and, of course, λ is unknown); that is, if X is the time to failure for one of these components, then X is an exponential random variable with mean $\mu = 1/\lambda$. As before,

$$P\left(\frac{X}{2.21} \leq \mu \leq \frac{X}{.11}\right) = .8.$$

Let us further suppose we place one of these components in service and observe that it fails at $X = 30$ hours. Then $x/2.21 = 30/2.21 = 13.57$, $x/.11 = 30/.11 = 272.73$ and in a sense we can say we are 80 percent sure (or confident) that the average time to failure for all components of this type ($\mu = 1/\lambda$) lies between 13.57 and 272.73 hours, based on this one observed failure time of $x = 30$ hours. What is the basis for this statement? Our relative frequency interpretation for probability says that

$$P\left(\frac{X}{2.21} \leq \mu \leq \frac{X}{.11}\right) = .8$$

means that 80 percent of the time μ will lie between $X/2.21$ and $X/.11$; equivalently, if we repeatedly were to take independent observations of X, we would find that in the long run 80 percent of the (random) intervals $[X/2.21, X/.11]$ cover the population mean μ and the other 20 percent do not. Thus we can claim to be 80% sure that our single interval $[13.57, 272.73]$ includes the unknown population mean μ, that it is in fact one of those in the 80% majority. After we have observed $x = 30$ hours we still do not know the value for μ, of course, but the preceding statement does have some inferential content regarding the value for μ.

As you undoubtedly noticed from this discussion, the interval $[13.57, 272.73]$ is quite wide, so the value for μ has not been pinned down very precisely. Could not we get more information (maybe a shorter interval or a larger probability) if we based our statement on a sample of 2 or 3 or perhaps some larger number n of observations? The answer to this query is yes, of course, as we will see. If we base our interval statement on a random sample of $n > 1$ observations from the population, we can expect to get a shorter interval (with the same probability) or a larger probability that the interval includes μ (for the same length interval) or, in general, both, when we use more observations. Statistical inference is concerned with using an assumed probability model, together with a set of observed data, to derive information about the population from which the sample was selected.

Recall from Section 5.7 that if $X_1, X_2, ..., X_n$ are independent, each with continuous distribution function $F(x)$, (thus this is a random sample of a continuous random variable X), the ordered values $X_{(1)} \leq X_{(2)} \leq \cdots \leq X_{(n)}$ are called the order statistics of the sample. Also, recall from the probability integral transform that if X is continuous with distribution function $F(x)$, then the random variable $U = F(X)$ is uniform on $(0, 1)$. Thus $U_1 = F(X_1), U_2 = F(X_2), ..., U_n = F(X_n)$ is a random sample of a uniform $(0, 1)$ random variable and $U_{(1)} = F(X_{(1)}), U_2 = F(X_{(2)}), ..., U_{(n)} = F(X_{(n)})$ are the order statistics of a random sample of a uniform random variable. Let $U_{(j)} = F(X_{(j)})$ be the jth uniform order statistic. Then the event $\{U_{(j)} \leq t\}$ occurs if and only if j or more of the uniform random variables are less than or equal to t; that is,

$$P(U_{(j)} \leq t) = \sum_{k=j}^{n} P(k \text{ uniforms are each no larger than } t).$$

Now for an individual uniform $(0, 1)$ random variable

$$P(U \le t) = t;$$

n independent uniform random variables define n independent Bernoulli trials, with parameter t, because each one either is no larger than t (with probability t) or not (with probability $1 - t$). The probability of exactly k successes (observed uniform no larger than t) in n trials then is binomial, n, t, and from the preceding information

$$P(U_{(j)} \le t) = \sum_{k=j}^{n} \binom{n}{k} t^k (1 - t)^{n-k}.$$

The density function for $U_{(j)}$ is

$$f(t) = \frac{d}{dt} \sum_{k=j}^{n} \binom{n}{k} t^k (1 - t)^{n-k}$$

$$= j \binom{n}{j} t^{j-1} (1 - t)^{n-j}, \qquad 0 < t < 1,$$

because of cancellations in taking the derivative of the sum. Thus the jth uniform order statistic $U_{(j)}$ is a beta random variable with parameters $a = j$ and $b = n - j + 1$. Thus the expected value for $U_{(j)}$ is $j/[1 + (n - j + 1)] = j/(n + 1)$. The expected difference between two successive order statistics thus is $(j + 1)/(n + 1) - j/(n + 1) = 1/(n + 1)$; we would expect the uniform order statistics to be equally spaced.

EXERCISE 6.2

1. Assume $X_1, X_2, \ldots, X_9$ is a random sample of a normal random variable X with unknown mean μ and variance $\sigma^2 = 1$.
    (a) Evaluate $P(|X - \mu| \le \frac{1}{2})$.
    (b) Evaluate $P(|\overline{X} - \mu| \le \frac{1}{2})$.
2. Assume X_1, X_2 is a random sample of size 2 from a normal population known to have mean 0 and variance σ^2. Evaluate
    (a) $P(X_1^2 + X_2^2 \le \sigma^2)$.
    (b) $P(X_1^2 + X_2^2 \le 2\sigma^2)$.
    [*Hint.* What is the probability law for $(X_1^2 + X_2^2)/\sigma^2$?]
*3. Suppose as in Exercise 6.2.2, a random sample of size 2 is selected from a normal population known to have mean 0 and variance σ^2; further assume $x_1 = -.75$, $x_2 = .16$. How sure (or confident) would you be that
    (a) $\sigma^2 \ge 1.0221$?
    (b) $\sigma^2 \ge .4242$?
    (c) $\sigma^2 \ge 2.7909$?

4. Consider the small business with 10 employees discussed in this section (assuming the numbers of employees and wages as given there). Assume 2 employees are selected at random without replacement from the 10 and X_1 is the hourly wage of the first employee selected whereas X_2 is the hourly wage of the second employee selected.

   (a) Find the joint probability law for X_1 and X_2 and use it to verify that both marginal probability laws are identical with $p_X(x)$ as claimed.

   (b) What is the probability the average of the wages of the two employees selected does not exceed $3.10?

   (c) What is the probability the average of the wages of the two employees selected is at least $5.50?

5. Repeat Exercise 6.2.4, assuming 2 employees are selected at random with replacement.

6. Assume $X_1, X_2, ..., X_n$ is a random sample of a random variable X with moments $m_1, m_2, ...$ The kth *sample moment* is the random variable (statistic) defined by

$$M_k = \frac{1}{n} \sum_{i=1}^{n} X_i^k ,$$

   the average of the kth powers of the elements of the sample, $k = 1, 2, 3, ...$ Show that $E[M_k] = m_k$, the kth population moment, for each $k = 1, 2, 3, ...$

7. Assume the scores made on a real estate broker examination are normally distributed with mean $\mu = 200$, $\sigma = 20$. Any person making a score in excess of 230 points on the examination becomes a certified real estate broker. Assume 10 persons take the exam simultaneously and that the scores they make, $X_1, X_2, ..., X_{10}$, are a random sample from this population of possible scores.

   (a) What is the probability that any one of these persons will be certified?

   (b) What is the probability that exactly one of the 10 will be certified?

   (c) What is the probability that at least one of the 10 will be certified?

8. In the discussion in the text we saw that

$$P\left(\frac{X}{2.21} \le \mu \le \frac{X}{.11} \right) = .8$$

   if X is exponential with parameter λ and $\mu = 1/\lambda$. There are many random intervals, of course, which have probability .8 of covering μ.

   (a) Find the value a such that

$$P\left(\frac{X}{a} \ge \mu \right) = P\left(X \ge \frac{a}{\lambda} \right) = .8.$$

(b) Find the value b such that

$$P\left(\frac{X}{b} \le \mu\right) = P\left(X \le \frac{b}{\lambda}\right) = .8.$$

*(c) Show that the interval in (a) is the shortest, and the interval in (b) is the longest, interval that could be constructed from $f(x)$ with probability .8 of covering $\mu = 1/\lambda$.

9. Assume, as in the text and in Exercise 6.2.8, X is an exponential random variable with mean $\mu = 1/\lambda$. With an observed value of $x = 30$, find three intervals, each of which you feel is 90 percent sure to include the value for $\mu = 1/\lambda$.

*10. For the 10 persons discussed in Exercise 6.2.7, let $\overline{X}$ be the average of their scores and let S^2 be the sample variance of their scores.
 (a) Evaluate $P(\overline{X} \ge 210)$ and $P(180 \le \overline{X} \le 195)$.
 (b) Evaluate $P(S^2 \le 371)$ and $P(185 \le S^2 \le 653)$. (*Hint.* What is the probability law for $(n-1)S^2/\sigma^2 = 9S^2/400$?)

*11. It was shown in Theorem 6.2.2 that $E[S^2] = \sigma^2$ when $X_1, X_2, \dots, X_n$ is a random sample of X. Does this imply that $E[S] = \sigma$, where $S = \sqrt{S^2}$ is the sample standard deviation? If not, evaluate $E[S]$ assuming X, the population random variable, is normal.

12. Assume we know the reading X a spectrometer will produce is a normal random variable with mean μ (true unknown level of iron contamination) and standard deviation $\sigma = 5$ ppm. How large a sample (how many small samples should be burned), n, do we need so that $|\overline{X} - \mu| \le 1$ with probability at least .9?

13. Assume $X_1, X_2, \dots, X_{10}$ is a random sample from a normal population. What is the probability that $|\overline{X} - \mu| \le S$, where $\overline{X} = \frac{1}{10}\sum X_i$, $S^2 = \frac{1}{9}\sum(X_i - \overline{X})^2$?

14. If $X_1, X_2, \dots, X_n$ is a random sample of an exponential random variable with parameter λ, what is the probability law for
 (a) $\sum_{i=1}^{n} X_i$?
 *(b) $2\lambda \sum X_i$? (*Hint.* Look at the moment generating function.)

*15. Assume a population of M numbers, where M_i have value y_i, $i = 1, 2, \dots, k$; thus $\sum_{i=1}^{k} M_i = M$. Thus the population mean is $\mu = \sum_{i=1}^{k} M_i y_i / M$ and the population variance is

$$\sigma^2 = \sum_{i=1}^{k} \frac{M_i(y_i - \mu)^2}{M}$$

Let $X_1, X_2, \dots, X_n$, $n \le M$, be n values selected at random, without replacement, from the population. Evaluate $E[\overline{X}]$, $\text{Var}[\overline{X}]$ and $E[S^2]$, where $\overline{X} = (1/n)\sum_{i=1}^{n} X_i$, $S^2 = (1/(n-1))\sum(X_i - \overline{X})^2$. (*Hint.* The marginal probability law for each X_i is the same as the population distribution; to evaluate $\text{Cov}(X_i, X_j)$, use $E[X_1X_2] = E[X_2E[X_1|X_2]].$)

16. Suppose $X_1, X_2, ..., X_n$ is a random sample of size n of a continuous random variable X with median $t_{.5}$.
    (a) What is the probability that all n sample values are less than $t_{.5}$?
    (b) What is the probability that all n sample values are greater than $t_{.5}$?
    *(c) Use (a) and (b) to evaluate

$$P(X_{(1)} \le t_{.5} \le X_{(n)}),$$

where $X_{(1)}$ is the smallest and $X_{(n)}$ is the largest of the X_i's. After the sample values are observed, this probability tells us how sure we can be that the population median lies between the smallest and largest sample values.

*17. Let $X_1, X_2, ..., X_n$ be a random sample of a continuous random variable with distribution function $F(x)$.
    (a) With n an odd number, find the order statistic $X_{(j)}$ such that the expected area to the left of $X_{(j)}$ is $\frac{1}{2}$.
    (b) With n even, find a linear combination of two order statistics such that the same linear combination of the areas to their left is $\frac{1}{2}$. Is this combination unique?

6.3 Summary

Descriptive statistics: The most commonly used indicators of the middle of a set of data are the mean, $\bar{x}$, and the median, the middle of the ranked data. The most commonly used measures of the variability of a set of data are s, the standard deviation, and the interquartile range. Histograms give a quick indication of both location and spread of a set of data.

Inferential statistics: If $X_1, X_2, ..., X_n$ is a random sample of a random variable X, then (a) $X_1, X_2, ..., X_n$ are independent, and (b) Each X_i has the same marginal probability law. If $X_1, X_2, ..., X_n$ is a random sample of a random variable with mean μ and variance σ^2, then (a) $E[\bar{X}] = \mu$, $\text{Var}[\bar{X}] = \sigma^2/n$, and (b) $E[S^2] = \sigma^2$, where $\bar{X} = \sum X_i/n$, $S^2 = \sum(X_i - \bar{X})^2/(n-1)$. If $X_1, X_2, ..., X_n$ is a random sample of a normal random variable with mean μ and variance σ^2, then (a) $\bar{X}$ is normal with mean μ, variance σ^2/n, (b) $(n-1)S^2/\sigma^2$ is a χ^2 random variable with $n-1$ degrees of freedom, (c) $\bar{X}$ and S are independent random variables, and (d) $(\sqrt{n})(\bar{X}-\mu)/S$ has the T distribution with $n-1$ degrees of freedom.

7

Estimation of Parameters

We will in this chapter begin studying problems of statistical inference. For historical reasons, many problems of inference (and their solutions) have been dichotomized into two areas: estimation of parameters and tests of hypotheses; we will examine the first of these two types in this chapter. Typically, in a problem of parameter estimation we assume we have available a random sample of a random variable X, whose probability law is assumed known, except for the values for the parameters of the probability law. The problem then is to use the observed numbers in the sample to guess (estimate) these unknown parameter values.

For example, an oil reservoir (say, the oil pan of an automobile) may have a certain level of iron contamination (μ parts per million, with μ unknown). A small sample of oil from the reservoir can be burned on a spectrometer, resulting in an observed number x, the spectrometer reading for the iron contamination. Because of the way the spectrometer operates, and other reasons, it may seem reasonable to assume the number produced by the spectrometer is the observed value for a normal random variable with mean μ and variance σ^2 (both unknown). Thus to analyze, say, n samples from this same source, we might assume we will observe $X_1, X_2, \ldots, X_n$, a random sample of a normal random variable X. What is a reasonable use of these values in estimating μ and σ^2? Indeed, what should one look or ask for in using observed values for random variables to estimate parameters?

As a second example, previous experience in testing the reliability (or useful lifetime) of electronic components produced in a given way shows that the times to failure for such components are exponential random variables. Suppose a new design is proposed for producing one of these components; our previous experience leads us to feel that the times to failure for items made according to this new design will again be exponential but with an unknown value for λ. If we construct n components according to this design and assume their times to failure $X_1, X_2, ..., X_n$ will be independent exponential random variables with parameter λ, what can be done with the observed times to estimate λ?

In both these situations we assumed we knew the form or type of probability law for X, the population random variable. The open question is what the value(s) for the parameter(s) may be. In this chapter we will consider the problem of point estimation of unknown parameters (what single number might be the best guess for the parameter's value?) and the question of interval estimation of unknown parameters (can we find an interval with known probability of covering the unknown parameter value?).

Generally, the *estimator* for an unknown parameter will be a function of $X_1, X_2, ..., X_n$ and then is itself a random variable with its own probability law. We will continue to use capital letters for such random variables, sometimes with a caret (^) or tilde (~) to distinguish different rationales of estimation employed. Once the observed sample values $x_1, x_2, ..., x_n$ are known, the observed value for the estimator can also be computed; this observed value for an estimator is called the *estimate* of the parameter (based on the observed sample values) and is denoted by a lowercase letter. Thus if γ represents the true unknown value for the parameter of a probability law, we could use $\hat{\Gamma}$ and $\tilde{\Gamma}$ to represent two different estimators for γ, whose observed values given a sample from the population are, respectively, $\hat{\gamma}$ and $\tilde{\gamma}$; it is important to recognize the difference between γ, $\hat{\Gamma}$, and $\hat{\gamma}$. The observed value of an estimator (the estimate $\hat{\gamma}$, say) will in general not be equal to γ, the true parameter value.

7.1 Estimation by Method of Moments and by Maximum Likelihood

Late in the nineteenth century the English statistician Karl Pearson proposed a system for essentially fitting density functions to large bodies of data. One could think of this procedure as one that "matches" a density function to the histogram of the data. The Pearson system utilizes roughly seven basic types of density functions; which particular type, and which one of that type, should be used with a given data set is determined by the values of the first four moments of the data set. Although we do not have the space to discuss the Pearson system itself for choosing which density "best represents" a body of observed data, we will discuss using the sample moments to estimate the parameters of an assumed population probability law. This procedure for generating estimates is called the *method of moments*. Let us begin by defining the sample moments.

DEFINITION 7.1.1.   Let $X_1, X_2, ..., X_n$ be a random sample of a random variable X. The average value of the kth powers of $X_1, X_2, ..., X_n$

$$M_k = \frac{1}{n} \sum_{i=1}^{n} X_i^k$$

is called the kth sample moment, for $k = 1, 2, 3, ...$ ∎

If we assume the population random variable has a known probability law, with unknown parameter θ, it is generally the case that the first moment of X,

$$m_1 = E[X] = g(\theta)$$

is some function of the unknown parameter. The method of moments then generates an estimator $\tilde{\theta}$ for θ by solving the equation

$$M_1 = g(\tilde{\Theta}),$$

that is, the value $\tilde{\Theta}$ that makes the first sample moment equal to $g(\tilde{\Theta})$ is the method of moments estimator for θ.

DEFINITION 7.1.2.   Given $X_1, X_2, ..., X_n$ is a random sample of a random variable X whose probability law depends on unknown parameters $\theta_1, \theta_2, ..., \theta_k$, say, the method of moments estimators for the parameters are given by setting sample moments equal to population moments and solving the resulting equations simultaneously.   ∎

The following examples illustrate this procedure.

EXAMPLE 7.1.1.   Suppose $T_1, T_2, ..., T_n$ represent independent times to failure for a piece of equipment, assumed to have an exponential lifetime with (unknown) parameter λ. The first moment of the population random variable is

$$E[X] = \frac{1}{\lambda}$$

and the first sample moment is $M_1 = (1/n) \sum_{i=1}^{n} X_i = \bar{X}$. The method of moments estimator for λ is given by solving

$$\bar{X} = \frac{1}{\tilde{\Lambda}},$$

which immediately gives $\tilde{\Lambda} = 1/\overline{X}$ as the estimator. If $n = 5$ and the observed times are 30.4, 7.8, 1.4, 13.1, 67.3 (hours), then $\overline{x} = \frac{120}{5} = 24$ and the estimate for λ is $\tilde{\lambda} = \frac{1}{24} = .042$. ∎

It is possible that the first moment of a probability law with unknown parameter θ does not depend on θ. In such a case the lowest order moments possible are used to find the method of moments estimator.

EXAMPLE 7.1.2.   Assume X is a uniform random variable on the interval $(-\theta, \theta)$, where θ is unknown. The mean for X then is 0, no matter what the value for θ may be. The second moment of X is

$$E[X^2] = \int_{-\theta}^{\theta} \frac{x^2}{2\theta}\,dx = \frac{x^3}{6\theta}\bigg|_{-\theta}^{\theta} = \frac{\theta^2}{3},$$

which is a function of θ. If $X_1, X_2, \ldots, X_n$ is a random sample of X, then, the method of moments estimator is given by solving

$$\frac{1}{n}\sum X_i^2 = M_2 = \frac{\tilde{\Theta}^2}{3},$$

which gives $\tilde{\Theta} = \sqrt{3M_2}$ as the estimator. If the observed values in a sample of size 4 are $-.808, 2.590, 2.314, -.268$, we have

$$\frac{1}{4}\sum x_i^2 = \frac{12.787}{4} = 3.197$$

and the estimate for θ is

$$\tilde{\theta} = \sqrt{3(3.197)} = 3.097. \qquad ∎$$

If a probability law has two or more unknown parameters, an equal number of sample and population moments must be equated, and solved simultaneously.

EXAMPLE 7.1.3.   Suppose the time X between orders for a given part, at a large warehouse, is a gamma random variable with parameters r and λ. As we know from Chapter 4,

$$E[X] = \frac{r}{\lambda}$$

$$E[X^2] = \frac{r(1 + r)}{\lambda^2}$$

so the method of moments estimators for the two parameters, given a random sample of n values of X, are specified by

$$\overline{X} = \frac{\tilde{R}}{\tilde{\Lambda}}, \qquad M_2 = \frac{\tilde{R}(1 + \tilde{R})}{\tilde{\Lambda}^2} = \frac{\overline{X}(1 + \tilde{R})}{\tilde{\Lambda}} = \frac{\overline{X}}{\tilde{\Lambda}} + \overline{X}^2$$

from which we find the estimators to be

$$\tilde{\Lambda} = \frac{\overline{X}}{M_2 - \overline{X}^2}, \qquad \tilde{R} = \frac{\overline{X}^2}{M_2 - \overline{X}^2}$$

If we found $n = 10$ observed values for X to be 15.5, 4.5, 6.8, 46.0, 34.5, 4.7, 20.9, 8.2, 14.9, 17.7, we have

$$\bar{x} = \frac{173.7}{10} = 17.37, \qquad \frac{1}{10}\sum x_i^2 = \frac{4674.43}{10} = 467.443$$

and the parameter estimates then are

$$\tilde{\lambda} = \frac{17.37}{467.443 - (17.37)^2} = .1048, \qquad \tilde{r} = \frac{(17.37)^2}{467.443 - (17.37)^2} = 1.821$$ ∎

The method of moments is one procedure for generating estimators of unknown parameters; it provides an attractive rationale and is generally quite easy to employ. In 1921 Sir R. A. Fisher proposed a different rationale for estimating parameters and pointed out a number of reasons that it might be preferable. The procedure proposed by Fisher is called *maximum likelihood* and is generally acknowledged to be superior to the method of moments (in those cases in which the two lead to two different estimators). Fisher's procedure had in fact been used earlier by Gauss and others for certain isolated problems. Fisher presented it as a unifying concept to cover a broad range of problems and it is generally accepted as the best rationale to apply in estimating unknown parameters, when one is willing to assume the form of the population probability law is known. The maximum likelihood procedure says one should examine the *likelihood function* of the sample values and take as the estimates of the unknown parameters those values that maximize this likelihood function. The procedure is spelled out in the following definition.

DEFINITION 7.1.3.   Assume $X_1, X_2, \ldots, X_n$ is a random sample of a random variable X whose probability law depends on an unknown parameter θ. The likelihood function is

$$\begin{aligned} L_X(\theta) &= p_X(x_1)\, p_X(x_2) \cdots p_X(x_n) && \text{if } X \text{ is discrete} \\ &= f_X(x_1) f_X(x_2) \cdots f_X(x_n) && \text{if } X \text{ is continuous.} \end{aligned}$$

The maximum likelihood estimate for θ is the value $\hat{\theta}$ which maximizes $L_X(\theta)$. ∎

It is important to examine the structure of the likelihood function itself. Suppose first that X is a discrete random variable and $X_1, X_2, \ldots, X_n$ is a random sample of X; $x_1, x_2, \ldots, x_n$ are the observed sample values. Then the probability of observing the values that did in fact occur in the sample is

$$P(X_1 = x_1, X_2 = x_2, \ldots, X_n = x_n) = P(X_1 = x_1) P(X_2 = x_2) \cdots P(X_n = x_n)$$
$$= p_X(x_1) p_X(x_2) \cdots p_X(x_n)$$
$$= L_X(\theta),$$

the likelihood function; that is, $L_X(\theta)$ expresses the probability of observing the numbers obtained as a function of θ. If X is a continuous random variable with density function $f_X(x)$, the probability of observing sample values in a neighborhood of those that occurred is

$$P(x_1 - \frac{\Delta x_1}{2} \le X_1 \le x_1 + \frac{\Delta x_1}{2}, x_2 - \frac{\Delta x_2}{2} \le X_2 \le x_2 + \frac{\Delta x_2}{2},$$

$$\ldots, x_n - \frac{\Delta x_n}{2} \le X_n \le x_n + \frac{\Delta x_n}{2}) \doteq f_X(x_1) f_X(x_2) \cdots f_X(x_n) \cdot \Delta x_1 \Delta x_2 \cdots \Delta x_n$$

$$= L_X(\theta) \Delta x_1 \Delta x_2 \cdots \Delta x_n.$$

The likelihood function is proportional to the probability of observing values in a neighborhood of $x_1, x_2, \ldots, x_n$. In either case the maximum likelihood estimate for θ is the value $\hat{\theta}$, which maximizes $L_X(\theta)$; thus $\hat{\theta}$ is, in a sense, the "most likely" (maximum probability) value for the unknown parameter θ.

Since $L_X(\theta)$ is (at least proportional to) the probability of observing the sample values, it will always be nonnegative (for any possible value for θ). Thus we could, if we like, equally well examine $K(\theta) = \ln L_X(\theta)$, the natural logarithm of the likelihood function. Because the natural logarithm is a monotonic increasing function the value of θ that maximizes $K(\theta)$ is identical with the value that maximizes $L_X(\theta)$. Maximizing $K(\theta)$ frequently leads to a simpler procedure (at least in appearance) than does directly maximizing $L_X(\theta)$. Let us examine some examples.

EXAMPLE 7.1.4.   Consider the case discussed in Example 7.1.1, the estimation of the parameter λ of an exponential probability law, given the observed values of a random sample of size n from the population. Suppose, as in Example 7.1.1, $n = 5$ and the five observed values are 30.4, 7.8, 1.4, 13.1, and 67.3 (hours). Then the likelihood function for the sample is

$$L_X(\lambda) = (\lambda e^{-30.4\lambda})(\lambda e^{-7.8\lambda})(\lambda e^{-1.4\lambda})(\lambda e^{-13.1\lambda})(\lambda e^{-67.3\lambda})$$
$$= \lambda^5 e^{-120\lambda},$$

a continuous function of λ, $\lambda > 0$. The natural logarithm of $L(\lambda)$ is

$$K(\lambda) = \ln(\lambda^5 e^{-120\lambda}) = 5 \ln \lambda - 120 \lambda.$$

To find the maximizing value

$$\frac{dK(\lambda)}{d\lambda} = \frac{5}{\lambda} - 120$$

and setting $dK(\lambda)/d\lambda = 0$ gives

$$\frac{5}{\lambda} - 120 = 0$$

from which we find

$$\hat{\lambda} = \frac{5}{120} = .042.$$

Since $d^2 K(\lambda)/d\lambda^2 = -5/\lambda^2 < 0$, $\hat{\lambda}$ does locate the maximum value for $K(\lambda)$ [as well as $L_X(\lambda)$]; this is the same value as $\tilde{\lambda}$, the method of moments estimate for λ. For the general case of observing $x_1, x_2, \dots, x_n$ in the random sample, the likelihood function is

$$L_X(\lambda) = \prod_{i=1}^{n} (\lambda e^{-\lambda x_i}) = \lambda^n e^{-\lambda \sum x_i},$$

$$K(\lambda) = \ln L(\lambda) = n \ln \lambda - \lambda \sum x_i,$$

$$\frac{dK(\lambda)}{d\lambda} = \frac{n}{\lambda} - \sum x_i,$$

from which we find

$$\frac{n}{\hat{\lambda}} = \sum x_i$$

or

$$\hat{\lambda} = \frac{n}{\sum x_i} = \frac{1}{\bar{x}},$$

the same as the method of moments estimate found in Example 7.1.1. ∎
    In Example 7.1.4 the likelihood function was a nice smooth continuous function of the parameter and had a single maximum at which its derivative was zero. It is possible for the likelihood function to have a shape like that pictured in Figure 7.1.1, in which case its derivative is not zero at its maximum value. The following example presents a case of this sort.

EXAMPLE 7.1.5. Assume X is a uniform random variable on the interval $(-\theta, \theta)$, and let $X_1, X_2, \ldots, X_n$ be a random sample of X. The likelihood function for the sample values then is

$$L_X(\theta) = \prod_{i=1}^{n} \frac{1}{2\theta} = \frac{1}{(2\theta)^n}, \qquad -\theta \le x_i \le \theta, \qquad i = 1, 2, \ldots, n.$$

The condition that each x_i lies between $-\theta$ and θ says the magnitude of each x_i does not exceed θ; that is $|x_i| \le \theta$, from which it follows that θ must be at least as large as the maximum $|x_i|$. The likelihood function equals $1/(2\theta)^n$ for $\theta \ge$ maximum $|x_i|$ (and equals 0 otherwise); this function has the shape pictured in Figure 7.1.1 from which it is easy to see that the maximum of $L(\theta)$ occurs with θ equal to its smallest possible value, which is the maximum $|x_i|$. Thus the maximum likelihood estimator for θ is

$$\hat{\theta} = \text{maximum} \, |X_i|, \qquad i = 1, 2, \ldots, n.$$

For the four observed values $-.808, 2.590, 2.314, -.268$, the largest magnitude is 2.590 so the maximum likelihood estimate for θ, based on these observed values, is $\hat{\theta} = 2.590$; this is not the same as the method of moments estimate discussed earlier. ■

In the two cases just discussed it was quite straightforward to find the parameter value that maximizes the likelihood function. For some probability laws it is not possible to express the maximum likelihood estimate as a simple closed-form function of the observed values; the equation defining the maximum likelihood estimate must be solved numerically. With modern computing equipment this is generally not a serious problem.

If the population probability law has two or more unknown parameters, then the likelihood function has an equal number of unknowns (the parameter values). To maximize a function of, say, two unknowns, θ_1, θ_2, we

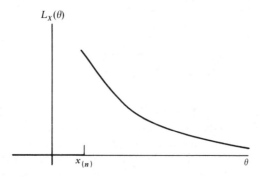

Figure 7.1.1

set the two partial derivatives

$$\frac{\partial L_X(\theta_1, \theta_2)}{\partial \theta_1}, \qquad \frac{\partial L_X(\theta_1, \theta_2)}{\partial \theta_2}$$

[or equivalently, the partial derivatives of $K(\theta_1, \theta_2) = \ln L_X(\theta_1, \theta_2)$] both equal to zero and solve the resulting equations simultaneously to determine the maximum likelihood estimates. This is illustrated in the following two examples.

EXAMPLE 7.1.6. Assume $X_1, X_2, \ldots, X_n$ is a random sample of a normal random variable X with mean μ and variance σ^2. For example, these values could be the repeated spectrometer readings for samples taken from the same oil reservoir with unknown true iron content μ. The likelihood function then has two unknowns, μ and σ^2, and is given by

$$L_X(\mu, \sigma^2) = \prod_{i=1}^{n} f_X(x_i) = \prod_{i=1}^{n} \frac{1}{\sigma\sqrt{2\pi}} \exp\left[-(x_i - \mu)^2/2\sigma^2\right]$$

$$= \left(\frac{1}{2\pi\sigma^2}\right)^{n/2} \exp\left[-\sum (x_i - \mu)^2/2\sigma^2\right].$$

The natural logarithm of $L_X(\mu, \sigma^2)$ is

$$K(\mu, \sigma^2) = \ln L_X(\mu, \sigma^2)$$

$$= -\frac{n}{2}\ln 2\pi - \frac{n}{2}\ln \sigma^2 - \frac{\sum(x_i - \mu)^2}{2\sigma^2}$$

from which we find

$$\frac{\partial K(\mu, \sigma^2)}{\partial \mu} = \frac{\sum(x_i - \mu)}{\sigma^2}$$

$$\frac{\partial K(\mu, \sigma^2)}{\partial \sigma^2} = -\frac{n}{2\sigma^2} + \frac{\sum(x_i - \mu)^2}{2(\sigma^2)^2}.$$

Setting these two partial derivatives equal to zero gives

$$\frac{\sum(x_i - \hat{\mu})}{\hat{\sigma}^2} = 0$$

$$\frac{\sum(x_i - \hat{\mu})^2}{2(\hat{\sigma}^2)^2} = \frac{n}{2\hat{\sigma}^2}.$$

From the first equation we must have

$$\sum(x_i - \hat{\mu}) = \sum x_i - n\hat{\mu} = 0$$

which gives $\hat{\mu} = (1/n)\sum x_i = \bar{x}$. Substituting this value into the second equation gives

$$\hat{\sigma}^2 = \frac{1}{n}\sum (x_i - \bar{x})^2.$$

The maximum likelihood estimators for μ and σ^2 thus are $\bar{X}$ and $(n-1)S^2/n$, respectively. ∎

EXAMPLE 7.1.7.   Let X be a gamma random variable with parameters r and λ, as in Example 7.1.3, and assume $X_1, X_2, \ldots, X_n$ is a random sample of X. The likelihood function for the sample is

$$L_X(r, \lambda) = \prod_{i=1}^{n} f_X(x_i) = \prod_{i=1}^{n} \frac{\lambda^r x_i^{r-1}}{\Gamma(r)} e^{-\lambda x_i}$$

$$= \frac{\lambda^{nr}}{[\Gamma(r)]^n} \exp(-\lambda \sum x_i)\left(\prod x_i\right)^{r-1}$$

and its natural logarithm is

$$K(r, \lambda) = \ln L_X(r, \lambda)$$
$$= nr \ln \lambda - n \ln \Gamma(r) - \lambda \sum x_i + (r-1)\ln\left(\prod x_i\right).$$

Then

$$\frac{\partial K(r, \lambda)}{\partial r} = n \ln \lambda - \frac{n}{\Gamma(r)}\frac{d\Gamma(r)}{dr} + \sum_{i=1}^{n} \ln x_i$$

$$\frac{\partial K(r, \lambda)}{\partial \lambda} = \frac{nr}{\lambda} - \sum x_i.$$

From the second equation [setting $\partial K(r, \lambda)/\partial \lambda = 0$] we easily find

$$\hat{\lambda} = \frac{n\hat{r}}{\sum x_i} = \frac{\hat{r}}{\bar{x}}$$

Substituting into the first equation $[\partial K(r, \lambda)/\partial r = 0]$ then gives

$$n \ln\left(\frac{\hat{r}}{\bar{x}}\right) + \sum \ln x_i = \frac{n}{\Gamma(\hat{r})}\frac{d\Gamma(r)}{dr}\bigg|_{r=\hat{r}}$$

to be solved for $\hat{r}$. This is an example of one of those cases alluded to earlier in which the solution cannot be expressed in closed form and numerical methods must be employed. ∎

Maximum likelihood estimators are invariant under transformations of the parameter. Suppose $\hat{\Gamma}$ is the maximum likelihood estimator of a parameter γ. If for some reason it is desired to estimate a function of γ, say, $\theta = h(\gamma)$, rather than γ itself, we could revert to the likelihood function $L_X(\gamma)$ and express it in terms of θ rather than γ [by substituting the values of θ corresponding to those of γ through the function $\theta = h(\gamma)$]. Then the maximum likelihood estimator of θ, $\hat{\Theta}$, would be determined by the value that maximizes L_X. The invariant property of maximum likelihood estimators says that this maximizing value $\hat{\theta}$ of L_X will yield the estimator $\hat{\Theta} = h(\hat{\Gamma})$; thus there is no need actually to express L_X in terms of θ to get the estimator if we already have $\hat{\Gamma}$.

For example, in Example 7.1.4 we found the maximum likelihood estimate for the parameter λ of an exponential probability law to be $\hat{\lambda} = 1/\bar{x}$. The invariance of maximum likelihood says the maximum likelihood estimate for the population mean $\mu = 1/\lambda$ is given by $\hat{\mu} = (1/\bar{x})^{-1} = \bar{x}$, and the maximum likelihood estimate for $P(X > a)$ is $e^{-a/\bar{x}}$ for any fixed a.

EXERCISE 7.1

1. Assume the amount of rainfall recorded at a certain station on a given date (for example, September 30) is uniformly distributed on the interval $(0, b)$. If a sample of 10 years' records shows that the following amounts were recorded on that date, compute the method of moments estimate of b: 0, 0, .7, 1, .1, 0, .2, .5, 0, .6. (Measurements are recorded in inches.)

2. Assume the amount of growth in height of 1-foot tall Monterey pine trees in a year is a normal random variable with unknown mean and unknown variance. The growths of five trees were recorded as: 3 feet, 5 feet, 2 feet, 1.5 feet, 3.5 feet. Compute the method of moments estimates for μ and σ^2.

3. The number of cars arriving at a supermarket parking lot per hour is assumed to be a Poisson random variable with parameter λ. The following numbers of cars were observed arriving during the hour from 9 to 10 A.M. on six successive days: 50, 47, 82, 91, 46, 64. Compute $\hat{\lambda}$.

4. X is a geometric random variable with parameter p. Given a random sample of n observations of X, what is the method of moments estimator of p?

5. Richard is allowed to shoot a basketball from the free-throw line until he makes a basket. He does this five times, with the number of shots necessary being 5, 1, 7, 4, and 9, respectively. Compute the method of moments estimate of p, the probability he will make a basket, shooting from the free-throw line. (Assume the number of shots necessary is a geometric random variable.)

6. X is a Bernoulli random variable with parameter p. Given a random sample of n observations of X, compute the method of moments estimator for p.

7. X is a binomial random variable with parameters n (known) and p. Given a random sample of N observations of X, compute the method of moments estimator for p.

8. Suppose X is a normal random variable with mean $\mu = 5$ and unknown variance σ^2. What is the method of moments estimator for σ^2, based on a random sample of n observations of X?

9. X is a binomial random variable with parameters n and p, both unknown. Given a random sample of N observations of X, compute the the method of moments estimators of n and p.

10. Suppose X is a normal random variable with mean $\mu = 10$ and variance σ^2 unknown. What is the maximum likelihood estimator for σ^2, based on a random sample of n observations of X?

11. Suppose you buy a dozen Valencia oranges, weigh each of them, and find

$$\sum_{i=1}^{12} x_i = 180, \qquad \sum_{i=1}^{12} x_i^2 = 2799$$

(in ounces). Compute the maximum likelihood estimates of μ and σ^2, assuming the 12 oranges you have purchased are a random sample of all Valencia oranges (and their weights are normally distributed).

12. Suppose X is a Poisson random variable with parameter μ. Given a random sample of n observations of X, what is the maximum likelihood estimator for μ?

13. Assume the number of new car sales, per day, that a particular dealer makes is a Poisson random variable with parameter μ. Given that in 20 days the total number of sales he made was 30 cars, what is the maximum likelihood estimate of μ?

14. If X is a geometric random variable with parameter p, what is the maximum likelihood estimator for p based on a random sample of n observations of X?

15. Suppose X is a binomial random variable with parameters n (known) and p. What is the maximum likelihood estimator for p based on a random sample of N observations of X?

*16. A used car salesman is willing to assume the number of sales he makes, per day, is a Poisson random variable with parameter μ. Over the past 30 days, he made 0 sales on 20 days and one or more sales on each of the remaining 10 days. What is the method of moments estimate for μ?

*17. Answer Exercise 7.1.16, using the maximum likelihood method.

*18. The lifetimes of a component are assumed to be exponential with parameter λ. Ten of these components were placed on test independently. The only data recorded were the number of components that had failed in less than 100 hours versus the number that had not failed. It was found that three had failed before 100 hours and the remaining seven had not. What is the maximum likelihood estimate for λ?

*19. Assume $(X_{11}, X_{12}), (X_{21}, X_{22}), ..., (X_{n1}, X_{n2})$ is a random sample of a bivariate normal vector (X_1, X_2). Evelute the method of moments and maximum likelihood estimators for $\mu_1, \mu_2, \sigma_1^2, \sigma_2^2$ and ρ.

20. In his first 100 baseball games in the 1980 season, a professional baseball player made 139 hits in 374 times at bat. Assuming he has the same probability of getting a hit each time at bat, and the attempts are independent, what is the maximum likelihood estimate of the probability he will get at least two hits in his next four times at bat.

*21. If $(X_1, X_2, ..., X_k)$ is a multinomial random vector with parameters $n, p_1, p_2, ..., p_k$, show that the maximum likelihood estimates for p_1, $p_2, ..., p_k$ are given by $\hat{p}_i = x_i/n$, $i = 1, 2, ..., k$. (Recall x_i is the number of times outcome i occurs in the n trials.)

7.2 Properties of Estimators

In the last section we studied two different, generally applicable methods for constructing estimators of unknown parameters. In many cases the two methods generate the same estimator, but in many important problems they do not lead to the same estimator. When we are faced with the choice of two or more estimators for the same parameter, it becomes important to develop criteria for comparing them; if we were able to develop some scale of "goodness" of estimators, then we would always want to use the estimator that was best for the given problem. Unfortunately, there is no universally appropriate scale of goodness that can be used in comparing estimators of the same parameter.

As seen, the estimator of an unknown parameter is a statistic, a random variable whose value can be observed on the basis of a sample. Because an estimator $\hat{\Gamma}$ is always a random variable, the particular value it takes on varies from one sample to another; thus we certainly would not expect to have the estimate $\hat{\gamma}$ equal to the true unknown value γ for every sample we take. If we are considering two estimators, $\tilde{\Gamma}$ and $\hat{\Gamma}$, for the same parameter γ, we can in theory derive the probability laws of $\tilde{\Gamma}$ and of $\hat{\Gamma}$ from the probability law of the random sample. Comparison of $\tilde{\Gamma}$ and $\hat{\Gamma}$ then reduces to a comparison of their two respective probability laws in some way. If, for example, $\tilde{\Gamma}$ was uniformly distributed from $\gamma - \frac{1}{4}$ to $\gamma + \frac{1}{4}$ while $\hat{\Gamma}$ was uniformly distributed on the interval from $\gamma - \frac{1}{16}$ to $\gamma + \frac{1}{16}$, we would obviously prefer $\hat{\Gamma}$ as an estimator of γ (and thus would use $\hat{\gamma}$ rather than $\tilde{\gamma}$ as our estimate for any particular sample). Unfortunately, the comparison of probability laws of estimators is generally not this straightforward; in most cases the probability density functions (or probability functions) of various estimators are of widely divergent types and there is then a variety of ways in which they could be compared. Perhaps the most obvious property to investigate of the probability law of an estimator Γ would be its mean value. As the following definition states, if the mean value of Γ is γ, then Γ is unbiased; otherwise, it is a biased estimator.

DEFINITION 7.2.1. An estimator Γ of an unknown parameter γ is *unbiased* if $E[\Gamma] = \gamma$, for all values of γ. The difference $B[\Gamma] = E[\Gamma] - \gamma$ is called the *bias* in Γ; if $B[\Gamma] \neq 0$ then Γ is a biased estimator. ■

Thus an unbiased estimator is a random variable whose expected value is the parameter being estimated; if we were repeatedly to take samples of size n and for each compute the observed value of Γ (the estimate for that sample outcome), then the average of these observed values would be γ, the parameter being estimated.

EXAMPLE 7.2.1. We saw in Chapter 6 that $E[\overline{X}] = \mu$, the population mean, no matter what the probability law from which we were sampling. Thus $\overline{X}$ is an unbiased estimator of μ in a normal density, of the parameter p for a Bernoulli random variable, and of the parameter μ for a Poisson random variable (since in each of these cases the parameter mentioned is the mean of the random variable being sampled). ■

The probability law for an estimator Γ, which is unbiased, is centered at γ, the true parameter value (in the sense that the mean equals γ). In any given case it is generally quite easy to find several unbiased estimators for the same parameter. This is illustrated in the following example.

EXAMPLE 7.2.2. Let $X_1, X_2, \ldots, X_n$ be a random sample of a Poisson random variable with parameter μ (remember the subscript simply identifies the order in which the random variables are selected). Then, since $E[X_i] = \mu$, $i = 1, 2, \ldots, n$, the first sample value, or the second, or indeed any one of X_1 through X_n is an unbiased estimator for μ. If we average the first two, the last two, any two or any three, or all n, each such average also provides an unbiased estimator for μ. In addition to these $2^n - 1$ unbiased estimators, recall that if X is Poisson with parameter μ then the variance of X is also μ; since

$$S^2 = \frac{1}{n-1} \sum_{i=1}^{n} (X_i - \overline{X})^2$$

is an unbiased estimator for the population variance (Theorem 6.2.2), S^2 is also an unbiased estimator for μ. Similarly, the statistic $X_1(X_1 - X_2)$, and others like it, provides an unbiased estimator for μ. In any given case there are many possible unbiased estimators for the same parameter. ■

The rather appealing property of unbiasedness is shared by many possible estimators. To choose between different unbiased estimators, one would reasonably also consider their variances. If both Γ_1 and Γ_2 are unbiased estimators of a parameter γ, and $\sigma_{\Gamma_1}^2 < \sigma_{\Gamma_2}^2$, then Γ_1 varies less about γ over repeated samples than does Γ_2 $\left(\text{in the sense that } E[(\Gamma_1 - \gamma)^2]\right.$ $< E[(\Gamma_2 - \gamma)^2]\right)$. It would then be logical to use $\hat{\gamma}_1$ instead of $\hat{\gamma}_2$ as the estimate for γ based on any given sample. Historically, if both Γ_1 and Γ_2 are unbiased estimators of a parameter γ, and $\sigma_{\Gamma_1}^2 < \sigma_{\Gamma_2}^2$, the estimator Γ_1 is called *more efficient* than Γ_2. But if one is to compare the "desirability" of estimators on the basis of the expected squared distance between the estimator and the parameter (the quantity $E[(\Gamma - \gamma)^2]$), there is really no reason to consider only unbiased estimators (those for which $E[\Gamma] = \gamma$). That is, let Γ represent a general estimator for γ, with expected value $E[\Gamma]$ (which may not necessarily be the same as γ). The quantity $E[(\Gamma - \gamma)^2]$ is called the mean square error of Γ, as an estimator for γ, as given in the following definition.

DEFINITION 7.2.2. Let $X_1, X_2, ..., X_n$ be a random sample of X, whose probability law depends on an unknown parameter γ, and let $\Gamma = h(X_1, X_2, ..., X_n)$ be any statistic (function only of the elements of the sample). Then the *mean square error* of Γ (as an estimator for γ) is

$$MSE[\Gamma] = E[(\Gamma - \gamma)^2].$$ ■

A couple of facts are immediately apparent regarding mean square errors and biases. Suppose Γ is an unbiased estimator for γ, and then $E[\Gamma] = \gamma$, which says that $B[\Gamma] = E[\Gamma] - \gamma = 0$ and

$$\begin{aligned} MSE[\Gamma] &= E[(\Gamma - \gamma)^2] \\ &= E[(\Gamma - E[\Gamma])^2] \\ &= \text{Var}[\Gamma]. \end{aligned}$$

Thus if Γ is an unbiased estimator for γ, its bias is zero and its mean square error equals its variance. For the general case it is easy to see that

$$\begin{aligned} MSE[\Gamma] &= E[(\Gamma - \gamma)^2] \\ &= E[((\Gamma - E[\Gamma]) + (E[\Gamma] - \gamma))^2] \\ &= E[(\Gamma - E[\Gamma])^2 + (E[\Gamma] - \gamma)^2 + 2(\Gamma - E[\Gamma])(E[\Gamma] - \gamma)] \\ &= \text{Var}[\Gamma] + (B[\Gamma])^2 + 2(E[\Gamma] - \gamma)E[(\Gamma - E[\Gamma])] \\ &= \text{Var}[\Gamma] + (B[\Gamma])^2, \end{aligned}$$

since $E[(\Gamma - E[\Gamma])] = 0$. Thus in any case the mean square error for Γ is equal to its variance plus the square of its bias.

A biased estimator has its probability law centered at $E[\Gamma] \neq \gamma$, which means that its long-term average value over repeated samples differs from γ, the true parameter value. But if its variance and bias are sufficiently small, it might still provide an attractive choice as an estimator for γ, in the sense that its mean square error might still be quite small (and thus the expected squared distance between Γ and γ may be smaller than that of some unbiased estimators). The following example illustrates a biased estimator that performs better (in terms of mean square error) than a good unbiased estimator.

EXAMPLE 7.2.3.   Let X be an exponential random variable with parameter λ; thus the density for X is

$$f(x) = \lambda e^{-\lambda x}, \qquad x > 0;$$

the mean value for X is $\mu = 1/\lambda$, and its variance is $\sigma^2 = 1/\lambda^2$. If X_1, X_2 is a random sample of X, then, as mentioned earlier, X_1, X_2, and $\overline{X} = (X_1 + X_2)/2$ are each unbiased estimators for $\mu = 1/\lambda$. Of these three, we would prefer $\overline{X}$ since its variance is $\sigma^2/2 = 1/2\lambda^2$, whereas the variances of both X_1 and X_2 are $1/\lambda^2$; since the variance of $\overline{X}$ is half as large as the variance of X_1 (or X_2), we would expect its observed value to be closer to $\mu = 1/\lambda$. But if we consider biased estimators, we can find one whose mean square error is smaller than $1/2\lambda^2$, meaning that its average squared distance from μ is smaller than that of $\overline{X}$. This estimator is $\sqrt{X_1 X_2}$, the geometric mean of X_1 and X_2. Since

$$E[\sqrt{X}] = \int_0^\infty \sqrt{x}\lambda e^{-\lambda x}\, dx = \frac{\Gamma\left(\frac{3}{2}\right)}{\sqrt{\lambda}} = \frac{1}{2}\sqrt{\frac{\pi}{\lambda}}$$

$$E[(\sqrt{X})^2] = E[X] = \frac{1}{\lambda},$$

we have

$$\operatorname{Var}[\sqrt{X_1 X_2}] = E[X_1 X_2] - (E[\sqrt{X_1 X_2}])^2$$
$$= E[X_1]E[X_2] - (E[\sqrt{X_1}]E[\sqrt{X_2}])^2$$
$$= \frac{1}{\lambda^2} - \left(\frac{\pi}{4\lambda}\right)^2 = \frac{16 - \pi^2}{16\lambda^2},$$

$$B[\sqrt{X_1 X_2}] = E[\sqrt{X_1 X_2}] - \frac{1}{\lambda}$$
$$= \frac{\pi}{4\lambda} - \frac{1}{\lambda} = \frac{\pi - 4}{4\lambda}$$

Thus

$$MSE[\sqrt{X_1 X_2}] = \frac{16 - \pi^2}{16\lambda^2} + \left(\frac{\pi - 4}{4\lambda}\right)^2$$

$$= \frac{4 - \pi}{2\lambda^2}$$

and since $4 - \pi < 1$, $MSE[\sqrt{X_1 X_2}] < \mathrm{Var}[\bar{X}]$. The average squared distance between $\sqrt{X_1 X_2}$ and μ is smaller than the average squared distance between $\bar{X}$ and μ, even though $\sqrt{X_1 X_2}$ is a biased estimator. (As you are asked to show in Exercise 7.2.5, one can actually do better than $\sqrt{X_1 X_2}$ in terms of average squared distance.) ∎

What can be said about the smallest possible value for the mean square error of an estimator? The Cramér-Rao inequality, which we will not be able to prove, gives a lower bound for the mean square error of any estimator of the parameter of a probability law that satisfies certain regularity conditions. These regularity conditions include: (a) the range of the random variable must be independent of the parameter and (b) the derivative of the density (or probability) function with respect to the parameter must be a continuous differentiable function of the parameter. The Cramér-Rao inequality is given in the following theorem.

Theorem 7.2.1. Assume $X_1, X_2, ..., X_n$ is a random sample of a random variable X whose density function $f_X(x)$ [or probability function $p_X(x)$] depends on an unknown parameter γ and which satisfies the preceding regularity conditions. Let $\Gamma = g(X_1, X_2, ..., X_n)$ be an estimator of γ with bias $B[\Gamma]$. Then if X is a continuous random variable, the mean square error of Γ must satisfy

$$MSE[\Gamma] \geq \frac{(1 + B'[\Gamma])^2}{nE\left[\left(\frac{\partial}{\partial \gamma} \ln f_X(X)\right)^2\right]}$$

where

$$\frac{d}{d\gamma} B[\Gamma] = B'[\Gamma].$$

If X is discrete,

$$MSE[\Gamma] \geq \frac{(1 + B'[\Gamma])^2}{nE\left[\left(\frac{\partial}{\partial \gamma} \ln p_X(X)\right)^2\right]}$$

∎

Notice immediately that this theorem also provides a bound on the variance of any unbiased estimator (which is its most frequent use). If Γ is unbiased, $B[\Gamma] = 0$, $MSE[\Gamma] = \text{Var}[\Gamma]$ and we have

$$\text{Var}[\Gamma] \geq \frac{1}{nE\left[\left(\dfrac{\partial}{\partial\theta}\ln f_X(X)\right)^2\right]}.$$

The principal use for the bound is in seeing whether or not a given estimator, say, Γ_1, has a mean square error equal to the bound; if it does (as long as the regularity conditions are satisfied), then Γ_1 is the best possible estimator (in terms of mean square error). This is illustrated in the following example.

EXAMPLE 7.2.4. Assume $X_1, X_2, ..., X_n$ is a random sample of a Poisson random variable with parameter μ. Thus

$$p_X(x) = \frac{\mu^x}{x!} e^{-\mu}, \qquad x = 0, 1, 2, ...,$$

so $dp_X(x)/d\mu$ is a continuous differentiable function of μ for all $\mu > 0$ and the range of X does not depend on μ. We know that $\bar{X}$ has mean μ and variance μ/n, so $\bar{X}$ is an unbiased estimator for μ, $B[\bar{X}] = 0$ and $B'[\bar{X}] = 0$. Now

$$\ln p_X(X) = X \ln \mu - \ln X! - \mu$$

$$\frac{\partial \ln p_X(X)}{\partial \mu} = \frac{X}{\mu} - 1 = \frac{X - \mu}{\mu},$$

$$E\left[\left(\frac{\partial}{\partial\mu}\ln p_X(X)\right)^2\right] = E\left[\frac{(X - \mu)^2}{\mu^2}\right]$$

$$= \frac{\text{Var}[X]}{\mu^2} = \frac{1}{\mu}.$$

The Cramér-Rao lower bound for the variance of any unbiased estimator of μ is thus $1/n(1/\mu) = \mu/n$, the same as the variance of $\bar{X}$. Thus $\bar{X}$ has the smallest possible variance of any unbiased estimator for μ, and there is no need to consider any other unbiased estimators. ∎

If the regularity conditions are not satisfied, of course, the Cramér-Rao bound does not apply and estimators can be found with smaller mean square errors. The following example discusses such a case.

EXAMPLE 7.2.5.   Assume $X_1, X_2, \ldots, X_n$ is a random sample of a uniform random variable X on the interval $(0, \theta)$. Thus the density for X is

$$f_X(x) = \frac{1}{\theta}, \qquad 0 < x < \theta$$

and the range for X depends on θ, so the regularity conditions for the Cramér-Rao bound are not satisfied. The maximum likelihood estimator for θ is $X_{(n)}$, the maximum sample value. Its density function is

$$f_{X_{(n)}}(x) = \frac{nx^{n-1}}{\theta^n}, \qquad 0 < x < \theta$$

from which we find

$$E[X_{(n)}] = \frac{n\theta}{n+1}$$

$$B[X_{(n)}] = \frac{n\theta}{n+1} - \theta = -\frac{\theta}{n+1}$$

$$E[X_{(n)}^2] = \frac{n\theta^2}{n+2}$$

so

$$\text{Var}[X_{(n)}] = \frac{n\theta^2}{n+2} - \left(\frac{n\theta}{n+1}\right)^2 = \frac{n\theta^2}{(n+2)(n+1)^2}$$

$$MSE[X_{(n)}] = \frac{n\theta^2}{(n+2)(n+1)^2} + \left(-\frac{\theta}{n+1}\right)^2$$

$$= \frac{2\theta^2}{(n+2)(n+1)}.$$

If the Cramér-Rao lower bound were applicable for this case, its value would be

$$\frac{[n/(n+1)]^2}{n(-1/\theta)^2} = \frac{1}{n}\left(\frac{n\theta}{n+1}\right)^2 > \frac{2\theta^2}{(n+2)(n+1)}$$

since

$$(1 + B'[X_{(n)}])^2 = \left(1 - \frac{1}{n+1}\right)^2 = \left(\frac{n}{n+1}\right)^2$$

and

$$\frac{\partial \ln f_X(X)}{\partial \theta} = \frac{\partial(-\ln \theta)}{\partial \theta} = -\frac{1}{\theta}.$$

$X_{(n)}$ has a considerably smaller mean square error than the value given by the bound, possible only because the regularity conditions are not satisfied. ∎

One further property that an estimator may have is called *consistency*. It is an asymptotic, or large sample, property since it actually describes a limiting property of the probability law of the estimator as the sample size n increases without limit.

DEFINITION 7.2.3. Let Γ be an estimator for γ, based on a random sample of size n. If

$$\lim_{n \to \infty} P(|\Gamma - \gamma| \geq \varepsilon) = 0, \qquad \text{for any} \qquad \varepsilon > 0,$$

then Γ is a *consistent* estimator for γ. ∎

To be a consistent estimator the sequence of probabilities that $|\Gamma - \gamma| \geq \varepsilon$ must converge to zero as the sample size n goes to ∞. In a sense then the probability law for Γ must get more and more concentrated in any neighborhood of γ, no matter how small, if Γ is consistent. From the Chebyshev inequality (Theorem 5.5.1) we know that

$$P(|\Gamma - \gamma| \geq \varepsilon) \leq \frac{1}{\varepsilon^2} E[(\Gamma - \gamma)^2]$$

$$= \frac{1}{\varepsilon^2} MSE[\Gamma].$$

It follows that if $MSE[\Gamma] \to 0$ as $n \to \infty$ (thus both the variance and bias of Γ go to zero as $n \to \infty$), then Γ is consistent.

EXAMPLE 7.2.6. Let $X_1, X_2, \ldots, X_n$ be a random sample of a random variable with mean μ and finite variance σ^2. Then, as seen several times, $\overline{X}$ is an unbiased estimator for μ and its variance is σ^2/n. Since $MSE[\overline{X}] = \sigma^2/n \to 0$ as $n \to \infty$, $\overline{X}$ is a consistent estimator for μ. ∎

There are many other properties that have been defined for estimators, which we do not have the space to discuss. The interested reader can consult *Introduction to the Theory of Statistics*, 3rd ed., A. M. Mood, F. A. Graybill, and D. C. Boes, McGraw-Hill, 1974, for a more complete discussion of parameter estimation and properties of estimators.

We described the maximum likelihood method of deriving estimators in Section 7.1. Let us close this section by discussing some of the properties of estimators generated by this method.

First, maximum likelihood estimators are not in general unbiased. As seen in Example 7.1.6, the maximum likelihood estimators for μ and σ^2, for a normal population, are $\overline{X} = (1/n) \sum X_i$ and $\hat{\sigma}^2 = (1/n) \sum (X_i - \overline{X})^2$, respectively. Of these two, $\overline{X}$ is unbiased and $\hat{\sigma}^2$ is biased. It is easy to correct the bias in $\hat{\sigma}^2$, though; as we know $S^2 = n\hat{\sigma}^2/(n-1)$ is unbiased, so simply multiplying the maximum likelihood estimator for σ^2 by the right constant makes it unbiased in this case.

Maximum likelihood estimators are in general consistent estimators, as long as the population probability law satisfies some weak regularity conditions and they are generally asymptotically unbiased (bias $\to 0$ as $n \to \infty$). In fact, subject to these same weak regularity conditions, the following theorem, which we will not prove, can be shown to be true.

Theorem 7.2.2. If $X_1, X_2, \ldots, X_n$ is a random sample of a random variable X whose density $f_X(x)$ [or probability function $p_X(x)$] depends on an unknown parameter γ, the probability law for the maximum likelihood estimator $\hat{\Gamma}$ is approximately normal with mean γ and variance

$$\frac{1}{nE\left[((\partial/\partial\gamma)\ln f_X(X))^2\right]} \qquad \text{if } X \text{ is continuous}$$

$$\frac{1}{nE\left[((\partial/\partial\gamma)\ln p_X(X))^2\right]} \qquad \text{if } X \text{ is discrete,}$$

for large n. ■

Notice that the variance of the approximating normal probability law is the Cramér-Rao lower bound. Thus the maximum likelihood estimator is generally asymptotically unbiased and its asymptotic probability law is normal with variance equal to the smallest possible value for unbiased estimators; thus it is called *asymptotically efficient.*

EXAMPLE 7.2.7. We have seen (Example 7.1.4) that the maximum likelihood estimator for the parameter λ of an exponential probability law is $1/\overline{X}$. This exponential probability law does satisfy the regularity conditions of Theorem 7.2.2, so the results given there apply. We have

$$\ln f_X(X) = \ln \lambda - \lambda X$$

$$\frac{\partial \ln f_X(X)}{\partial \lambda} = \frac{1}{\lambda} - X$$

so

$$E\left[\left(\frac{\partial}{\partial\lambda}\ln f_X(X)\right)^2\right] = E\left[\left(\frac{1}{\lambda} - X\right)^2\right] = \frac{1}{\lambda^2},$$

and $\hat{\Lambda} = 1/\overline{X}$ is approximately normal with mean λ and variance λ^2/n

for large values of n. Thus $(\hat{\Lambda} - \lambda)\sqrt{n}/\lambda$ is approximately standard normal, for large n, making it easy to (approximately) evaluate probability statements about the relative error in $\hat{\lambda}$ as an estimate of λ. For example, if n is large,

$$P\left(\frac{|\hat{\Lambda} - \lambda|\sqrt{n}}{\lambda} \le 1.96\right) \doteq .95,$$

which implies

$$P\left(\frac{|\hat{\Lambda} - \lambda|}{\lambda} \le \frac{1.96}{\sqrt{n}}\right) = .95$$

so with $n = 100$ we can be 95 percent sure the relative error in $\hat{\lambda}$ does not exceed .196.   ∎

EXERCISE 7.2

1. Suppose $X_1, X_2, ..., X_6$ is a random sample of a normal random variable with mean μ and variance σ^2. Determine C such that

    $$C[(X_1 - X_2)^2 + (X_3 - X_4)^2 + (X_5 - X_6)^2]$$

    is an unbiased estimator of σ^2.

2. Let $X_1, X_2, ..., X_n$ be a random sample of a Poisson random variable with parameter μ and define

    $$M = d\{X_1(X_1 - X_2) + X_2(X_2 - X_3) + \cdots + X_{n-1}(X_{n-1} - X_n)\}.$$

    Find the value for the constant d that makes M an unbiased estimator for μ.

3. $X_1, X_2, ..., X_n$ is a random sample of a Bernoulli random variable. Find the value of the constant a that makes

    $$a\{X_1 + X_1^2 + X_2 + X_2^2 + \cdots + X_n + X_n^2\}$$

    an unbiased estimator for p.

4. $X_1, X_2, ..., X_n$ is a random sample of a geometric random variable with parameter p. Find the value of the constants a and b such that

    $$b[X_1(aX_2 - X_1) + X_2(aX_3 - X_2) + \cdots + X_{n-1}(aX_n - X_{n-1})]$$

    is an unbiased estimator for $1/p$.

5. In Example 7.2.3 we saw that $MSE[\sqrt{X_1 X_2}]$ is smaller than $MSE[\bar{X}]$ if X_1, X_2 is a random sample of an exponential random variable. Find the constant a such that

    $$MSE[a\bar{X}] < MSE[\sqrt{X_1 X_2}].$$

6. Let $X_1, X_2, ..., X_n$ be a random sample of a Bernoulli random variable with parameter p. Compute

$$MSE\left[\frac{\sum X_i}{n + 1}\right].$$

7. If $X_1, X_2, ..., X_n$ is a random sample of an exponential random variable with parameter λ, find the value for a that minimizes $MSE[a\bar{X}]$ as an estimator for $\mu = 1/\lambda$.

8. What is the smallest possible variance for an unbiased estimator of p, the parameter of a Bernoulli distribution, based on a random sample of size n? Can you find an estimator that has this variance?

9. What is the smallest possible variance for an unbiased estimator of $\mu = 1/p$, the mean of a geometric distribution, based on a random sample of size n? Can you find an unbiased estimator that has this variance?

10. Evaluate the Cramér-Rao lower bound for the variance of an unbiased estimator of $\mu = 1/\lambda$, the mean of an exponential probability law. Can you find an unbiased estimator that has this variance?

11. Assume X is a gamma random variable with known parameter r and unknown λ. Let $X_1, X_2, ..., X_n$ be a random sample of X; evaluate the Cramér-Rao lower bound for the variance of an unbiased estimator of $v = 1/\lambda$. Can you find an unbiased estimator that has this variance?

*12. Let X be a random variable with density

$$f_X(x) = e^{\alpha - x}, \qquad x > \alpha.$$

(a) What is the maximum likelihood estimator for α?

(b) Evaluate the Cramér-Rao lower bound for the variance of an unbiased estimator for α, based on a random sample of size n. Does this density satisfy the regularity conditions for the Cramér-Rao bound?

(c) Make the estimator in (a) unbiased, if necessary, and evaluate its variance.

13. Suppose Γ_1 and Γ_2 are independent unbiased estimators of a parameter γ, with variances σ_1^2 and σ_2^2, respectively.

(a) Show that $a\Gamma_1 + (1 - a)\Gamma_2$ is also an unbiased estimator for γ.

(b) Find the value for a that minimizes the variance of $a\Gamma_1 + (1 - a)\Gamma_2$.

14. Assume X is uniform on the interval $(0, \gamma)$. Based on a random sample of n observations, the maximum likelihood estimator is $\hat{\Gamma} = \max(X_1, X_2, ..., X_n)$ whereas the method of moments estimator is $\tilde{\Gamma} = 2\bar{X}$. Compute $E[\hat{\Gamma}]$ and $E[\tilde{\Gamma}]$ and the mean square errors of the two estimators. Which would you prefer? Are they consistent?

*15. The *Best Linear Unbiased Estimator* (BLUE) of a parameter γ, based on a random sample $X_1, X_2, ..., X_n$, is that linear function $\Gamma = \sum_{i=1}^{n} a_i X_i$ such that
(a) $E[\Gamma] = \gamma$.
(b) $\text{Var}[\Gamma] \le \text{Var}[\tilde{\Gamma}]$, where $\tilde{\Gamma} = \sum_{i=1}^{n} b_i X_i$ is any other unbiased estimator of γ.
If $X_1, X_2, ..., X_n$ is a random sample of a random variable X with mean μ, variance σ^2, show that $\overline{X} = (1/n) \sum X_i$ is the BLUE estimator for μ.

*16. Let $X_1, X_2, ..., X_n$ be a random sample of a normal random variable with mean μ, variance σ^2. The Cramér-Rao lower bound applies as stated in Theorem 7.2.1 for each of the two parameters. Evaluate this bound for unbiased estimators of μ and of σ^2. Can you find an unbiased estimator of σ^2 that attains its bound?

*17. Consider estimators of the form $a \sum_{i=1}^{n} (X_i - \overline{X})^2$ for σ^2, the variance of a normal distribution with unknown μ and σ^2, given a random sample of size n. Find the value for a that minimizes $MSE[a \sum (X_i - \overline{X})^2]$.

18. $X_1, X_2, ..., X_{40}$ is a random sample of a geometric random variable with parameter p. What is the approximate probability law for $1/\overline{X} = 40/\sum X_i$?

19. Let $X_1, X_2, ..., X_n$ be a random sample of X, a Poisson random variable with parameter μ. What is the approximate probability law for $\overline{X} = (1/n) \sum X_i$?

7.3   Confidence Intervals

We have studied methods of generating estimators of unknown parameters and some general properties that such estimators may possess. An estimate of an unknown parameter, computed from the values observed in a random sample, is simply a number; if a good procedure (such as maximum likelihood) was used to produce the estimator, then it may very well have some excellent properties. Suppose, for example, the manufacturer of a processed food, say, a frozen dinner, prints on his packages "net weight 16 ounces"; suppose further the distribution of weights of these packages is normal with mean μ and variance σ^2, both unknown to us as consumers. If we select a random sample of, say, 10 of these packages and find their average weight to be $\bar{x} = \frac{1}{10} \sum x_i = 15.9$ ounces, then we have computed the observed value of the minimum variance unbiased estimator ($\overline{X}$) for μ, the average weight of all packages made by this manufacturer. Since $\overline{X}$ has the smallest variance of all unbiased estimators for μ, we would, of course, expect its observed value (15.9 ounces) to be close to μ (in the sense of smallest variance). But we still do not know exactly how close $\bar{x}$ may be to μ, nor indeed do we know the probability that $\bar{X}$ will be within a given distance of μ for a sample of size 10. We will in this section study *confidence intervals* for unknown parameters. A confidence interval for an unknown parameter gives both an indication of the numerical value of an unknown parameter, as well as a measure of how confident we are of that numerical value. The definition of a confidence interval for an unknown parameter follows.

DEFINITION 7.3.1. Suppose X is a random variable whose probability law depends on an unknown parameter θ. Given a random sample of X, X_1, $X_2, \ldots, X_n$, the two statistics L_1 and L_2 form a $100 (1 - \alpha) \%$ *confidence interval* for θ if

$$P(L_1 \leq \theta \leq L_2) \geq 1 - \alpha,$$

no matter what the unknown value of θ. (Either of L_1 or L_2 may be constants, including $-\infty$ or ∞.) $1 - \alpha$ is called the *confidence coefficient* of the interval (L_1, L_2). ∎

Note, first, that the interval (L_1, L_2) might be termed a *random interval*, since at least one of its end points is a random variable and hence will vary from one sample to another. Once we have taken the sample, we can compute the observed values for L_1 and L_2 and then can say that we are $100(1 - \alpha) \%$ confident that the interval bracketed by their observed values contains the unknown value of θ. For any given sample, of course, either the interval contains the true parameter value or it does not, and we do not know which of these possibilities is true for our computed interval. If repeated samples of the same size were taken, and the observed (L_1, L_2) computed for each, the proportion $1 - \alpha$ of the computed intervals would contain the true parameter value in the long run. This leads to our saying we are $100(1 - \alpha) \%$ confident that our single interval contains the true parameter value.

One of the most frequently used confidence intervals is for the mean of a normal distribution. Assume $X_1, X_2, \ldots, X_n$ is a random sample of a normal random variable with mean μ and variance σ^2. Then, as we know from Theorem 5.7.3, $\sqrt{n}(\bar{X} - \mu)/S$ has the T-distribution with $n - 1$ degrees of freedom, where $\bar{X} = (1/n) \sum X_i$, $S^2 = \sum (X_i - \bar{X})^2/(n - 1)$. Let $t_{1 - \alpha/2}$ denote the value for which the area under this T-density, to its left, is equal to $1 - \alpha/2$ (see Figure 7.3.1). Because of the symmetry of this density about 0, the area to the left of $-t_{1 - \alpha/2}$ is $\alpha/2$ (see Figure 7.3.2), as is the area to the right of $t_{1 - \alpha/2}$. Thus the area under the density between $-t_{1 - \alpha/2}$ and

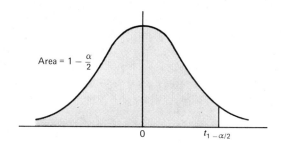

Figure 7.3.1

$t_{1-\alpha/2}$ is $1 - \alpha$ so we know that

$$P\left(-t_{1-\alpha/2} \leq \frac{\sqrt{n}(\overline{X} - \mu)}{S} \leq t_{1-\alpha/2} \right) = 1 - \alpha.$$

The inequality

$$\frac{\sqrt{n}(\overline{X} - \mu)}{S} \leq t_{1-\alpha/2}$$

is equivalent to

$$\overline{X} - \mu \leq \frac{S}{\sqrt{n}} t_{1-\alpha/2}$$

(since $S/\sqrt{n} > 0$) and to

$$-\mu \leq -\overline{X} + \frac{S}{\sqrt{n}} t_{1-\alpha/2}$$

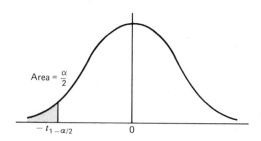

Figure 7.3.2

and to

$$\overline{X} - \frac{St_{1-\alpha/2}}{\sqrt{n}} \leq \mu \qquad \text{(multiply by } -1\text{)}.$$

It is also easy to see that

$$-t_{1-\alpha/2} \leq \frac{\sqrt{n}(\overline{X} - \mu)}{S}$$

is equivalent to

$$\mu \leq \overline{X} + \frac{St_{1-\alpha/2}}{\sqrt{n}}$$

by going through the same steps. But this shows that the event $\{-t_{1-\alpha/2} \leq \sqrt{n}(\overline{X} - \mu)/S \leq t_{1-\alpha/2}\}$ is equivalent to the event

$$\left\{ \overline{X} - \frac{St_{1-\alpha/2}}{\sqrt{n}} \leq \mu \leq \overline{X} + \frac{St_{1-\alpha/2}}{\sqrt{n}} \right\}.$$

Equivalent events must have the same probability so

$$P(L_1 \leq \mu \leq L_2) = 1 - \alpha,$$

where

$$L_1 = \overline{X} - \frac{St_{1-\alpha/2}}{\sqrt{n}},$$

$$L_2 = \overline{X} + \frac{St_{1-\alpha/2}}{\sqrt{n}}$$

and (L_1, L_2) is a $100(1 - \alpha)\%$ confidence interval for μ. This proves the following theorem.

Theorem 7.3.1. Let $X_1, X_2, \ldots, X_n$ be a random sample of a normal random variable with mean μ and variance σ^2. Then the interval (L_1, L_2) with

$$L_1 = \overline{X} - \frac{St_{1-\alpha/2}}{\sqrt{n}}$$

$$L_2 = \overline{X} + \frac{St_{1-\alpha/2}}{\sqrt{n}}$$

is a $100(1 - \alpha)\%$ confidence interval for μ. ∎

EXAMPLE 7.3.1. Suppose, as discussed earlier, the weight of a frozen food package produced by a given manufacturer is a normal random variable with mean μ and variance σ^2 (both unknown to us). We weigh $n = 10$ of these packages, selected at random and independently from those produced by this manufacturer and suppose we find

$$\sum_{i=1}^{10} x_i = 159, \qquad \sum_{i=1}^{10} x_i^2 = 2531;$$

the estimates for μ and σ then are $\bar{x} = 15.9$, $s = .57$. To compute a 90% confidence interval for μ, we want $1 - \alpha = .9$ so $1 - \alpha/2 = .95$; for a 95% confidence interval, $1 - \alpha = .95$, $1 - \alpha/2 = .975$. From Table 3 in the Appendix, with $d = 9$ degrees of freedom, we find $t_{.95} = 1.833$, $t_{.975} = 2.262$. The observed 90% confidence limits for μ then are

$$\bar{x} - \frac{1.833s}{\sqrt{n}} = 15.57, \qquad \bar{x} + \frac{1.833s}{\sqrt{n}} = 16.23$$

and we are 90 percent sure the interval $(15.57, 16.23)$ covers the true value for μ, based on this sample. The observed 95% confidence limits for μ are

$$\bar{x} - \frac{2.262s}{\sqrt{n}} = 15.49, \qquad \bar{x} + \frac{2.262s}{\sqrt{n}} = 16.31$$

so we are 95 percent sure the interval $(15.49, 16.31)$ covers the true value for μ, based on this sample. As is generally the case, to be more confident that our interval includes the true parameter value, the length of the confidence interval increases. ∎

It is also quite straightforward to compute a confidence interval for the variance of a normal random variable, given a random sample of size n. Let $X_1, X_2, \ldots, X_n$ be a random sample of a normal random variable with mean μ and variance σ^2. Then, as seen in Theorem 6.2.3,

$$\frac{\sum (X_i - \bar{X})^2}{\sigma^2} = \frac{(n-1)S^2}{\sigma^2}$$

has a χ^2 probability law with $n - 1$ degrees of freedom. From Table 2 in the Appendix, with 30 or fewer degrees of freedom, we can find the two numbers $\chi^2_{\alpha/2}$ and $\chi^2_{1-\alpha/2}$ such that

$$P\left(\chi^2_{\alpha/2} \leq \frac{(n-1)S^2}{\sigma^2} \leq \chi^2_{1-\alpha/2} \right) = 1 - \alpha.$$

(If the degrees of freedom exceed 30 we can approximate $\chi^2_{\alpha/2}$ and $\chi^2_{1-\alpha/2}$

as discussed in Section 5.6.) The event

$$\left\{ \chi^2_{\alpha/2} \le \frac{(n-1)S^2}{\sigma^2} \le \chi^2_{1-\alpha/2} \right\}$$

is equivalent to the event

$$\left\{ \frac{(n-1)S^2}{\chi^2_{1-\alpha/2}} \le \sigma^2 \le \frac{(n-1)S^2}{\chi^2_{\alpha/2}} \right\}$$

and thus

$$P(L_1 \le \sigma^2 \le L_2) = 1 - \alpha,$$

with

$$L_1 = \frac{(n-1)S^2}{\chi^2_{1-\alpha/2}},$$

$$L_2 = \frac{(n-1)S^2}{\chi^2_{\alpha/2}},$$

so (L_1, L_2) is a $100(1 - \alpha)\%$ confidence interval for σ^2. This proves the following theorem.

Theorem 7.3.2. Let $X_1, X_2, \ldots, X_n$ be a random sample of a normal random variable with mean μ and variance σ^2. Then the interval (L_1, L_2), with

$$L_1 = \frac{(n-1)S^2}{\chi^2_{1-\alpha/2}},$$

$$L_2 = \frac{(n-1)S^2}{\chi^2_{\alpha/2}},$$

is a $100(1 - \alpha)\%$ confidence interval for σ^2. ∎

EXAMPLE 7.3.2. Let us assume the same sample data as given in Example 7.3.1 and compute 95 and 99% confidence limits for σ^2. For that sample we had $n = 10$, $\sum x_i = 159$, $\sum x_i^2 = 2531$ so $(n-1)s^2 = \sum (x_i - \bar{x})^2$ $= 2.90$. From Table 2, with $d = 9$ df, we find $\chi^2_{.005} = 1.73$, $\chi^2_{.025} = 2.70$, $\chi^2_{.975} = 19.0$, $\chi^2_{.995} = 23.6$ so the 95% confidence limits for σ^2 are

$$\frac{2.90}{19} = .15 \qquad \frac{2.90}{2.70} = 1.07$$

and the 99% confidence limits are

$$\frac{2.90}{23.6} = .12 \qquad \frac{2.90}{1.73} = 1.68.$$

Based on this sample we are 95 percent sure the interval $(.15, 1.07)$ covers σ^2 and 99 percent sure that the interval $(.12, 1.68)$ covers σ^2. ∎

In many cases we may want confidence limits for a function of a parameter, rather than for the parameter itself; in Example 7.3.2 we might like confidence limits for σ rather than for σ^2, for example. It is quite straightforward to derive confidence limits for any monotonic (increasing or decreasing) function of a parameter, granted we have confidence limits for the parameter itself. Let θ be an unknown parameter and let (L_1, L_2) be a $100(1 - \alpha)\%$ confidence interval for θ; we desire confidence limits for $g(\theta)$, a monotonic function of θ. The event

$$\{L_1 \leq \theta \leq L_2\}$$

is equivalent to the event

$$\{g(L_1) \leq g(\theta) \leq g(L_2)\}$$

as long as g is a monotonic increasing function. Thus $(g(L_1), g(L_2))$ is a $100(1 - \alpha)\%$ confidence interval for $g(\theta)$. If g is monotonic decreasing $(g(L_2), g(L_1))$ is a $100(1 - \alpha)\%$ confidence interval for $g(\theta)$.

EXAMPLE 7.3.3.   We saw in Example 7.3.2 that 95% confidence limits for σ^2 were $(.15, 1.07)$, based on the sample observed. The 95% confidence limits for σ, then, based on the same sample, are $(\sqrt{.15}, \sqrt{1.07}) = (.39, 1.03)$; 99% confidence limits are given by $(\sqrt{.12}, \sqrt{1.68}) = (.35, 1.30)$. ∎

Suppose $X_1, X_2, \ldots, X_n$ is a random sample of an exponential random variable X with parameter λ. Then, as we know, $\sum X_i$ is a gamma random variable with parameters n and λ so the moment generating function for $\sum X_i$ is

$$m_{\sum X_i}(t) = \left(\frac{\lambda}{\lambda - t}\right)^n = \frac{1}{(1 - t/\lambda)^n}.$$

The moment generating function for $2\lambda \sum X_i$ then is

$$m_{\sum X_i}(2\lambda t) = \frac{1}{(1 - 2\lambda t/\lambda)^n} = \frac{1}{(1 - 2t)^n},$$

the moment generating function for a χ^2 random variable with $2n$ degrees of freedom; thus $2\lambda \sum X_i$ is a χ^2 random variable with $2n$ degrees of freedom, which allows us to easily find confidence limits for λ. Again, from Table 2 (for 30 or fewer degrees of freedom) or the approximation, we can find the two values $\chi^2_{\alpha/2}$, $\chi^2_{1-\alpha/2}$, such that

$$P(\chi^2_{\alpha/2} \leq 2\lambda \sum X_i \leq \chi^2_{1-\alpha/2}) = 1 - \alpha.$$

The event

$$\{\chi^2_{\alpha/2} \leq 2\lambda \sum X_i \leq \chi^2_{1-\alpha/2}\}$$

is equivalent to

$$\left\{ \frac{\chi^2_{\alpha/2}}{2\sum X_i} \leq \lambda \leq \frac{\chi^2_{1-\alpha/2}}{2\sum X_i} \right\}$$

which proves the following theorem.

Theorem 7.3.3. Let $X_1, X_2, \ldots, X_n$ be a random sample of an exponential random variable with parameter λ. Then (L_1, L_2) is a $100(1-\alpha)\%$ confidence interval for λ, where

$$L_1 = \frac{\chi^2_{\alpha/2}}{2\sum X_i},$$

$$L_2 = \frac{\chi^2_{1-\alpha/2}}{2\sum X_i}.$$

χ^2 with $2n$ degrees of freedom

■

EXAMPLE 7.3.4.   The time to failure for an electronic device is assumed to be an exponential random variable with (unknown) parameter λ. Assume 10 of these devices are placed on test (independently) and their observed times to failure are 607.5, 1947.0, 37.6, 129.9, 409.5, 529.5, 109.0, 582.4, 499.0, 188.1 hours, respectively. Granted these are the observed values of a random sample of an exponential random variable X, let us illustrate the computation of a 90% confidence interval for λ, the unknown parameter of the exponential probability law. We have $\sum x_i = 5039.5$, and, with $2(10) = 20$ degrees of freedom, $\chi^2_{.05} = 10.9$, $\chi^2_{.95} = 31.4$ so the observed 90% confidence limits for λ are

$$\frac{10.9}{2(5039.5)} = .00108,$$

$$\frac{31.4}{2(5039.5)} = .00312.$$

We can also use these sample values to evaluate confidence limits for functions of λ, which might be of interest. For example, the mean time to failure for this type of device is $\mu = 1/\lambda$; we can be 90 percent sure this mean time to failure lies between $\frac{1}{.00312} = 320.5$ and $\frac{1}{.00108} = 925.9$ hours, based on this sample.

The probability one of these devices will work at least t hours without failure is $P(X > t) = e^{-\lambda t}$; this is called the *reliability* of the device for a period of t hours. We can be 90% sure the reliability of this device for a 100-hour period $(e^{-100\lambda})$ lies between $e^{-100(.00312)} = .732$ and $e^{-100(.00108)} = .898$, based on this sample.   ∎

We have seen how to construct confidence intervals for the mean and variance of a normal population, as well as for the parameter of an exponential distribution. Each of these cases involves a parameter of a continuous probability law. Now let us consider confidence intervals for the parameter p of a Bernoulli probability law. Assume $X_1, X_2, ..., X_n$ is a random sample of a Bernoulli random variable X; then $Y = \sum_{i=1}^{n} X_i$ is binomial with parameters n and p.

Suppose we want a confidence interval for p with confidence coefficient $1 - \alpha$. Then for any assumed value for $0 < p < 1$ we can find the largest integer $h_1(p)$ such that $P(Y \le h_1(p)) \le \alpha/2$ and the smallest integer $h_2(p)$ such that $P(Y \ge h_2(p)) \le \alpha/2$. Because of the discreteness of the binomial probability law, we cannot in general make these probabilities exactly equal $\alpha/2$ for all p. Then since the events $\{Y \le h_1(p)\}$ and $\{Y \ge h_2(p)\}$ are mutually exclusive, we have

$$P(Y \le h_1(p) \text{ or } Y \ge h_2(p)) \le \frac{\alpha}{2} + \frac{\alpha}{2} = \alpha,$$

from which it follows that the complement of this event has probability at least $1 - \alpha$; that is,

$$P(h_1(p) < Y < h_2(p)) = 1 - P(Y \le h_1(p) \text{ or } Y \ge h_2(p))$$
$$\ge 1 - \alpha,$$

no matter what the unknown value for p. The two functions $h_1(p)$ and $h_2(p)$ are monotonic and nondecreasing (and also discontinuous step functions). Thus we have no trouble defining the inverse functions and the event $\{h_1(p) < Y < h_2(p)\}$ is equivalent to $\{h_2^{-1}(Y) < p < h_1^{-1}(Y)\}$ so

$$P(h_2^{-1}(Y) < p < h_1^{-1}(Y)) \ge 1 - \alpha,$$

since the probabilities of equivalent events are equal, that is, $(h_2^{-1}(Y), h_1^{-1}(Y))$ is a $100(1 - \alpha)\%$ confidence interval for p. Recall that $h_1(p)$ is the integer such that

$$P(Y \le h_1(p)) \le \frac{\alpha}{2},$$

that is,

$$\sum_{j=0}^{h_1(p)} \binom{n}{j} p^j (1 - p)^{n-j} \le \frac{\alpha}{2}.$$

But if we have observed $Y = y$, $h_1^{-1}(y)$ is the upper confidence limit for p, and $h_1(h_1^{-1}(y)) = y$ so we want the value of p such that

$$\sum_{j=0}^{y} \binom{n}{j} p^j (1 - p)^{n-j} = \frac{\alpha}{2};$$

this is the upper confidence limit for p. Similarly, $h_2(h_2^{-1}(y)) = y$ so the lower confidence limit is the value for p such that

$$\sum_{j=y}^{n} \binom{n}{j} p^j (1 - p)^{n-j} = \frac{\alpha}{2}.$$

Tables and graphs giving the solutions of these two equations (the two confidence limits) are available for various values of n, the number of trials, and y, the observed number of successes. For moderate values of n it is not difficult to find the solutions on a programmable calculator.

Theorem 7.3.4. Let Y be a binomial random variable with parameters n and p and let p_1 and p_2 be such that

$$\sum_{j=0}^{y} \binom{n}{j} p_2^j (1 - p_2)^{n-j} = \frac{\alpha}{2}$$

$$\sum_{j=y}^{n} \binom{n}{j} p_1^j (1 - p_1)^{n-j} = \frac{\alpha}{2},$$

for any given observed value y. Then (p_1, p_2) is a $100(1 - \alpha)\%$ confidence interval for p. ■

EXAMPLE 7.3.5. Assume there is a constant probability p that a person entering a store will make a purchase and that persons selected at random leaving the store constitute a random sample of a Bernoulli random variable (success = purchase made, failure = no purchase). Of $n = 10$ persons selected, it was found that $y = 4$ had made a purchase. Then a 90% confidence interval for p is determined by solving

$$\sum_{j=0}^{4} \binom{10}{j} p^j (1 - p)^{10-j} = .05$$

$$\sum_{j=4}^{10} \binom{10}{j} p^j (1 - p)^{10-j} = .05$$

for p (the first equation gives the upper limit and the second equation gives the lower limit). The value for p (to 3 decimal places) that satisfies the first equation is .696 and the value that satisfies the second is .150, as is easily verified with a hand-held calculator. Thus if a random sample of 10 independent Bernoulli random variables gives $Y = 4$ successes, the 90% confidence interval for p is (.150, .696). ∎

The same procedure discussed for deriving a confidence interval for a Bernoulli parameter works for other discrete populations, as well as for populations with a continuous probability law. The following example applies this procedure to a continuous probability law.

EXAMPLE 7.3.6.   Assume $X_1, X_2, \ldots, X_n$ is a random sample of a uniform random variable X on $(0, \theta)$. The maximum likelihood estimator for θ is $X_{(n)}$, the maximum sample value, with distribution function

$$F_{X_{(n)}}(t) = \frac{t^n}{\theta^n}, \qquad 0 \le t \le \theta.$$

Suppose we desire a $100(1 - \alpha)\%$ confidence interval for θ. Clearly,

$$P\big(X_{(n)} \le h_1(\theta)\big) = \frac{[h_1(\theta)]^n}{\theta^n} = \frac{\alpha}{2}$$

gives

$$h_1(\theta) = \theta \left(\frac{\alpha}{2}\right)^{1/n}$$

and

$$P\big(X_{(n)} \ge h_2(\theta)\big) = 1 - \frac{[h_2(\theta)]^n}{\theta^n} = \frac{\alpha}{2}$$

gives

$$h_2(\theta) = \theta \left(1 - \frac{\alpha}{2}\right)^{1/n}.$$

Thus we have

$$P\big(h_1(\theta) < X_{(n)} < h_2(\theta)\big) = 1 - \alpha$$

and the event

$$\{h_1(\theta) < X_{(n)} < h_2(\theta)\} = \left\{\theta\left(\frac{\alpha}{2}\right)^{1/n} < X_{(n)} < \theta\left(1 - \frac{\alpha}{2}\right)^{1/n}\right\}$$

is equivalent to the event

$$\left\{ \frac{X_{(n)}}{(1 - \alpha/2)^{1/n}} < \theta < \frac{X_{(n)}}{(\alpha/2)^{1/n}} \right\}$$

so

$$\left(\frac{X_{(n)}}{(1 - \alpha/2)^{1/n}}, \frac{X_{(n)}}{(\alpha/2)^{1/n}} \right)$$

provides a $100(1 - \alpha)\%$ confidence interval for θ. For example, if the largest value in a sample of $n = 40$ yields $X_{(40)} = 36.9$ and we desire a 95% confidence interval for θ, we find $\alpha/2 = .025$, $(\alpha/2)^{1/40} = .9119$, $(1 - \alpha/2)^{1/40} = .9994$ and the 95% limits for θ are

$$\frac{36.9}{.9994} = 36.92, \qquad \frac{36.9}{.9119} = 40.46.$$

We are 95 percent sure that θ lies between 36.92 and 40.46, based on this sample (assuming independence and uniformity). ∎

The large sample distribution for maximum likelihood estimators (Theorem 7.2.2) also proves useful in deriving confidence intervals for unknown parameters. Recall that (as long as the regularity conditions are satisfied) the maximum likelihood estimator, Γ, of a parameter γ is, for a "large" sample size n, approximately normally distributed with mean γ and variance

$$\sigma^2 = \frac{1}{nE\left[\left((\partial/\partial\gamma) \ln f_X(X)\right)^2\right]}$$

$$\left(\text{or} \quad \frac{1}{nE\left[\left((\partial/\partial\gamma) \ln p_X(X)\right)^2\right]} \quad \text{if } X \text{ is discrete} \right).$$

It follows then that $(\Gamma - \gamma)/\sigma$ is approximately standard normal so

$$P\left(-z_{1-\alpha/2} \leq \frac{\Gamma - \gamma}{\sigma} \leq z_{1-\alpha/2} \right) = 1 - \alpha,$$

where $z_{1-\alpha/2}$ comes from the standard normal table. The event

$$\left\{ -z_{1-\alpha/2} \leq \frac{\Gamma - \gamma}{\sigma} \leq z_{1-\alpha/2} \right\}$$

then is equivalent to $\{\Gamma - \sigma z_{1-\alpha/2} \leq \gamma \leq \Gamma + \sigma z_{1-\alpha/2}\} = \{g_1(\Gamma) \leq \gamma \leq g_2(\Gamma)\}$, so the observed values for $g_1(\Gamma)$ and $g_2(\Gamma)$ provide confidence limits for γ. This is illustrated in the following example, again for the Bernoulli case.

EXAMPLE 7.3.7. Suppose we have a large random sample, $X_1, X_2,$ ..., X_n, of a Bernoulli random variable X with parameter p. The maximum likelihood estimator for p is $\bar{X} = (1/n) \sum X_i$; also we have

$$\frac{\partial}{\partial p} \ln p_X(X) = \frac{X}{p} - \frac{1-X}{1-p} = \frac{X-p}{p(1-p)}$$

so

$$\frac{1}{nE[((\partial/\partial p) \ln p_X(X))^2]} = \frac{p(1-p)}{n}$$

and $\bar{X}$ is approximately normal with mean p and variance

$$\sigma^2 = \frac{p(1-p)}{n}.$$

Thus we have

$$P\left(-z_{1-\alpha/2} \leq \frac{\bar{X}-p}{\sqrt{p(1-p)/n}} \leq z_{1-\alpha/2}\right) \doteq 1-\alpha$$

and we want to find an event of the form $\{g_1(\bar{X}) \leq p \leq g_2(\bar{X})\}$, which is equivalent to

$$\left\{-z_{1-\alpha/2} \leq \frac{\bar{X}-p}{\sqrt{p(1-p)/n}} \leq z_{1-\alpha/2}\right\} = \left\{\frac{(\bar{X}-p)^2}{p(1-p)/n} \leq z_{1-\alpha/2}^2\right\}.$$

The equation

$$(\bar{X}-p)^2 \leq \frac{p(1-p)z_{1-\alpha/2}^2}{n}$$

has the solutions

$$g_1(\bar{X}) = \frac{\bar{X} + (z^2/2n) - (z/\sqrt{n})\sqrt{\bar{X}(1-\bar{X}) + z^4/4n}}{1 + (z^2/n)}$$

$$g_2(\bar{X}) = \frac{\bar{X} + (z^2/2n) + (z/\sqrt{n})\sqrt{\bar{X}(1-\bar{X}) + z^4/4n}}{1 + (z^2/n)},$$

where $z = z_{1-\alpha/2}$. Especially for n large (and reasonable $1-\alpha$), z^2/n should be essentially zero so the more frequently used large sample limits are

$$\bar{X} \pm \frac{z}{\sqrt{n}}\sqrt{\bar{X}(1-\bar{X})}.$$

If in a sample of $n = 100$ we find the proportion of successes to be $\bar{x} = .65$, we can be approximately 95 percent sure $(z_{.975} = 1.96)$ that the true value for p lies between

$$.65 - \frac{1.96}{\sqrt{100}}\sqrt{.65(.35)} = .56$$

and

$$.65 + \frac{1.96}{\sqrt{100}}\sqrt{.65(.35)} = .74.$$

It is interesting to compare this large sample approximate procedure with the exact procedure of Example 7.3.5. There we saw that observing $Y = 4$ successes in $n = 10$ trials gives $(.150, .696)$ as the exact 90% confidence limits for p. Since $z_{.95} = 1.64$, $\sqrt{\bar{X}(1 - \bar{X})/n} = \sqrt{.4(.6)/10} = .155$, the large sample 90% limits are

$$.4 \pm 1.64(1.55) = .4 \pm .254 = (.146, .654),$$

surprisingly close to the exact values since $n = 10$ is hardly large. The large sample values are so accurate mainly because the estimated value for p is .4 and the exact binomial distribution with $n = 10$, $p = .4$ is fairly close to being symmetric. As the estimate for p gets closer to 0 or 1, a larger value for n is required for large sample confidence limits to be this close to the exact values. ∎

EXERCISE 7.3

1. Assume a government agency desires to estimate the miles per gallon that a particular new (make and model) car is capable of attaining. To do this, the agency acquires one of these cars, fills the gas tank, and then a trained driver drives the car for 100 miles; the gas tank is then refilled and the same driver again drives the car 100 mi, at which time it is again refilled and driven 100 mi, and so on. This operation is performed a total of $n = 10$ times; the numbers of gallons needed to fill the gas tank these 10 times are:

    4.78, 4.42, 3.94, 4.15, 4.90, 3.92, 3.94, 4.68, 4.32, 4.23.

    Assuming these values are a random sample from a normal population with mean μ and variance σ^2,
    (a) Compute estimates for μ and σ.
    (b) Compute 90% confidence limits for μ, the number of gallons needed to drive this car 100 miles.
    (c) Compute 90% confidence limits for $m = 100/\mu$, the miles per gallon one might expect from this make and model.

2. Use the data in Exercise 7.3.1 to compute 95% confidence limits for σ^2 and for σ, the population variance and standard deviation.

3. Frequently, "one-sided" confidence limits are derived for unknown parameters. Using a random sample of size n of a normal random variable with parameters μ and σ^2, find a statistic L such that $P(L \leq \mu) = 1 - \alpha$. L is called a lower confidence limit for μ. Use the data in Exercise 7.3.1 to evaluate a lower 95% confidence limit for μ, the number of gallons needed to drive that car 100 miles. Also, find a 95% upper confidence limit for $100/\mu$, the miles per gallon one could expect from the car. (U is a 95% upper confidence limit if $P(100/\mu \leq U) = .95$).

4. The time needed to diagnose the fault and to make the repair for large pieces of equipment is frequently assumed to be an exponential random variable. Assume the time necessary for diagnosing and repairing a transmission problem for a 1981 Datsun automobile is an exponential random variable with parameter λ. The observed times for diagnosing and repairing $n = 9$ of these transmissions were:

$$1.7, .9, 3.0, 3.6, .5, 7.3, 3.2, .3, 6.1$$

hours, respectively. Use these values to construct 95% (two-sided) confidence limits for the mean time to diagnose and repair one of these transmissions.

5. Assume the repair shop doing the (diagnosing and) repair in Exercise 7.3.4 charges $30 per hour. Find 95% confidence limits for the expected amount it will cost to make the repair for one of these transmissions.

6. Assume there is a constant probability p that a towel manufactured by machine in a factory will be declared a "second" because of some defect. If in 1000 towels selected at random from one day's output, it is found that 30 are "seconds," find 95% (two-sided) confidence limits for p.

7. Suppose the factory mentioned in Exercise 7.3.6 selects a sample of 1000 towels for inspection each working day; thus each day a 95% confidence interval is computed for p. In a $5(52) = 260$ day work year, then, 260 confidence intervals have been computed.
   (a) What number of these intervals would you expect covered the true value for p?
   (b) What is the (approximate) probability that at least 95 percent of the computed intervals include the true value for p?

*8. Each hour a radio station broadcasts a "beep" marking the hour. The station has a quartz timepiece that is used to trigger the instant of the beep. Unless the quartz used is very high quality, and maintained under carefully controlled conditions, it is not 100 percent accurate. Assume the difference between the time the beep starts and the "exact" time of the hour is a uniform random variable X on the interval $(-\theta, \theta)$, using microseconds as a unit. A random sample of $n = 15$ observed values for X were:

$$221, \quad 265, \quad -140, \; 327, \; -401, \; 308, \; -317, \; 447,$$
$$-137, \; -228, \; -477, \quad 69, \quad 475, \quad 56, \; -101.$$

Compute 99% confidence limits for θ, the magnitude of the most extreme error in timing for the beep.

*9. (a) Find the expected length of a 90% confidence interval for the mean of a normal population μ, using the result in Theorem 7.3.1, based on a sample of size $n = 16$.

   (b) What is the probability that the length of this confidence interval is less than σ?

*10. (a) Find the expected length of a 90% confidence interval for the standard deviation σ of a normal population, using the result of Theorem 7.3.2, for a sample of size $n = 16$.

   (b) What is the probability the length of this confidence interval is less than σ?

11. Suppose $X_1, X_2, \ldots, X_n$ is a random sample of a gamma random variable with $r = 2$ and λ unknown.

   (a) What is the probability law for $Y = \sum X_i$?

   (b) Find the constant k (which may involve λ) such that kY is a χ^2 random variable.

   (c) Use the result of (b) to construct $100(1 - \alpha)\%$ confidence limits for λ and for μ, the population mean.

12. Assume the data for repair times given in Exercise 7.3.4 come from a gamma distribution with $r = 2$ and λ unknown. Use these numbers to compute 95% confidence limits for μ, the mean time needed to repair a Datsun transmission.

13. Use the large sample distribution for the maximum likelihood estimator for μ, the mean of a Poisson distribution, to construct approximate $100(1 - \alpha)\%$ confidence limits for μ.

14. The number of new houses sold, per week for 15 weeks, by an active real estate firm, were

$$3, \; 3, \; 4, \; 6, \; 2, \; 4, \; 4, \; 3, \; 1, \; 2, \; 0, \; 5, \; 7, \; 1, \; 4$$

respectively. Assuming these are the observed values for a random sample of size 15 of a Poisson random variable with parameter μ, use the procedure in Exercise 7.3.13 to compute 95% confidence limits for μ.

*15. If $X_1, X_2, \ldots, X_n$ is a random sample of a Poisson random variable

X with parameter μ, $Y = \sum X_i$ is again Poisson with parameter $n\mu$. An exact confidence interval for μ can be based on finding two functions, $h_1(\mu)$ and $h_2(\mu)$, such that

$$P\big(Y \le h_1(\mu)\big) \le \frac{\alpha}{2}$$

$$P\big(Y \ge h_2(\mu)\big) \le \frac{\alpha}{2},$$

so we would have

$$P\big(h_1(\mu) < Y < h_2(\mu)\big) \ge 1 - \alpha.$$

Again, $\{h_1(\mu) < Y < h_2(\mu)\}$ is equivalent to $\{h_2^{-1}(Y) < \mu < h_1^{-1}(Y)\}$ and the confidence interval for μ is given by solving

$$\sum_{k=y}^{\infty} \frac{(n\mu)^k}{k!} e^{-n\mu} = \frac{\alpha}{2}$$

$$\sum_{k=0}^{y} \frac{(n\mu)^k}{k!} e^{-n\mu} = \frac{\alpha}{2}$$

for μ. Apply this procedure to the data given in Exercise 7.3.14, and compare the resulting 95% confidence limits with the large sample limits.

7.4   Two-Sample Procedures

We have seen procedures for estimating unknown parameters, some ways of comparing estimators and how to construct confidence intervals for certain parameters. In every case discussed we assumed we had a random sample of a random variable and were concerned with inferences about the unknown parameter(s) of the probability law for the random variable. Many practical problems are concerned with comparing the parameters of two or more probability laws. For example, we might assume the miles per gallon that we would achieve driving car A a fixed distance is a normal random variable X with parameters μ_X, σ_X^2. We might also assume the miles per gallon achieved driving car B the same distance is a normal random variable Y with parameters μ_Y, σ_Y^2. If we were considering the purchase of either car A or car B, it would be of some interest to estimate $\mu_X - \mu_Y$, the expected difference in miles per gallon to be achieved by the two cars. Or suppose an overweight person would like to lose some weight; the amount she would lose (over a 6-month period) with diet 1 is a random variable X_1 with mean μ_1, whereas the amount of weight she would lose (in the same period) with diet 2 is a random variable X_2 with mean μ_2. This person would undoubtedly be interested in the value for $\mu_1 - \mu_2$ in choosing which diet she might prefer to use. We will in this section consider problems involving independent random samples of two random variables.

Suppose $X_1, X_2, \ldots, X_n$ is a random sample of a random variable X whose probability law depends on an unknown parameter θ and let $Y_1, Y_2, \ldots, Y_m$ be an independent random sample of a random variable Y whose probability law depends on an unknown parameter λ (the sample of Y values is independent of the sample of X values means all the random variables in the full set $X_1, X_2, \ldots, X_n, Y_1, Y_2, \ldots, Y_m$ are independent). The independence of the two samples then implies that the likelihood function for the $m + n$ random variables is

$$L_{X,Y}(\theta, \lambda) = L_X(\theta)\, L_Y(\lambda),$$

the product of the likelihood functions of the individual samples. But then the values for θ and λ that maximize $L_{X,Y}(\theta, \lambda)$, the joint likelihood function, are identical with the values that maximize the individual likelihood functions, $L_X(\theta)$ and $L_Y(\lambda)$. That is, if $\hat{\theta}$ maximizes $L_X(\theta)$ and $\hat{\lambda}$ maximizes $L_Y(\lambda)$, then $L_{X,Y}(\hat{\theta}, \hat{\lambda})$ is the maximum value for the joint likelihood function. Then the maximum likelihood estimate for a function, $g(\lambda, \theta)$, of the parameters of the two probability laws, is $g(\hat{\lambda}, \hat{\theta})$ from (an extension of) the invariance principle of maximum likelihood.

EXAMPLE 7.4.1. If $n = 5$ test drives of car A (each of 100 miles) gives an average mile per gallon estimate of $\bar{x} = 22.1$ and $m = 3$ test drives of car B (also each of 100 miles) gives $\bar{y} = 24.3$ miles per gallon, the maximum likelihood estimate for $\mu_Y - \mu_X = 24.3 - 22.1 = 2.2$ miles per gallon. If $n = 10$ ladies of the same build and initial weight lose an average of $\bar{x} = 16$ pounds on diet 1 and $m = 8$ ladies (again same build and initial weight) lose an average of $\bar{y} = 19$ pounds on diet 2, the maximum likelihood estimate for $\mu_1 - \mu_2$ is $\bar{x} - \bar{y} = -3$ pounds. ∎

EXAMPLE 7.4.2. At times models are adopted that link the probability laws for random variables whose observed values are produced under different conditions. Accelerated life testing provides examples of this kind. Suppose the time to failure for an electronic component, in standard use, is an exponential random variable X with parameter λ. If this type of component is used in an environment with an elevated (specified) temperature and is subjected to vibrations (with a given frequency), assume its time to failure is again an exponential random variable Y, but now with parameter 4λ (these conditions describe an increased stress level, leading to a shorter expected time to failure). Assume $X_1, X_2, \ldots, X_n$ is a random sample of X

and $Y_1, Y_2, \ldots, Y_m$ is an independent random sample of Y. Then

$$L_X(\lambda) = \prod_{i=1}^{n} (\lambda e^{-\lambda x_i}) = \lambda^n e^{-\lambda \Sigma x_i}$$

$$L_Y(\lambda) = \prod_{i=1}^{m} (4\lambda e^{-4\lambda y_i}) = 4^m \lambda^m e^{-4\lambda \Sigma y_i}$$

and the joint likelihood function for all $m + n$ observations is

$$L_{X,Y}(\lambda) = (\lambda^n e^{-\lambda \Sigma x_i})(4^m \lambda^m e^{-4\lambda \Sigma y_i})$$
$$= 4^m \lambda^{n+m} \exp\left[-\lambda\left(\sum x_i + 4 \sum y_i\right)\right]$$

and

$$K(\lambda) = \ln L_{X,Y}(\lambda)$$
$$= m \ln 4 + (n + m) \ln \lambda - \lambda\left(\sum x_i + 4 \sum y_i\right),$$

$$\frac{dK(\lambda)}{d\lambda} = \frac{n + m}{\lambda} - \left(\sum x_i + 4 \sum y_i\right)$$

from which we find

$$\hat{\lambda} = \frac{n + m}{\sum x_i + 4 \sum y_i}$$

is the combined maximum likelihood estimate for λ, using the data from both sorts of tests. The maximum likelihood estimate of the mean time to failure for one of these components in standard use ($\mu = 1/\lambda$) is

$$\hat{\mu} = \frac{1}{\hat{\lambda}} = \frac{\sum x_i + 4 \sum y_i}{m + n} = \frac{n}{n + m} \bar{x} + 4 \frac{m}{n + m} \bar{y},$$

a linear combination of $\bar{x}$ and $4\bar{y}$. The multiplier of 4 times $\bar{y}$ occurs, of course, because of the assumption

$$E[Y] = \frac{1}{4\lambda} = \frac{1}{4} E[X] = \frac{\mu}{4}.$$

What would be the maximum likelihood estimate of the mean time to failure of one of these components in the increased stress environment? ∎

We can also use some of our earlier results to find confidence intervals for certain functions of the parameters of two probability laws. Suppose, for example, $X_1, X_2, \ldots, X_n$ is a random sample of a normal random variable X with parameters μ_X and σ^2, and $Y_1, Y_2, \ldots, Y_m$ is an independent random sample of a normal random variable Y with parameters μ_Y, σ^2 (note the two variances are assumed equal). Then, as we know, $\bar{X} - \bar{Y}$ is a normal random variable with mean $\mu_X - \mu_Y$ and variance $\sigma^2(1/m + 1/n) = (m + n)\sigma^2/mn$, so it follows that

$$Z = \frac{\sqrt{mn/(m + n)}\,((\bar{X} - \bar{Y}) - (\mu_X - \mu_Y))}{\sigma}$$

is a standard normal random variable. You can easily verify that the maximum likelihood estimate for σ^2, the common variance of both populations, is

$$\hat{\sigma}^2 = \frac{1}{m + n}\left(\sum(x_i - \bar{x})^2 + \sum(y_j - \bar{y})^2\right)$$

and that

$$S_p^2 = \frac{1}{m + n - 2}\left(\sum(X_i - \bar{X})^2 + \sum(Y_j - \bar{Y})^2\right)$$

is an unbiased estimator for σ^2. S_p^2 is called a "pooled" estimator for σ^2, because it adds (pools) together the sums of squares of deviations of the individual observations about the mean, for the two samples. Clearly, $(m + n - 2)S_p^2/\sigma^2$ is a χ^2 random variable with $m + n - 2$ degrees of freedom, and S_p^2 is independent of the standard normal random variable Z. But then

$$T = \frac{\sqrt{mn/(m + n)}\,((\bar{X} - \bar{Y}) - (\mu_X - \mu_Y))}{\sigma}\bigg/\sqrt{\frac{(m + n - 2)S_p^2}{\sigma^2(m + n - 2)}}$$
$$= \frac{\sqrt{mn/(m + n)}\,((\bar{X} - \bar{Y}) - (\mu_X - \mu_Y))}{S_p}$$

has the T-distribution with $m + n - 2$ degrees of freedom. If $t_{1-\alpha/2}$ is the $100(1 - \alpha/2)$th quantile of this T-distribution, we have

$$P\left(-t_{1-\alpha/2} \le \frac{\sqrt{mn/(m + n)}\,[(\bar{X} - \bar{Y}) - (\mu_X - \mu_Y)]}{S_p} \le t_{1-\alpha/2}\right) = 1 - \alpha;$$

the event

$$\left\{-t_{1-\alpha/2} \le \frac{\sqrt{mn/(m + n)}\,[(\bar{X} - \bar{Y}) - (\mu_X - \mu_Y)]}{S_p} \le t_{1-\alpha/2}\right\}$$

is equivalent to

$$\left\{\bar{X} - \bar{Y} - t_{1-\alpha/2}S_p\sqrt{\frac{m + n}{mn}} \le \mu_X - \mu_Y \le \bar{X} - \bar{Y} + t_{1-\alpha/2}S_p\sqrt{\frac{m + n}{mn}}\right\}$$

so

$$L_1 = \bar{X} - \bar{Y} - t_{1-\alpha/2} S_p \sqrt{\frac{m+n}{mn}},$$

$$L_2 = \bar{X} - \bar{Y} + t_{1-\alpha/2} S_p \sqrt{\frac{m+n}{mn}}$$

are $100(1 - \alpha)\%$ confidence limits for $\mu_X - \mu_Y$. This proves the following theorem.

Theorem 7.4.1. If $X_1, X_2, \dots, X_n$ and $Y_1, Y_2, \dots, Y_m$ are independent samples from normal distributions with means μ_X, μ_Y, respectively, and variance σ^2, then

$$L_1 = \bar{X} - \bar{Y} - t_{1-\alpha/2} S_p \sqrt{\frac{m+n}{mn}},$$

$$L_2 = \bar{X} - \bar{Y} + t_{1-\alpha/2} S_p \sqrt{\frac{m+n}{mn}}$$

are $100(1 - \alpha)\%$ confidence limits for $\mu_X - \mu_Y$.   ∎

EXAMPLE 7.4.3.   On $n = 5$ 100-mile test drives of car A, the observed miles per gallon achieved were 20.7, 24.0, 23.2, 21.6, 21.0 and on $m = 3$ 100-mile test drives of car B, the observed miles per gallon achieved were 25.8, 23.7, 23.4. Letting $x_1, x_2, \dots, x_5$ represent the observations of car A and y_1, y_2, y_3 the observations for car B, and making the assumptions given in Theorem 7.4.3, we can evaluate, say, 90% confidence limits for $\mu_X - \mu_Y$, the difference in expected miles per gallon for the two cars. If the confidence interval includes the value 0, then it would appear likely that the expected miles per gallon are the same for the two cars (since $\mu_X - \mu_Y = 0$, that is, $\mu_X = \mu_Y$ is one of the values covered by the confidence interval). On the other hand, if the confidence interval does not include 0, it would seem likely that the expected miles per gallon are not the same for the two cars (and one might want to eliminate the lower mile per gallon car from consideration of purchase). For the observed values given,

$$\sum x_i = 110.5, \qquad \sum x_i^2 = 2450.29, \qquad \sum y_j = 72.9, \qquad \sum y_j^2 = 1774.89,$$

so

$$\bar{x} = 22.1, \qquad \sum (x_i - \bar{x})^2 = 8.24, \qquad \bar{y} = 24.3, \qquad \sum (y_j - \bar{y})^2 = 3.42$$

so $s_p = 1.394$, $t_{.95} = 1.943$ and the 90% confidence limits for $\mu_X - \mu_Y$ are:

$$22.1 - 24.3 - (1.943)(1.394)\sqrt{8/15} = -4.18$$
$$22.1 - 24.3 + (1.943)(1.394)\sqrt{8/15} = -0.22.$$

Based on these sample values (and the assumed model), we can be 90 percent sure $\mu_X - \mu_Y$ lies between -4.18 and -0.22; in particular, then, unless an event with probability .1 has occurred, zero is not one of the possible values for $\mu_X - \mu_Y$ and it would appear we could expect better gas mileage with car B than with car A. ∎

The confidence limits given in Theorem 7.4.1 are based on the assumption the variances of the X and Y values are equal, that $\sigma_X^2 / \sigma_Y^2 = 1$. If this assumption is not made there is no standard, simple procedure that is commonly used to derive confidence limits for $\mu_X - \mu_Y$.

With $X_1, X_2, ..., X_n$ and $Y_1, Y_2, ..., Y_m$ independent samples of normal random variables with means μ_X, μ_Y, respectively, and variances σ_X^2, σ_Y^2, respectively, there is a simple procedure one can use to evaluate confidence limits for σ_Y^2 / σ_X^2. Again,

$$\frac{\sum (X_i - \bar{X})^2}{\sigma_X^2}$$

is a χ^2 random variable with $n - 1$ degrees of freedom and

$$\frac{\sum (Y_j - \bar{Y})^2}{\sigma_Y^2}$$

is an independent χ^2 random variable with $m - 1$ degrees of freedom. It follows then that the ratio

$$\frac{\sum (X_i - X)^2}{(n-1)\sigma_X^2} \bigg/ \frac{\sum (Y_j - Y)^2}{(m-1)\sigma_Y^2} = \frac{\sigma_Y^2 S_X^2}{\sigma_X^2 S_Y^2}$$

has an F-distribution with $n - 1$ and $m - 1$ degrees of freedom, and

$$P\left(F_{\alpha/2} \le \frac{\sigma_Y^2 S_X^2}{\sigma_X^2 S_Y^2} \le F_{1-\alpha/2} \right) = 1 - \alpha,$$

where $F_{\alpha/2}$, $F_{1-\alpha/2}$ are quantiles from this F-distribution. But then we also have

$$P\left(\frac{S_Y^2}{S_X^2} F_{\alpha/2} \le \frac{\sigma_Y^2}{\sigma_X^2} \le \frac{S_Y^2}{S_X^2} F_{1-\alpha/2} \right) = 1 - \alpha$$

so

$$L_1 = \frac{S_Y^2}{S_X^2} F_{\alpha/2},$$

$$L_2 = \frac{S_Y^2}{S_X^2} F_{1-\alpha/2}$$

are $100(1 - \alpha)\%$ confidence limits for σ_Y^2/σ_X^2, the ratio of the two population variances. This establishes the following theorem.

Theorem 7.4.2. Let $X_1, X_2, \ldots, X_n$ be a random sample of a normal random variable X with parameters μ_X, σ_X^2 and let $Y_1, Y_2, \ldots, Y_m$ be an independent random sample of a normal random variable Y with parameters μ_Y, σ_Y^2. Then

$$L_1 = \frac{S_Y^2}{S_X^2} F_{\alpha/2},$$

$$L_2 = \frac{S_Y^2}{S_X^2} F_{1-\alpha/2}$$

are $100(1 - \alpha)\%$ confidence limits for σ_Y^2/σ_X^2, where

$$S_X^2 = \frac{1}{n-1} \sum (X_i - \bar{X})^2, \qquad S_Y^2 = \frac{1}{m-1} \sum (Y_j - \bar{Y})^2,$$

and $F_{\alpha/2}$, $F_{1-\alpha/2}$ are the $100(\alpha/2)$th and $100(1 - \alpha/2)$th quantiles of the F-distribution with $n - 1$, $m - 1$ degrees of freedom, respectively. ∎

EXAMPLE 7.4.4.   With the data given in Example 7.4.3, we have $n = 5$, $\sum (x_i - \bar{x})^2 = 8.24$, $s_X^2 = 2.06$, $m = 3$, $\sum (y_j - \bar{y})^2 = 3.42$, $s_Y^2 = 1.71$. To evaluate 90% confidence limits for σ_Y^2/σ_X^2, we find

$$F_{.95}(4, 2) = 19.25, \qquad F_{.05}(4, 2) = \frac{1}{F_{.95}(2, 4)} = \frac{1}{6.94} = .1441$$

so the 90% confidence limits for σ_Y^2/σ_X^2 are:

$$\tfrac{1.71}{2.06} (.1441) = .1196, \qquad \tfrac{1.71}{2.06} (19.25) = 15.98.$$

The 90% confidence limits for σ_Y/σ_X are

$$\sqrt{.1196} = .346, \qquad \sqrt{15.98} = 3.997. \qquad ∎$$

We can also derive confidence limits for the ratio of two exponential parameters, using a procedure very similar to that described in Theorem

7.4.2. Suppose $X_1, X_2, \ldots, X_n$ is a random sample of an exponential random variable with parameter λ_X and $Y_1, Y_2, \ldots, Y_m$ is an independent random sample of an exponential random variable with parameter λ_Y. Then $2\lambda_X \sum X_i$ is a χ^2 random variable with $2n$ degrees of freedom and $2\lambda_Y \sum Y_j$ is an independent χ^2 random variable with $2m$ degrees of freedom. Thus the ratio

$$\frac{2\lambda_X \sum X_i}{2n} \bigg/ \frac{2\lambda_Y \sum Y_j}{2m} = \frac{\lambda_X \overline{X}}{\lambda_Y \overline{Y}}$$

has the F-distribution with $2n$ and $2m$ degrees of freedom and

$$L_1 = \frac{\overline{Y}}{\overline{X}} F_{\alpha/2},$$

$$L_2 = \frac{\overline{Y}}{\overline{X}} F_{1-\alpha/2}$$

are $100(1 - \alpha)\%$ confidence limits for λ_X/λ_Y. This establishes the following theorem.

Theorem 7.4.3. Let $X_1, X_2, \ldots, X_n$ be a random sample of an exponential random variable with parameter λ_X, $Y_1, Y_2, \ldots, Y_m$ be an independent random sample of an exponential random variable with parameter λ_Y, and let $F_{\alpha/2}$, $F_{1-\alpha/2}$ be the $100(\alpha/2)$th, $100(1 - \alpha/2)$th quantiles of the F-distribution with $2n$, $2m$ degrees of freedom. Then

$$L_1 = \frac{\overline{Y}}{\overline{X}} F_{\alpha/2},$$

$$L_2 = \frac{\overline{Y}}{\overline{X}} F_{1-\alpha/2}$$

are $100(1 - \alpha)\%$ confidence limits for λ_X/λ_Y, or equivalently, for μ_Y/μ_X, the ratio of the two means. ∎

EXAMPLE 7.4.5. It is frequently assumed that the amount of time needed by a repair shop to identify the difficulty with and to repair a piece of equipment is an exponential random variable with parameter λ. The confidence limits in Theorem 7.4.3 can be used to evaluate confidence limits for the ratio of the mean times required for two different repair shops to complete similar jobs. Assume repair shop 1 required 2.2, 6.4, 1.3, 7.1, 5.1 hours, respectively, to identify and repair $n = 5$ similar equipment breakdowns, whereas repair shop 2 required 2.3, 0.1, 4.5, 0.2, 2.4 hours, respectively, to identify and repair $m = 5$ similar equipment breakdowns. Assuming the times for repair shop 1 are independent, exponential with parameter λ_1 and mean $\mu_1 = 1/\lambda_1$, whereas the times for repair shop 2 are independent, exponential with parameter λ_2 and mean $\mu_2 = 1/\lambda_2$, we have $\sum x_i = 22.1$, $\bar{x} = 4.42$ (for shop 1), $\sum y_j = 9.5$, $\bar{y} = 1.90$ (for shop 2). To evaluate 90% confidence limits for μ_1/μ_2, we find $F_{.05} = \frac{1}{2.98} = .336$, $F_{.95} = 2.98$ (both with 10, 10 degrees of freedom) so we can be 90 percent sure that μ_1/μ_2 lies between $\frac{4.42}{1.90}(.336) = .781$ and $\frac{4.42}{1.90}(2.98) = 6.93$, based on these sample values. ∎

It is frequently of interest to compare two Bernoulli parameters. Assume $X_1, X_2, \ldots, X_n$ is a random sample of a Bernoulli random variable with parameter p_1 and $Y_1, Y_2, \ldots, Y_m$ is an independent random sample of a Bernoulli random variable with parameter p_2. The maximum likelihood estimates for p_1 and p_2 are

$$\hat{p}_1 = \frac{1}{n}\sum x_i = \bar{x}, \qquad \hat{p}_2 = \frac{1}{m}\sum y_j = \bar{y}$$

and for m and n both large (say, 20 or more, as long as p_1 and p_2 are neither too close to zero or to one), we know from Theorem 7.2.2 that both $\bar{X}$ and $\bar{Y}$ are approximately normally distributed. But then the difference $\bar{X} - \bar{Y}$ is approximately normal with mean $p_1 - p_2$ and variance $p_1 q_1/n + p_2 q_2/m$. As in Example 7.3.7, this true variance should be well approximated by

$$\frac{\bar{X}(1 - \bar{X})}{n} + \frac{\bar{Y}(1 - \bar{Y})}{m},$$

so

$$Z = \frac{\bar{X} - \bar{Y} - (p_1 - p_2)}{\sqrt{\bar{X}(1 - \bar{X})/n + \bar{Y}(1 - \bar{Y})/m}}$$

is approximately standard normal. Then

$$P(-z_{1-\alpha/2} \leq Z \leq z_{1-\alpha/2}) \doteq 1 - \alpha$$

and the event

$$\{-z_{1-\alpha/2} \leq Z \leq z_{1-\alpha/2}\}$$

is equivalent to

$$\left\{\bar{X} - \bar{Y} - z_{1-\alpha/2}\sqrt{\frac{\bar{X}(1 - \bar{X})}{n} + \frac{\bar{Y}(1 - \bar{Y})}{m}} \leq p_1 - p_2\right.$$

$$\left. \leq \bar{X} - \bar{Y} + z_{1-\alpha/2}\sqrt{\frac{\bar{X}(1 - \bar{X})}{n} + \frac{\bar{Y}(1 - \bar{Y})}{m}}\right\}$$

so

$$L_1 = \bar{X} - \bar{Y} - z_{1-\alpha/2} \sqrt{\frac{\bar{X}(1-\bar{X})}{n} + \frac{\bar{Y}(1-\bar{Y})}{m}}$$

$$L_2 = \bar{X} - \bar{Y} + z_{1-\alpha/2} \sqrt{\frac{\bar{X}(1-\bar{X})}{n} + \frac{\bar{Y}(1-\bar{Y})}{m}}$$

are approximate $100(1-\alpha)\%$ confidence limits for $p_1 - p_2$. This establishes the following theorem.

Theorem 7.4.4. Let $X_1, X_2, \ldots, X_n$ be a random sample of a Bernoulli random variable with parameter p_1 and let $Y_1, Y_2, \ldots, Y_m$ be an independent random sample of a Bernoulli random variable with parameter p_2. Then

$$L_1 = \bar{X} - \bar{Y} - z_{1-\alpha/2} \sqrt{\frac{\bar{X}(1-\bar{X})}{n} + \frac{\bar{Y}(1-\bar{Y})}{m}}$$

$$L_2 = \bar{X} - \bar{Y} + z_{1-\alpha/2} \sqrt{\frac{\bar{X}(1-\bar{X})}{n} + \frac{\bar{Y}(1-\bar{Y})}{m}},$$

where $z_{1-\alpha/2}$ is the $100(1-\alpha/2)$th quantile of the standard normal probability law, are approximate $100(1-\alpha)\%$ confidence limits for $p_1 - p_2$. ∎

EXAMPLE 7.4.6. A manufacturing concern has its choice of two different suppliers of the same small part, at the same price. Each part supplied by A is defective with probability p_1 and each part supplied by B is defective with probability p_2; if we inspect n parts supplied by A, the total number of defectives observed, $\sum x_i$, is binomial with parameters n and p_1. The total number of defectives observed, $\sum y_j$, of m parts inspected from B, is binomial with parameters m and p_2. Suppose of $n = 100$ parts from A, we find $\sum x_i = 10$ defectives, whereas of $m = 150$ parts from B we find $\sum y_i = 11$ defectives. Thus $\bar{x} = .10$, $\bar{y} = .073$, $z_{.95} = 1.64$ so we can be 90 percent sure that $p_1 - p_2$ lies between

$$.10 - .073 - 1.64\sqrt{\frac{.1(.9)}{100} + \frac{(.927)(.073)}{150}} = -.033$$

$$.10 - .073 + 1.64\sqrt{\frac{.1(.9)}{100} + \frac{(.927)(.073)}{150}} = .087$$

based on these sample values.   ∎

EXERCISE 7.4

1. If $X_1, X_2, ..., X_n$ is a random sample of a normal random variable with parameters μ_X and σ_X^2, and $Y_1, Y_2, ..., Y_m$ is an independent random sample of a normal random variable with parameters μ_Y and σ_Y^2, what are the maximum likelihood estimates for μ_X, μ_Y, σ_X^2, σ_Y^2?

*2. If, in Exercise 7.4.1, we assume $\mu_Y = 2\mu_X$ and $\sigma_Y^2 = \frac{1}{2}\sigma_X^2$, what are the maximum likelihood estimates for μ_X and σ_X^2?

3. The times to failure for 10 items tested under standard conditions were 6.1, 1.5, 0.3, 2.6, 13.5, 2.5, 19.5, 3.5, 2.1, 0.7 weeks, respectively, whereas the times to failure for 8 items tested under "severe" conditions were 4.8, 1.8, 0.2, 5.4, 1.2, 1.8, 0.1, 5.6 weeks, respectively. Assume the standard condition failure times are observed values of exponential random variables with parameter λ, whereas the severe condition failure times are observed values of exponential random variables with parameter 2λ. What is the maximum likelihood estimate for λ based on these sample values?

4. Assume the numbers of pounds lost by $n = 12$ ladies on diet 1 were:

    12.8, 14.8, 13.5, 16.0, 18.1, 15.1, 11.4, 15.2, 10.2, 14.6, 15.5, 17.1

    whereas the numbers of pounds lost by $m = 10$ ladies on diet 2 were:

    9.3, 11.5, 14.6, 10.7, 13.2, 12.6, 16.7, 10.6, 13.1, 12.7.

    If these are independent observations from two normal populations, compute 90% confidence limits for σ_X/σ_Y (X represents diet 1, Y diet 2).

5. Use the data in Exercise 7.4.4 to evaluate 95% confidence limits for $\mu_X - \mu_Y$.

*6. Assume the standard-condition failure times given in Exercise 7.4.3 are observed values for a random sample of an exponential random variable X with parameter λ_1, whereas the severe condition failure times given there are the observed values of an independent sample of an exponential random variable Y with parameter λ_2. Based on these observed values, what is the maximum likelihood estimate for $P(X \leq Y)$?

7. Use the data given in Example 7.4.5, making the same assumptions, to compute 90% confidence limits for σ_1^2/σ_2^2, where σ_1^2, σ_2^2 are the variances of the repair times for the two shops.

8. The number of days of service provided by $n = 7$ watch batteries of brand A were:

    307, 355, 324, 384, 434, 362, 274

    whereas the number of days of service provided by $m = 9$ watch batteries of brand B were:

$$295, 336, 422, 329, 389, 374, 473, 326, 386.$$

Use these data to compute 90% confidence limits for the difference in the two expected lifetimes, assuming normality and that the two variances are equal.

9. A candidate for a national political office commissioned opinion polls in two different states to estimate the proportion of voters in each state who would vote for him on election day. In state A it was found that 112 of 250 eligible voters said they would vote for the candidate and, in state B, 207 of 400 eligible voters said they would vote for the candidate in question. Idealize the situation slightly by assuming the eligible voters contacted in state A are independent Bernoulli trials with probability p_1 of success (vote for the candidate) and do the same for state B, except the probability of success is p_2. Compute 90% confidence limits for $p_1 - p_2$. Based on these data, does it appear the candidate will do equally well in the two states?

*10. Explain why it is necessary that $\sigma_X^2 = \sigma_Y^2$ in the confidence interval for $\mu_X - \mu_Y$ discussed in Theorem 7.4.1. Could we get a confidence interval for $\mu_X - \mu_Y$ if we assume $\sigma_X^2 = a\sigma_Y^2$, where a is known?

7.5 Summary

Estimator for an unknown parameter: Random variable that is a function of the elements of a random sample; observed value used to guess the numerical parameter value.

Estimate: Observed value for an estimator.

Sample moments: The kth sample moment is the average of the kth powers of the elements of a random sample.

Method of moments estimators: Derived by equating sample moments to population moments.

Likelihood function: Probability of observing sample values in a neighborhood of those observed; it is a function of the unknown parameter(s) of the probability law of the random variable sampled.

Maximum likelihood estimates: Value(s) for the unknown parameter(s) that maximize(s) the likelihood function.

Maximum likelihood estimator(s): Function(s) of the elements of the random sample whose functional form is determined by the maximum likelihood estimates.

Unbiased estimator: One whose expected value is the parameter estimated.

Bias in an estimator: Difference between its expected value and the parameter estimated; $B[\Gamma] = E[\Gamma] - \gamma$.

More efficient estimator: Of two unbiased estimators of the same parameter, the one with the smaller variance.

Mean square error of an estimator: Expected squared distance between the estimator and the parameter:

$$MSE[\Gamma] = E[(\Gamma - \gamma)^2] = \text{Var}[\Gamma] + (B[\Gamma])^2.$$

Cramér-Rao lower bound: Gives smallest possible value for the mean square error of an estimator (variance of an unbiased estimator), Theorem 7.2.1. Not always achieved.

Consistent estimator: One such that $\lim_{n \to \infty} P(|\Gamma - \gamma| \geq \varepsilon) = 0$, for any $\varepsilon > 0$.

Confidence interval for an unknown parameter: Random interval that has a known probability of covering the unknown parameter value.

Confidence coefficient: Probability that a confidence interval covers an unknown parameter.

Confidence intervals for

Mean of normal: Theorem 7.3.1.

Variance of normal: Theorem 7.3.2.

Exponential parameter: Theorem 7.3.3.

Bernoulli parameter: Theorem 7.3.4.

Difference of two normal means: Theorem 7.4.1.

Ratio of two normal variances: Theorem 7.4.2.

Ratio of two exponential parameters: Theorem 7.4.3.

Difference of two Bernoulli parameters: Theorem 7.4.4.

Large sample confidence limits: Example 7.3.7.

8

Tests of Hypotheses

In Chapter 7 we considered problems of estimation; in this chapter we will study what are generally called tests of hypotheses. A common dictionary definition of the word hypothesis states a hypothesis is "an unproved theory, proposition, supposition." For our purposes we will be concerned with hypotheses about probability laws; by observing values of the random variable whose probability law is affected, we gather evidence regarding the truth of the hypothesis.

For example, in Example 7.4.2 we assumed the time to failure for a component in "standard use" to be an exponential random variable with parameter λ, and assumed if this same component were tested in a more severe environment, its time to failure is again exponential, but with parameter 4λ. This situation could be cast as defining a hypothesis: Probability law for time to failure, standard use, is exponential, parameter λ_1. Probability law for time to failure, severe environment, is exponential, parameter λ_2. It could be the case that $\lambda_2 = 4\lambda_1$ (this is the hypothesis). If we were to observe n independent times to failure, $X_1, X_2, ..., X_n$, for the standard use and independent times to failure, $Y_1, Y_2, ..., Y_m$, in the severe environment, the observed values of the random variables should contain information we could use to decide whether or not the hypothesis $(\lambda_2 = 4\lambda_1)$ appears to be true. We will examine some of the classical methodology for accomplishing this in the present chapter.

By defining a "test of a hypothesis," we will simply mean we have specified a rule that, for any possible collection of observed values for the random variables, tells us whether or not to accept the hypothesis. There are, of course, any number of possible rules (tests) that could be employed for any given problem. We will seek to find the rule that may be best in some sense. For example, in the life-testing case, recall that $\bar{X}$ estimates $1/\lambda_1$ (the mean time to failure, standard use) whereas $\bar{Y}$ estimates $1/\lambda_2$ (mean time to failure in the severe environment). If in actual fact it is true that $\lambda_2 = 4\lambda_1$, then we would expect to find $\bar{X}/\bar{Y}$ equal to approximately $(1/\lambda_1)/(1/4\lambda_1) = 4$. But since both $\bar{X}$ and $\bar{Y}$ are (continuous) random variables, and both would vary from one collection of possible observed values to another, we realize that we will not find $\bar{x}/\bar{y} = 4$ for all possible samples, even if $\lambda_2 = 4\lambda_1$. Thus the thought that might immediately occur, that we accept the hypothesis $\lambda_2 = 4\lambda_1$, only if we find $\bar{x}/\bar{y} = 4$, really is not totally practical. The variability in both $\bar{X}$ and $\bar{Y}$ makes a range of values for $\bar{x}/\bar{y}$ possible, when $\lambda_2 = 4\lambda_1$; thus we probably should settle on some rule (test) like "Accept the hypothesis $\lambda_2 = 4\lambda_1$ if $\bar{x}/\bar{y}$ is close to 4," where it remains to be seen what might be meant by the phrase "close to 4." As we will see, this is in fact the best type of rule from several points of view.

The observed values of random samples contain information about the probability law, but they do not contain "perfect" information, because the same set of observed values could come from many different probability laws. We might expect (and it will be so) that the likelihood function proves useful in distinguishing between the different possible probability laws that could have generated the same sample values. One point to understand clearly is that we will generally be unable to "prove" a hypothesis is either true or false, in the deductive sense, based on observed values of random variables. Again, referring to the life-testing example, even if we observe sample values that yield $\bar{x}/\bar{y} = 4$, it is quite possible that $\lambda_2 = 4.1\lambda_1$, or $\lambda_2 = 3.8\lambda_1$, and so on. Observing $\bar{x}/\bar{y} = 4$ is a long way from "proving" that $\lambda_2 = 4\lambda_1$. We will use the standard terminology of *accepting* the hypothesis (if the observed sample values seem consistent with it) or *rejecting* the hypothesis (if they are not).

We will begin our discussion at the simplest possible point, one that in general is unrealistic but which allows a clear exposition of the issues involved and the statistical methodology employed to resolve them. Throughout this chapter we will assume the *form* of the probability laws generating our random samples is known. For example, our observed times to failure are a random sample of an exponential random variable, or the spectrometric readings we observe are a random sample of a normal random variable, and so on. What we do not know (which is the concern of the hypothesis to be tested) is the value(s) of the parameter(s) of the probability law. For this reason the techniques we will discuss are frequently called *parametric* tests.

8.1 Simple Hypotheses

The simplest possible situation is one in which we have observed a random sample, $X_1, X_2, \ldots, X_n$, of a random variable X and want to choose between two distinct, completely specified probability laws for X. A hypothesis that completely specifies the probability law for X is called *simple*.

DEFINITION 8.1.1. A *simple hypothesis* H is any statement that completely specifies the probability law for a random variable X. A hypothesis that is not simple is called *composite*. A *test* of a hypothesis, H, is any rule that tells us whether to accept H or reject H, for every possible observed random sample of X. ∎

For example, if we have found $n = 4$ observed spectrometer readings to be 61, 47, 53, 58, and assume these are the observed values of a random sample of a random variable X, then

$$H: \ X \text{ is normal}, \mu = 60, \sigma = 5$$
$$H: \ X \text{ is normal}, \mu = 50, \sigma = 10$$
$$H: \ X \text{ is normal}, \mu = 100, \sigma = 2$$
$$H: \ X \text{ is exponential}, \lambda = .02$$

are each simple hypotheses. Any statement, such as, "X is normal, $\mu = 60$" or "X is normal, $\sigma = 5$" or "X is exponential," which does not *completely* specify the probability law for X is a composite, not a simple, hypothesis. We will consider only simple hypotheses in this section.

In the straightforward (admittedly unrealistic) case of deciding between two simple hypotheses, it is fairly easy to find the best possible *test* (rule for deciding which of the two hypotheses should be accepted). To distinguish between the two hypotheses considered, we will call one of them the *null hypothesis*, denoted by H_0, and the other, the *alternative hypothesis*, denoted by H_1. When we test a simple H_0 versus a simple H_1, our rule must, for any possible observed sample values, tell us which of the two hypotheses to accept; thus "accept H_0" is equivalent to "reject H_1" and vice versa. To avoid confusion, we will always apply the terminology accept–reject to the *null* hypothesis.

A little thought immediately reveals that we could make two different possible errors in testing a simple H_0 versus a simple H_1. These are called "type I error" and "type II error," as defined in Table 8.1.1. Any test of H_0 will tell us *either* to accept H_0 or reject H_0, based on the observed sample values. Thus it is not possible to commit both errors simul-

Table 8.1.1

	H_0 is true	H_0 is false
Accept H_0	No error	Type II error
Reject H_0	Type I error	No error

taneously. We will define

$$\alpha = P(\text{type I error})$$
$$= P(\text{reject } H_0, \text{ given } H_0 \text{ true})$$
$$= P(\text{reject } H_0 | H_0 \text{ true})$$

$$\beta = P(\text{type II error})$$
$$= P(\text{accept } H_0 | H_1 \text{ true}).$$

Every test of H_0 has values for the pair (α, β) associated with it. It would seem ideal if we could find the test that simultaneously minimizes both α and β, but this is not possible. Since each of α and β is a probability, we know $\alpha \geq 0$ and $\beta \geq 0$; that is, 0 is the minimum value for each. No matter what H_0 and H_1 state and what observed values occur in the sample, we could use the rule (test): Accept H_0. With this test we would never commit a type I error, since we would not reject H_0, no matter what the sample values were. Thus for this test we would have $\alpha = 0$, its smallest possible value (and this test has $\beta = 1$, its largest possible value). The converse of this test, which would always reject H_0, gives $\beta = 0$, $\alpha = 1$. Neither of these tests is desirable, because they maximize one of the two probabilities of error while minimizing the other. The following example illustrates the type of trade-off that typically exists between α and β.

EXAMPLE 8.1.1. Assume $X_1, X_2, ..., X_9$ is a random sample of a normal random variable whose variance σ^2 is known to equal 1, and we want to test $H_0: \mu = 2$ versus $H_1: \mu = 3$. Since the two simple hypotheses specify values for μ, and $\bar{X} = \frac{1}{9} \sum X_i$ is the minimum variance unbiased estimator for μ, it would seem reasonable to use a test that recommends acceptance or rejection of H_0, based on the observed value for $\bar{X}$. More specifically, if H_0 is true, we would expect the observed value for $\bar{X}$ to be close to 2 rather than to 3; if H_0 is not true, we would expect to find the observed value for $\bar{X}$ close to 3 rather than to 2. Thus we might consider a test that accepts H_0 if $\bar{x} \leq c$ and which rejects H_0 if $\bar{x} > c$, where c is a constant to be chosen. Figure 8.1.1 graphs the density for $\bar{X}$ for the two

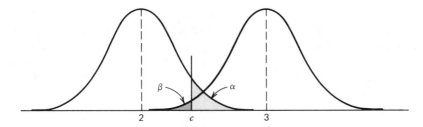

Figure 8.1.1

possible cases and indicates the values of α and β for a particular c. We have

$$\alpha = P(\text{reject } H_0 | H_0 \text{ true})$$
$$= P(\overline{X} > c | H_0 \text{ true})$$
$$= 1 - N\left(\frac{c - 2}{1/3}\right)$$
$$= N(3(2 - c))$$

since $\overline{X}$ is normal with mean 2, variance $\frac{1}{9}$ if H_0 is true. Similarly,

$$\beta = P(\text{accept } H_0 | H_0 \text{ false})$$
$$= P(\overline{X} \le c | H_0 \text{ false})$$
$$= N\left(\frac{c - 3}{1/3}\right) = N(3(c - 3))$$

since $\overline{X}$ is normal with mean 3, variance $\frac{1}{9}$ if H_0 is false (meaning H_1 is true). Table 8.1.2 gives four choices for c, with the resulting values of α and β. As we make c larger, the value for α decreases but the value for β increases.

Table 8.1.2

c	α	β
2.2	.2743	.0082
2.4	.1151	.0359
2.6	.0359	.1151
2.8	.0082	.2743

The trade-off in magnitude of α and β, as illustrated in Table 8.1.2, is typical. What then is a reasonable procedure to use in choosing a test? In order to discuss this choice a little more exactly, let us assume we want to test $H_0: \theta = \theta_0$ versus $H_1: \theta = \theta_1$, where θ_0 and θ_1 are specified values for the parameter of the probability law for a discrete random variable X. We will have a random sample, $X_1, X_2, \ldots, X_n$, of X to use in making our decision to accept or reject H_0. Let us use

$$\underset{\sim}{x} = (x_1, x_2, \ldots, x_n)$$

to represent the observed sample values and let S be the sample space for $\underset{\sim}{x}$ (collection of all possible $\underset{\sim}{x}$ values that might occur). The likelihood function for the sample then has only two possible values, $L_X(\theta_0)$ or $L_X(\theta_1)$, because we are assuming that these are the only possible values for θ. Now recall that $L_X(\theta_0)$ is the product of the probability functions evaluated at the observed sample values, with θ_0 the value for the parameter. Thus $L_X(\theta_0)$ actually gives the probability of observing $\underset{\sim}{x}$, assuming θ_0 is the correct parameter value and $L_X(\theta_1)$ gives the probability of observing $\underset{\sim}{x}$, assuming θ_1 is the correct parameter value.

Any test of H_0 versus H_1 must give the decision "accept H_0" or the decision "reject H_0" for any possible observed $\underset{\sim}{x}$. Thus we can think of any test as providing a partition of S, the sample space for $\underset{\sim}{x}$, into two parts.

$$A = \{\underset{\sim}{x}: \text{we accept } H_0\}$$
$$R = \{\underset{\sim}{x}: \text{we reject } H_0\}.$$

We will call A the *acceptance region* for the test and R the *rejection* (or *critical*) *region* for the test, and since they are a partition of S, $A \cup R = S$ and $A \cap R = \varnothing$ (these simply say we must make exactly one of the two decisions for every possible $\underset{\sim}{x} \in S$). Thus every test can be thought of as a partition of S or, equivalently, the class of all possible tests of H_0 versus H_1 is defined by all possible partitions of S. For the two trivial tests mentioned earlier we have

$$\text{Always accept } H_0: A = S \qquad R = \varnothing$$
$$\text{Always reject } H_0: A = \varnothing \qquad R = S,$$

so each of these provides a (trivial) partition of S.

Since $L_X(\theta_0)$ and $L_X(\theta_1)$ each provide the probability of observing $\underset{\sim}{x}$, note that

$$\sum_{\underset{\sim}{x} \in S} L_X(\theta_0) = 1$$

$$\sum_{\underset{\sim}{x} \in S} L_X(\theta_1) = 1,$$

the sum of the probabilities of all possible observed x must, of course, equal 1 in either case. But then for any specified test (partition of S into A, R)

$$\sum_{x \in S} L_x(\theta_0) = \sum_{x \in A} L_x(\theta_0) + \sum_{x \in R} L_x(\theta_0)$$

$$= P(\text{accept } H_0 | \theta_0) + P(\text{reject } H_0 | \theta_0)$$

$$= (1 - \alpha) + \alpha$$

$$\sum_{x \in S} L_x(\theta_1) = \sum_{x \in A} L_x(\theta_1) + \sum_{x \in R} L_x(\theta_1)$$

$$= P(\text{accept } H_0 | \theta_1) + P(\text{reject } H_0 | \theta_1)$$

$$= \beta + (1 - \beta),$$

so the probabilities of the two types of error are given by summing the appropriate likelihood function over the correct region:

$$\alpha = \sum_{x \in R} L_x(\theta_0), \qquad \beta = \sum_{x \in A} L_x(\theta_1).$$

Any reasonable test should accept H_0 for any outcome **x** such that $L_x(\theta_0) > L_x(\theta_1)$, because if this inequality is true, it says this **x** is more likely to occur if θ_0 is the true parameter value than if θ_1 is the true parameter value. Let $k > 0$ be an arbitrary constant and consider the particular test that has as its acceptance region

$$A = \{x: L_x(\theta_0) > kL_x(\theta_1)\};$$

the rejection region for this test then is

$$R = \{x: L_x(\theta_0) \leq kL_x(\theta_1)\}.$$

Now suppose we choose the value for k so that

$$\alpha = \sum_{x \in R} L_x(\theta_0)$$

is fixed (say, at .01 or .05 or whatever value we choose), and consider any other test (with partition A^*, R^*), which has the same probability of type I error, that is,

$$\alpha = \sum_{x \in R^*} L_x(\theta_0)$$

as well. Then the two tests together partition S into four parts: $A \cap A^*$,

$A \cap R^*$, $R \cap A^*$, $R \cap R^*$, and we know

$$0 = \alpha - \alpha = \sum_{\underset{x \in R}{}} L_X(\theta_0) - \sum_{\underset{x \in R^*}{}} L_X(\theta_0)$$

$$= \left(\sum_{\underset{x \in R \cap R^*}{}} L_X(\theta_0) + \sum_{\underset{x \in R \cap A^*}{}} L_X(\theta_0) \right)$$

$$- \left(\sum_{\underset{x \in A \cap R^*}{}} L_X(\theta_0) + \sum_{\underset{x \in R \cap R^*}{}} L_X(\theta_0) \right)$$

$$= \sum_{\underset{x \in R \cap A^*}{}} L_X(\theta_0) - \sum_{\underset{x \in A \cap R^*}{}} L_X(\theta_0)$$

so it follows that

$$\sum_{\underset{x \in R \cap A^*}{}} L_X(\theta_0) = \sum_{\underset{x \in A \cap R^*}{}} L_X(\theta_0).$$

But for all $x \in R$ (and thus those in $R \cap A^*$ in particular) we have

$$L_X(\theta_0) \le k L_X(\theta_1)$$

and for all $x \in A$ (and in $A \cap R^*$)

$$L_X(\theta_0) > k L_X(\theta_1).$$

Thus

$$\sum_{\underset{x \in A \cap R^*}{}} L_X(\theta_0) > k \sum_{\underset{x \in A \cap R^*}{}} L_X(\theta_1)$$

and

$$- \sum_{\underset{x \in R \cap A^*}{}} L_X(\theta_0) \ge -k \sum_{\underset{x \in R \cap A^*}{}} L_X(\theta_1);$$

adding these two inequalities gives

$$0 \ge k \sum_{\underset{x \in A \cap R^*}{}} L_X(\theta_1) - k \sum_{\underset{x \in R \cap A^*}{}} L_X(\theta_1)$$

or

$$\sum_{\underset{x \in R \cap A^*}{}} L_X(\theta_1) \ge \sum_{\underset{x \in A \cap R^*}{}} L_X(\theta_1).$$

Adding

$$\sum_{\underset{x \in A \cap A^*}{}} L_X(\theta_1)$$

to both sides gives

$$\sum_{\underline{x} \in R \cap A^*} L_X(\theta_1) + \sum_{\underline{x} \in A \cap A^*} L_X(\theta_1) \geq \sum_{\underline{x} \in A \cap R^*} L_X(\theta_1) + \sum_{\underline{x} \in A \cap A^*} L_X(\theta_1),$$

that is,

$$\sum_{\underline{x} \in A^*} L_X(\theta_1) \geq \sum_{\underline{x} \in A} L_X(\theta_1).$$

This equation says the probability of accepting H_0 when $\theta = \theta_1$, using the test with partition (A^*, R^*), must be at least as large as the probability of accepting H_0 (again, when $\theta = \theta_1$), using the test with A, R as previously defined. That is, if we consider any other test with the same probability of type I error, α, its probability of type II error must be at least as large as β for the test with rejection region

$$R = \{\mathbf{x}: L_X(\theta_0) \leq kL_X(\theta_1)\}$$

so the test with this rejection region minimizes β for α fixed. This establishes the following theorem, named after J. Neyman and E. Pearson who first published it in the 1930s. It forms the basis for choosing the "best" test of a simple H_0 versus a simple H_1.

Theorem 8.1.1. To test the simple $H_0: \theta = \theta_0$ versus the simple $H_1: \theta = \theta_1$, based on a random sample of size n of a random variable whose probability law depends on θ, the test with critical region

$$R = \{\underline{x}: L_X(\theta_0) \leq kL_X(\theta_1)\}$$

has the smallest possible value for

$$\beta = \sum_{\underline{x} \in A} L_X(\theta_1)$$

among all tests with the same value for

$$\alpha = \sum_{\underline{x} \in R} L_X(\theta_0). \qquad \blacksquare$$

This theorem gives a very simple way of finding the partition of S that will give the smallest possible probability of type II error, for any chosen value for α. One simply uses the likelihood function to assign sample values to the rejection (or critical) region R and adjusts the value for k to give the desired value for α. Exactly the same reasoning can be employed with samples of continuous random variables, using integration rather than summation. Thus we again define the critical region by comparing the values of the likelihood function and adjust k to give the desired value for α.

For all the standard probability laws, this best partitioning of S, the sample space for $\underset{\sim}{x}$, reduces to an equivalent partition of the possible range for some statistic $g(\underset{\sim}{x})$, a function of the observed values. Thus the test can equally well, and more simply, be expressed in terms of $g(\underset{\sim}{x})$, the observed value for $g(\underset{\sim}{X})$. We will refer to $g(\underset{\sim}{X})$ as being the *test statistic* for the given hypotheses. The following examples apply this theorem to some standard cases.

EXAMPLE 8.1.2.   Let us reconsider the case discussed in Example 8.1.1 and contrast what we did there to construct a test of $H_0 : \mu = 2$ versus $H_1 : \mu = 3$, with this best procedure. Given a random sample of size n of a normal random variable with mean μ and known $\sigma^2 = 1$ (so the hypotheses are simple), the two values for the likelihood function are

$$L_X(2) = \left(\frac{1}{2\pi} \right)^{n/2} \exp\left[\frac{-\sum (x_i - 2)^2}{2} \right]$$

$$L_X(3) = \left(\frac{1}{2\pi} \right)^{n/2} \exp\left[\frac{-\sum (x_i - 3)^2}{2} \right].$$

The best test then has critical region

$$R = \left\{ \underset{\sim}{x} : \left(\frac{1}{2\pi} \right)^{n/2} \exp\left[\frac{-\sum (x_i - 2)^2}{2} \right] \right.$$

$$\left. \leq k \left(\frac{1}{2\pi} \right)^{n/2} \exp\left[\frac{-\sum (x_i - 3)^2}{2} \right] \right\}$$

which is equivalent to

$$R = \left\{ \underset{\sim}{x} : \sum (x_i - 2)^2 \geq -2 \ln k + \sum (x_i - 3)^2 \right\}$$

and to

$$R = \left\{ \underset{\sim}{x} : \sum x_i \geq -\ln k + \frac{5n}{2} \right\}$$

and to

$$R = \left\{ \underset{\sim}{x} : \bar{x} \geq -\frac{1}{n} \ln k + \frac{5}{2} = c \right\},$$

the same critical region we employed. Thus $\bar{X}$ is called the test statistic (since only its value is needed to decide whether to accept or reject H_0) and we can choose c to set α at any value we like (you can verify that $c = 2 + \frac{1}{3} z_{1-\alpha}$). There is no real interest in finding the value for k (although we could if we want), because it is much simpler to compare the observed value for $\overline{X}$ with c rather than evaluating the likelihood functions, $L_X(2)$ and $L_X(3)$ and checking to see whether $L_X(2) \le kL_X(3)$. ∎

EXAMPLE 8.1.3. Suppose we assume the time to failure X for a piece of equipment is an exponential random variable with parameter λ. Given a random sample $X_1, X_2, \ldots, X_n$ of lifetimes we want to test $H_0: \lambda = .01$ versus $H_1: \lambda = .04$. The likelihood functions for the two cases then are

$$L_X(.01) = (.01)^n e^{-.01 \Sigma x_i}$$
$$L_X(.04) = (.04)^n e^{-.04 \Sigma x_i}$$

and the best critical region is defined by

$$R = \left\{ \underline{x} : (.01)^n e^{-.01 \Sigma x_i} \le k(.04)^n e^{-.04 \Sigma x_i} \right\},$$

which you can verify is equivalent to

$$R = \left\{ \underline{x} : \overline{x} \le c \right\}$$

so again, $\overline{X}$ is the test statistic. Since if $H_0: \lambda = .01$ is true, $2(.01) n\overline{X}$ is a χ^2 random variable with $2n$ degrees of freedom, to have any desired value for α we need

$$\alpha = P(\overline{X} \le c \,|\, H_0 \text{ true})$$
$$= P(2(.01) n\overline{X} \le (.02) nc \,|\, H_0 \text{ true})$$
$$= P(\chi^2 \le (.02) nc)$$

so

$$\chi^2_\alpha = .02nc$$

and we use

$$c = \frac{50\chi^2_\alpha}{n}.$$

Thus if we take a sample of $n = 8$ lifetimes and want $\alpha = .1$, we find (for $2n = 16$ degrees of freedom),

$$\chi^2_{.1} = 9.31, \qquad c = \frac{(50)(9.31)}{8} = 58.19$$

and we should reject H_0 if we find $\overline{x} \le 58.19$. This test has the smallest possible β among all those with $\alpha = .1$. ∎

We can, of course, also employ this procedure to find the best test for simple hypotheses about the parameter of a discrete probability law. Since the test statistic then is in general a function of the discrete sample values, it is itself a discrete random variable. Because of this discreteness, only a discrete collection of values for α are attainable. Nonetheless, if we select k to have any attainable α value, the Neyman–Pearson critical region still gives the test with the smallest possible β among all those with the same α (or any smaller α, actually).

EXAMPLE 8.1.4.   Assume a large lot of items is received, each of which is either defective or nondefective; the lot is large enough so that it is reasonable to assume individual items selected at random without replacement are independent Bernoulli trials with parameter p, where p is the proportion of defectives in the lot. Suppose $n = 50$ items are selected at random, tested, and $\sum_{i=1}^{50} X_i$ is the number of defectives found (in the n tested); we want to test $H_0: p = .1$ versus $H_1: p = .2$. The two likelihood functions then are

$$L_X(.1) = (.1)^{\Sigma x_i}(.9)^{50 - \Sigma x_i}$$
$$L_X(.2) = (.2)^{\Sigma x_i}(.8)^{50 - \Sigma x_i}$$

and the best critical region is

$$R = \left\{ \mathbf{x}: (.1)^{\Sigma x_i}(.9)^{50 - \Sigma x_i} \leq k(.2)^{\Sigma x_i}(.8)^{50 - \Sigma x_i} \right\},$$

which is equivalent to

$$R = \left\{ \underline{x}: \sum x_i \geq c \right\},$$

so $Y = \sum_{i=1}^{50} X_i$ is the test statistic. Y is a binomial random variable with parameters 50 and p and, if H_0 is true,

$$P(Y \geq c \,|\, p = .1) = \sum_{j=c}^{50} \binom{50}{j}(.1)^j(.9)^{50-j}$$

so the possible values for α are limited by the values that can be achieved by this discrete sum. The values given in Table 8.1.3 can be verified without difficulty on a hand-held calculator.

Table 8.1.3

c	$P(Y \geq c \,|\, p = .10)$
7	.230
8	.122
9	.058
10	.025
11	.009

Thus with a sample of $n = 50$, the best test of H_0: $p = .1$ versus H_1: $p = .2$ says we should reject H_0 if $y \geq 9$, if we want $\alpha = .058$, and reject H_0 if $y \geq 10$, if we want $\alpha = .025$. In either case we are using the test with the smallest β among all those whose probability of type I error does not exceed the α value chosen. The choice of which α value to use, of course, depends on the person applying the test and her desired probability of committing a type I error. ∎

There is in actual fact a rather close connection between tests of hypotheses and estimation of parameters. In general, any statistic that provides a good estimator for an unknown parameter will also be the test statistic for testing hypotheses about the same parameter. The form of the best critical region (as defined by Theorem 8.1.1) can also generally be surmised by considering the values of the estimator (test statistic) that are more consistent with H_1 than with H_0.

In testing a simple H_0 versus a simple H_1, we see from Theorem 8.1.1 how, for a fixed sample size n, to find the test (critical region) that gives the smallest value for β for any given value of α. There are three basic quantities involved in such a test: the sample size n, α, and β. In testing simple hypotheses about the mean of a normal distribution, we can select, quite straightforwardly, desired values for α and β, and then find the sample size n that, together with the best test (partition of all possible observed samples), gives the desired α and β. This is illustrated in the following example.

EXAMPLE 8.1.5. Assume a normal random variable X has unknown mean μ and known variance σ^2 and we want to test H_0: $\mu = \mu_0$ versus H_1: $\mu = \mu_1$, where μ_0 and $\mu_1 > \mu_0$ are any desired constants. Further suppose we would like to fix both α and β, the probabilities of the two types of error. How large should n, the sample size, be? As we know from Example 8.1.2, the best critical region is one in which we reject H_0 if $\bar{x} \geq c$, as long as $\mu_1 > \mu_0$ (if $\mu_1 < \mu_0$, the best critical region is specified by $\bar{x} \leq c$). Given a random sample of size n of X, we know that $\bar{X}$ is normal with mean μ_0, variance σ^2/n if H_0 is true, and $\bar{X}$ is normal with mean μ_1, variance σ^2/n if H_1 is true. Thus with $\mu_1 > \mu_0$,

$$\alpha = P(\text{type I error})$$

$$= P(\bar{X} \geq c \mid \mu = \mu_0)$$

$$= P\left(\frac{(\bar{X} - \mu_0)\sqrt{n}}{\sigma} \geq \frac{(c - \mu_0)\sqrt{n}}{\sigma} \,\middle|\, \mu = \mu_0 \right)$$

$$= 1 - N\left(\frac{(c - \mu_0)\sqrt{n}}{\sigma}\right)$$

$$= N\left(\frac{\sqrt{n}(\mu_0 - c)}{\sigma}\right)$$

so we require

$$\frac{\sqrt{n}(\mu_0 - c)}{\sigma} = z_\alpha.$$

Similarly, the probability of a type II error is

$$\beta = P(\overline{X} \le c \,|\, \mu = \mu_1)$$

$$= P\left(\frac{(\overline{X} - \mu_1)\sqrt{n}}{\sigma} \le \frac{(c - \mu_1)\sqrt{n}}{\sigma} \,\Big|\, \mu = \mu_1\right)$$

$$= N\left(\frac{(c - \mu_1)\sqrt{n}}{\sigma}\right)$$

so we also need

$$\frac{\sqrt{n}(c - \mu_1)}{\sigma} = z_\beta.$$

Thus we have two equations

$$\mu_0 - c = z_\alpha \frac{\sigma}{\sqrt{n}}$$

$$c - \mu_1 = z_\beta \frac{\sigma}{\sqrt{n}}$$

in two unknowns, c and $\sqrt{n}$. It is easy to see that the solutions are given by

$$\sqrt{n} = \frac{\sigma(z_\alpha + z_\beta)}{\mu_0 - \mu_1}, \qquad c = \frac{z_\alpha \mu_1 + z_\beta \mu_0}{z_\alpha + z_\beta},$$

so a sample of size

$$n = \frac{\sigma^2(z_\alpha + z_\beta)^2}{(\mu_0 - \mu_1)^2},$$

employed with the best test, gives the desired values for α and β (this value for n is not likely to be an integer; the conservative approach then is to round up to the next higher integer).

For example, suppose we assume X is normal with $\sigma = 2$ and we would like to find the sample size n that, using the best test, will give $\alpha = .01$, $\beta = .05$ in testing H_0: $\mu = 2$ versus H_1: $\mu = 5$. We have $z_\alpha = z_{.01} = -2.33$, $z_\beta = z_{.05} = -1.64$, so the preceding solution gives

$$n = \frac{2^2(-2.33 - 1.64)^2}{(2 - 5)^2} = 7.005$$

so a sample of $n = 8$ will be sufficient. With $n = 8$, if we want to hold $\alpha = .01$, we find c from

$$\frac{\sqrt{8}(2 - c)}{2} = z_{.01} = -2.33,$$

which gives

$$c = 2 + \frac{4.66}{\sqrt{8}} = 3.65.$$

The value for β then would be

$$P(\bar{X} \leq 3.65 \,|\, \mu = 5) = N\left(\frac{(3.65 - 5)\sqrt{8}}{2}\right)$$

$$= N(-1.91) = .0281,$$

smaller than the desired $\beta = .05$ because we rounded n up to 8. On the other hand, we could keep $\beta = .05$ by using the value for c determined from

$$\frac{\sqrt{8}(c - 5)}{2} = z_\beta = -1.64,$$

which gives

$$c = 5 - \frac{3.28}{\sqrt{8}} = 3.84;$$

the value for α then is

$$P(\bar{X} \geq 3.84 \,|\, \mu = 2) = 1 - N\left(\frac{(3.84 - 2)\sqrt{8}}{2}\right)$$

$$= 1 - N(2.60) = .0047.$$

If we round n up, as previously recommended, either α or β, or both, will be smaller than the values initially specified. ∎

It is, of course, extremely rare that one would have a random sample from a normal population whose variance σ^2 is known but whose mean μ is unknown. Thus the procedure discussed in the preceding example is not frequently employed per se. It is, however, a useful prototype for cases in which the test statistic, used in the best test of a simple H_0 versus a simple H_1, has a distribution that is well approximated by the normal for large n. This is illustrated in the following example.

EXAMPLE 8.1.6.   Assume, as in Example 8.1.4, a large lot of items contains the proportion p of defectives; we want to test H_0: $p = .1$ versus H_1: $p = .2$ with $\alpha = \beta = .01$, say (at least approximately). How large should the sample size n be? Again, as long as the number of items in the lot is large enough that inspection of individual items can be well approximated by independent Bernoulli trials, each with parameter p, the number of defectives Y in a random sample of size n is binomial with parameters n and p. For large n, Y is approximately normal, $\mu = np$, $\sigma^2 = npq$, so to have $\alpha = .01$, we want

$$.01 = P(Y \geq c \,|\, p = .1)$$
$$= 1 - P(Y \leq c - 1 \,|\, p = .1)$$
$$\doteq 1 - N\left(\frac{c - 1 - .1n}{\sqrt{n(.1)(.9)}} \right)$$

which says then we want

$$\frac{c - 1 - .1n}{\sqrt{.09n}} \doteq 2.33.$$

To have $\beta = .01$, we want

$$.01 = P(Y \leq c - 1 \,|\, p = .2)$$
$$\doteq N\left(\frac{c - 1 - .2n}{\sqrt{n(.2)(.8)}} \right)$$

so we also require

$$\frac{c - 1 - .2n}{\sqrt{.16n}} \doteq -2.33.$$

Solving these two equations simultaneously gives $n = (7 \times 2.33)^2 = 266.02$, so a sample of $n = 267$ would be sufficient; again, because we have rounded n up to the next larger integer, we can take α or β (or both) smaller than

.01. Since n has turned out to be "large," the normal approximation used should be very accurate and, with $n = 267$, we can hold both α and β to being roughly .01 in testing H_0: $p = .1$ versus H_1: $p = .2$.   ∎

EXERCISE 8.1

1. Suppose that X is a Bernoulli random variable with parameter p. We take a random sample of four observations of X and want to test H_0: $p = \frac{1}{4}$ versus H_1: $p = \frac{3}{4}$. If we reject H_0 only if we get four successes in the sample, compute the values of α and β.

2. Given that X is a uniform random variable on the interval $(0, \theta)$, we might test H_0: $\theta = 1$ versus the alternative H_1: $\theta = 2$ by taking a sample of 2 observations of X and rejecting H_0 if $\overline{X} > .99$. Compute α and β for this test.

3. Assume we have a random sample of size n of a continuous random variable X and want to test the simple hypothesis that the density for X is

    $$f_X(x) = 2x, \qquad 0 < x < 1$$

    versus the simple alternative that the density for X is

    $$f_X(x) = 1, \qquad 0 < x < 1.$$

    Find the best test for this hypothesis.

4. Five oil samples are removed from the same oil reservoir and analyzed on a spectrometer for their iron content. The iron readings produced are assumed to be normal with mean μ (unknown true iron contamination in the reservoir) and variance $\sigma^2 = 3$. Granted

    $$\sum_{i=1}^{5} x_i = 265,$$

    would you accept H_0: $\mu = 50$, with $\alpha = .05$, in testing versus H_1: $\mu = 55$? if the alternative were H_1: $\mu = 45$?

5. A random variable X is known to be normal with $\mu = 5$, σ^2 unknown. What is the test statistic and best critical region in testing H_0: $\sigma^2 = 10$ versus H_1: $\sigma^2 = 20$ for a sample of size n? How does your answer change if the alternative is H_1: $\sigma^2 = 5$?

6. The times for an auto repair shop to diagnose and repair a certain problem are assumed to be exponential with parameter λ and mean $\mu = 1/\lambda$, with units of hours. Six cars with the same problem required 1.8, 5.2, 0.4, 5.1, 0.6, 3.5 hours, respectively, to be successfully re-

paired. Based on these observed values would you accept H_0: $\mu = 4$ with $\alpha = .1$, with the alternative H_1: $\mu = 3$?

7. Assume X is a Poisson random variable with parameter μ. What is the test statistic and best critical region for testing H_0: $\mu = 2$ versus H_1: $\mu = 1$, based on a random sample of n observations of X?

8. If meteorites strike the surface of the moon "at random," the number of meteorite craters per unit of area should be a Poisson random variable with parameter μ (expected number per unit area). If $n = 10$ units of area are examined and the total number of meteorite craters found is 8, would you accept H_0: $\mu = \frac{1}{2}$, with $\alpha = .068$, with the alternative H_1: $\mu = 1$? Evaluate β for this test.

*9. Theorem 8.1.1 can also be used to find the best test that a random variable X has one completely specified probability law versus another. Based on a random sample of n observations of a continuous random variable, describe the best critical region for testing H_0: X is normal, $\mu = 100$, $\sigma = 20$ versus H_1: X is exponential, $\lambda = .01$.

10. Use the large sample (normal approximation) methodology to evaluate how large a sample, n, is required to test H_0: $\mu = 150$ versus H_1: $\mu = 200$, with $\alpha = .01$, $\beta = .02$, assuming the random variable is exponential with mean μ. Would you accept or reject H_0 if for the sample size determined, you found $\bar{x} = 175$?

11. Repeat Exercise 8.1.10 if the random variable is Poisson with parameter μ, testing H_0: $\mu = 2$ versus H_1: $\mu = 1.5$, with $\alpha = .01$, $\beta = .02$. Would you accept or reject H_0 if you found $\bar{x} = 1.75$?

12. Each of $n = 10$ persons used the same instrument to measure the same object; the true value was then subtracted from each of these. These 10 differences are assumed to be a random sample of a normal random variable X with mean 0 and variance σ^2 and are used to test H_0: $\sigma^2 = 2$ versus H_1: $\sigma^2 = 4$. The critical region used for the test is

$$R = \{\underset{\sim}{x}: \sum x_i^2 \geq 32\}.$$

Evaluate α and β for this test.

13. To test H_0: $p = .01$ versus H_1: $p = .005$, $n = 1000$ flashbulbs will be ignited, where p is the probability a flashbulb is defective (and will not flash). The rejection region is

$$R = \{\underset{\sim}{x}: \sum x_i \leq 6\},$$

where $x_i = 0$ if the ith bulb flashes, $x_i = 1$ if it does not. Evaluate α and β for this test.

8.2 Composite Hypotheses

In most applications unknown parameters of probability laws are not restricted to either of just two possible values. The probability of selecting a defective item from a large lot is not restricted to $p = .1$ or $p = .2$; indeed, if the lot contains N items, the proportion defective must be one of the values $0, 1/N, 2/N, \ldots, (N - 1)/N, 1$, which if we idealize to Bernoulli trials, actually is represented by the interval $[0, 1]$. The possible parameter values for a normal probability law are $\{(\mu, \sigma) \colon -\infty < \mu < \infty, \sigma > 0\}$, a set that is continuous in two dimensions. Thus in most applications one is interested in cases in which H_0 or H_1 (or both) is a composite, not simple, hypothesis. We will discuss tests of composite hypotheses in this section.

To keep our discussion concrete, let us first describe the case in which we have a random sample of a random variable X, whose probability law depends on a single unknown parameter θ, which can equal any value in a continuous interval on the real line (for example, X is Bernoulli and its parameter p lies in the interval $[0, 1]$, or X is exponential and its parameter λ is positive). The collection of possible values for the unknown parameter will be called the *parameter space* and will be denoted by Ω; for a Bernoulli X, the parameter space is $\Omega = \{p \colon 0 \le p \le 1\}$ and for X exponential the parameter space is $\Omega = \{\lambda \colon \lambda > 0\}$. A probability law with one unknown parameter then is said to have a one-dimensional parameter space Ω.

One of the most frequently occurring composite-hypothesis situations is referred to as testing a one-sided alternative. If θ_0 is any known, fixed value in Ω, and we want to test $H_0 \colon \theta \le \theta_0$ versus the alternative $H_1 \colon \theta > \theta_0$, we have a one-sided alternative; similarly, in testing $H_0 \colon \theta \ge \theta_0$ versus $H_1 \colon \theta < \theta_0$, we again have a one-sided alternative. In testing one-sided alternatives, we generally find a range of values for θ for which H_0 is true, as well as a range of values for θ for which H_1 is true. Thus if we assume H_0 true, we do not have a single unique probability law for the random variable X; the same is true for H_1. This contrasts with the simple versus simple case, in which assuming either H_0 or H_1 true does uniquely specify the probability law for X. The real impact of H_0 and H_1 being composite is that we no longer have single specific numbers that we can call the probabilities of type I and type II errors. When H_0 is composite, $P(\text{reject } H_0 | H_0 \text{ true})$ depends on which particular value, of all those specified by H_0, we assume to be the true value for θ; actually, $P(\text{reject } H_0 | H_0 \text{ true})$ has now become a function of θ, defined for all θ specified by H_0. Similarly, with H_1 composite, $P(\text{accept } H_0 | H_1 \text{ true})$ is actually a function defined for all θ specified by H_1. Both these functions can be evaluated from the *operating characteristic* (OC) function or the *power function* of the test, defined as follows.

DEFINITION 8.2.1. The *operating characteristic* function (frequently called the OC curve) of a test of a hypothesis H_0 about a parameter θ is

$$C(\theta) = P(\text{accept } H_0 | \theta).$$

The complementary function

$$Q(\theta) = 1 - C(\theta)$$

is called the *power function* of the test. ∎

Different tests (rules for accepting or rejecting H_0) have different OC curves or, equivalently, different power functions. Because the value of an OC curve is a probability for any θ, we know immediately that $0 \leq C(\theta) \leq 1$ for all θ; similarly, $0 \leq Q(\theta) \leq 1$, the power function is also bounded by 0 and 1.

Suppose $\Omega = \{\theta : \theta > 0\}$ and we want to test $H_0 : \theta \leq \theta_0$ versus $H_1 : \theta > \theta_0$, where θ_0 is some specific value in Ω. The ideal OC curve (pictured in Figure 8.2.1) then would be

$$\begin{aligned} C(\theta) &= 1, & \theta \leq \theta_0 \\ &= 0, & \theta > \theta_0. \end{aligned}$$

If θ is really unknown, and all we can do is select a random sample of values for X on which to base our test, this ideal OC curve can not be realized. The following examples illustrate the computation of OC curves.

EXAMPLE 8.2.1. Suppose we have a random sample of size 10 of an exponential random variable X and want to test $H_0 : \lambda \leq 1$ versus $H_1 : \lambda > 1$.

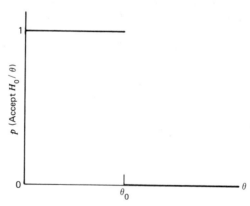

Figure 8.2.1

The critical region for our test is (rather arbitrarily)

$$R = \{\underline{x}: \bar{x} \le .545\}.$$

The OC curve for this test then is

$$
\begin{aligned}
C(\lambda) &= P(\bar{X} > .545 \,|\, \lambda) \\
&= P(20\lambda\bar{X} > 10.9\lambda) \\
&= P(\chi^2(20) > 10.9\lambda) \\
&= 1 - F_{\chi^2}(10.9\lambda),
\end{aligned}
$$

one minus the χ^2 distribution function, 20 degrees of freedom, evaluated at 10.9λ, since $2\lambda n\bar{X}$ is a χ^2 random variable with $2n$ degrees of freedom. From Table 2 in the Appendix, we can evaluate certain quantiles of the χ^2 distribution function with 20 degrees of freedom and, from these, the OC curve for this test. For example, we find $F_{\chi^2}(8.26) = .01$, $10.9\lambda = 8.26$ gives $\lambda = .758$ so

$$C(.758) = 1 - F_{\chi^2}(8.26) = .99.$$

Similarly,

$$
\begin{aligned}
C(1) &= 1 - F_{\chi^2}(10.9) = .95 \\
C(1.771) &= 1 - F_{\chi^2}(19.3) = .5 \\
C(2.881) &= 1 - F_{\chi^2}(31.4) = .05, \text{ and so on.}
\end{aligned}
$$

We can directly use the χ^2 quantiles to evaluate points on the OC curve for this test. Since the $\chi^2(20)$ distribution function is monotonic increasing, the OC curve is monotonic decreasing. The OC curve for this test is pictured in Figure 8.2.2; this shape is typical for the commonly used tests. Notice that for all $\lambda \le 1$ (those values specified by H_0) we have $C(\lambda) \ge .95$ and thus

$$
\begin{aligned}
Q(\lambda) &= 1 - C(\lambda) \\
&= P(\text{reject } H_0 \,|\, \lambda) \\
&\le .05 \quad \text{for} \quad \lambda \le 1.
\end{aligned}
$$

Thus $Q(\lambda) = 1 - C(\lambda) = P(\text{reject } H_0 \,|\, \lambda)$ is, for λ specified by H_0, actually, the function whose values give the probability of type I error. Similarly,

$$C(\lambda) = P(\text{accept } H_0 \,|\, \lambda) = 1 - Q(\lambda)$$

is, for λ specified by H_1, the function whose values give the probability of type II error. The largest value of $Q(\lambda)$, for all λ specified by H_0, is called the *size* of the test and is denoted by α. Thus the size of this test is $\alpha = .05$.

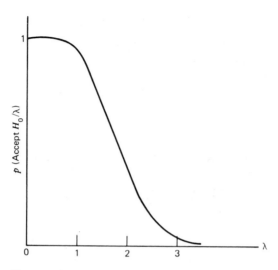

Figure 8.2.2

EXAMPLE 8.2.2. A flashbulb manufacturer claims that p, the probability any one of its bulbs is defective and will not work, is no larger than .01. Suppose we formally wish to test this claim. We want to test $H_0: p \leq .01$ versus $H_1: p > .01$; to do so we will fire 10 of the bulbs. If X, the number that do not work, is 1 or more we will reject H_0; otherwise (if $X = 0$), we will accept H_0. Granted there is a constant probability p that any one of the bulbs will not fire, and the bulbs used are independent trials, X is binomial, $n = 10$, p. The OC curve for this test is

$$C(p) = P(\text{accept } H_0 \,|\, p)$$
$$= P(X = 0 \,|\, p)$$
$$= (1 - p)^{10},$$

again, a monotonic decreasing function of p. The size of the test then is $Q(.01) = 1 - C(.01) = 1 - (.99)^{10} = .096$; if in fact $p = .05$, the probability of a type II error is $C(.05) = (.95)^{10} = .599$, whereas if $p = .1$, the probability of a type II error is $C(.1) = (.9)^{10} = .349$ with this test. ∎

Comparing tests of composite hypotheses thus involves the comparison of functions, either comparing OC functions or equivalently, power functions for competing tests; this comparison of functions is a more complex task than is involved in finding the best test of a simple H_0 versus a simple H_1. Again, in the composite case the trivial rules "always accept H_0" or "always reject H_0" could be employed, ignoring the observed sample values; you can easily verify that each of these is the best rule, for some $\theta \in \Omega$, and the worst for others, so we would not want to use either one. In the simple H_0–simple H_1 case we decided to consider all possible tests with the same value for $\alpha = P(\text{type I error})$ and, among these, called the one with the smallest value for $\beta = P(\text{type II error})$ the best test.

The same reasoning applied to the composite situation leads us to consider all tests of the same size α (α = maximum value of $Q(\theta)$ for θ values specified by H_0) and, among these, to seek the test that has the highest value for $Q(\theta)$ [equivalently, the one with the smallest value for $C(\theta) = 1 - Q(\theta)$] for all θ values specified by H_1. Such a test, if it exists, is called *uniformly most powerful*, because it maximizes the power function for all θ specified by H_1. Uniformly most powerful tests do exist for many important cases; these can frequently be found by again using the Neyman–Pearson Theorem, 8.1.1. Recall that in testing H_0: $\theta = \theta_0$ versus H_1: $\theta = \theta_1$ (for convenience assume $\theta_1 > \theta_0$) the best test is one with critical region

$$R = \{\underline{x}: L_X(\theta_0) \leq k L_X(\theta_1)\},$$

where k is selected to make

$$P(\text{type I error}) = P(\underline{X} \in R \,|\, \theta = \theta_0) = \alpha;$$

also recall that we can in general simplify the implementation of the rule by finding the statistic, say, $g(\mathbf{x})$, such that

$$\{\underline{x}: L_X(\theta_0) \leq k L_X(\theta_1)\} \qquad \text{and} \qquad \{\underline{x}: g(\underline{x}) \geq c\}$$

are equivalent (now c is chosen to fix the value for α). Suppose in testing H_0: $\theta \leq \theta_0$ versus H_1: $\theta > \theta_0$, the likelihood function $L_X(\theta)$ is such that

$$\{\underline{x}: L_X(\theta_0) \leq k L_X(\theta_1)\}$$

is equivalent to

$$\{\underline{x}: g(\underline{x}) \geq c\},$$

for *every* $\theta_1 > \theta_0$, and is equivalent to

$$\{\underline{x}: g(\underline{x}) \leq c\}$$

for every $\theta_1 < \theta_0$ (both these inequalities may be reversed). Then it would follow that

1. The maximum value for $Q(\theta)$, for all $\theta \leq \theta_0$ (those specified by H_0) is $\alpha = Q(\theta_0)$, so the test is of size α.
2. The probability of a type II error is the smallest, for each $\theta > \theta_0$, among all tests of the same size (equivalently, the power is the greatest for all $\theta > \theta_0$).

That is, this test would be uniformly most powerful of size α. The following example should help clarify this reasoning.

　　EXAMPLE 8.2.3.   Let $X_1, X_2, ..., X_n$ be a random sample of an exponential random variable; we want to test $H_0: \lambda \leq \lambda_0$ versus $H_1: \lambda > \lambda_0$, where $\lambda_0 > 0$ is some specified constant. The inequality

$$L_X(\lambda_0) \leq k L_X(\lambda_1)$$

is

$$\lambda_0^n e^{-\lambda_0 \Sigma x_i} \leq k \lambda_1^n e^{-\lambda_1 \Sigma x_i},$$

which is equivalent to

$$(\lambda_1 - \lambda_0) \sum x_i \leq \ln k + n \ln \frac{\lambda_1}{\lambda_0}.$$

Then since $\lambda_1 - \lambda_0 > 0$ for all $\lambda_1 > \lambda_0$ and $\lambda_1 - \lambda_0 < 0$ for all $\lambda_1 < \lambda_0$, this inequality is equivalent to

$$\bar{x} \leq \frac{\ln k + n \ln \lambda_1/\lambda_0}{n(\lambda_1 - \lambda_0)} = c \quad \text{if} \quad \lambda_1 > \lambda_0$$

$$\bar{x} \geq \frac{\ln k + n \ln \lambda_1/\lambda_0}{n(\lambda_1 - \lambda_0)} = c \quad \text{if} \quad \lambda_1 < \lambda_0.$$

But then the test whose critical region is specified by $R = \{x: \bar{x} \leq c\}$, where $P(\bar{X} \leq c | \lambda = \lambda_0) = \alpha$, is the uniformly most powerful test of size α for $H_0: \lambda \leq \lambda_0$ versus $H_1: \lambda > \lambda_0$, since it is of size α [maximum $Q(\lambda) = Q(\lambda_0)$ for $\lambda \leq \lambda_0$] and it gives the smallest probability of type II error for *every* $\lambda_1 > \lambda_0$. The test in Example 8.2.1 is in fact the best (of size .05) one can achieve in testing $H_0: \lambda \leq 1$ versus $H_1: \lambda > 1$.　　■

　　The reasoning previously employed also gives uniformly most powerful tests for testing.

1.　One-sided alternatives about p, the Bernoulli parameter.
2.　One-sided alternatives about μ, the parameter of a Poisson probability law.
3.　One-sided alternatives about μ, the mean of a normal probability law with σ known.
4.　One-sided alternatives about σ, the standard deviation of a normal probability law with μ known.

You are asked to verify these statements in the exercises.

A more general methodology is called for in testing hypotheses about parameters of probability laws whose parameter space Ω has two or more dimensions (that is, the probability law has two or more unknown parameters). Let θ be the vector of parameter values of the probability law; for generality let k be the number of components in θ. Ω then is the collection of all possible values for θ and is a k-dimensional space or set. One of the most frequently employed methodologies is based on the generalized likelihood ratio. Suppose we want to test a hypothesis H_0 that specifies values or ranges for some one or more parameters of the probability law, versus the alternative H_1 that simply states H_0 is false. It is instructive to think of H_0 then as saying the parameter values lie in some subset $\omega \subset \Omega$; that is, $H_0 \colon \theta \in \omega$ and the alternative is $H_1 \colon \theta \notin \omega$. The likelihood function is $L_X(\theta)$. We can maximize the likelihood function by finding that $\theta \in \Omega$, which makes $L_X(\theta)$ as large as possible; the values in θ that do this, of course, are the maximum likelihood estimates, $\hat{\theta}$, of the components of θ, and $L_X(\hat{\theta})$ is the achieved maximum of $L_X(\theta)$ over the whole parameter space Ω. We can also maximize $L_X(\theta)$ over the values specified by $H_0 \colon \theta \in \omega$; let $L_X(\hat{\omega})$ denote the largest value for $L_X(\theta), \theta \in \omega$. Of course, $L_X(\hat{\omega}) \leq L_X(\hat{\theta})$, since in constraining the possible values for (θ) to only those in $\omega \subset \Omega$, the achieved maximum in the restricted space must be no larger than the maximum in Ω. Thus we would always have

$$l = \frac{L_X(\hat{\omega})}{L_X(\hat{\theta})} \leq 1.$$

This ratio,

$$l = \frac{L_X(\hat{\omega})}{L_X(\hat{\theta})}.$$

is called the *generalized likelihood ratio*. If in fact $H_0 \colon \theta \in \omega$ is true, we would expect $L_X(\hat{\omega})$ to be roughly equal to $L_X(\hat{\theta})$ and the ratio l of the two maxima should be close to 1. If H_0 is not true, we might expect $L_X(\hat{\omega})$ to be considerably smaller than $L_X(\hat{\theta})$. The generalized likelihood ratio test criterion, for testing $H_0 \colon \theta \in \omega$ versus $H_1 \colon \theta \notin \omega$, uses the critical region

$$R = \left\{ x \colon l = \frac{L_X(\hat{\omega})}{L_X(\hat{\theta})} \leq k \right\},$$

where again $k < 1$ is chosen to make the size of the test equal to α. It can be shown (you are asked to do this in the exercises) that this generalized likelihood ratio test is identical with the Neyman–Pearson best test (Theorem 8.1.1) of simple H_0 versus simple H_1 if $\Omega = \{\theta_0, \theta_1\}$ contains only two points. The generalized likelihood ratio test criterion thus reduces to the best test in the simple versus simple case and in general gives good tests in other more complicated cases. The commonly employed tests about the parameters of a normal probability law (with the other parameter unspecified) are particular cases of this generalized likelihood ratio test criterion. The following example, and theorems, cover these cases.

EXAMPLE 8.2.4.   Many light bulbs currently sold have a statement like "Average Life—750 Hours" printed on the package containing them. It would seem reasonable (for pretested bulbs) that the lifetime of bulbs made by a given manufacturer should be normal in form; the statement on the package then is referring to the mean value of that normal distribution. Suppose n bulbs are purchased and turned on until they burn out; their lifetimes could be assumed to be a random sample of size n from this normal distribution. Let us derive the likelihood ratio test of the hypothesis $H_0: \mu = 750$ versus the alternative $H_1: \mu \neq 750$. The full parameter space is $\Omega = \{(\mu, \sigma): -\infty < \mu < \infty, \sigma > 0\}$ and the restricted space specified by H_0 is $\omega = \{(750, \sigma): \sigma > 0\}$ (this is called a two-sided alternative). The generalized likelihood function for the sample is

$$L_X(\mu, \sigma^2) = \frac{\exp\left(-\sum(x_i - \mu)^2/2\sigma^2\right)}{(2\pi\sigma^2)^{n/2}}$$

To determine $L_X(\hat{\omega})$, we must assume H_0 is true and find the maximum value for L_X, that is, we want to maximize $L_X(750, \sigma^2)$ with respect to σ^2. Recalling the maximization of L_X from Section 7.1 it is easily seen that the maximizing value is

$$\hat{\sigma}^2 = \sum \frac{(x_i - 750)^2}{n}$$

and thus

$$L_X(\hat{\omega}) = e^{-n/2}\left(\frac{n}{2\pi \sum (x_i - 750)^2}\right)^{n/2}$$

To evaluate $L_X(\hat{\theta})$, recall that the maximum likelihood estimators for μ and σ^2 are

$$\hat{\mu} = \bar{x}, \qquad \hat{\sigma}^2 = \sum (x_i - \bar{x})^2/n.$$

We find then that

$$L_X(\hat{\theta}) = e^{-n/2}\left(\frac{n}{2\pi \sum (x_i - \bar{x})^2}\right)^{n/2}$$

and the generalized likelihood ratio is

$$l = \frac{L_X(\hat{\omega})}{L_X(\hat{\Theta})} = \left(\frac{\sum (x_i - \bar{x})^2}{\sum (x_i - 750)^2} \right)^{n/2}$$

after canceling common factors. It is easy to verify the identity

$$\sum (x_i - 750)^2 = \sum [(x_i - \bar{x}) + (\bar{x} - 750)]^2$$
$$= \sum [(x_i - \bar{x})^2 + 2(x_i - \bar{x})(\bar{x} - 750) + (\bar{x} - 750)^2]$$
$$= \sum (x_i - \bar{x})^2 + n(\bar{x} - 750)^2,$$

since

$$\sum 2(x_i - \bar{x})(\bar{x} - 750) = 2(\bar{x} - 750) \sum (x_i - \bar{x}) = 0.$$

If we substitute this in the denominator of the ratio l, and divide both numerator and denominator by $\sum (x_i - \bar{x})^2$ we have

$$l^{2/n} = 1 \left/ \left(1 + \frac{n(\bar{x} - 750)^2}{\sum (x_i - \bar{x})^2} \right) \right.$$

We should reject H_0 if l is small (k is chosen to make our probability of type I error equal to α). Notice that if l (or $l^{2/n}$) is small, then $n(\bar{x} - 750)^2 / \sum (x_i - \bar{x})^2$ is large; that is, the rejection region is defined as those sample values such that

$$\frac{n(\bar{x} - 750)^2}{\sum (x_i - \bar{x})^2} > d$$

or, equivalently, those such that

$$\frac{|\bar{x} - 750| \sqrt{n}}{\sqrt{\sum (x_i - \bar{x})^2}} > \sqrt{d}.$$

We recall that

$$\frac{(\bar{x} - 750) \sqrt{n(n - 1)}}{\sqrt{\sum (x_i - \bar{x})^2}}$$

is the observed value of a T random variable if in fact we have a random sample of n observations of a normal random variable with $\mu = 750$. Thus the generalized likelihood ratio test criterion critical region of $H_0: \mu = 750$ versus $H_1: \mu \neq 750$ is equivalent to

$$|t| > \sqrt{d(n - 1)};$$

thus if we are to have a probability of type I error equal to α we should reject H_0 only if

$$\frac{|\bar{x} - 750| \sqrt{n(n - 1)}}{\sqrt{\sum (x_i - \bar{x})^2}} > t_{1 - \alpha/2},$$

where $t_{1-\alpha/2}$ is the $100(1 - \alpha/2)$th percentile of the t distribution with $n - 1$ degrees of freedom.   ∎

The preceding example is a special case of the following theorem, whose proof is left to the reader.

Theorem 8.2.1. $X_1, X_2, \ldots, X_n$ is a random sample of a normal random variable with unknown mean μ and unknown variance σ^2. Then the generalized likelihood ratio test criterion critical region R for a test of size α of H_0 versus H_1 is specified as follows for the stated H_0 and H_1.

Test	H_0	H_1	R
1.	$\mu \leq \mu_0$	$\mu > \mu_0$	$\bar{x} > \mu_0 + \dfrac{s}{\sqrt{n}} t_{1 - \alpha}$
2.	$\mu \geq \mu_0$	$\mu < \mu_0$	$\bar{x} < \mu_0 - \dfrac{s}{\sqrt{n}} t_{1 - \alpha}$
3.	$\mu = \mu_0$	$\mu \neq \mu_0$	$\left| \dfrac{\bar{x} - \mu_0}{s/\sqrt{n}} \right| > t_{1 - \alpha/2}$

∎

The following example derives the generalized likelihood ratio test for the hypothesis that the variance of a normal random variable has a specified value.

EXAMPLE 8.2.5.   The manufacturer of a precision scale claims that the standard deviation of measurements made by his machine will not exceed .02 mg. Assuming repeated measurements made with this machine are normally distributed, let us derive the likelihood ratio test of H_0: $\sigma \leq .02$ versus the alternative H_1: $\sigma > .02$. The likelihood function of the sample is, again,

$$L_X(\mu, \sigma^2) = \left(\frac{1}{2\pi\sigma^2} \right)^{n/2} \exp\left[\frac{-\sum (x_i - \mu)^2}{2\sigma^2} \right].$$

To find $L_X(\hat{\omega})$, we must maximize L_X under the assumption $\sigma \le .02$. The maximizing value for μ is $\bar{x}$ and if

$$\hat{\sigma}^2 = \sum \frac{(x_i - \bar{x})^2}{n} \le (.02)^2,$$

then $\hat{\sigma}$ is the maximizing value for σ; if

$$\sum \frac{(x_i - \bar{x})^2}{n} > (.02)^2,$$

then the maximizing value for σ is .02 (under the restriction $\sigma \le .02$). Thus

$$L_X(\hat{\omega}) = e^{-n/2} \left(\frac{n}{2\pi \sum (x_i - \bar{x})^2} \right)^{n/2} \qquad \text{if } \sum \frac{(x_i - \bar{x})^2}{n} \le (.02)^2$$

$$= \left(\frac{1}{2\pi(.02)^2} \right) \exp \left[-\sum \frac{(x_i - \bar{x})^2}{2(.02)^2} \right] \quad \text{if } \sum \frac{(x_i - \bar{x})^2}{n} > (.02)^2.$$

Again, the values that maximize L_X are

$$\hat{\mu} = \bar{x}, \qquad \hat{\sigma}^2 = \sum \frac{(x_i - \bar{x})^2}{n}$$

and

$$L_X(\hat{\theta}) = \left(\frac{n}{2\pi \sum (x_i - \bar{x})^2} \right)^{n/2} e^{-n/2}.$$

Then we have

$$l = \frac{L_X(\hat{\omega})}{L_X(\hat{\theta})} = 1 \qquad \text{if } \sum \frac{(x_i - \bar{x})^2}{n} \le (.02)^2$$

$$= \left(\frac{\sum (x_i - \bar{x})^2}{n(.02)^2} \right)^{n/2} \exp \left[\frac{n}{2} - \frac{\sum (x_i - \bar{x})^2}{2(.02)^2} \right] \quad \text{if } \sum \frac{(x_i - \bar{x})^2}{n} > (.02)^2.$$

Figure 8.2.3 gives the graph of λ versus $\sum (x_i - \bar{x})^2/n(.02)^2$. Notice that $l < k$ is equivalent to $\sum (x_i - \bar{x})^2/n(.02)^2 > c$; if in fact H_0 is true, then $\sum (X_i - \bar{X})^2/(.02)^2$ is a χ^2 random variable with $n - 1$ degrees of freedom and

$$P\left(\sum (X_i - \bar{X})^2/n(.02)^2 > c \right) = P(\chi^2(n - 1) > cn).$$

Thus to have a probability of type I error equal to α we should choose $c = \chi^2_{1-\alpha}/n$, where $\chi^2_{1-\alpha}$ is the $100(1 - \alpha)$th percentile of the χ^2 distribution with

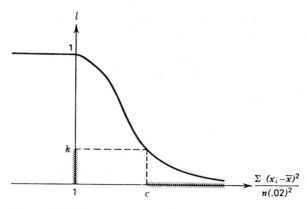

Figure 8.2.3

$n - 1$ degrees of freedom. If, for example, we used the machine $n = 10$ times to weigh the same object and we wanted to test $H_0: \sigma \leq .02$ versus $H_1: \sigma > .02$ with $\alpha = .1$, then $\chi^2_{.9} = 14.7$ (with 9 df), and we should reject H_0 if

$$\sum (x_i - \bar{x})^2 > 14.7(.02)^2 = 0.00588,$$

where $x_1, x_2, \ldots, x_{10}$ are the 10 observed measurements. ∎

Theorem 8.2.2 gives the generalized likelihood ratio test criterion critical regions for testing hypotheses about the variance of a normal random variable with unknown mean.

Theorem 8.2.2. $X_1, X_2, \ldots, X_n$ is a random sample of a normal random variable X whose mean μ is unknown. Then the generalized likelihood ratio test criterion critical region R for a test of size α of H_0 versus H_1 is specified as follows for the stated H_0 and H_1.

Test	H_0	H_1	R
1.	$\sigma^2 \leq \sigma_0^2$	$\sigma^2 > \sigma_0^2$	$\sum (x_i - \bar{x})^2 > \sigma_0^2 \chi^2_{1-\alpha}$
2.	$\sigma^2 \geq \sigma_0^2$	$\sigma^2 < \sigma_0^2$	$\sum (x_i - \bar{x})^2 < \sigma_0^2 \chi^2_{\alpha}$
3.	$\sigma^2 = \sigma_0^2$	$\sigma^2 \neq \sigma_0^2$	$\sum (x_i - \bar{x})^2 < \sigma_0^2 \chi^2_{\alpha/2}$
		and	$\sum (x_i - \bar{x})^2 > \sigma_0^2 \chi^2_{1-\alpha/2}$

∎

There is a direct relationship between confidence intervals for unknown parameters and tests of hypotheses about the same parameters. In fact, a $100(1 - \alpha)\%$ confidence interval for an unknown parameter θ can be used to define a test of size α (one minus the confidence coefficient) for a hypothesis about θ and vice versa. To illustrate this, recall from Section 7.3 that the interval from

$$\bar{x} - t_{1-\alpha/2} \frac{s}{\sqrt{n}}$$

to

$$\bar{x} + t_{1-\alpha/2} \frac{s}{\sqrt{n}}$$

is a $100(1 - \alpha)\%$ confidence interval for μ, based on a random sample of size n of a normal random variable X. Suppose, in testing $H_0: \mu = \mu_0$ versus the two-sided alternative $H_1: \mu \neq \mu_0$, we use the rule: Construct the preceding $100(1 - \alpha)\%$ confidence limits for μ. If μ_0 (the hypothesized value) falls in the confidence interval, accept H_0, and if it does not, reject H_0. That is, with this rule we would accept $H_0: \mu = \mu_0$ as long as

$$\bar{x} - t_{1-\alpha/2} \frac{s}{\sqrt{n}} \leq \mu_0 \leq \bar{x} + t_{1-\alpha/2} \frac{s}{\sqrt{n}},$$

which is easily seen to be equivalent to

$$\frac{|\bar{x} - \mu_0|}{s/\sqrt{n}} \leq t_{1-\alpha/2},$$

the same acceptance region as the generalized likelihood ratio test of size α for this hypothesis. This rule then is the same test.

Conversely, the generalized likelihood ratio test of $H_0: \mu = \mu_0$ versus $H_1: \mu \neq \mu_0$ has acceptance region (Theorem 8.2.1) defined by

$$\frac{|\bar{x} - \mu_0|}{s/\sqrt{n}} \leq t_{1-\alpha/2},$$

which is equivalent to

$$\bar{x} - t_{1-\alpha/2} \frac{s}{\sqrt{n}} \leq \mu_0 \leq \bar{x} + t_{1-\alpha/2} \frac{s}{\sqrt{n}}.$$

So if we use this test of size α of $H_0: \mu = \mu_0$ versus $H_1: \mu \neq \mu_0$ and, based on our observed sample, define the set of μ_0 values such that we would accept $H_0: \mu = \mu_0$, this resulting set of μ_0 values is identical with the $100(1 - \alpha)\%$ confidence interval for μ, the population mean. A confidence interval for an unknown parameter can in general be translated into a test about the value for that parameter and vice versa.

EXERCISE 8.2

1. Assume the annual rainfall at a certain recording station is a normal random variable with mean μ and standard deviation 2 inches. The rainfall recorded (in inches) in each of 5 years was 18.6, 20.4, 17.3, 15.1 and 22.6. Test the hypothesis that $\mu \geq 21$ versus the alternative $\mu < 21$ with $\alpha = .1$.

2. A producer of frozen fish is being investigated by the Bureau of Fair Trades. Each package of fish that this producer markets carries the claim that it contains 12 ounces of fish; a complaint has been registered that this claim is not true. The bureau acquires 100 packages of fish marketed by this company and, letting x_i be the observed weight (in ounces) of the ith package, $i = 1, 2, ..., 100$, they find

$$\sum x_i = 1150, \qquad \sum x_i^2 = 13{,}249.75.$$

It would seem reasonable to assume the true weights of packages that they market are normally distributed with mean μ and variance σ^2, neither of which is known. With $\alpha = .01$, would the bureau accept or reject $H_0: \mu \geq 12$ versus $H_1: \mu < 12$, based on this sample?

3. In deciding whether a certain type of plant would be appropriate for hedges, it is of some importance that individual plants exhibit small variability in the amounts they will grow in a year (at the same age). Specifically, we might assume that the growth made by a plant of a specific type and age (for given climatic conditions) is a normal random variable with mean μ and variance σ^2. Then to decide whether the plant would be appropriate for hedges, we might like to test $H_0: \sigma^2 \geq \frac{1}{4}$ versus $H_1: \sigma^2 < \frac{1}{4}$ with $\alpha = .05$ (measurements made in feet). Suppose we record the growth of five plants of this type for 1 year and find them to be 1.9, 1.1, 2.7, 1.6 and 2.0 feet. Should we accept H_0?

*4. A manufacturer of insulated copper wire claims that his process will coat the wire so well that defects in insulation occur at a rate of no more than 1 per 100 feet. Assume ten 1000-foot rolls are examined, starting at the beginning of the roll, and the distance, X, to the first defect found is measured for each roll. If these defects occur "at random" at a constant rate of λ per foot, X is then exponential with parameter λ. Granted the 10 observed values of X were (in feet) 24, 32, 98,

35, 20, 108, 77, 54, 8, 50, would you accept his claim with $\alpha = .01$?

*5. Assume the same manufacturer's claim as in Exercise 8.2.4, except now assume each of the 1000-foot rolls is inspected over its total length and the number of defects Y in 1000 feet is counted.

(a) If the defects occur at random at a constant rate of λ per foot, what is $E[Y]$?

(b) Granted the numbers of defects found in the 10 rolls were 12, 16, 10, 9, 14, 3, 11, 15, 10, 15, would you accept the manufacturer's claim with $\alpha = .1$? (*Hint.* Recall the normal approximation to the Poisson probability law.)

6. Assume each of $n = 13$ ladies followed the same diet for a period of 2 months. The amounts of weight lost, in kilograms, were 3.9, 4.3, 5.6, 5.6, 4.1, 6.5, 3.7, 5.9, 4.3, 3.7, 4.4, 5.9, 5.0, respectively. Assuming these are the observed values of a random sample of a normal random variable, would you accept $H_0: \mu \geq 4.5$ (versus $H_1: \mu < 4.5$) with $\alpha = .1$?

7. Using the data in Exercise 8.2.6, and with the normal assumption, would you accept $H_0: \sigma \leq .8$ (versus $H_1: \sigma > .8$), with $\alpha = .05$, where σ is the standard deviation of the normal distribution?

8. Show that the Neyman–Pearson Theorem (8.1.1) and the reasoning employed in Example 8.2.3 do lead to the uniformly most powerful tests in cases (1) through (4), listed after that example.

9. Show that with the parameter space $\Omega = \{\theta_0, \theta_1\}$, the generalized likelihood ratio critical region for testing $H_0: \theta = \theta_0$ versus $H_1: \theta \neq \theta_0$, is identical with the best test of Theorem 8.1.1.

10. The number of telephone requests for a particular song received by a radio station, per day, is a Poisson random variable with parameter μ. On 10 successive days the numbers of such requests received were 7, 1, 8, 3, 7, 8, 7, 3, 6, 5, respectively. Would you accept $H_0: \mu \geq 6$ (versus $H_1: \mu < 6$) with $\alpha = .05$, based on these data? (See the hint following Exercise 8.2.5.)

*11. A sportfishing boat takes 10 people fishing, each day, during the salmon season. The ticket seller tells each prospective customer that his probability is at least .6 of catching at least one salmon, each day. On 12 successive days the boat took 10 people fishing. The numbers of people who caught at least one salmon, on these 12 days, were 6, 7, 4, 5, 2, 5, 4, 7, 4, 7, 9, 4, respectively. Assume, for each day, the persons on the boat are independent Bernoulli trials with probability p of catching at least one salmon. Would you accept the ticket seller's claim, with $\alpha = .05$, based on these data?

12. During the 1980 U.S. presidential campaign, a nationwide political polling group announced that of 1200 randomly selected respondents,

584 would vote to reelect the incumbent. Letting p represent the proportion of voters in the whole electorate who would say they would reelect the incumbent, would you accept $H_0: p \geq .5$, based on this sample, with $\alpha = .1$?

13. The lubricating oil in an aircraft engine was changed on a given day; the new oil that was put into the engine contained 30 ppm iron. After 25 flight hours, $n = 11$ small samples of the oil were removed and burned on a spectrometer to estimate the current iron contamination level. The observed spectrometer readings were 34.9, 37.4, 40.1, 39.2, 34.4, 25.1, 40.7, 34.5, 30.6, 33.2, 34.0. Assuming these are the observed values of a normal random variable, would you accept $H_0: \mu = 30$ (versus $H_1: \mu \neq 30$), with $\alpha = .05$?

14. Make the same assumptions as those mentioned in Exercise 8.2.13 and use the data given there to test $H_0: \sigma \leq 4$ (versus $H_1: \sigma > 4$), with $\alpha = .1$.

*15. The evaluation of the power of the T-test of Theorem 8.2.1 involves the noncentral T-distribution, which we have not had the space to discuss. In testing $H_0: \mu \leq \mu_0$ versus $H_1: \mu > \mu_0$, we reject H_0 if

$$\frac{\sqrt{n}(\bar{x} - \mu_0)}{s} > t_{1-\alpha}.$$

Show why the T-distribution cannot be used to evaluate

$$P\left(\frac{\sqrt{n}(\bar{X} - \mu_0)}{S} > t_{1-\alpha} \Big| \mu = \mu_1 \right),$$

where μ_1 is any value greater than μ_0.

8.3 Some Two-Sample Tests

In Section 7.4 we discussed sampling from two different probability laws and using the observed sample values to derive confidence limits for certain functions of the parameters of the two probability laws. In many cases it is of interest to test hypotheses about the parameters of two probability laws. For example, we might assume that a standard diet recommended for weight reduction generates normal random variables (amounts of weight lost by people using it) with mean μ_1 and variance σ^2; some new proposed diet also leads to normally distributed weight losses, say, with mean μ_2 and variance σ^2. How might we use observed sample values to test $H_0: \mu_1 \leq \mu_2$ versus $H_1: \mu_1 > \mu_2$? Or suppose the number of automobile accidents on a certain highway, per day, is assumed to be a Poisson random variable X with parameter μ_1, when the speed limit is set at 104 kilometers/hour (65 mph); if the speed limit is changed to 80 kph (50 mph), we might assume the number of accidents per day Y to be Poisson with parameter μ_2. How could we use observed values of X and Y to test $H_0: \mu_1 \leq \mu_2$ versus $H_1: \mu_1 > \mu_2$? We will discuss some of the commonly used methodology for making this type of test in this section.

As you might expect after reading Section 8.2, the generalized likelihood ratio test criterion is frequently the basis for tests of hypotheses regarding the parameters of two different probability laws. We will go through this rationale in some detail for one test and then simply discuss the tests commonly used for several other cases (and whether they come from the generalized likelihood ratio test criterion).

Suppose $x_1, x_2, ..., x_n$ are the observed values of a random sample of a normal random variable X with mean μ_1, variance σ^2, and $y_1, y_2, ..., y_m$ are the observed values of an independent random sample of a normal random variable Y with mean μ_2, variance σ^2 (note the two variances are assumed equal). We want to test $H_0: \mu_1 = \mu_2$ versus $H_1: \mu_1 \neq \mu_2$ and will employ the generalized likelihood ratio test criterion. The vector of parameters (for the combined sample of $m + n$ values) then is $\theta = (\mu_1, \mu_2, \sigma^2)$ and the parameter space is $\Omega = \{(\mu_1, \mu_2, \sigma^2): -\infty < \mu_1 < \infty, -\infty < \mu_2 < \infty, \sigma^2 > 0\}$. If $H_0: \mu_1 = \mu_2$ is assumed true, the constrained parameter space is

$$\omega = \{(\mu, \mu, \sigma^2): -\infty < \mu < \infty, \sigma^2 > 0\}.$$

The likelihood function is

$$
L_X(\mu_1, \sigma^2)\, L_Y(\mu_2, \sigma^2) = \left(\frac{1}{2\pi\sigma^2}\right)^{n/2} \exp\left[-\frac{\sum (x_i - \mu_1)^2}{2\sigma^2}\right]
$$

$$
\cdot \left(\frac{1}{2\pi\sigma^2}\right)^{m/2} \exp\left[-\frac{\sum (y_j - \mu_2)^2}{2\sigma^2}\right]
$$

$$
= \left(\frac{1}{2\pi\sigma^2}\right)^{m+n/2}
$$

$$
\cdot \exp\left[-\frac{\left(\sum (x_i - \mu_1)^2 + \sum (y_j - \mu_2)^2\right)}{2\sigma^2}\right].
$$

To find the overall maximum of this likelihood function (the maximum likelihood estimates), define

$$
K = \ln L_X(\mu_1, \sigma)\, L_Y(\mu_2, \sigma)
$$

$$
= -\frac{(m+n)}{2} \ln 2\pi - \frac{(m+n)}{2} \ln \sigma^2 - \frac{\sum (x_i - \mu_1)^2}{2\sigma^2} - \frac{\sum (y_j - \mu_2)^2}{2\sigma^2}.
$$

Then

$$\frac{\partial K}{\partial \mu_1} = \frac{\sum (x_i - \mu_1)}{\sigma^2}$$

$$\frac{\partial K}{\partial \mu_2} = \frac{\sum (y_j - \mu_2)}{\sigma^2}$$

$$\frac{\partial K}{\partial \sigma^2} = -\frac{m + n}{2\sigma^2} + \frac{\sum (x_i - \mu_1)^2 + \sum (y_j - \mu_2)^2}{2(\sigma^2)^2} ;$$

setting these partial derivatives equal to zero and solving the resulting equations simultaneously easily gives the estimates

$$\hat{\mu}_1 = \bar{x}, \qquad \hat{\mu}_2 = \bar{y}, \qquad \hat{\sigma}^2 = \frac{\sum (x_i - \bar{x})^2 + \sum (y_j - \bar{y})^2}{m + n}.$$

Thus the unconstrained maximum value for the likelihood function is

$$L(\hat{\theta}) = \left(\frac{m + n}{2\pi \left(\sum (x_i - \bar{x})^2 + \sum (y_j - \bar{y})^2 \right)} \right)^{(m+n)/2} e^{-(m+n)/2}.$$

With H_0 assumed true, the likelihood function becomes

$$L(\omega) = \left(\frac{1}{2\pi\sigma^2} \right)^{(m+n)/2} \exp \left[-\frac{\left(\sum (x_i - \mu)^2 + \sum (y_j - \mu)^2 \right)}{2\sigma^2} \right],$$

exactly the likelihood function for a random sample of size $m + n$ of a normal random variable with mean μ and variance σ^2; thus the maximizing values are

$$\hat{\mu} = \frac{\sum x_i + \sum y_j}{m + n} = \frac{n\bar{x} + m\bar{y}}{m + n}$$

$$\hat{\sigma}^2 = \frac{\sum (x_i - \hat{\mu})^2 + \sum (y_j - \hat{\mu})^2}{m + n}$$

and

$$L(\hat{\omega}) = \left(\frac{m + n}{2\pi \left(\sum (x_i - \hat{\mu})^2 + \sum (y_j - \hat{\mu})^2 \right)} \right)^{(m+n)/2} e^{-(m+n/2}.$$

The generalized likelihood ratio test criterion then is

$$l = \frac{L(\hat{\omega})}{L(\hat{\theta})} = \left(\frac{\sum (x_i - \bar{x})^2 + \sum (y_j - \bar{y})^2}{\sum (x_i - \hat{\mu})^2 + \sum (y_j - \hat{\mu})^2} \right)^{(m+n)/2}$$

after canceling common factors. Now recalling that

$$\hat{\mu} = \frac{n\bar{x} + m\bar{y}}{m + n},$$

we can write

$$\sum (x_i - \hat{\mu})^2 + \sum (y_j - \hat{\mu})^2$$
$$= \sum (x_i - \bar{x} + \bar{x} - \hat{\mu})^2 + \sum (y_j - \bar{y} + \bar{y} - \hat{\mu})^2$$
$$= \sum (x_i - \bar{x})^2 + \sum (y_j - \bar{y})^2 + n(\bar{x} - \hat{\mu})^2 + m(\bar{y} - \hat{\mu})^2$$
$$= \sum (x_i - \bar{x})^2 + \sum (y_j - \bar{y})^2 + \frac{mn(\bar{x} - \bar{y})^2}{m + n}.$$

Dividing both numerator and denominator by $\sum (x_i - \bar{x})^2 + \sum (y_j - \bar{y})^2$ then gives

$$l = \frac{1}{(1 + a)^{(m+n)/2}},$$

where

$$a = \frac{mn(\bar{x} - \bar{y})^2}{m + n} \Bigg/ \left(\sum (x_i - \bar{x})^2 + \sum (y_j - \bar{y})^2 \right)$$
$$= \frac{t^2}{(m + n - 2)};$$

note that

$$t = (\bar{x} - \bar{y}) \sqrt{\frac{mn}{m + n}} \Bigg/ \sqrt{\frac{\sum (x_i - \bar{x})^2 + \sum (y_j - \bar{y})^2}{m + n - 2}}$$

is the observed value of a T random variable with $m + n - 2$ degrees of freedom if $H_0: \mu_1 = \mu_2$ is true. The critical region for the generalized likelihood ratio test criterion is defined by $l \leq k$; but $l \leq k$ is equivalent to $a \geq c$, which in turn is equivalent to $|t| \geq d$. Thus to have probability of type I error equal to α we should reject H_0 if

$$|\bar{x} - \bar{y}| \sqrt{\frac{mn}{m + n}} \Bigg/ \sqrt{\frac{\sum (x_i - \bar{x})^2 + \sum (y_j - \bar{y})^2}{m + n - 2}} \geq t_{1 - \alpha/2},$$

where the quantile is chosen from the T-distribution with $m + n - 2$ degrees of freedom. This establishes the two-sided alternative portion of Theorem 8.3.1; the one-sided alternative results also follow from the generalized likelihood ratio test criterion.

Theorem 8.3.1. Let $X_1, X_2, \ldots, X_n$ be a random sample of a normal random variable with mean μ_1, variance σ^2 and let $Y_1, Y_2, \ldots, Y_m$ be an independent random sample of a normal random variable with mean μ_2, variance σ^2. Define

$$S_p^2 = \frac{\sum (X_i - \bar{X})^2 + \sum (Y_j - \bar{Y})^2}{m + n - 2},$$

$$T = (\bar{X} - \bar{Y})\sqrt{\frac{mn}{m+n}} \bigg/ S_p.$$

Then the generalized likelihood ratio test of size α of the following hypotheses is as indicated, where the t quantiles are selected from the T distribution with $m + n - 2$ degrees of freedom.

H_0	H_1	Rejection Region
$\mu_1 = \mu_2$	$\mu_1 \neq \mu_2$	$\|T\| \geq t_{1-\alpha/2}$
$\mu_1 \leq \mu_2$	$\mu_1 > \mu_2$	$T > t_{1-\alpha}$
$\mu_1 \geq \mu_2$	$\mu_1 < \mu_2$	$T < t_\alpha$

EXAMPLE 8.3.1. Given that $n = 8$ 60-watt light bulbs of brand G provided 686, 784, 769, 848, 728, 739, 757, 743 hours of service, respectively, and that $m = 10$ 60-watt bulbs of brand W provided 762, 783, 763, 749, 806, 783, 831, 784, 790, 750 hours of service, respectively, let us use these data to illustrate the preceding theorem. Thus we assume these are two independent samples of normal random variables, both with the same variance; we want to test $H_0: \mu_1 = \mu_2$ (that the two mean lifetimes are the same) versus $H_1: \mu_1 \neq \mu_2$, with $\alpha = .05$. To make this two-sided test, then, we find $t_{.975} = 2.120$ (with $m + n - 2 = 16$ degrees of freedom), and we find from these data, $\bar{x} = 756.75$, $\bar{y} = 780.1$, $\sum(x_i - \bar{x})^2 = 15{,}555.5$, $\sum(y_j - \bar{y})^2 = 5884.9$, so the pooled (unbiased) estimate for σ^2 is

$$s_p^2 = \frac{15{,}555.5 + 5884.9}{16} = 1340.025.$$

Thus the observed T statistic is

$$t = (756.75 - 780.1)\frac{\sqrt{80/18}}{\sqrt{1340.025}}$$

$$= -1.345;$$

since $|t| = 1.345 < 2.120 = t_{.975}$, we accept H_0.

As mentioned at the end of Section 8.2, confidence intervals can be translated into acceptance regions for tests of hypotheses. Clearly, the test(s) described in Theorem 8.3.1 actually correspond directly to the confidence interval for $\mu_1 - \mu_2$, the difference of two normal means, discussed in Section 7.4. If you use the data given in Example 8.3.1 to evaluate a 95% two-sided confidence interval for $\mu_1 - \mu_2$, you will find the computed interval includes 0; that is, one of the possible values for $\mu_1 - \mu_2$ is 0, which is equivalent to $\mu_1 = \mu_2$. Thus the same data necessarily lead to accepting $H_0: \mu_1 = \mu_2$, with probability of type I error equal to one minus the confidence coefficient.

In Section 7.4 we also discussed $100(1 - \alpha)\%$ confidence intervals for σ_1^2/σ_2^2, the ratio of two normal variances. Let us convert this type of interval into a test of $H_0: \sigma_1^2 = \sigma_2^2$ versus $H_1: \sigma_1^2 \neq \sigma_2^2$. Again, then, let $X_1, X_2, ..., X_n$ be a random sample of a normal random variable X with mean μ_1 and variance σ_1^2 and let $Y_1, Y_2, ..., Y_m$ be an independent random sample of a normal random variable Y with mean μ_2 and variance σ_2^2. A $100(1 - \alpha)\%$ two-sided confidence interval for σ_2^2/σ_1^2 then has end points

$$\frac{S_Y^2}{S_X^2} F_{\alpha/2}(n - 1, m - 1)$$

and

$$\frac{S_Y^2}{S_X^2} F_{1-\alpha/2}(n - 1, m - 1).$$

The rejection region for testing $H_0: \sigma_1^2 = \sigma_2^2$ versus $H_1: \sigma_1^2 \neq \sigma_2^2$ then consists of those sample outcomes *not* covered by the confidence interval; that is, we reject $H_0: \sigma_1^2 = \sigma_2^2$ if the confidence interval does not include the point $\sigma_2^2/\sigma_1^2 = 1$. The confidence interval will not include 1 if the lower limit exceeds 1,

$$\frac{S_Y^2}{S_X^2} F_{\alpha/2}(n - 1, m - 1) > 1,$$

that is,

$$\frac{S_X^2}{S_Y^2} < F_{\alpha/2}(n - 1, m - 1),$$

of if the upper limit is smaller than 1,

$$\frac{S_Y^2}{S_X^2} F_{1-\alpha/2}(n - 1, m - 1) < 1,$$

that is

$$\frac{S_X^2}{S_Y^2} > F_{1-\alpha/2}(n-1, m-1).$$

Translating one-sided confidence limits for σ_1^2/σ_2^2 leads to tests of one-sided alternatives, summarized in Theorem 8.3.2. It can be shown that each of these tests is also given by applying the generalized likelihood ratio test criterion.

Theorem 8.3.2. Let $X_1, X_2, \ldots, X_n$ be a random sample of a normal random variable X with mean μ_1, variance σ_1^2 and let $Y_1, Y_2, \ldots, Y_m$ be an independent random sample of a normal random variable Y with mean μ_2, variance σ_2^2. Define

$$S_X^2 = \frac{1}{n-1}\sum(X_i - \overline{X})^2, \qquad S_Y^2 = \frac{1}{m-1}\sum(Y_j - \overline{Y})^2;$$

the generalized likelihood ratio tests (of size α) of the following hypotheses are as listed.

H_0	H_1	Rejection Region		
$\sigma_1^2 \leq \sigma_2^2$	$\sigma_1^2 > \sigma_2^2$	$\dfrac{S_X^2}{S_Y^2} > F_{1-\alpha}$		
$\sigma_1^2 \geq \sigma_2^2$	$\sigma_1^2 < \sigma_2^2$	$\dfrac{S_X^2}{S_Y^2} < F_{\alpha}$		
$\sigma_1^2 = \sigma_2^2$	$\sigma_1^2 \neq \sigma_2^2$	$\dfrac{S_X^2}{S_Y^2} < F_{\alpha/2}$	or	$\dfrac{S_X^2}{S_Y^2} > F_{1-\alpha/2}$

All quantiles used are from the F-distribution with $n-1$, $m-1$ degrees of freedom. ∎

EXAMPLE 8.3.2.   We will again use the data given in Example 8.3.1. To test that the two mean lifetimes were equal, we actually assumed the two variances to be equal. If this assumption were in doubt, we might first use the data to test the equality of the variances and, if we accept $H_0: \sigma_1^2 = \sigma_2^2$, then the previous test of equality of the two means can be made. Thus suppose we want to test $H_0: \sigma_1^2 = \sigma_2^2$, versus $H_1: \sigma_1^2 \neq \sigma_2^2$ with $\alpha = .02$.

Using the data from 8.3.1, we have

$$s_X^2 = \frac{15555.5}{7} = 2222.21, \qquad s_Y^2 = \frac{5884.9}{9} = 653.88$$

so

$$\frac{s_X^2}{s_Y^2} = 3.40;$$

from the F table with 7 and 9 degrees of freedom we find $F_{.01} = .149$, $F_{.99} = 5.613$, so we accept $H_0: \sigma_1^2 = \sigma_2^2$. Even though there is a sizable difference in the two estimated variances, it is not sufficiently great for us to reject equality with $\alpha = .02$. ∎

If in Example 8.3.2 we had used $\alpha = .10$, we find $F_{.95} = 3.29$ and we would reject $H_0: \sigma_1^2 = \sigma_2^2$ (the probability we are wrong in rejecting H_0 is .1). But then we would not be able to use the T statistic of Example 8.3.1 in testing $H_0: \mu_1 = \mu_2$ because we have decided the two population variances are unequal. What does one do in this case to test that two normal means are equal, when their variances are not? This is a famous problem, called the Behrens–Fisher problem, one for which there is not unanimity regarding the best test to employ. One can examine the generalized likelihood ratio test for $H_0: \mu_1 = \mu_2$ versus $H_1: \mu_1 \neq \mu_2$, but unfortunately the probability law for the ratio l in this case depends on σ_1^2/σ_2^2; if we do not know the value for σ_1^2/σ_2^2, then we cannot use the generalized likelihood ratio to find the value k that sets $P(\text{reject } H_0 | \mu_1 = \mu_2) = \alpha$.

One appealing approximate test, which is commonly used in this case, was first proposed by Welch, in the British journal *Biometrika* in 1937. Let us discuss his approximate procedure and apply it to the data in Example 8.3.1. Again, assume we have independent samples of sizes n and m, respectively, from two normal populations; the population means are μ_1, μ_2, respectively, and the population variances are σ_1^2, σ_2^2, not assumed equal. We want to test $H_0: \mu_1 = \mu_2$ versus $H_1: \mu_1 \neq \mu_2$. Welch reasoned that the difference, $\overline{X} - \overline{Y}$, has variance

$$\frac{\sigma_1^2}{n} + \frac{\sigma_2^2}{m}$$

and that

$$S_X^2 = \frac{1}{n-1}\sum(X_i - \overline{X})^2, \qquad S_Y^2 = \frac{1}{m-1}\sum(Y_j - \overline{Y})^2,$$

respectively, give unbiased estimates for the two variances. Thus the magnitude of

$$U = \frac{(\bar{X} - \bar{Y})}{\sqrt{S_X^2/n + S_Y^2/m}}$$

would be a reasonable quantity to use in deciding whether to accept or reject H_0: $\mu_1 = \mu_2$. (If $m = n$ and $\sigma_1^2 = \sigma_2^2$, note that U has the T-distribution used in Theorem 8.3.1; with $\sigma_1^2 \neq \sigma_2^2$, U does not have a T-distribution.) Welch also then gives some further reasoning to show it is reasonable to approximate the true distribution for U by a T-distribution with degrees of freedom

$$d = (S_X^2/n + S_Y^2/m)^2 \left/ \left(\frac{S_X^4}{n^2(n-1)} + \frac{S_Y^4}{m^2(m-1)} \right) \right. ;$$

thus if we find $|u| > t_{1-\alpha/2}$, with degrees of freedom d, we reject H_0. The approximate probability of a type I error then is α, with this test. Since the degrees of freedom, d, will most likely not be an integer, we find $t_{1-\alpha/2}$ by interpolation. The procedure is illustrated in Example 8.3.3.

EXAMPLE 8.3.3. Suppose we had rejected H_0: $\sigma_1^2 = \sigma_2^2$ using the data from Example 8.3.1 and still want to test H_0: $\mu_1 = \mu_2$. We will employ Welch's procedure to do this. For the data given we have $n = 8$, $m = 10$, $\bar{x} = 756.75$, $\bar{y} = 780.1$, $s_X^2 = 2222.21$, $s_Y^2 = 653.88$; the observed value for the test statistic is

$$u = \frac{(756.75 - 780.1)}{\sqrt{2222.21/8 + 653.88/10}}$$

$$= -1.26,$$

slightly smaller in magnitude than the observed T statistic of Example 8.3.1. The degrees of freedom for the approximate T-distribution is

$$d = \left(\frac{2222.21}{8} + \frac{653.88}{10} \right)^2 \left/ \left(\frac{(2222.21)^2}{64(7)} + \frac{(653.88)^2}{100(9)} \right) \right.$$

$$= 10.24;$$

with 10 degrees of freedom, $t_{.975} = 2.228$, and with 11 degrees of freedom, $t_{.975} = 2.201$, so the interpolated value is

$$t_{.975} = 2.228 + .24(2.201 - 2.228) = 2.222.$$

We would still accept H_0: $\mu_1 = \mu_2$ with $\alpha = .05$ (at least approximately). If one examines the formula that determines d, the degrees of freedom, for this approximate test, it will always lie between $m + n - 2$ (the appropriate degrees of freedom with the assumption that $\sigma_1^2 = \sigma_2^2$) and the smaller of $n - 1$ and $m - 1$. Because the t quantile (with fixed area) decreases with increasing degrees of freedom, we must always accept H_0 with the Welch approximate test whenever we accept with the test given in Theorem 8.3.1 (for the same α). In cases in which we reject H_0: $\mu_1 = \mu_2$, using the test of Theorem 8.3.1, it is prudent to check whether one would still reject using the Welch approximate test whenever one is uncomfortable with the assumption $\sigma_1^2 = \sigma_2^2$.   ∎

In many applications one may have dependent samples of two random variables, the dependence sometimes occurring purposefully to more carefully control extraneous factors that may affect the comparison of interest. For example, suppose one were interested in investigating the effect of alcohol consumption (at a specified rate) on some reaction time (say, the length of time needed to hit a brake pedal of an automobile). One way to investigate such an effect would be to select, say, n people and measure their reaction times. Then we could also independently select a group of m people, have each person consume alcohol at the specified rate, and measure their reaction times. We would then have two independent samples and, assuming normality, could use Theorem 8.3.1 to test hypotheses of interest about μ_1 and μ_2, the two average reaction times. There are a number of facts one could criticize in this procedure. Perhaps one of the most important criticisms is that, by chance, the n people selected initially might all have naturally slow reaction times and, again by chance, the people in the group of m might all have naturally fast reaction times. This could lead us to accept H_0: $\mu_1 \geq \mu_2$, for example, simply because of the use of two independent groups. There are other criticisms that could equally be made if the preceding procedure were followed; our main interest here is simply to motivate a technique that is frequently employed in experimental work to avoid this criticism (and others as well), and to see that a reasonable model can be made incorporating dependent samples.

In the previous discussion of reaction times it probably occurred to you that a clearer investigation of the effect of alcohol consumption on reaction time could be made if we selected n people and measured their reaction time, as before. Rather than selecting a second independent group of people to consume the alcohol, we could use the same n people in the second part; each of the n consumes the alcohol and then we measure each of their reaction times a second time. This would result in n paired measurements, two reaction time measurements for n people, one made before and one made after the alcohol consumption, say, X_i = reaction time before, Y_i = reaction time after, for person i. But now it would seem reasonable that X_i and Y_i are correlated random variables, because they are reaction times for the same individual. If individual i has a natural fast reaction time, we might expect both X_i and Y_i to be above their respective averages (mean reaction times before and after). If we assume the X_i's to be normal with mean μ_1 and the Y_i's to be normal with mean μ_2, we cannot use Theorem 8.3.1 to test $H_0: \mu_1 \geq \mu_2$, for example, if we assume the two samples are correlated. Another assumption of Theorem 8.3.1 is that the two variances are equal. It is quite conceivable that the variances of reaction times before and after are not equal, again negating the use of Theorem 8.3.1.

What then might be a reasonable model, and does it lead to an easy way of testing hypotheses about the values for μ_1 and μ_2? The following model is very frequently used in such situations. We have n pairs, (X_i, Y_i), $i = 1, 2, \ldots, n$; we assume they are a random sample from a *bivariate* normal population with parameters μ_1, μ_2, σ_1^2, σ_2^2, ρ. Recall then that any linear function of X_i and Y_i is again normal; in particular, if we define $D_i = X_i - Y_i$, $i = 1, 2, \ldots, n$, the differences in the two reaction times for individual i, the D_i's are independent, normal, mean $\mu_D = \mu_1 - \mu_2$, variance $\sigma_D^2 = \sigma_1^2 + \sigma_2^2 - 2\rho\sigma_1\sigma_2$. Note then that $\mu_D = 0$ is equivalent to $\mu_1 = \mu_2$ and we can use the T-test of Theorem 8.2.1 to test hypotheses about $\mu_D = \mu_1 - \mu_2$. This test is called the *paired T-test* because of the natural pairings of the observations. The result is described in the following theorem.

Theorem 8.3.3. (Paired T-Test).   Assume (X_i, Y_i), $i = 1, 2, \ldots, n$, is a random sample of a bivariate normal vector (X, Y) with parameters μ_1, μ_2, σ_1^2, σ_2^2, ρ; define $D_i = X_i - Y_i$, $i = 1, 2, \ldots, n$, $\mu_D = \mu_1 - \mu_2$,

$$\bar{D} = \frac{1}{n}\sum D_i, \qquad S_D^2 = \frac{1}{n-1}\sum (D_i - \bar{D})^2.$$

Then $T = (\bar{D} - \mu_D)\sqrt{n}/S_D$ has the T-distribution with $n - 1$ degrees of freedom. This distribution can be used to test the following hypotheses, with $P(\text{type I error}) = \alpha$, as follows.

H_0	H_1	Rejection Region
$\mu_1 \leq \mu_2$	$\mu_1 > \mu_2$	$\bar{d}\sqrt{n}/s_D > t_{1-\alpha}$
$\mu_1 \geq \mu_2$	$\mu_1 < \mu_2$	$\bar{d}\sqrt{n}/s_D < t_\alpha$
$\mu_1 = \mu_2$	$\mu_1 \neq \mu_2$	$|\bar{d}|\sqrt{n}/s_D > t_{1-\alpha/2}$

EXAMPLE 8.3.4.   The following reaction times were gathered from $n = 10$ volunteers; the units used are milliseconds. For each individual the x value is the first reaction time (before consumption of beverage) and the y value is the second reaction time (after consumption) for the same individual; $d = x - y$.

Individual	1	2	3	4	5	6	7	8	9	10
$x =$	469	563	693	737	706	595	634	511	620	496
$y =$	697	814	850	933	821	788	818	761	792	763
$d =$	−228	−251	−157	−196	−115	−193	−184	−250	−172	−267

We will use these data to test $H_0: \mu_1 \geq \mu_2$ versus $H_1: \mu_1 < \mu_2$ with $\alpha = .01$. We find $\sum d_i = -2013$, $\sum d_i^2 = 425753$, $\bar{d} = -201.3$, $s_d = 47.77$, so

$$\frac{\bar{d}\sqrt{10}}{s_d} = -13.33;$$

with 9 degrees of freedom, $t_{.01} = -2.821$, so we reject H_0. One might wonder why we chose to test $H_0: \mu_1 \geq \mu_2$. Surely it is expected that if any difference in average reaction time exists, the later reaction time should be greater than the former, meaning we would expect $\mu_1 < \mu_2$ (as we have in fact concluded). Frequently, in experimental work null hypotheses are expressed so that hopefully the data collected will lead to rejection of H_0, mainly because if we do reject H_0, the only possible error we could commit is the type I, whose maximum value is α (.01 in this case). This gives an easy way of controlling the probability that an error has been committed. Had we tested (and accepted) $H_0: \mu_1 \leq \mu_2$, the only possible error we could have committed is a type II that (as can be shown) is as small as possible, given α, for all alternatives, but we do not have so clear an indication of its maximum value for reasonable alternatives. ∎

In Section 7.4 (Theorem 7.4.3) we also discussed a confidence interval for the ratio, λ_X/λ_Y, of two exponential parameters. This interval is easily converted into tests of hypotheses about the equality of λ_X and λ_Y. Assume as before $X_1, X_2, \ldots, X_n$ is a random sample of an exponential random variable with parameter λ_X, whereas $Y_1, Y_2, \ldots, Y_m$ is an independent random sample of an exponential random variable with parameter λ_Y. The $100(1 - \alpha)\%$ confidence limits for λ_X/λ_Y then are

$$\frac{\bar{Y}}{\bar{X}} F_{\alpha/2}, \qquad \frac{\bar{Y}}{\bar{X}} F_{1-\alpha/2},$$

where both quantiles are from the F-distribution with degrees of freedom $2n$, $2m$. If $H_0: \lambda_X = \lambda_Y$ is true, the ratio $\lambda_X/\lambda_Y = 1$, and the confidence interval for λ_X/λ_Y will not cover 1 if the lower limit exceeds 1 or if the upper limit is smaller than 1; thus the confidence interval does not include 1 if

$$\frac{\bar{Y}}{\bar{X}} F_{\alpha/2} > 1, \qquad \text{that is,} \qquad \frac{\bar{X}}{\bar{Y}} < F_{\alpha/2},$$

or if

$$\frac{\overline{Y}}{\overline{X}} F_{1-\alpha/2} < 1, \qquad \text{that is,} \qquad \frac{\overline{X}}{\overline{Y}} > F_{1-\alpha/2},$$

which defines the rejection region. It can be shown that this test is again the same as the generalized likelihood ratio test.

Theorem 8.3.4. Let $X_1, X_2, ..., X_n$ be a random sample of an exponential random variable with parameter λ_X and let $Y_1, Y_2, ..., Y_m$ be an independent random sample of an exponential random variable with parameter λ_Y. The generalized likelihood ratio tests of the following hypotheses are as listed; all quantiles are from the F-distribution with $2n$ and $2m$ degrees of freedom.

H_0:	H_1:	Rejection Region
$\lambda_X \le \lambda_Y$	$\lambda_X > \lambda_Y$	$\dfrac{\overline{X}}{\overline{Y}} < F_\alpha$
$\lambda_X \ge \lambda_Y$	$\lambda_X < \lambda_Y$	$\dfrac{\overline{X}}{\overline{Y}} > F_{1-\alpha}$
$\lambda_X = \lambda_Y$	$\lambda_X \ne \lambda_Y$	$\dfrac{\overline{X}}{\overline{Y}} < F_{\alpha/2}$ or
		$\dfrac{\overline{X}}{\overline{Y}} > F_{1-\alpha/2}$

EXAMPLE 8.3.5. A computer was used to generate 13 independent, exponential random variables. The first 7 values generated were 2.542, 3.508, 5.593, 5.746, .054, .243, .002 and the last 6 were 1.371, 7.655, 2.866, 2.966, 7.276, 6.144, respectively. If this generator is working as it should, these values should be just like two independent samples of sizes $n = 7$, $m = 6$, respectively, of exponential random variables with the same parameter. Let us use these values to test $H_0: \lambda_X = \lambda_Y$ versus $H_1: \lambda_X \ne \lambda_Y$ with $\alpha = .1$; the $n = 7$ values are a random sample of X, whereas the $m = 6$ values are a random sample of Y. We have $\bar{x} = \frac{17.688}{7} = 2.527$, $\bar{y} = \frac{28.278}{6} = 4.713$, so $\bar{x}/\bar{y} = .536$. With $2n = 14$, $2m = 12$ df, $F_{.05} = .395$, $F_{.95} = 2.637$ and since $\bar{x}/\bar{y}$ lies between these two values, we accept H_0; the total of the first 7 is fairly small compared to the total of the last 6, but the difference is not sufficiently great to warrant rejecting $H_0: \lambda_X = \lambda_Y$ with $\alpha = .1$ (as indeed it should not be). ∎

Frequently, one may want to test hypotheses regarding the values of two Bernoulli parameters. Let us discuss a simple way of doing this for large samples from both populations; the particular test we will describe here is actually identical with a "contingency table" test that we will discuss in Chapter 10. Let $X_1, X_2, \ldots, X_n$ be a random sample of a Bernoulli random variable with parameter p_1 and let $Y_1, Y_2, \ldots, Y_m$ be an independent random sample of a Bernoulli random variable with parameter p_2. How might we test $H_0: p_1 = p_2$ versus $H_1: p_1 \neq p_2$, with a specified value for α (at least approximately)? Define $\overline{X} = (1/n) \sum X_i$, $\overline{Y} = (1/m) \sum Y_j$ and, as we know, $\overline{X}$ then is approximately normal with mean p_1, variance $p_1(1-p_1)/n$, $\overline{Y}$ is approximately normal with mean p_2, variance $p_2(1-p_2)/m$ and the two are independent. The difference $\overline{X} - \overline{Y}$ then is approximately normal with mean $p_1 - p_2$ and variance $p_1(1-p_1)/n + p_2(1-p_2)/m$; if $H_0: p_1 = p_2$ is true, the mean of this difference is 0 and the variance is $p(1-p) \cdot (1/n + 1/m) = (m + n) p(1 - p)/mn$, where p is the common value for p_1 and p_2. Under the assumption $p_1 = p_2 = p$, the maximum likelihood estimate for p is

$$\hat{p} = \frac{\sum x_i + \sum y_j}{n + m} = \frac{n\overline{x} + m\overline{y}}{n + m}$$

and, still making this assumption,

$$Z = \frac{\overline{X} - \overline{Y}}{\sqrt{\dfrac{m + n}{mn} \left(\dfrac{n\overline{X} + m\overline{Y}}{m + n} \right) \left(1 - \dfrac{n\overline{X} + m\overline{Y}}{m + n} \right)}}$$

is approximately a standard normal random variable. Thus if we reject $H_0: p_1 = p_2$ when $|Z| > z_{1 - \alpha/2}$, our probability of type I error is approximately α.

This test is not equivalent to the confidence interval for $p_1 - p_2$, discussed in Section 7.4, although it is close; it is informative to see where the difference lies. The test is also not equivalent to the generalized likelihood ratio test for this hypothesis. As already mentioned, it is equivalent to a contingency table test, whose motivation will be discussed in Chapter 10.

EXAMPLE 8.3.6. Assume of $n = 50$ television sets of brand R that were sold by a given store, 20 required a service call within their one-year warranty period, whereas of $m = 40$ sets of brand Z, 12 required a service call within their one-year warranty period. Assuming the $n = 50$ R sets represent independent Bernoulli trials with $p_1 =$ probability a service call is required, and the $m = 40$ Z sets represent independent Bernoulli trials with $p_2 =$ probability a service call is required, should we accept $H_0: p_1 =$

p_2 (versus $H_1: p_1 \neq p_2$) with $\alpha = .1$? We can use the test just discussed to make this decision. We have $\bar{x} = \frac{20}{50} = .4$, $\bar{y} = \frac{12}{40} = .3$,

$$\frac{m+n}{mn}\left(\frac{n\bar{x}+m\bar{y}}{m+n}\right)\left(1 - \frac{n\bar{x}+m\bar{y}}{m+n}\right) = \frac{90}{40(50)}\left(\frac{20+12}{90}\right)\left(1 - \frac{20+12}{90}\right)$$

$$= .0103$$

so

$$z = \frac{.4 - .3}{\sqrt{.0103}} = .98.$$

Since $z_{.95} = 1.64$, we accept H_0, based on these samples and conclude that the two brands appear to be equally reliable during their warranty period. ■

We have been studying the classical Neyman–Pearson system for testing hypotheses. Within this system, the basic philosophy is to consider all possible tests of the same size α, where α is to be chosen in advance; the test procedure used then is the one that minimizes the probability of type II error, or equivalently, which has maximum power, granted one can find such a test. The sample sizes involved are fixed and the test leads to either of two decisions: Accept H_0 or reject H_0.

Many statistical practitioners employ a procedure that is a slight variation on this approach, frequently called the *tests of significance* methodology (the Neyman–Pearson approach is called the methodology of *tests of hypotheses*). Let us discuss some numerical values to illustrate the tests of significance procedure. Suppose a random sample of size $n = 20$ of a normal random variable is selected and we want to test $H_0: \mu \leq 2$ versus $H_1: \mu > 2$, using the T-test of Theorem 8.3.1 with $\alpha = .05$. The rule there says then that we should compute the observed t value and reject H_0 if $t \geq 1.729$, the 95th quantile of the T-distribution with 19 degrees of freedom. Thus, if the observed t value were, say, 1.730 or 2.623 or 100 we should reject H_0, whereas if the observed t value were 1.728 or 1.7284, we should accept H_0, employing the rule as stated. The Neyman–Pearson system results in a go–no go decision, without regard to how close to rejecting (but a little short) the outcome was or by how much the critical value was exceeded. Proponents of the tests of significance approach feel that this go–no go discrete approach is wasteful of information in not giving a more continuous indication of how well the observed sample values agreed or disagreed with the stated hypothesis.

The tests of significance approach to testing a hypothesis, say, $H_0: \mu \leq 2$ versus $H_1: \mu > 2$ as before, does not set the value for α, the probability of type I error, in advance. In general, proponents will use the same test statistic (the observed t in this case) and then compute the probability of getting an observed test statistic this extreme (or more extreme) if H_0 is true; that is, for the previous case, they will observe t and then compute $\alpha' = P(T \geq t)$, the area under the T density with 19 degrees of freedom, as pictured in Figure 8.3.1. If α' is "sufficiently small," they will then proceed as if H_0 is false, otherwise, they proceed as if H_0 is true. For example, if with 19 degrees of freedom we observe $t = 1.623$ in testing $H_0: \mu \leq 2$, we find $\alpha' = .061$; this says that the probability is (no greater than) .061 of observing a T value this extreme if $H_0: \mu \leq 2$ is true. Whether H_0 is then accepted or rejected depends on the practitioner; in some applications he might reject H_0 with this α' and in others he might not.

In testing two-sided alternatives, the observed tail area must be doubled, because in that case either observed test statistics that are too large or too small would be inconsistent with H_0. For example, assume again a random sample of $n = 20$ of a normal random variable, to be used to test $H_0: \mu = 2$ versus $H_1: \mu \neq 2$. If the observed T value is $t = 1.623$, now we compute

$$\alpha' = 2P(T \geq 1.623) = .122$$

as the probability of getting an observed value this extreme (or more so); this is done because an observed $t = -1.623$ or smaller is just as extreme, if $\mu = 2$, as is the assumed $t = 1.623$ or larger.

It is, of course, easy for the tests of significance advocate to see what the tests of hypothesis advocate would decide with any fixed α. If $\alpha' \leq \alpha$, the tests of hypothesis advocate would reject H_0 and if $\alpha' > \alpha$, he or she would not. It is not possible to go the other way, though, because of the subjective element used by the tests of significance advocate. That is, knowing that H_0 was rejected with $\alpha = .10$, say, does not reveal the value for α' (except we know $\alpha' \leq .10$) and the α' user might or might not have rejected H_0 with the same observed data.

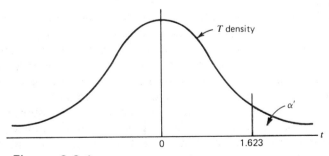

Figure 8.3.1

EXERCISE 8.3

1. To compare gas mileage capabilities of two new cars (comparable models) $n = 7$ cars of make D were driven by the same driver over the same course. The observed mileages were 22.8, 26.0, 25.6, 24.0, 25.3, 23.8, 24.5; $m = 11$ cars of make T were handled in the same way resulting in observed mileages of 19.7, 40.9, 17.2, 25.7, 40.0, 18.1, 24.5, 16.9, 26.8, 26.8, 42.8. Would you accept $H_0: \mu_D \geq \mu_T$ versus $H_1: \mu_D < \mu_T$, making the assumptions of Theorem 8.3.1?

2. Using the data of Exercise 8.3.1, would you accept the hypothesis that the variances in mileage are the same for these two makes of cars, with $\alpha = .02$? Does this result affect your conclusion in Exercise 8.3.1?

3. Ten randomly selected recent graduates of University C were selected and given an IQ test. Their scores were 120, 101, 87, 120, 107, 110, 118, 119, 112, 104. Ten recent graduates of University P were also given the same IQ test; their scores were 130, 133, 119, 123, 125, 124, 133, 120, 126, 126. Would you accept the hypothesis that the average IQ score is the same for graduates of these two universities, with $\alpha = .1$? Make the assumptions of Theorem 8.3.1.

4. Use the data of Exercise 8.3.3 to test the hypothesis that the variances in IQ scores for graduates of these two universities are equal, with $\alpha = .10$. Does the result of this test change your conclusion in Exercise 8.3.3?

5. It is assumed the number of days between earthquakes of magnitude 4.0 or more is an exponential random variable with parameter λ. On fault A the observed numbers of days between quakes of this magnitude, were 2.036, .753, .048, 5.816, 6.067, 1.449, 1.448, 1.604, respectively, for the most recent 9 earthquakes. On fault P the observed numbers of days between quakes of this magnitude were 1.972, 4.054, 2.801, 2.227, 3.826, 2.984, 1.193, 1.996, 0.982, 2.325, 3.404, respectively, for the most recent 12 earthquakes. Would you accept the hypothesis that earthquakes of this magnitude are occurring at the same rate for both faults?

6. Of 215 teenage girls admitted to a New York hospital during one month, it was found that 17 were unknowingly pregnant. Of 208 teenage girls admitted to a southern California hospital during the same month, it was found that 19 were unknowingly pregnant. Does it appear that the rate of unknowing pregnancies among teenage girls is the same in the two areas served by these hospitals?

*7. A large corporation operates two factories. It is assumed that the number of personnel accidents at each, per year, is a Poisson random variable. If there were 47 such accidents at one of the plants (in a year) and 68 accidents at the other plant (in the same year), does it appear that the two plants have the same expected number of accidents per year? (*Hint.* Develop a test based on approximate normality.)

8. The ease (or difficulty) with which one loses weight may very well be heavily dependent on the person's genetic background. To compare two different weight-loss diets, and to control the genetic contribution to how well one diet appeared versus the other, 12 identical twins were located; each person in each pair was overweight by roughly the same amount. Within each pair of twins, one of the two was selected at random and placed on diet P for 3 months; the other was placed on diet S for 3 months. At the end of this time the weight lost (in kilograms) was recorded. The results follow.

Twin Number

Diet	1	2	3	4	5	6	7	8	9	10	11	12
P	12.4	10.3	6.8	11.5	10.4	9.8	5.7	9.5	9.8	8.0	7.1	10.9
S	12.8	10.0	8.7	11.9	10.6	9.7	7.9	10.8	11.6	8.8	9.0	11.1

Use two different T statistics to test $H_0: \mu_P = \mu_S$ versus $H_1: \mu_P \neq \mu_S$, where μ_P, μ_S are the expected weight losses for the two diets.

9. Of 200 families watching television at a given time in New York, it was found that 45 were watching network A; of 110 families watching television at the same time in New Jersey, it was found that 32 were watching network A. Assuming these are the results of random samples, would you accept the hypothesis that network A is equally popular in both states (at this time)?

10. Evaluate the power of the test in Exercise 8.3.2 if $10\sigma_D^2 = \sigma_T^2$.

11. Using $\alpha = .1$ in Exercise 8.3.5, evaluate the power of your test if $\lambda_A = 2\lambda_P$.

*12. Use the data in Exercise 8.3.3 to test $H_0: \mu_P = \mu_C + 20$ versus $H_1: \mu_P \neq \mu_C + 20$ where μ_P, μ_C are the expected IQ scores for graduates of the two universities.

*13. Use the data in Exercise 8.3.1 to test $H_0: \sigma_T^2 = 2\sigma_D^2$.

8.4 Summary

Hypothesis: Statement about a probability law.

Simple hypothesis: Statement that uniquely identifies a probability law.

Composite hypothesis: One that is not simple.

Test of a hypothesis: Rule for deciding whether to accept or reject a hypothesis.

Critical region of test: Collection of possible observations that lead to rejection of the hypothesis.

Rejection region of test: Same as critical region.

Acceptance region of test: Complement of critical region.

Type I error: Rejecting a true hypothesis.

Type II error: Accepting a false hypothesis.

Best Neyman–Pearson test: Rule that minimizes $\beta = P(\text{type II error})$ for any fixed $\alpha = P(\text{type I error})$, when testing simple H_0 versus simple H_1 (see Theorem 8.1.1).

Equivalence of confidence intervals and tests: The values covered by the confidence interval are those such that H_0 is accepted.

Parameter space Ω: Collection of all possible values of the parameters of a probability law.

Operating characteristic function of a test: $C(\theta) = P(\text{accept } H_0 \mid \text{value for the parameter})$.

Power function of a test: $Q(\theta) = P(\text{reject } H_0 \mid \text{value for the parameter})$.

Size of a test: Maximum probability of rejecting H_0, assuming H_0 is true.

Uniformly most powerful test: One that maximizes the power function among all tests of the same size.

Generalized likelihood ratio test criterion: Test based on the value of $l = L_X(\hat{\omega})/L_X(\hat{\theta})$, where $L_X(\hat{\omega})$, $L_X(\hat{\theta})$ are the maximum values of the likelihood function, assuming H_0 is true, ignoring H_0, respectively.

Standard tests:

   Mean of normal, σ^2 unknown, Theorem 8.2.1

   Variance of normal, Theorem 8.2.2

   Equality of two normal means, Theorem 8.3.1

   Equality of two normal variances, Theorem 8.3.2

   Paired T test, Theorem 8.3.3

   Equality of two exponential parameters, Theorem 8.3.4.

Significance tests: Procedure in which the size is not fixed in advance; acceptance or rejection is based on the probability of observing sample as extreme, or more so, if the hypothesis is true.

9

Least Squares and Regression

We have studied the method of moments and the maximum likelihood method for estimating unknown parameters of probability laws. In this chapter we will study a third procedure for estimating parameters, called least squares estimation. This procedure is especially applicable to models that involve the values for two or more variables; least squares is employed to estimate unknown parameters in the assumed relationships between the variables. As we will see, the least squares estimators have surprisingly strong properties, based on a small number of assumptions.

9.1 Least Squares Estimation

The entrance exam scores (x) and graduating grade point averges (y) for 10 recent graduates of University A are given in Table 9.1.1

Table 9.1.1

Student number	1	2	3	4	5
Entrance exam x	355	448	351	528	434
Graduating grade point y	2.72	2.71	2.70	3.73	3.09
Student number	6	7	8	9	10
Entrance exam x	471	300	422	421	490
Graduating grade point y	3.20	2.00	2.98	2.69	2.94

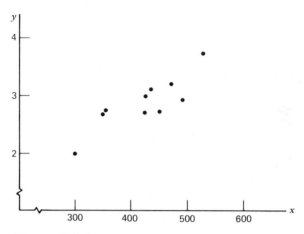

Figure 9.1.1

One might expect that those students with the higher entrance exam scores (*x*) would also have higher graduating grade points (*y*). To see how true this may be, let us plot the points from Table 9.1.1 in the (*x*, *y*) plane, as in Figure 9.1.1. Thus we have one point representing each student and can see that there is a general tendency for higher *y* values to be associated with higher *x* values. In fact, there appears to be an approximate straight line relationship between *y* and *x*, where "approximate straight line" is certainly a general, ill-defined term. Let us use these data to discuss a particular class of models that is used frequently in many areas.

For any given entrance exam score *x* it is easy to imagine a population or distribution of graduating grade points *y*. That is, we could in effect assume an essentially unlimited number of persons, each of whom makes the same score *x* on the entrance exam, and each of whom goes on to graduate. Their graduating grade points form the population of *y* values for this given entrance score *x*. If the plot in Figure 9.1.1 is representative, it would seem reasonable that the population of *y* values is shifting higher in some sense as *x* increases. The simplest possible way in which we could model this shift in the populations as *x* increases is to assume in particular that the mean or expected value of the distribution increases with *x*. Thus if Y_x is the population random variable, for those scoring *x* on the entrance exam, we might assume $E[Y_x] = g(x)$, where $g(x)$ is an increasing function. If, as suggested by Figure 9.1.1, we wanted to assume a linear increase, then we could take $g(x) = a + bx$, a general linear function, so we could assume $E[Y_x] = a + bx$, where the two constants *a* and *b* are, of course, unknown. For this simplest case, let us also assume the variance of the population of *y* values, for a given *x*, does not depend on *x*. Thus we also assume $V[Y_x] = \sigma^2$ is a constant for all values of x (the variance of graduating grade point averages among those scoring $x = 300$ is the same as for those scoring $x = 500$ or any other possible value).

Now suppose we pick certain x values, say, $x_1, x_2, \ldots, x_n$, and for each of these we select one (or more) value(s) at random from the corresponding population of y values; we will denote these observed values by $y_1, y_2, \ldots, y_n$. (Our notation is being slightly abused here; actually, y_i is the observed value for Y_{x_i}, but subscripting the y values by x_i leads to a very cumbersome system so we will simply use y_i as the value observed from the population with entrance exam score x_i and will use Y_i to represent the population random variable.) Assuming we select values at random from the populations of y values, it follows that $Y_1, Y_2, \ldots, Y_n$ are independent random variables, and $E[Y_i] = a + bx_i$, $i = 1, 2, \ldots, n$. We certainly would not expect the observed value selected from the population of y values with entrance exam score x_i to be equal to $a + bx_i$, the mean of that population. It is convenient to let e_i represent the deviation between the value selected, Y_i, and the population mean $a + bx_i$, that is,

$$e_i = Y_i - (a + bx_i)$$

from which we can write

$$Y_i = a + bx_i + e_i, \qquad i = 1, 2, \ldots, n.$$

These quantities $e_1, e_2, \ldots, e_n$ are sometimes called measurement errors or errors of observation; they are unobservable random variables, denoted by lowercase letters to be consistent with the notation that is historically used for this model. Note then that

$$E[e_i] = E[Y_i - (a + bx_i)] = 0$$

and

$$\text{Var}[e_i] = \text{Var}[Y_i] = \sigma^2 \qquad \text{for all} \qquad i.$$

These e_i's account for the fact that the observed values will not lie on a perfect straight line when plotted as in Figure 9.1.1, although the population mean values are assumed to fall on a straight line. The model we have just described is called a *simple linear regression* model. To summarize,

1.   We have a population of y values for each x; the population ran-random variable corresponding to x_i is Y_i.
2. $E[Y_i] = a + bx_i$ for each x_i.
3. $\text{Var}[Y_i] = \sigma^2$ for each x_i.
4.   The errors of observation, $e_i = Y_i - a - bx_i$ are uncorrelated (as

we have described the model, they are actually independent, but we do not really need such a strong assumption).

This simple linear regression model has three unknown parameters, a, b, and σ^2. By far the most frequently used method for estimating the unknown parameters is called *least squares*. This procedure is described in the following definition.

DEFINITION 9.1.1. Let $Y_1, Y_2, ..., Y_n$ be uncorrelated, $E[Y_i] = g(x_i)$, $\text{Var}[Y_i] = \sigma^2$, $i = 1, 2, ..., n$, where the x_i's are constants and the function $g(x_i)$ involves unknown parameters. The least squares estimators of the unknown parameters in $g(x)$ are those values that minimize

$$Q = \sum_{i=1}^{n} (Y_i - g(x_i))^2 = \sum_{i=1}^{n} e_i^2.$$

The estimator for σ^2 is

$$S^2 = k \sum (Y_i - \hat{g}(x_i))^2,$$

where $\hat{g}(x_i)$ is the least squares estimator for $E[Y_i] = g(x_i)$ and k is chosen to make S^2 unbiased. ∎

In Figure 9.1.2 the 10 entrance exam-graduating grade point values are plotted again. The least squares line is the one that minimizes the sum of squares of vertical deviations (dashed lines) between the observed points and itself, as pictured. This estimated line then is an estimate of $a + bx$, the true underlying straight line on which we assume the population means lie.

It is quite straightforward to find the values that minimize

$$Q = \sum (y_i - g(x_i))^2 = \sum (y_i - a - bx_i)^2$$

for our straight line case $(g(x_i) = a + bx_i)$. Since Q is a quadratic function of a and b, it is a differentiable function of both and the minimizing values are those that satisfy $\partial Q/\partial a = \partial Q/\partial b = 0$. Thus we have

$$\frac{\partial Q}{\partial a} = -2\sum (y_i - a - bx_i) = -2\sum y_i + 2na + 2b\sum x_i$$

$$\frac{\partial Q}{\partial b} = -2\sum x_i(y_i - a - bx_i) = -2\sum x_iy_i + 2a\sum x_i + 2b\sum x_i^2;$$

to distinguish the values that make these partial derivatives zero, we will put carats over a and b. Thus the least squares estimates for a and b must

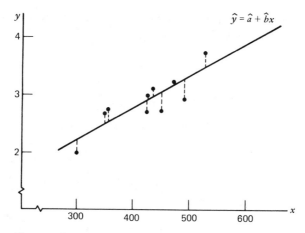

Figure 9.1.2

satisfy

$$-2 \sum y_i + 2n\hat{a} + 2\hat{b} \sum x_i = 0$$
$$-2 \sum x_i y_i + 2\hat{a} \sum x_i + 2\hat{b} \sum x_i^2 = 0,$$

which is equivalent to

$$n\hat{a} + \hat{b} \sum x_i = \sum y_i$$
$$\hat{a} \sum x_i + \hat{b} \sum x_i^2 = \sum x_i y_i.$$

These latter equations are frequently called the *normal* equations, the equations whose solutions determine the least squares estimates. From the first normal equation we have

$$\hat{a} = \bar{y} - \hat{b}\bar{x},$$

where

$$\bar{y} = \frac{1}{n} \sum y_i, \qquad \bar{x} = \frac{1}{n} \sum x_i$$

and substituting this in the second gives

$$(\bar{y} - \hat{b}\bar{x}) \sum x_i + \hat{b} \sum x_i^2 = \sum x_i y_i$$

from which we find

$$\hat{b} = \frac{\sum x_i y_i - \bar{y} \sum x_i}{\sum x_i^2 - \bar{x} \sum x_i}.$$

There are many equivalent ways to write this equation for $\hat{b}$, making use of various computational formulas. As we know,

$$\sum (x_i - \bar{x})^2 = \sum x_i^2 - \frac{(\sum x_i)^2}{n}$$

$$= \sum x_i^2 - \bar{x} \sum x_i$$

so the denominator can be written $\sum (x_i - \bar{x})^2$. Similarly,

$$\sum (x_i - \bar{x})(y_i - \bar{y}) = \sum (x_i y_i - x_i \bar{y} - \bar{x} y_i + \bar{x}\bar{y})$$

$$= \sum x_i y_i - \bar{y} \sum x_i - \bar{x} \sum y_i + n\bar{x}\bar{y})$$

$$= \sum x_i y_i - \bar{y} \sum x_i$$

$$= \sum (x_i - \bar{x}) y_i$$

since $n\bar{y} = \sum y_i$. Thus the least squares estimate for b can be written

$$\hat{b} = \frac{\sum (x_i - \bar{x})(y_i - \bar{y})}{\sum (x_i - \bar{x})^2},$$

which is the formula most frequently written for $\hat{b}$. Actually, in terms of computational accuracy it is always best to divide last, so the ratio of

$$\sum x_i y_i - \frac{(\sum x_i)(\sum y_i)}{n}$$

to

$$\sum x_i^2 - \frac{(\sum x_i)^2}{n}$$

provides the most accuracy. As earlier, the estimators are the same functions of the sample random variables and are represented by capital letters; thus the least squares estimators are $\hat{A} = \bar{Y} - \hat{B}\bar{x}$,

$$\hat{B} = \frac{\sum (x_i - \bar{x})(Y_i - \bar{Y})}{\sum (x_i - \bar{x})^2},$$

and the estimator for the line is $\hat{Y} = \hat{A} + \hat{B}x$. The unbiased estimator for σ^2 is

$$S^2 = \frac{1}{n-2} \sum (Y_i - \hat{Y}_i)^2$$

$$= \frac{1}{n-2} \sum (Y_i - \hat{A} - \hat{B}x_i)^2$$

$$= \frac{1}{n-2} \left[\sum (Y_i - \bar{Y})^2 - \hat{B}^2 \sum (x_i - \bar{x})^2 \right]$$

as you are asked to verify in the subsequent exercises. This establishes the following theorem.

Theorem 9.1.1. Let $Y_1, Y_2, \ldots, Y_n$ be uncorrelated, $E[Y_i] = a + bx_i$, $\text{Var}[Y_i] = \sigma^2$, where $x_1, x_2, \ldots, x_n$ are constants. The least squares estimators for a and b are

$$\hat{A} = \bar{Y} - \hat{B}\bar{x},$$

$$\hat{B} = \frac{\sum (x_i - \bar{x})(Y_i - \bar{Y})}{\sum (x_i - \bar{x})^2}$$

and the unbiased estimator for σ^2 is

$$S^2 = \frac{1}{n-2}\left[\sum (Y_i - \bar{Y})^2 - \hat{B}^2 \sum (x_i - \bar{x})^2\right]. \qquad \blacksquare$$

If $Y_1, Y_2, \ldots, Y_n$ are independent random variables with the same mean μ and variance σ^2, we know that $\bar{Y}$ is a good estimate for μ and $\sum (Y_i - \bar{Y})^2/(n-1)$ is unbiased for σ^2. The divisor is $n-1$ here because only one quantity ($\bar{y}$) is needed to estimate the mean for each y_i. The divisor used in defining S^2 in Theorem 9.1.1 is $n-2$ because in our simple linear regression model two quantities ($\hat{A}$ and $\hat{B}$) are needed to estimate the means for the Y_i's. In a sense one degree of freedom is used up in estimating each of a and b, leaving $n-2$ for estimating σ^2.

EXAMPLE 9.1.1.   Let us apply the foregoing results to the entrance exam-graduating grade point data of Table 9.1.1. We find $\sum x_i = 4220$, $\sum x_i^2 = 1824336$, $\sum y_i = 28.76$, $\sum y_i^2 = 84.4936$, $\sum x_i y_i = 12375.73$; these are the basic statistics needed to evaluate $\hat{a}$, $\hat{b}$, and s^2. Some of the modern hand-held calculators will provide all these values after all 10 pairs (x_i, y_i) are entered in the calculator once. Indeed, most of the same calculators have built-in algorithms also to evaluate $\hat{a}$ and $\hat{b}$; some will compute s^2 automatically (but check the formula used for s^2; there are occasional differences in the denominator used). We find, then, $\bar{x} = 422$, $\bar{y} = 2.876$,

$$\sum (x_i - \bar{x})^2 = 1{,}824{,}336 - \frac{(4220)^2}{10} = 43{,}496$$

$$\sum (x_i - \bar{x})(y_i - \bar{y}) = 12{,}375.73 - \frac{(4220)(28.76)}{10} = 239.01$$

$$\sum (y_i - \bar{y})^2 = 84.4936 - \frac{(28.76)^2}{10} = 1.77984$$

so

$$\hat{b} = \frac{239.01}{43,496} = .00549,$$

$$\hat{a} = 2.876 - (.00549)\,422 = .559$$

(if you use full 10-digit decimal accuracy in $\hat{b}$, you will get $\hat{a} = .557$ rounded to 3 decimals). Thus the estimated least squares line, from these data, is $\hat{y} = .559 + .00549x$. On the average, an increase of 10 points on the entrance exam score adds about .05 to the graduating grade point average. This estimated line is plotted in Figure 9.1.2. The unbiased estimate for σ^2 is

$$s^2 = \tfrac{1}{8}(1.77984 - (.00549)^2\,(43496)) = .0586$$

(if $\hat{b}$ is carried to 10 decimals you will get .0583 for this value), and the estimate for σ is $s = \sqrt{.0586} = .242$. Many of the formulas we will discuss for least squares are quite sensitive to rounding errors. A good rule is to carry as many decimals as possible and do no rounding until the final answer. Thus we will use $\hat{b} = .00549$, $\hat{a} = .557$, $s^2 = .0583$ in later discussion of this example. ∎

The least squares estimators derive their name from the fact that they minimize a sum of squares. Let us investigate some of their properties. First, note that $\hat{A}$ and $\hat{B}$ are linear functions of $Y_1, Y_2, ..., Y_n$. In fact, we can write

$$\hat{B} = \frac{\sum (x_i - \bar{x})\,Y_i}{\sum (x_j - \bar{x})^2} = \sum c_i Y_i$$

$$\hat{A} = \bar{Y} - \hat{B}\bar{x} = \frac{1}{n}\sum Y_i - \bar{x}\sum c_i Y_i$$

$$= \sum d_i Y_i,$$

where

$$c_i = \frac{x_i - \bar{x}}{\sum (x_j - \bar{x})^2}, \qquad d_i = \frac{1}{n} - \bar{x}c_i.$$

Then from Corollary 5.4.1 we know that

$$E[\hat{B}] = \sum c_i(a + bx_i)$$
$$= a\sum c_i + b\sum c_i x_i = b,$$

since

$$\sum c_i = 0, \qquad \sum c_i x_i = \frac{\sum (x_i - \bar{x}) x_i}{\sum (x_j - \bar{x})^2} = \frac{\sum (x_i - \bar{x})^2}{\sum (x_j - \bar{x})^2} = 1,$$

$$E[\hat{A}] = \sum d_i(a + bx_i)$$
$$= a \sum d_i + b \sum d_i x_i = a,$$

since

$$\sum d_i = \sum \left(\frac{1}{n} - \bar{x}c_i \right) = 1 - \bar{x} \sum c_i = 1,$$

$$\sum d_i x_i = \sum x_i \left(\frac{1}{n} - \bar{x}c_i \right) = \frac{1}{n} \sum x_i - \bar{x} \sum c_i x_i = \bar{x} - \bar{x} = 0.$$

Thus $\hat{A}$ and $\hat{B}$ are unbiased estimators for a and b, respectively. Since we are assuming the Y_i's are uncorrelated, $\text{Var}[Y_i] = \sigma^2$ for all i, we also know from Corollary 5.4.1 that

$$\text{Var}[\hat{B}] = \sigma^2 \sum c_i^2 = \sigma^2 \frac{\sum (x_i - \bar{x})^2}{[\sum (x_j - \bar{x})^2]^2}$$

$$= \frac{\sigma^2}{\sum (x_j - \bar{x})^2}$$

$$\text{Var}[\hat{A}] = \sigma^2 \sum d_i^2$$

$$= \frac{\sigma^2 \sum x_i^2}{n \sum (x_j - \bar{x})^2}$$

and from Theorem 5.4.2

$$\text{Cov}[\hat{A}, \hat{B}] = \sigma^2 \sum c_i d_i = \frac{-\sigma^2 \bar{x}}{\sum (x_i - \bar{x})^2};$$

you are asked in the exercises to verify the algebra in these last two expressions. If $\bar{x} > 0$, the covariance (and correlation) between $\hat{A}$ and $\hat{B}$ is negative; this means that $\hat{A}$ tends to decrease as $\hat{B}$ increases and vice versa. It is easy to see the reason for this. With $x = \bar{x}$, the value on the least squares line is

$$\hat{Y} = \hat{A} + \hat{B}\bar{x} = (\bar{Y} - \hat{B}\bar{x}) + \hat{B}\bar{x} = \bar{Y};$$

that is, this line always passes through the point whose coordinates are $\bar{x}, \bar{Y}$. Now $\hat{A}$ is the estimate of the y intercept and $\hat{B}$ is the estimate of the slope. Because the line must always pass through $(\bar{x}, \bar{Y})$, the fitted least squares line pivots on this point in a sense; as long as $\bar{x} > 0$, increasing the slope must decrease the y intercept and vice versa, which leads to the negative covariance between $\hat{A}$ and $\hat{B}$. Note as well that $\hat{A}$ and $\hat{B}$ are uncorrelated if and only if $\bar{x} = 0$. This is summarized in the following theorem.

Theorem 9.1.2. If $\hat{A}$, $\hat{B}$ are the least square estimators for the y intercept and slope, respectively, in a simple linear regression model, then both $\hat{A}$ and $\hat{B}$ are unbiased; their variances and covariances are

$$\text{Var}[\hat{A}] = \frac{\sigma^2 \sum x_i^2}{n \sum (x_j - \bar{x})^2}$$

$$\text{Var}[\hat{B}] = \frac{\sigma^2}{\sum (x_j - \bar{x})^2}$$

$$\text{Cov}[\hat{A}, \hat{B}] = \frac{-\sigma^2 \bar{x}}{\sum (x_j - \bar{x})^2} .$$ ∎

EXAMPLE 9.1.2. For the entrance exam-graduating grade point data discussed in Example 9.1.1, we assumed the variance of the y values for any given x is the (unknown) constant σ^2, the standard simple linear regression assumption. Then, since $\sum x_i^2 = 1,824,336$, $\sum (x_i - \bar{x})^2 = 43,496$, $\bar{x} = 422$, Theorem 9.1.2 shows that

$$\text{Var}[\hat{A}] = \frac{\sigma^2 (1824336)}{10(43496)} = 4.194\sigma^2$$

$$\text{Var}[\hat{B}) = \frac{\sigma^2}{43,496} = .0000230\sigma^2$$

$$\text{Cov}[\hat{A}, \hat{B}] = \frac{-\sigma^2 (422)}{43,496} = -.00970\sigma^2;$$

that is, over repeated samples of the same size (with the same x_i values), the variances and covariances of the two estimators are the preceding multiples of σ^2, the unknown variance of the Y values for any given x. Because the same data can be used to estimate σ^2 as well, we can also estimate the variances and covariance of these two estimators. We will let $s_{\hat{A}}^2$ and $s_{\hat{B}}^2$ be the estimated variances of these two estimators; since $s^2 =$

.0583 estimates σ^2 from Example 9.1.1, we have

$$s_{\hat{A}}^2 = 4.194(.0583) = .245$$
$$s_{\hat{B}}^2 = .0000230(.0583) = .00000134.$$

The estimated covariance between $\hat{A}$ and $\hat{B}$ is $-.00970(.0583) = -.000566$. These estimated variances for $\hat{A}$ and $\hat{B}$ (and their square roots, the estimated standard deviations) play an important role in problems of statistical inference, as we will see. ∎

A major reason that least squares is so frequently employed in estimating unknown parameters is that the variances of the estimators, given in Theorem 9.1.2, are within the framework of the assumed model the smallest possible among all linear unbiased estimators; for this reason they are called *best linear unbiased*, a term used earlier in Exercise 7.2.15. The proof of this fact is straightforward, but somewhat tedious; let us simply sketch one way of establishing this property for $\hat{B}$ (it also holds for $\hat{A}$). Recall we can write $\hat{B} = \sum c_i Y_i$, where

$$c_i = \frac{x_i - \bar{x}}{\sum (x_j - \bar{x})^2} \quad \text{and} \quad \sum c_i = 0, \quad \sum c_i x_i = 1,$$

so as we saw earlier, $\hat{B}$ is unbiased. A general linear function of the Y_i's can be written $k + \sum (c_i - h_i) Y_i$, where k and the h_i's are totally arbitrary constants (the c_i's are as previously defined). The expected value for this linear function is (with the assumptions of our simple linear regression model)

$$k + \sum (c_i - h_i)(a + bx_i)$$
$$= k + a \sum c_i + b \sum c_i x_i - a \sum h_i - b \sum h_i x_i$$
$$= k + b - a \sum h_i - b \sum h_i x_i;$$

if this is to equal b, no matter what the unknown values for a and b may be (so that it is unbiased for b), it follows that $k = 0, \sum h_i = 0, \sum h_i x_i = 0$. Thus if the linear function is to be unbiased for b, it suffices to examine possible values for the h_i's in $\sum (c_i - h_i) Y_i$, such that $\sum h_i = 0, \sum h_i x_i = 0$. Recall that

$$c_i = \frac{x_i - \bar{x}}{\sum (x_j - \bar{x})^2}$$

and thus

$$\sum c_i h_i = \frac{\sum x_i h_i}{\sum (x_j - \bar{x})^2} - \bar{x} \frac{\sum h_i}{\sum (x_j - \bar{x})^2} = 0$$

because of these two unbiased requirements. But then the variance of

$\sum (c_i - h_i) Y_i$ is

$$\sigma^2 \sum (c_i - h_i)^2 = \sigma^2 \sum (c_i^2 - 2c_i h_i + h_i^2)$$
$$= \sigma^2 \sum (c_i^2 + h_i^2);$$

since the c_i's are fixed, this is clearly minimized by $h_1 = h_2 = \cdots = h_n = 0$. But this says the linear unbiased estimator of b with the smallest variance is

$$\sum (c_i - 0) Y_i = \sum c_i Y_i = \hat{B},$$

our least squares estimator. This same line of reasoning can be used, essentially unchanged, to show that $\hat{A}$ is also the best linear unbiased estimator for a. In fact, if $E[Y_i] = g(x_i)$, and the unknown parameters enter $g(\ )$ linearly, it can be shown that the least squares estimates for these parameters are best linear unbiased, a surprisingly strong statement from so few assumptions. This result is called the Gauss–Markov Theorem and is stated as:

Theorem 9.1.3. Assume $Y_1, Y_2, \ldots, Y_n$ are uncorrelated, $E[Y_i] = g(x_i)$, $\text{Var}[Y_i] = \sigma^2$ for all i, where $g(x_i)$ is a linear function of $k \leq n$ parameters. Then each of the least squares estimators of the unknown parameters in $g(x_i)$ is best linear unbiased. ∎

In the case we have been considering $g(x_i) = a + bx_i$ is a linear function of two unknown parameters and $\hat{A}$ and $\hat{B}$ are best linear unbiased for a and b. The result in Theorem 9.1.3 applies to more general cases as well. Suppose we had assumed $g(x_i) = a + bx_i + cx_i^2$ or $g(x_i) = a + bx_i + cx_i^2 + dx_i^3$, each of which is linear in unknown parameters (a, b, c, in one case, a, b, c, d, in the other). In each of these two cases the least squares estimators would again be best linear unbiased. Be aware, though, that the least squares estimators themselves can change as one assumes different models, a fact that our notation may partially conceal. Suppose for our entrance exam score-graduating grade point data we were interested in using the assumption

$$E[Y_i] = a + bx_i + cx_i^2,$$

say. Then the least squares estimators for a, b, and c are those values that minimize

$$Q(a, b, c) = \sum (y_i - a - bx_i - cx_i^2)^2;$$

to determine the estimates $\hat{a}, \hat{b}, \hat{c}$, we would solve three equations (given by $\partial Q/\partial a = \partial Q/\partial b = \partial Q/\partial c = 0$), simultaneously. The values for $\hat{a}$ and $\hat{b}$ that result are in general different from the values we have derived that came from minimizing

$$Q(a, b) = \sum (y_i - a - bx_i)^2.$$

The reason, of course, is that if a_0 and b_0, say, minimize $Q(a, b)$, it does not follow that we will find the minimum value of $Q(a, b, c)$ by restricting ourselves to having $a = a_0$ and $b = b_0$ and allowing only the value for c to be determined. Looking at it another way, consider the two assumptions.

(i) $E[Y_i] = a + bx_i$
(ii) $E[Y_i] = a + bx_i + cx_i^2$

Even though we have used a and b in both statements, they do not really represent the same concept in both; in (i) we consider all possible straight lines and choose the one that minimizes $Q(a, b)$, whereas in (ii) we consider all possible parabolas and choose the one that minimizes $Q(a, b, c)$. The following example should help clarify the point.

EXAMPLE 9.1.3. Suppose we wanted to check a precision electronic scale designed to weigh objects whose true weight x does not exceed, say, 100 milligrams (about 3.5 ounces). Also, assume we have available $n = 10$ objects whose exact weights are 10, 20, 30, ..., 100 milligrams. We will place each of these objects on the scale once; define $x_i = 10i$, $i = 1, 2, ..., 10$, and let Y_i be the scale reading when the object with weight x_i is on the scale. If the scale were absolutely perfect we would find $Y_i = x_i$ for each i; the scale reading would agree perfectly with the known weight. With such small weights, though, and a scale that reads to, say, $\frac{1}{100}$th of a milligram, we might expect the scale reading actually to vary a little if we repeatedly weighed the same object, that is, we might assume a population of readings, any one of which might occur, when object x_i is on the scale. In short, let us assume the scale reading Y_i is a random variable and $\text{Var}[Y_i] = \sigma^2$ for all i. If the scale were accurate, then, it should be true that $E[Y_i] = x_i$, the population means should lie on a straight line through the origin. To allow for the possibility that the scale is not accurate, we will assume $E[Y_i] = bx_i$, where b is unknown and to be estimated. The least squares estimate is determined by minimizing

$$Q(b) = \sum (y_i - bx_i)^2.$$

We have

$$\frac{\partial Q}{\partial b} = -2 \sum x_i(y_i - bx_i)$$

$$= -2 \sum x_i y_i + 2b \sum x_i^2$$

and to make this quantity zero, we have

$$\hat{b} = \frac{\sum x_i y_i}{\sum x_i^2}.$$

Note that this does differ from the estimator of the slope of a straight line that is not forced through the origin. Because only one quantity, b, needs to be estimated to give estimates for the means of $Y_1, Y_2, ..., Y_n$, we lose only one degree of freedom and the unbiased estimator for σ^2 is

$$S^2 = \frac{1}{n-1} \sum (Y_i - \hat{B} x_i)^2$$

$$= \frac{1}{n-1} \left(\sum Y_i^2 - \hat{B}^2 \sum x_i^2 \right). \qquad \blacksquare$$

If $E[Y] = g(x)$ is a linear function of 3 or more unknown parameters, the least squares estimates of these parameters and of σ^2 can be very simply represented using matrix notation; without matrix notation, their representation becomes increasingly cumbersome. The basic idea, though, remains as we have described; the least squares estimators are the values that minimize the sum of squares of observation errors, $\sum e_i^2$. Because we have not assumed knowledge of matrix theory we will not discuss these models with 3 or more unknown parameters any further.

We have seen that the least squares estimators are best linear unbiased, within the framework of the assumptions; in particular, we assumed all the variances of the Y_i's to be equal. Suppose we were to assume $Y_1, Y_2, ..., Y_n$ are uncorrelated, $E[Y_i] = g(x_i)$, but now $\text{Var}[Y_i] = \sigma^2 c_i^2$, where the $c_1^2, c_2^2, ..., c_n^2$ values are known. What then would be reasonable estimators for the unknown parameters in $g(x)$? A very simple transformation allows us actually to use our previous results. If we divide Y_i by c_i, we have

$$\text{Var}\left[\frac{Y_i}{c_i} \right] = \frac{1}{c_i^2} \text{Var}[Y_i] = \frac{1}{c_i^2}(\sigma^2 c_i^2) = \sigma^2;$$

thus

$$E\left[\frac{Y_i}{c_i} \right] = \frac{g(x_i)}{c_i},$$

and since the Y_i/c_i's have constant variance, Theorem 9.1.3 assures us that the best linear unbiased estimates for the unknown parameters in $g(x_i)$ are given by least squares, that is, the values that minimize

$$\sum \left(\frac{y_i}{c_i} - \frac{g(x_i)}{c_i} \right)^2 = \sum \frac{(y_i - g(x_i))^2}{c_i^2}.$$

Thus if the y_i's have unequal variances, the best estimators come from minimizing a weighted sum of squares of the observation errors $e_i = Y_i - g(x_i)$. Those observation errors with the biggest variances (biggest c_i^2) then get the least weight, because they are divided by c_i^2 and those with the smallest variances get the largest weight, an intuitively reasonable result. This establishes the following corollary.

Corollary 9.1.3. Assume $Y_1, Y_2, \ldots, Y_n$ are uncorrelated, $E[Y_i] = g(x_i)$, $\text{Var}[Y_i] = \sigma^2 c_i^2$, where $g(x_i)$ is a linear function of $k \leq n$ parameters. Then the best linear unbiased estimators for these unknown parameters are the values that minimize

$$\sum_{i=1}^{n} \frac{(Y_i - g(x_i))^2}{c_i^2}.$$ ∎

EXAMPLE 9.1.4. Suppose the scale discussed in Example 9.1.3 is such that the variance of the reading increases with the weight of the object; thus the variance of the readings in repeatedly weighting, say, the 100 milligrams object is greater than the variance of the readings with the 10 milligrams object. More specifically, let us assume the variance increases linearly with x_i. Thus suppose again $E[Y_i] = bx_i$, where x_i is the true weight, the Y_i's are uncorrelated, but now $\text{Var}[Y_i] = \sigma^2 x_i$. What is the best linear unbiased estimate for b? According to the preceding corollary, we should find the value for b that minimizes

$$Q = \sum \frac{(y_i - bx_i)^2}{x_i}.$$

Thus

$$\frac{dQ}{db} = -2 \sum \frac{x_i(y_i - bx_i)}{x_i}$$
$$= -2 \sum y_i + 2b \sum x_i,$$

and we have

$$\hat{b} = \frac{\sum y_i}{\sum x_i} = \frac{\frac{1}{n}\sum y_i}{\frac{1}{n}\sum x_i} = \frac{\bar{y}}{\bar{x}},$$

the ratio of the two averages. The variance assumption has an effect on the best estimator for the unknown parameter in the mean. ∎

In the simple linear regression model the variable Y_i is called the *dependent* variable and x_i is called the *independent* variable. The dependent variable is a random variable, whereas the independent variable values are treated as given constants, in deriving the optimal properties of the least squares estimators in Theorem 9.1.3. What can be said about the situation in which both the dependent and independent variables are random variables? This case would occur, for example, if in our entrance exam score–graduating grade point example the 10 records used (listed in Table 9.1.1) had been selected at random from all the records available; this random selection from all the records available would make the x values that occur observed values of random variables. In this case we can reason conditionally, conditioning on the values that occur for the independent variable. The least squares estimators for a and b are best linear unbiased, given the values that occurred for the independent variable.

The material we have discussed in this section falls in the realm of what is frequently called curve fitting; given observed values (x_i, y_i), find the best fitting curve for the data. Our least squares approach is constructive, in a sense; to apply least squares we must first specify the functional form $g(x)$, which we want to use. Least squares then picks out the particular parameter values of the function $g(x)$ that best fit the data. Given a set of observed data, it is always wise to plot the observed points as in Figure 9.1.1; this can be very helpful in choosing the mathematical form of $g(x)$ to be used. If one has several observed y values for each x value, plotting the data can also be very helpful in determining reasonable assumptions about $\text{Var}[Y_i]$, which itself has an effect on the least squares estimates as we have seen. Once one has chosen a particular form $g(x)$, decided on the variance assumption, and used least squares, it is useful to examine the *residuals* about the fitted equation, the differences

$$\hat{e}_i = y_i - \hat{y}_i = y_i - \widehat{g(x_i)}.$$

These residuals should be patternless (if the right model has been chosen) in the sense that they are estimates of the e_i's that are assumed uncorrelated, mean 0, variance σ^2. In particular, plotting $\hat{e}_i$ versus x_i and $\hat{e}_i$ versus y_i should reveal no particular pattern; if there appears to be a pattern (such as $\hat{e}_i$ increases with x_i or y_i), one might want to change the assumed mathematical form for $g(x)$ or change the assumptions about $\text{Var}[Y_i]$ or both. Curve fitting is still (and always will be) an art form, calling for both imagination and knowledge in its practice.

EXERCISE 9.1

1. Use the grade point data of Table 9.1.1 to find the best fitting straight line that passes through the origin. Plot this line versus the observed data (and the line of Example 9.1.1). Which model bests fits the observed data?

2. Spectrometric oil analysis is used as a diagnostic tool in many industries. In a gasoline combustion engine the lubricating oil comes into contact with the moving engine parts. As wear of these parts occurs, traces of the metals from the parts may be detectable when the oil is analyzed on a spectrometer. The following data were gathered by sampling the engine oil of an automobile; the y values are the parts per million (ppm) of iron as read from the spectrometer and the x values are the number of miles driven since the oil was changed.

y	16	21	28	31	34	47	47	51	58	61
x	480	1005	1650	1989	2560	3023	3566	4011	4500	5002

Make the assumptions of the simple linear regression model and estimate a, b, and σ^2. Compute the residuals $\hat{e}_i$. Plot the data and the fitted line.

3. Reread Example 9.1.3. The following data were recorded using such a scale.

y	9.03	19.31	28.94	39.84	49.12	59.05	68.69	78.32	87.79	97.70
x	10	20	30	40	50	60	70	80	90	100

(a) Evaluate the estimate $\hat{b}$ for the line through the origin and compute s^2.

(b) Use these data to estimate a and b for the simple linear regression model and compute s^2.

*4. For the line through the origin model of Example 9.1.3, show that

$$E\left[\sum_{i=1}^{n} (Y_i - \hat{B}x_i)^2 \right] = (n-1)\sigma^2.$$

(Show $\sum (Y_i - \hat{B}x_i)^2 = \sum Y_i^2 - \hat{B}^2 \sum x_i^2$ and evaluate

$$E\left[\sum Y_i^2 \right] - \sum x_i^2 E\left[\hat{B}^2 \right].)$$

5. Use the data in Exercise 9.1.3 and make the assumptions of Example 9.1.4. Estimate this new line.

*6. Assume only two values are used for the independent variable: x_1 and

x_2. A sample of n_1 y values is taken with $x = x_1$ and a sample of n_2 y values is taken with $x = x_2$. With the simple linear regression model assumptions, show that the least squares line passes through the two points $(x_1, \bar{y}_1)(x_2, \bar{y}_2)$, where $\bar{y}_1$, $\bar{y}_2$ are the two average y values corresponding to x_1, x_2, respectively.

*7.   Verify that S^2 in Theorem 9.1.1 is an unbiased estimator for σ^2. [*Hint.* Write $(Y_i - \bar{Y})$ as $\sum f_j Y_j$, $\hat{B} = \sum c_j Y_j$, and use Theorem 5.4.2 and Corollary 5.4.1.]

*8.   Verify the values given for $\text{Var}[\hat{A}]$ and $\text{Cov}[\hat{A}, \hat{B}]$ given in Theorem 9.1.2.

9.   In describing the simple linear regression model, we have tacitly assumed there are at least two different values represented among $x_1, x_2, \ldots, x_n$, the values of the independent variable. What goes wrong if this is not true?

10.   In the linear regression through the origin, $E[Y_i] = bx_i$, find the best linear unbiased estimator for b if $\text{Var}[Y_i] = \sigma^2 x_i^2$, that is, the standard deviation of the Y_i's increases linearly with x_i.

11.   In the linear regression through the origin (Example 9.1.3), the estimated or fitted line is $\hat{y} = \hat{b}x$.
   (a) Does this line pass through the point with coordinates $(\bar{x}, \bar{y})$?
   (b) Is there a weighted regression approach such that this line through the origin goes through $(\bar{x}, \bar{y})$?

12.   In the simple linear regression model, the fitted line is $\hat{y} = \hat{a} + \hat{b}x$ and, for any specific x^*, $\hat{a} + \hat{b}x^*$ estimates the mean of the population of y values with $x = x^*$. What is the variance of this estimator? For which value of x^* is this variance the smallest?

13.   An onboard (electronic) aircraft navigation aid uses a known starting point, wind speed, direction and aircraft heading to compute the aircraft's location continuously during flight. To check the accuracy of this device an aircraft was flown 12 times. On each flight the aircraft left a known position and flew visually to another known location. When arriving at the known location, the navigation aid's location was read and used to compute a radial error, the straight line distance between where the aircraft was known to be and where the navigation aid said it was. In the following table $x =$ time of flight (in minutes) and $y =$ radial error (in meters).

x	10	11	33	59	57	38	9	37	47
y	696	1693	1123	9668	10333	2848	206	1472	14257

x	46	67	21
y	13133	16725	9300

(a) Assume the straight line through the origin model and compute $\hat{b}$, $s_{\hat{b}}$ (the estimated standard derivation for $\hat{b}$) and estimate the mean radial error for a $x = 60$-minute flight.

(b) Repeat (a) assuming $\text{Var}[Y_i] = \sigma^2 x_i$.

(c) Repeat (a) assuming $\text{Var}[Y_i] = \sigma^2 x_i^2$.

14. The number of errors in a binary bit $(0, 1)$ message of x characters is assumed to be (approximately) a Poisson random variable Y with mean $\mu = \lambda x$, where λ is unknown (and small). If n messages, of lengths $x_1, x_2, \ldots, x_n$, are examined and the number of errors, $Y_1, Y_2, \ldots, Y_n$, are counted, then

$$E[Y_i] = \lambda x_i, \qquad \text{Var}[Y_i] = \lambda x_i.$$

What is the best linear unbiased estimator for λ?

*15. Let $Y_1, Y_2, \ldots, Y_n$ and $x_1, x_2, \ldots, x_n$ be as defined in Exercise 9.1.14. Since

$$E\left[\frac{Y_i}{\sqrt{x_i}}\right] = \lambda \sqrt{x_i},$$

$$\Lambda = \sum \frac{Y_i}{\sqrt{x_i}} \bigg/ \sum \sqrt{x_i}$$

is also an unbiased estimator for λ. Show that $\text{Var}[\Lambda]$ is at least as large as $\text{Var}[\hat{\Lambda}]$ derived in Exercise 9.1.14.

*16. Show that $\text{Cov}[\hat{B}, Y_i - \hat{A} - \hat{B}x_i] = 0$ for each i. (*Hint.* $\text{Cov}[\hat{B}, Y_i - \hat{A} - \hat{B}x_i] = \text{Cov}[\hat{B}, Y_i] - \text{Cov}[\hat{A}, \hat{B}] - x_i \text{Var}[\hat{B}]$.)

9.2 Interval Estimation and Tests of Hypotheses

The least squares estimators are best linear unbiased, no matter what the probability law(s) for the Y_i's may be, as long as $E[Y_i] = g(x_i)$, $\text{Var}[Y_i] = \sigma^2$ and the Y_i's are uncorrelated. It is rather remarkable that such strong properties can be established with so few assumptions. Of course, if we want to compute confidence intervals or test hypotheses, it is necessary to be more specific about the probability law(s) for the Y_i values. Many different inference problems can be easily addressed if the Y_i values are all normal. We will add that assumption now and discuss a number of distributional results. Thus throughout this section we will now assume

1. $Y_1, Y_2, \ldots, Y_n$ are independent, normal.
2. $E[Y_i] = g(x_i)$, where the x_i's are known constants and $g(x)$ is a linear function of $k < n$ unknown parameters.
3. $\text{Var}[Y_i] = \sigma^2$ for all i.

With the assumption of normality, we can then investigate other methods of estimation, in particular, maximum likelihood. With $g(x_i) = a + bx_i$, the general straight-line form, the likelihood function for the sample is

$$L_Y(a, b, \sigma^2) = \prod \frac{1}{\sigma\sqrt{2\pi}} \exp\left[-\frac{(y_i - a - bx_i)^2}{2\sigma^2} \right]$$

$$= \left(\frac{1}{2\pi\sigma^2}\right)^{n/2} \exp\left[-\frac{\sum(y_i - a - bx_i)^2}{2\sigma^2} \right]$$

Note that the parameters a and b occur only in the exponent; furthermore, since the exponent has a negative sign, maximizing L_Y with respect to a and b is accomplished by minimizing $\sum(y_i - a - bx_i)^2/2\sigma^2$ with respect to a and b. But this is exactly (apart from the divisor $2\sigma^2$, which only affects the relative magnitude) the function we minimized in deriving the least squares estimators for a and b; thus the maximum likelihood estimates for a and b are identical with the least squares values $\hat{a}$ and $\hat{b}$, whose values we know from Section 9.1. Using these values, we easily see that $\partial L_Y/\partial\sigma^2 = 0$ with

$$\hat{\sigma}^2 = \frac{1}{n}\sum(y_i - \hat{a} - \hat{b}x_i)^2.$$

This establishes the following theorem.

Theorem 9.2.1. With assumptions (1), (2), (3), the maximum likelihood estimators for the unknown parameters in $g(x)$ are identical with the least squares estimators. The maximum likelihood estimate for σ^2 is

$$\hat{\sigma}^2 = \frac{1}{n}\sum(y_i - \hat{a} - \hat{b}x_i)^2. \qquad \blacksquare$$

The same estimators, then, occur whether least squares or maximum likelihood is employed as a rationale, making them doubly attractive. Because we have not changed any of the assumptions about $E[Y_i]$ or $Var[Y_i]$, in adding the assumption of normality, we also know immediately that $\hat{A}$, $\hat{B}$, and $S^2 = \sum(Y_i - \hat{A} - \hat{B}x_i)^2/(n - 2)$ are unbiased estimators for a, b, and σ^2, respectively, and that $Var[\hat{A}]$, $Var[\hat{B}]$, $Cov[\hat{A}, \hat{B}]$ are as given in Theorem 9.1.2. We will continue to use s^2, with a divisor of $n - 2$, rather than $\hat{\sigma}^2$ as our estimate for σ^2. But we can also easily establish the following facts.

1. $\hat{A} = \sum d_i Y_i$ and $\hat{B} = \sum c_i Y_i$ are both linear functions of the same normal random variables. Thus the probability law for $(\hat{A}, \hat{B})$ is bivariate normal with means a, b and variances and covariance as given in Theorem 9.1.2. In particular, then, the marginal probability laws for both $\hat{A}$ and $\hat{B}$ are normal.

2. Exercise 9.1.16 sketches a method of showing that $\text{Cov}[\hat{A}, Y_i - \bar{Y} - \hat{B}(x_i - \bar{x})] = \text{Cov}[\hat{B}, Y_i - \bar{Y} - \hat{B}(x_i - \bar{x})] = 0$, from which it follows that each of $\hat{A}$ and $\hat{B}$ is independent of $(Y_i - \bar{Y} - \hat{B}(x_i - \bar{x}))$, as well as of $\sum (Y_i - \bar{Y} - \hat{B}(x_i - \bar{x}))^2 = (n - 2) S^2$.

3. An argument similar to that employed with Theorem 6.2.3 shows that $(n - 2) S^2 / \sigma^2$ is a χ^2 random variable with $n - 2$ degrees of freedom.

These facts are summarized in the following theorem.

Theorem 9.2.2. Let $Y_1, Y_2, \ldots, Y_n$ be independent, normal, with $E[Y_i] = a + bx_i$, $\text{Var}[Y_i] = \sigma^2$, where the x_i's are known constants. Then the least squares estimators $\hat{A}$ and $\hat{B}$ are jointly normal (parameters previously specified) and each is independent of

$$\frac{(n - 2) S^2}{\sigma^2} = \frac{\sum (Y_i - \bar{Y} - \hat{B}(x_i - \bar{x}))^2}{\sigma^2},$$

which itself has a χ^2 probability law with $n - 2$ degrees of freedom. (The number of degrees of freedom is n less the number of unknown parameters in $E[Y] = g(x)$.) ■

Recall from Section 5.7 that ratios of standard normal random variables to square roots of independent χ^2 random variables, over their degrees of freedom, have the T distribution. This fact proves useful in constructing confidence intervals for, and testing hypotheses about, both a and b. Define, as in Example 9.1.2,

$$S_{\hat{A}}^2 = \frac{S^2 \sum x_i^2}{n \sum (x_i - \bar{x})^2}, \qquad S_{\hat{B}}^2 = \frac{S^2}{\sum (x_i - \bar{x})^2},$$

the estimators for

$$\sigma_{\hat{A}}^2 = \frac{\sigma^2 \sum x_i^2}{n \sum (x_i - \bar{x})^2}, \qquad \sigma_{\hat{B}}^2 = \frac{\sigma^2}{\sum (x_i - \bar{x})^2},$$

respectively. Then

$$\frac{\hat{A} - a}{\sigma_{\hat{A}}} \qquad \text{and} \qquad \frac{\hat{B} - b}{\sigma_{\hat{B}}}$$

are each standard normal and independent of $(n - 2) S^2/\sigma^2$. It follows then that

$$\frac{\hat{A} - a}{\sigma_{\hat{A}}} \Bigg/ \sqrt{\frac{(n - 2) S^2}{\sigma^2(n - 2)}} = \frac{\hat{A} - a}{S_{\hat{A}}}$$

and

$$\frac{\hat{B} - b}{\sigma_{\hat{B}}} \Bigg/ \sqrt{\frac{(n - 2) S^2}{\sigma^2(n - 2)}} = \frac{\hat{B} - b}{S_{\hat{B}}}$$

each have the T-distribution with $n - 2$ degrees of freedom, as summarized in Corollary 9.2.2.

Corollary 9.2.2.   With the assumptions stated in Theorem 9.2.2, and $S_{\hat{A}}, S_{\hat{B}}$ as previously defined, both $(\hat{A} - a)/S_{\hat{A}}$ and $(\hat{B} - b)/S_{\hat{B}}$ have the T distribution with $n - 2$ degrees of freedom.   ∎

These T-distributions can be used to construct confidence intervals for a and b in exactly the same manner as discussed in Chapter 7 for getting confidence intervals for the mean of a normal random variable. They can also be used to test hypotheses about a and b (and are actually generalized likelihood ratio tests) as discussed in Chapter 8. The following examples illustrate these applications.

EXAMPLE 9.2.1.   Table 9.1.1 presented entrance exam scores and graduating grade point averages for $n = 10$ graduates of University A. For the data given we found in Example 9.1.1, $\hat{a} = .557$, $\hat{b} = .00549$, $s^2 = .0583$, and, in Example 9.1.2, using the same data, $s_{\hat{A}}^2 = .245$, $s_{\hat{B}}^2 = .00000134$ so $s_{\hat{A}} = \sqrt{.245} = .495$, $s_{\hat{B}} = .00116$. For the T distribution with $n - 2 = 8$ degrees of freedom, we find $t_{.95} = 1.860$. If we assume the distribution of graduating grade point averages is normal for each x (perhaps not a very plausible assumption), 90% confidence limits for a are given by

$$\hat{a} \pm 1.860 s_{\hat{A}} = (-.364, 1.478).$$

The 90% confidence limits for b are

$$\hat{b} \pm 1.860 s_{\hat{B}} = (.00333, .00765).$$ ∎

EXAMPLE 9.2.2.   In Exercise 9.1.3 data are given for scale readings (Y_i) versus actual true weights (x_i) of 10 objects. As discussed in Example 9.1.3, the regression of Y on x should pass through the origin, if the scale is accurate and working properly. To test that the line passes through the origin, we would assume $E[Y_i] = a + bx_i$ and then test $H_0: a = 0$ versus $H_1: a \neq 0$. For the data given in the exercise, $\sum x_i = 550$, $\sum x_i^2 = 38500$, $\sum y_i = 537.79$, $\sum y_i^2 = 36883.5557$, $\sum x_i y_i = 37682.3$, from which we find $\hat{a} = -.247$, $\hat{b} = .982$, $s^2 = .182$, $s_{\hat{A}} = .292$, $s_{\hat{B}} = .0047$, using the formulas for an arbitrary straight line. Since $|\hat{a}|/s_{\hat{A}} = .85$ is about equal to $t_{.79}$, with 8 degrees of freedom, we would accept H_0 (if we reject, we would have $\alpha \doteq 2(.21) = .42$). Having accepted H_0: $a = 0$ our model becomes $E[Y_i] = bx_i$, and b, s^2, and $s_{\hat{B}}$ must be reestimated. We find, with the same data, $\hat{b} = .979$, $s^2 = .177$, $s_{\hat{B}} = .00214$. If the scale is accurate, we would expect $b = 1$; to test $H_0: b = 1$, we compute $|\hat{b} - 1|/s_{\hat{B}} = 9.92$, which is approximately equal to $t_{.999995}$, with 9 degrees of freedom, so we would reject H_0. It would appear that this scale has a bias that gets more severe for larger x (since $\hat{b}$ is significantly smaller than 1). We could test $H_0: b = 1$ using either the assumption $E[Y_i] = bx_i$, as illustrated, or $E[Y_i] = a + bx_i$. In either case you can easily see we would reject H_0, but the magnitude of the t statistic differs in the two cases, because both $\hat{b}$ and $s_{\hat{B}}$ depend on whether the constant term a is included in the model. ∎

There are other quantities besides a and b individually about which one may want to test hypotheses or for which one might want to compute confidence intervals. For example, in the entrance exam-graduating grade point average data, the quantity $a + bx^*$ represents the mean graduating grade point average for all students who score x^* on the entrance exam; in the preceding scale example, $a + bx^*$ represents the expected scale reading for all objects whose true weight is x^*. The estimator for $a + bx^*$ is $\hat{Y}_{x^*} = \hat{A} + \hat{B}x^*$, which, because both $\hat{A}$ and $\hat{B}$ have normal distributions, is again a normal random variable for any fixed x^*. Its variance is

$$\text{Var}[\hat{Y}_{x^*}] = \text{Var}[\hat{A}] + x^{*^2}\text{Var}[\hat{B}] + 2x^*\text{Cov}[\hat{A}, \hat{B}]$$

$$= \left[\frac{1}{n} + \frac{(x^* - \bar{x})^2}{\sum(x_i - \bar{x})^2}\right]\sigma^2 = \sigma_{x^*}^2,$$

as is easily verified. Since both $\hat{A}$ and $\hat{B}$ are independent of S^2, it follows that $\hat{Y}_{x^*}$ is independent of S^2 as well. The estimator for $\sigma_{x^*}^2$ is

$$S_{x^*}^2 = \left[\frac{1}{n} + \frac{(x^* - \bar{x})^2}{\sum(x_i - \bar{x})^2}\right]S^2,$$

and using the same reasoning employed to establish Corollary 9.2.2, we find that

$$\frac{\hat{Y}_{x^*} - (a + bx^*)}{S_{x^*}}$$

is then a T random variable with $n - 2$ degrees of freedom. This statistic can be used to test hypotheses about, or construct confidence intervals for, $a + bx^*$, the assumed population mean with $x = x^*$. The following example illustrates this.

EXAMPLE 9.2.3. Suppose you have just taken the entrance exam at University A and scored $x^* = 350$ points. Making our usual simple linear regression assumptions, plus normality, and using our previous data (Examples 9.2.1, 9.1.1), the best estimate of your graduating grade point average then is

$$\hat{a} + 350\hat{b} = .557 + (.00549)\,350 = 2.48.$$

We can use the fact that

$$\frac{\hat{Y}_{350} - (a + 350b)}{S_{350}}$$

has the T-distribution with 8 degrees of freedom to construct confidence intervals for the mean value of the graduating grade point averages of all students with entrance exam score $x = 350$. We have

$$s^2_{350} = \left(\frac{1}{10} + \frac{(350 - 422)^2}{43496} \right)(.0583)$$

$$= .0128$$

and thus $s_{350} = \sqrt{.0128} = .113$. With 8 degrees of freedom, $t_{.95} = 1.860$, $t_{.90} = 1.397$, so the 90% two-sided confidence limits for the mean graduating grade point average of all students scoring $x = 350$ are:

$$2.48 \pm (1.860)(.113) = (2.27,\ 2.69),$$

while the 90% upper and lower limits are, respectively,

$$2.48 + (1.397)(.113) = 2.64$$
$$2.48 - (1.397)(.113) = 2.32. \qquad \blacksquare$$

The confidence interval just described is for the mean of the population of y values, with $x = x^*$. In many cases we may want an interval that has a known probability of including a particular y value selected from the population with $x = x^*$. In the preceding example we were 90% sure the interval from 2.27 to 2.69 includes the mean value for all students who scored $x^* = 350$ on the entrance exam; if you are one of those students, the interval from 2.27 to 2.69 may or may not include your graduating grade point (the probability it does is smaller than .9). A 90% confidence interval for an individual y, with $x^* = 350$, is also easily derived; this interval is frequently called a *prediction interval*. Let Y_{x^*} be the individual y value that will occur. Assuming it (like all the earlier observed values) is selected at random from the full population, Y_{x^*} is independent of $\hat{Y}_{x^*}$, the estimator of the mean of the population. Thus

$$E[Y_{x^*} - \hat{Y}_{x^*}] = 0$$
$$\text{Var}[Y_{x^*} - \hat{Y}_{x^*}] = \text{Var}[Y_{x^*}] + \text{Var}[\hat{Y}_{x^*}]$$
$$= \sigma^2 + \sigma_{x^*}^2$$
$$= \left(\frac{n+1}{n} + \frac{(x^* - \bar{x})^2}{\sum(x_i - \bar{x})^2} \right) \sigma^2$$
$$= \sigma_{x^*+e}^2,$$

say, which we estimate by

$$S_{x^*+e}^2 = \left(\frac{n+1}{n} + \frac{(x^* - \bar{x})^2}{\sum(x_i - \bar{x})^2} \right) S^2.$$

It then follows that

$$\frac{Y_{x^*} - \hat{Y}_{x^*}}{S_{x^*+e}}$$

also has the T-distribution with $n-2$ degrees of freedom. This fact can be used to get confidence intervals about Y_{x^*} and to test hypotheses about Y_{x^*}.

EXAMPLE 9.2.4. Exercise 9.1.2 briefly discussed the use of spectrometric oil analysis as a diagnostic tool and presented the results of 10 such analyses. In a normally operating engine, noticeable traces of the metals being lubricated are found in the lubricating oil. Burning a sample of oil on a spectrometer gives an estimate of the level of these metallic contaminants at the time the sample is taken. As the engine is used, the level of contamination can only increase, as long as no new oil is added. Thus if samples are analyzed regularly (more or less), as with the data in Exercise 9.1.2, one can estimate the relationship between the level of iron contamination, say, and the number of miles the car was driven. The 10 data points of Exercise 9.1.2 are plotted in Figure 9.2.1, along with the least squares line. For the data from Exercise 9.1.2, you will find $\sum y_i = 394$, $\sum y_i^2 = 17{,}702$, $\sum x_i = 27{,}786$, $\sum x_i^2 = 97{,}685{,}656$, $\sum x_i y_i = 1{,}304{,}050$, $\bar{y} = 39.4$, $\bar{x} = 2778.6$, $\sum(x_i - \bar{x})^2 = 20{,}479{,}476.4$ $\sum(y_i - \bar{y})^2 = 2178.4$, $\sum(x_i - \bar{x})(y_i - \bar{y}) = 209{,}281.6$, so $\hat{b} = .0102$, $\hat{a} = 11.005$, and the least squares line is $\hat{y} = 11.005 + .0102x$. You can easily verify as well that $s^2 = 4.967$. This least

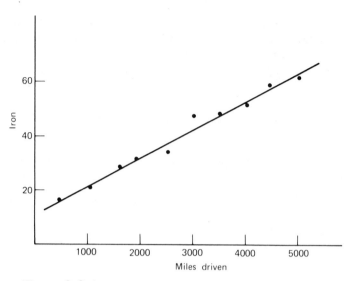

Figure 9.2.1

squares line, and s^2, describe the way that iron builds up in this engine when it is working properly. The basic idea in using oil analyses as a diagnostic aid is to continue taking oil samples regularly, say, every thousand miles. Then the "normal" performance as described by the least squares line and s^2 can be used to set limits for the observed iron content, for any number of hours, x^*, after the oil was changed. If an oil sample shows an iron content in excess of the normal limit, it may indicate that something is wearing faster than usual and perhaps some special maintenance is called for. For example, suppose an oil sample is taken from this engine at $x^* = 2200$ miles after the oil was changed. Then

$$s^2_{2200+e} = \left(1.1 + \frac{(2200 - 2778.6)^2}{20,479,476.4} \right) 4.967$$

$$= 5.544$$

so $s_{2200+e} = 2.355$. With 8 degrees of freedom, $t_{.95} = 1.860$. The estimated mean iron content at 2200 miles (if the engine is working properly) is $\hat{y}_{2200} = 11.005 + 2200(.0102) = 33.5$; with probability .95 the spectrometer iron reading, with $x = 2200$, should be no more than

$$33.5 + (1.860) \, 2.355 = 37.9.$$

If we analyze this sample, and find a reading of 38 or more, either an event with probability .05 has occurred or the engine is no longer working as it was in the "normal" phase; perhaps some abnormal wear is occurring that should be corrected. This type of computation can be made for any number x^* of hours since the oil was changed, assuming the relationship is linear for all x. The probability used, .95, can also be easily varied if desired, to give a smaller or larger false alarm rate, which is the probability of claiming the engine is not working properly when it really is. ∎

One should, of course, be aware that it can be dangerous to make predictions for individual Y values, or for $E[Y]$, whose x values lie outside the interval represented in the sample. For the oil analysis data of Exercise 9.1.2 the smallest x is 480 miles and the largest is 5002; the plot of (x, y) values appears quite linear for x in this range. Because we have not observed any Y values for $x = 9000$ miles, say, we could not be completely comfortable in extrapolating the observed straight line relationship out to $x = 9000$. Such extrapolation makes the added assumption that the observed relation also holds outside the range observed, a statement that may or may not be true, and one about which the observed data contains no information.

EXERCISE 9.2

1. Use the oil analysis data of Exercise 9.1.2 and Example 9.2.4 to evaluate 95% two-sided confidence limits for a and b. Compute 95% two-sided confidence limits for σ^2.

2. Assume you score $x^* = 400$ on the entrance exam for University A. Use the data of Example 9.2.3 to compute a 90% lower confidence limit for your graduate grade point.

3. Eight executives took the same 12-hour speed-reading course. At the first meeting each took a reading exam to measure their initial reading speed, x. At the final meeting each took another reading exam to measure their final achieved speed, y. The data recorded are:

y	356	364	361	413	401	421	460	438
x	134	132	233	290	225	104	295	111

   (a) Use these data to estimate a, b, σ^2 in the simple linear regression model.
   (b) Making the appropriate normal assumption, test $H_0: b = 0$ versus $H_1: b \neq 0$.

4. The initial weight (x) and the amount of weight lost from using a diet for one month (y) (both in pounds) for 12 people are:

y	31	9	22	30	27	17	14	21	31	28	27	15
x	214	168	176	159	173	163	157	182	209	196	170	176

Assuming a simple linear regression model, with normality, does it appear a person's initial weight affects the amount of weight lost when using this diet?

5. (a) If your initial weight is 170 pounds, how much would you expect to lose if you go on the diet described in Exercise 9.2.4, for one month?

(b) What is the smallest amount you will lose, with probability .9?

6. The term regression was apparently first used by F. Galton in describing laws of inheritance. He stated "Each peculiarity in a man is shared by his kinsman, but on the average in a less degree." This has been interpreted to mean that the laws of inheritance cause extremes to regress toward the overall mean. To illustrate the point, the English statistician Karl Pearson collected a large number of records of family member's heights. The following data are typical of what he found, where y = son's height, x = father's height (inches).

y	63.6	65.2	66	65.5	66.9	67.1	67.4	68.3	70.1	70
x	60	62	64	65	66	67	68	70	72	74

Notice that taller fathers do tend to have taller sons but the son's heights are not so extreme as the father's. Fit a simple linear regression to these data and test $H_0: b \geq 1$ versus $H_1: b < 1$, with $\alpha = .01$. The fact that $\hat{b} < 1$ tends to reinforce Galton's regression statement.

*7. Least squares and maximum likelihood estimators may coincide for other probability laws in addition to the normal. Assume an electronic component has an exponential time to failure with mean μ; also assume $\mu = a + bx$, where x is the number of times the equipment has been switched on (each turn-on sends a power surge through the equipment). Suppose each of n_1 components of this type has been switched on x_1 times, and $y_1, y_2, \ldots, y_{n_1}$ are the resulting times to failure; each of n_2 components of this type has been switched on $x_2 > x_1$ times, resulting in $z_1, z_2, \ldots, z_{n_2}$ as their times to failure.

(a) Evaluate the maximum likelihood estimates for a and b. Compare with Exercise 9.1.6.

(b) If we had 3 different values for the number of switch-ons, $x_1 < x_2 < x_3$, would the maximum likelihood estimators coincide with least squares?

8. (a) For the father–son height data of Exercise 9.2.6, what is your estimate for the mean height of sons whose fathers are 6 feet tall?
   (b) Construct a 90% two-sided confidence interval for the mean height of sons whose fathers are 6 feet tall.

9. It is possible to regress x on y, given any set of observations. If, as in the text, we assume $E[Y] = a + bx$, we also have $x = (E[Y] - a)/b = c + dE[Y]$. If we do regress x on Y and use least squares to estimate c and d, show that $\hat{c} \neq -\hat{a}/\hat{b}$, $\hat{d} \neq 1/\hat{b}$.

10. When renting a car from a major car rental agency, the customer must pay for the gasoline consumed. Thus when the car is returned the customer must pay for the gasoline needed to fill the tank. Rather than attempting to fill the tank when the customer returns the car, and causing possibly long delays, the customer is told to read the gas gauge to the nearest $\frac{1}{8}$ tank; his final bill is based on this reading. Later, when the car is prepared for rerental, the actual level of the tank is known from the amount of gasoline required to fill it up.
    (a) Let x represent the customer's report of the gas gauge reading and let y be the actual tank level (computed from (tank capacity − gallons needed to fill)/tank capacity). If all the gas gauges were accurate, and accurately read, and all customers were honest, what should be the relation between y and x?
    (b) Based on the following observed data, estimate the relation between y and x.
    (c) Test any hypotheses you find of interest using these data.

y	x
0.099	0.25
0.08	0.25
0.241	0.25
0.119	0.25
0.286	0.375
0.455	0.375
0.312	0.375
0.517	0.5
0.293	0.5
0.567	0.625
0.825	0.625
0.555	0.625
0.604	0.75
0.582	0.75

9.3 The Analysis of Variance

In Chapter 8 we saw how to test hypotheses about the equality of means of two normal populations, using a T statistic. If one has samples from 3 or more normal populations and wants to test that all their means are equal, then a new test procedure is called for. For example, suppose we wanted to test the hypothesis that the expected miles per gallon we could achieve is the same for each of 3 different automobiles, say A, B, and C. One could use a T statistic to compare A with B, then B with C, and finally C with A. Actually, the third comparison would be redundant, because the first two T statistics already involve $\bar{y}_A - \bar{y}_B$, $\bar{y}_B - \bar{y}_C$, and s, the pooled estimate of the common standard deviation, so the value for the T statistic could actually be constructed (quite simply) from the first two. More importantly, these first two T statistics are not independent and each would give rise to an observed significance level. It is not straightforward to translate these two observed significance levels into a single significance level for testing H_0: $\mu_A = \mu_B = \mu_C$.

The procedure used to test equality of means of several normal populations is called an analysis of variance, which sounds like a misnomer. As we will see, the procedure involves splitting a "total" variance into pieces (analyzing it) and then deciding whether to accept or reject equality of the population means based on the relative magnitudes of these pieces.

To keep track of which population a sample observation came from, and to distinguish between the several observations that came from the same population, we will use a double subscripting system. Suppose we have independent samples from each of k different normal populations; the sample from population i is of size n_i, $i = 1, 2, ..., k$. The sample random variables are denoted by Y_{ij}, $i = 1, 2, ..., k$, $j = 1, 2, ..., n_i$. We assume the Y_{ij}'s are independent, normal, $E[Y_{ij}] = \mu_i$, $\text{Var}[Y_{ij}] = \sigma^2$ and want to test H_0: $\mu_1 = \mu_2 = \cdots = \mu_k$ versus H_1: H_0 is false. Note that we assume the population variances are equal, as we did in using the T test for equality of two means.

It is convenient to represent the k population means in a slightly different fashion. So far we have assumed we have samples of size n_i from populations with means μ_i, $i = 1, 2, ..., k$. The sum of all the expected values of the sample random variables then is

$$\sum_i \sum_j E[Y_{ij}] = \sum_i \sum_j \mu_i = \sum_i n_i \mu_i$$

and the "average" expected value is this total divided by the complete sample size

$$\mu = \frac{\sum n_i \mu_i}{\sum n_i}.$$

Now let τ_i represent the deviation of the ith population mean from this average, $\tau_i = \mu_i - \mu$, and we have

$$\sum n_i \tau_i = \sum n_i(\mu_i - \mu)$$
$$= \sum n_i \mu_i - \mu \sum n_i = 0,$$

the weighted sum of these deviations equals zero. Our original hypothesis, that $\mu_1 = \mu_2 = \cdots = \mu_k$ is equivalent to $\mu_1 - \mu = \mu_2 - \mu = \cdots = \mu_k - \mu$, that is, $\tau_1 = \tau_2 = \cdots = \tau_k$, which, together with $\sum n_i \tau_i = 0$ actually implies then that our hypothesis of equal population means becomes $H_0: \tau_1 = \tau_2 = \cdots = \tau_k = 0$, and we can write $E[Y_{ij}] = \mu + \tau_i$, $\text{Var}[Y_{ij}] = \sigma^2$. It is also convenient to introduce errors of observation e_{ij}, deviations between the sample random variables and their expected values,

$$e_{ij} = Y_{ij} - (\mu + \tau_i)$$

as in Section 9.1. We now have our notation in consonance with that which is frequently used with analysis of variance discussions.

Let us assume we observe $\sum n_i$ random variables, Y_{ij}, where

$$Y_{ij} = \mu + \tau_i + e_{ij}, \qquad i = 1, 2, \ldots, k, \qquad j = 1, 2, \ldots, n_i,$$

$\mu, \tau_1, \tau_2, \ldots, \tau_k$ are unknown constants, the e_{ij}'s are normal and independent, with mean 0 and variance σ^2.

This model is called the *completely randomized* model in the literature of design of experiments. Independent samples are selected completely at random from each of k different normal populations. With the explicit introduction of the observation errors, a natural method of estimation of μ and the τ_i's is given by least squares (and because we have assumed normality, this is again equivalent to using maximum likelihood). That is, we want the values for $\mu, \tau_1, \tau_2, \ldots, \tau_k$ such that

$$Q = \sum_i \sum_j e_{ij}^2$$
$$= \sum_i \sum_j (Y_{ij} - \mu - \tau_i)^2$$

is minimized (and recall that $\sum n_i \tau_i = 0$, which simplifies things). We have

$$\frac{\partial Q}{\partial \mu} = -2 \sum_i \sum_j (Y_{ij} - \mu - \tau_i)$$
$$= -2Y_{..} + 2n\mu + 2\sum n_i \tau_i = -2Y_{..} + 2n\mu$$

$$\frac{\partial Q}{\partial \tau_i} = -2 \sum_j (Y_{ij} - \mu - \tau_i)$$
$$= -2Y_{i.} + 2n_i \mu + 2n_i \tau_i, \qquad i = 1, 2, \ldots, k,$$

where a dotted subscript indicates summation over that subscript:

$$Y_{i.} = \sum_j Y_{ij}, \qquad Y_{..} = \sum_i Y_{i.} = \sum_i \sum_j Y_{ij}, \qquad n_{.} = \sum_i n_i.$$

The equations to be solved (again the normal equations) for the estimators are (after canceling 2's)

$$n_{.}\hat{\mu} = Y_{..}$$
$$n_i\hat{\mu} + n_i\hat{\tau}_i = Y_{i.}, \qquad i = 1, 2, \ldots, k$$

and the solutions are easily found to be

$$\hat{\mu} = \overline{Y}_{..}, \quad \hat{\tau}_i = \overline{Y}_{i.} - \overline{Y}_{..}, \qquad i = 1, 2, \ldots, k,$$

where super bars indicate averaging:

$$\overline{Y}_{..} = \frac{1}{n_{.}} Y_{..}, \qquad \overline{Y}_{i.} = \frac{1}{n_i} Y_{i.}.$$

Not too surprisingly, the least squares (maximum likelihood) estimator of μ is $\overline{Y}_{..}$, the overall mean, and the estimator for τ_i is $\overline{Y}_{i.} - \overline{Y}_{..}$, the deviation between the ith sample mean and the overall mean. The unbiased estimator for σ^2 is

$$S^2 = \frac{1}{n_{.} - k} \sum \sum (Y_{ij} - \hat{\mu} - \hat{\tau}_i)^2$$

$$= \frac{1}{n_{.} - k} \sum \sum (Y_{ij} - \overline{Y}_{i.})^2,$$

where the divisor is $n_{.} - k$, the total sample size ($n_{.} = \sum n_i$) less the number of (independent) parameters specifying the population means (which is k). The estimators $\overline{Y}_{..}$ (for μ) and $\overline{Y}_{i.} - \overline{Y}_{..}$ (for τ_i) are both linear functions of normal random variables and thus are themselves normal. It is straightforward to verify that $\text{Cov}(\overline{Y}_{..}, \overline{Y}_{i.} - \overline{Y}_{..}) = 0$ for each i, using Theorem 5.4.2. It is also straightforward to verify that $\text{Cov}(\overline{Y}_{..}, Y_{ij} - \overline{Y}_{i.}) = 0$, and that $\text{Cov}(\overline{Y}_{i.} - \overline{Y}_{..}, Y_{ij} - \overline{Y}_{i.}) = 0$, for each i and j. This shows that $\overline{Y}_{..}$ is independent of both $\overline{Y}_{i.} - \overline{Y}_{..}$ and S^2 and that $\overline{Y}_{i.} - \overline{Y}_{..}$ is independent of S^2.

EXAMPLE 9.3.1.   Five identical models of automobile make A, five of make B, and five of make C were used to compare the mileages achievable by the three makes. The data gathered are given in Table 9.3.1.

Table 9.3.1

Make	A	B	C
	21.7	22.2	19.6
	26.5	24.2	23.9
	24.9	23.9	24.4
	23.9	23.7	22.2
	25.0	25.3	23.3

Let us assume these observations came from the completely randomized model and estimate the unknown parameters. That is, assume we have observed values for Y_{ij}, where

$$Y_{ij} = \mu + \tau_i + e_{ij}, \qquad i = 1, 2, 3, \qquad j = 1, 2, \dots, 5;$$

μ is an overall average mileage value for all three makes, τ_i is the average deviation of make i from μ, $i = 1, 2, 3$ (we have replaced A, B, C by 1, 2, 3, respectively) and the e_{ij}'s are independent, normal observation errors, with mean 0 and variance σ^2. We have $\bar{y}_{1.} = \frac{122}{5} = 24.4$, $\bar{y}_{2.} = \frac{119.3}{5} = 23.86$, $\bar{y}_{3.} = \frac{113.4}{5} = 22.68$, $\bar{y}_{..} = \frac{354.7}{15} = 23.647 = \hat{\mu}$, so $\hat{\tau}_1 = 24.4 - 23.647 = .753$, $\hat{\tau}_2 = 23.86 - 23.647 = .213$, $\hat{\tau}_3 = 22.68 - 23.647 = -.967$, and apart from rounding error, $\sum n_i \hat{\tau}_i = 0$, as must be the case, since

$$\sum n_i \hat{\tau}_i = \sum n_i (\bar{y}_{i.} - \bar{y}_{..}) = \sum y_{i.} - \bar{y}_{..} \sum n_i$$
$$= y_{..} - y_{..} = 0.$$

To estimate σ^2, we compute

$$s^2 = \frac{1}{15 - 3} \sum \sum (y_{ij} - \bar{y}_{i.})^2.$$

Notice that

$$\sum_j (y_{1j} - \bar{y}_{1.})^2 = 12.560$$

is simply the sum of squares of the 5 observations for car A about the mean for car A; similarly, this same quantity for cars B and C is

$$\sum_j (y_{2j} - \bar{y}_{2.})^2 = 4.972,$$

$$\sum_j (y_{3j} - \bar{y}_{3.})^2 = 14.548,$$

respectively. The unbiased estimator for σ^2 is the sum of these divided by 12,

$$s^2 = \frac{32.080}{12} = 2.673,$$

a pooled estimate like the one we used in Chapter 7, except now it is pooled over all 3 samples instead of just 2. With a sample of size 5 from population 1 we have $5 - 1 = 4$ degrees of freedom for estimating σ^2 (and 4 from each of populations 2 and 3 as well). The divisor in s^2 is the sum of these degrees of freedom from the 3 populations. ∎

Now let us return to testing the hypothesis that the population means are all equal, for the completely randomized model. The generalized likelihood ratio test criterion can be shown to lead to the test we will describe, although motivating it in a different way. The total sum of squares of all the observations about the overall observed mean is

$$\sum_i \sum_j (Y_{ij} - \overline{Y}_{..})^2.$$

This quantity can be partitioned as follows.

$$\sum_i \sum_j (Y_{ij} - \overline{Y}_{i.} + \overline{Y}_{i.} - \overline{Y}_{..})^2$$

$$= \sum_i \sum_j \{(Y_{ij} - \overline{Y}_{i.})^2 + (\overline{Y}_{i.} - \overline{Y}_{..})^2 + 2(Y_{ij} - \overline{Y}_{i.})(\overline{Y}_{i.} - \overline{Y}_{..})\}$$

$$= \sum_i \sum_j (Y_{ij} - \overline{Y}_{i.})^2 + \sum_i \sum_j (\overline{Y}_{i.} - \overline{Y}_{..})^2,$$

since

$$2\sum_i \sum_j (Y_{ij} - \overline{Y}_{i.})(\overline{Y}_{i.} - \overline{Y}_{..}) = 2\sum_i (\overline{Y}_{i.} - \overline{Y}_{..})\sum_j (Y_{ij} - \overline{Y}_{i.})$$

$$= 2\sum_i (\overline{Y}_{i.} - \overline{Y}_{..}) \cdot 0 = 0.$$

Also,

$$\sum_i \sum_j (\overline{Y}_{i.} - \overline{Y}_{..})^2 = \sum_i n_i (\overline{Y}_{i.} - \overline{Y}_{..})^2$$

because the summand does not depend on j. This gives us the basic algebraic identity.

$$\sum_i \sum_j (Y_{ij} - \overline{Y}_{..})^2 = \sum_i \sum_j (Y_{ij} - \overline{Y}_{i.})^2 + \sum_i n_i (\overline{Y}_{i.} - \overline{Y}_{..})^2.$$

The first quantity on the right is the sum of squares of the residuals

$$\hat{e}_{ij} = Y_{ij} - \hat{\mu} - \hat{\tau}_i,$$

the numerator of S^2, used to estimate σ^2 regardless of whether the population means are equal; divided by σ^2 it is in fact a χ^2 random variable with $\sum_i (n_i - 1) = n_. - k$ degrees of freedom. The second quantity on the right is actually $\sum_i n_i \hat{\tau}_i^2$ and is independent of S^2 (since each $\hat{\tau}_i$ is independent of S^2); if all the τ_i's are in fact equal to 0, it is not difficult to show that

$$\frac{1}{\sigma^2} \sum n_i (\overline{Y}_{i.} - \overline{Y}_{..})^2$$

is a χ^2 random variable with $k - 1$ degrees of freedom. Note as well that if all the τ_i's are 0, we would expect the $|\hat{\tau}_i|$'s to be small (because the least squares estimators $\hat{\tau}_i$ are unbiased) and thus $\sum n_i \hat{\tau}_i^2$ should be relatively small; when the τ_i's are unequal, $\sum n_i \hat{\tau}_i^2$ tends to get larger. In short,

$$\frac{\sum\sum(Y_{ij} - \overline{Y}_{i.})^2}{\sigma^2} \quad \text{and} \quad \frac{\sum n_i(\overline{Y}_{i.} - \overline{Y}_{..})^2}{\sigma^2}$$

are independent χ^2 random variables so the ratio

$$\frac{\sum n_i(\overline{Y}_{i.} - \overline{Y}_{..})^2}{k - 1} \Big/ \frac{\sum\sum(Y_{ij} - \overline{Y}_{i.})^2}{n_. - k}$$

has an F-distribution with $k - 1$, $n_. - k$ degrees of freedom if $\tau_1 = \tau_2 = \cdots = \tau_k = 0$. We reject $H_0: \tau_1 = \tau_2 = \cdots = \tau_k = 0$ if this ratio exceeds $F_{1-\alpha}$, the $100(1 - \alpha)$th quantile of the F-distribution with $k - 1$, $n_. - k$ degrees of freedom, to have $P(\text{Type I error}) = \alpha$; as mentioned earlier, this is in fact the generalized likelihood ratio test for H_0.

The quantities used in testing $H_0: \tau_1 = \tau_2 = \cdots = \tau_k = 0$ are frequently presented in tabular form called an analysis of variance. Table 9.3.2 presents the analysis of variance table for the completely randomized model, based on samples of size $n_1, n_2, \ldots, n_k$ from k populations. Because we are in fact partitioning a variance into pieces, the first column identifies the source of the variation, the second column gives the degrees of freedom for the different sources (which sum to the total degrees of freedom), the third column gives the sums of squares for the sources (these add up to the total sum of squares) and the final column gives the mean squares (ratio of sums of squares over degrees of freedom). The ratio of the between populations mean square to the residual mean square is the F statistic whose value is compared with the appropriate F quantile in deciding whether to reject $H_0: \tau_1 = \tau_2 = \cdots = \tau_k = 0$. Table 9.3.2 presents the sums of squares in the preceding forms as we derived them. As with computing a sample variance, there are simpler formulas for computational use, such as:

Total sum of squares:

$$\sum\sum(Y_{ij} - \overline{Y}_{..})^2 = \sum\sum Y_{ij}^2 - \frac{(Y_{..})^2}{n_.} = SS_T$$

Among populations sum of squares

$$\sum n_i(\overline{Y}_{i.} - \overline{Y}_{..})^2 = \sum \frac{Y_{i.}^2}{n_i} - \frac{(Y_{..})^2}{n_.} = SS_A$$

Residual sum of squares

$$\sum\sum \hat{e}_{ij}^2 = \sum\sum (Y_{ij} - \overline{Y}_{i.})^2 = \sum\sum Y_{ij}^2 - \sum \frac{(Y_{i.})^2}{n_i} = SS_R$$

It is quite easy to see from these computational sums of squares that the basic identity, Total sum of squares = Among populations plus Residual sums of squares, must hold. You are asked to verify these computational formulas in the following exercises.

Table 9.3.2

Source	Degrees of Freedom	Sums of Squares	Mean Squares
Among populations	$k - 1$	$SS_A = \sum n_i(\overline{Y}_{i.} - \overline{Y}_{..})^2$	$SS_A/(k - 1)$
Residual	$n_. - k$	$SS_R = \sum\sum (Y_{ij} - \overline{Y}_{i.})^2$	$SS_R/(n_. - k)$
Total	$n_. - 1$	$SS_T = \sum\sum (Y_{ij} - \overline{Y}_{..})^2$	

The following example computes the analysis of variance for testing equality of the mileages attained by the three car makes discussed in Example 9.3.1.

EXAMPLE 9.3.2. For the car data discussed in Example 9.3.1, we found $y_{1.} = 122$, $y_{2.} = 119.3$, $y_{3.} = 113.4$, $y_{..} = 354.7$. Thus we have $\sum y_{i.}^2/n_i = 8395.21$, $y_{..}^2/n_. = 8387.473$, $\sum\sum y_{ij}^2 = 8427.29$, from which we find $\sum y_{i.}^2/n_i - y_{..}^2/n_. = 7.737$, $\sum\sum y_{ij}^2 - \sum y_{i.}^2/n_i = 32.080$, $\sum\sum y_{ij}^2 - y_{..}^2/n_. = 39.817$; these are summarized in the following analysis of variance (see Table 9.3.3).

Table 9.3.3

Source	Degrees of Freedom	Sums of Squares	Mean Squares
Car makes	2	7.737	3.869
Residuals	12	32.080	2.673
Total	14	39.817	

To test the hypothesis that the expected mileages with the three car makes are equal, we compute the ratio of the car make mean square to the residual mean square, $3.869/2.673 = 1.447$; with 2 and 12 degrees of freedom, $F_{.95} = 3.886$ so we would accept H_0: Equal mileages, with $\alpha = .05$. ∎

The completely randomized model is in a number of ways the simplest model for samples from several populations. There are many more complicated models that have proved useful in various situations, which also lead to an analysis of variance. We will examine one further model and its motivation.

In the completely randomized model it is assumed that the sample values are selected at random from each population; thus all of the values selected from the same population have the same expected value ($\mu + \tau_i$ in the completely randomized model). In much experimental work, by design or otherwise, it may be that the observations from the same population have differing expected values, which one wants to remove from the comparison of the populations. For example, Table 9.3.1 presented the achieved miles per gallon from 15 different cars, five cars from each of three different makes. Nothing was said about whether all the cars were or were not driven under the same conditions. Suppose the first car listed (for each make) was driven in city traffic; the second, in flat open country at a slow speed (say, 35 miles per hour and less); the third, in flat country at a fast speed (say, at 55 mph); the fourth, in hilly country slow, and the last, in hilly country fast. It is quite possible that these different driving conditions affect the miles per gallon achieved by the cars; thus the five cars of the same make would have different expected values. It is certainly desirable to compare the car makes over a variety of driving conditions, but on the other hand, we would like to remove the effects of these different conditions from the comparison of the car makes. We will now describe the standard model used for achieving this—called a *randomized block* model or a *two-way classification-model without interaction*.

For the completely randomized model we assumed

$$Y_{ij} = \mu + \tau_i + e_{ij},$$

where μ was an overall mean parameter and τ_i represented the expected deviation from μ caused by the population sampled from (car make). A natural extension of this model is

$$Y_{ij} = \mu + \tau_i + \beta_j + e_{ij},$$

where μ and τ_i are as before; the parameters β_j represent the expected deviation caused by an extraneous condition (the different driving conditions in our example). The ranges on the two subscripts are $i = 1, 2, \ldots, k$ (as before), $j = 1, 2, \ldots, b$. That is, we assume with this model we have samples from each of k populations and the sample size from each population now is b (each $n_i = b$). For this case the model is nicely balanced and the analysis is quite simple; it is, of course, possible to have differing numbers of observations from the various populations, but this can complicate the analysis considerably; we do not have the space to discuss this case.

As in the completely randomized model, the τ_i's are deviations from the overall mean caused by the different populations so again

$$\sum n_i \tau_i = \sum b \tau_i = b \sum \tau_i = 0,$$

that is, $\sum \tau_i = 0$; similarly, since the β_j's represent deviations from the overall mean caused by the extraneous factor, $\sum \beta_j = 0$. There is a very important assumption hidden in this assumed model. We have

$$E[Y_{11}] = \mu + \tau_1 + \beta_1$$
$$E[Y_{12}] = \mu + \tau_1 + \beta_2$$
$$E[Y_{21}] = \mu + \tau_2 + \beta_1$$
$$E[Y_{22}] = \mu + \tau_2 + \tau_2$$

so

$$E[Y_{11} - Y_{12}] = (\mu + \tau_1 + \beta_1) - (\mu + \tau_1 + \beta_2)$$
$$= \beta_1 - \beta_2,$$
$$E[Y_{21} - Y_{22}] = (\mu + \tau_2 + \beta_1) - (\mu + \tau_2 + \beta_2)$$
$$= \beta_1 - \beta_2.$$

That is, the expected difference between observations 1 and 2 for population 1 is the same as this expected difference for population 2. Indeed,

$$E[Y_{pc} - Y_{pd}] = \beta_c - \beta_d = E[Y_{qc} - Y_{qd}]$$

for any possible p, q, c, d. This says that the extraneous factor affects the expected value in exactly the same way, for all populations; in our car make example this model assumes the expected change in mileage between, say, city driving and country slow driving is exactly the same for all three car makes, which may or may not be true. The model assumes there is no *interaction* between the populations sampled and the extraneous condition. To summarize, the assumptions for the *randomized block* model (two-way classification without interaction) are:

We have observations for bk random variables Y_{ij} such that

$$Y_{ij} = \mu + \tau_i + \beta_j + e_{ij}, \qquad i = 1, 2, \ldots, k,$$
$$j = 1, 2, \ldots, b;$$

μ, the τ_i's and β_j's are unknown constants such that $\sum \tau_i = \sum \beta_j = 0$, and the e_{ij}'s are independent, normal, random variables each with mean 0 and variance σ^2.

As with the completely randomized model, a dotted subscript means summation over that subscript and super bars indicate averages. Again, least squares and maximum likelihood estimators for μ, the τ_i's and β_j's are coincident. To get the normal equations, we minimize

$$Q = \sum\sum e_{ij}^2 = \sum\sum(Y_{ij} - \mu - \tau_i - \beta_j)^2;$$

we have

$$\frac{\partial Q}{\partial \mu} = -2\sum\sum(Y_{ij} - \mu - \tau_i - \beta_j) = -2Y_{..} + 2kb\mu$$

$$\frac{\partial Q}{\partial \tau_i} = -2\sum_j(Y_{ij} - \mu - \tau_i - \beta_j) = -2Y_{i.} + 2b\mu + 2b\tau_i, \qquad i = 1, 2, ..., k,$$

$$\frac{\partial Q}{\partial \beta_j} = -2\sum_i(Y_{ij} - \mu - \tau_i - \beta_j) = -2Y_{.j} + 2k\mu + 2k\beta_j, \qquad j = 1, 2, ..., b,$$

so the normal equations are

$$bk\hat{\mu} = Y_{..}$$
$$b\hat{\mu} + b\hat{\tau}_i = Y_{i.}, \qquad i = 1, 2, ..., k$$
$$k\hat{\mu} + k\hat{\beta}_j = Y_{.j}, \qquad j = 1, 2, ..., b.$$

The unbiased estimator for σ^2 is

$$S^2 = \frac{1}{(b-1)(k-1)}\sum\sum(Y_{ij} - \bar{Y}_{i.} - \bar{Y}_{.j} + \bar{Y}_{..})^2,$$

again, the sum of squares of all the residuals, $\hat{e}_{ij}$, divided by the total sample size less the number of parameters to be estimated; this last number is $bk - (b + k - 1) = (b-1)(k-1)$, since we have one μ, $k - 1$ τ_i's and $b - 1$ β_j's to be estimated.

Again, with this randomized block model, it is generally of interest to test $H_0: \tau_1 = \tau_2 = \cdots = \tau_k = 0$, and also the quantities needed to accomplish this are generally displayed in an analysis of variance. With the randomized block model, though, one additional source of variability, the extraneous factor, can be identified and, in a sense, corrected for in making the test. The basic identity of the completely randomized model still holds.

$$\sum\sum(Y_{ij} - \bar{Y}_{..})^2 = \sum\sum(Y_{ij} - \bar{Y}_{i.})^2 + b\sum(\bar{Y}_{i.} - \bar{Y}_{..})^2.$$

Now, though, the sum of squares that was the residuals can itself be partitioned into two parts,

$$\sum\sum(Y_{ij} - \bar{Y}_{i.})^2 = \sum\sum(Y_{ij} - \bar{Y}_{i.} - \bar{Y}_{.j} + \bar{Y}_{..})^2 + k\sum(\bar{Y}_{.j} - \bar{Y}_{..})^2$$

as is easily verified. Notice that the first term on the right is the sum of squares of the residuals for the randomized block model and is the numerator for S^2 as already noted; the second term is

$$k \sum \hat{\beta}_j^2 = k \sum (\bar{Y}_{.j} - \bar{Y}_{..})^2 = \sum \frac{Y_{.j}^2}{k} - \frac{Y_{..}^2}{bk} = SS_B$$

the sum of squares of the estimates of the extraneous factor values that we will call the block effects. The randomized block analysis of variance is given in Table 9.3.4.

Table 9.3.4

Source	Degrees of Freedom	Sums of Squares	Mean Squares
Among populations	$k - 1$	$SS_A = b \sum (\bar{Y}_{i.} - \bar{Y}_{..})^2$	$SS_A/(k - 1)$
Blocks	$b - 1$	$SS_B = k \sum (\bar{Y}_{.j} - \bar{Y}_{..})2$	$SS_B/(b - 1)$
Residual	$(b - 1)(k - 1)$	$SS_R = SS_T - SS_A - SS_B$	$SS_R/(b - 1)(k - 1)$
Total	$bk - 1$	$SS_T = \sum \sum (Y_{ij} - \bar{Y}_{..})^2$	

The ratio of the Among populations mean square to the Residual mean square has an F-distribution with $k - 1, (b - 1)(k - 1)$ degrees of freedom if $\tau_1 = \tau_2 = \cdots = \tau_k = 0$; again, H_0 is rejected if this ratio exceeds $F_{1-\alpha}$ to have $P(\text{type I error}) = \alpha$. Because the β_j parameters enter the model in exactly the same way, we can also test the hypothesis that $\beta_1 = \beta_2 = \cdots = \beta_b = 0$ by computing the ratio of the block mean square to the residual mean square; this equality of the β_j's is rejected if the ratio exceeds the $100(1 - \alpha)$th quantile of the F-distribution with $b - 1, (b - 1)(k - 1)$ degrees of freedom.

EXAMPLE 9.3.3. The data from Table 9.3.1 are repeated in Table 9.3.5, now assuming that five different driving conditions were involved, as listed. Since this is the same set of data, we again have $\hat{\mu} = 23.647$, $\hat{\tau}_1 = .753$, $\hat{\tau}_2 = .213$, $\hat{\tau}_3 = -.967$. Now, however, we also have to estimate the β_j's, the parameters representing the various driving conditions. The five means for the different conditions are: $\bar{y}_{.1} = 21.167$, $\bar{y}_{.2} = 24.867$, $\bar{y}_{.3} = 24.4$, $\bar{y}_{.4} = 23.267$, $\bar{y}_{.5} = 24.533$; subtracting $\hat{\mu}$ from each gives the estimates of the β_j's, $\hat{\beta}_1 = -2.480$, $\hat{\beta}_2 = 1.220$, $\hat{\beta}_3 = .753$, $\hat{\beta}_4 = -.380$, $\hat{\beta}_5 = .887$. For the analysis of variance to test $H_0: \tau_1 = \tau_2 = \tau_3 = 0$, we already have the

Table 9.3.5

| | Make | | |
Conditions	A	B	C
City	21.7	22.2	19.6
Flat Country, slow	26.5	24.2	23.9
Flat country, fast	24.9	23.9	24.4
Hilly country, slow	23.9	23.7	22.2
Hilly country, fast	25.0	25.3	23.3

sum of squares due to car makes, $SS_A = 7.737$, from Example 9.3.2, and the residual sum of squares for the completely randomized model, 32.080. The only new computations required are the sums of squares for blocks SS_B (the driving conditions) and SS_R, the residual sum of squares for this two-way model. We find

$$SS_B = \tfrac{1}{3} \sum y_{.j}^2 - y_{..}^2/15 = 8414.883 - 8387.473 = 27.411,$$
$$SS_R = 32.080 - 27.411 = 4.669.$$

The two-way analysis of variance for these data is given in Table 9.3.6.

Table 9.3.6

Source	Degrees of Freedom	Sums of Squares	Mean Squares
Car makes	2	7.737	3.869
Conditions	4	27.411	6.853
Residual	8	4.669	0.584
Total	14	39.817	

Notice that there has been a big reduction in the residual sum of squares, because of the large differences between the driving conditions; this has considerably reduced the residual mean square, our estimate of σ^2, the population variance (compare with Table 9.3.3, Example 9.3.2). Now we test $H_0: \tau_1 = \tau_2 = \tau_3 = 0$ by computing $3.869/.584 = 6.63$, which exceeds $F_{.95} = 4.459$ with 2 and 8 degrees of freedom, so now we would reject H_0 with $\alpha = .05$. Recognizing and removing the variability due to the different driving conditions has given a more sensitive test for differences in car makes. From the same analysis of variance we also have the computations necessary for testing that the driving conditions are equal in the effects on mileage; to test $H_0: \beta_1 = \beta_2 = \cdots = \beta_5 = 0$, we compute $6.853/.584 = 11.74$, which exceeds $F_{.95} = 3.838$ with 4 and 8 degrees of freedom.   ■

The analyses of variance we have discussed, and the regression models of Sections 9.1 and 9.2, are special cases of a general class of models called *linear statistical models*, ones that assume observed values of random variables, each of whose expected values are linear functions of unknown parameters. These linear statistical models assume a wide array of particular forms and have been used extensively in the literature of many subject matter areas. (The interested reader can consult S. R. Searle, *Linear Models*, Wiley, 1971, for a more complete introduction to this important topic.)

EXERCISE 9.3

1. The following table records the amounts of weight lost by 19 men, put on one of 3 different diets for 1 month.

Diet 1	12.8	8.6	10.6	6.0	10.5	10.3	
Diet 2	11.3	11.3	12.4	9.5	12.9	14.2	15.9
Diet 3	12.8	17.8	14.0	12.8	12.7	12.6	

(a) Assume a completely randomized model and estimate $\mu, \tau_1, \tau_2, \tau_3$, and σ^2.

(b) Would you accept $H_0: \tau_1 = \tau_2 = \tau_3 = 0$ with $\alpha = .05$?

2. To check their calibration techniques, each of 3 different spectrometers (of the same make and model) analyzed the same oil sample repeatedly. Their results (in ppm) are summarized in the following table.

Spectrometer 1	52	49	47	51	51	50	48	52	53
Spectrometer 2	49	52	51	50	51	50			
Spectrometer 3	50	49	51	50	50				

(a) Assume these are values selected from 3 populations, and a completely randomized model. Estimate $\mu, \tau_1, \tau_2, \tau_3, \sigma^2$.

(b) Would you accept $H_0: \tau_1 = \tau_2 = \tau_3$ with $\alpha = .1$?

3. Four different marine paints were compared for their ability to protect ships in a sea-going environment. A total of 16 ships were used, each

painted with one of four paints. Then each of the ships was deployed for 6 months. On the ships' return a score (bigger is better) was assigned to each based on the amount of chipping, peeling, and average remaining paint thickness. The scores are:

Paint 1	108	69	72	85
Paint 2	79	70	100	91
Paint 3	95	80	80	92
Paint 4	108	119	116	106

(a) Assume a completely randomized model and estimate the unknown parameters.

(b) Would you accept the hypothesis that the paints are equally effective?

4. Four different chemicals were compared for their effectiveness against mildew on each of six different types of surface. The numbers recorded are a measure of suppression ability over a period of time, higher numbers indicating better suppression. The data recorded are given in the following table.

Surface Type	Chemical			
	1	2	3	4
1	6.8	6.8	13.9	18.6
2	12.6	7.3	17.2	14.4
3	13.1	9.6	21.3	18.0
4	15.4	11.9	21.4	18.5
5	10.4	10.8	14.2	18.2
6	11.0	12.4	17.5	19.1

(a) Assume these data were generated by a randomized block model and estimate the parameters in the model.

(b) With the randomized block model would you accept the hypothesis that the 4 chemicals are equally effective in suppressing mildew?

5. If the four columns in the data in Exercise 9.3.3 represent four different geographic areas in which the ships were deployed, would your conclusions regarding differences in the paints be any different?

6. For the completely randomized model, show that $\bar{Y}_{..}$, $\bar{Y}_{i.} - \bar{Y}_{..}$, $Y_{ij} -$

$\overline{Y}_{i.}$ are all mutually independent. (*Hint.* Look at the covariances of the various terms.)

7. For a completely randomized model with all $n_i = n$, show that $\overline{Y}_{i.}$ is normal with mean $\mu + \tau_i$, variance σ^2/n. Use this to show that

$$\sum \frac{n(\overline{Y}_{i.} - \overline{Y}_{..})^2}{\sigma^2}$$

is a χ^2 random variable with $k - 1$ degrees of freedom, if the τ_i's are all equal.

8. The paired T test (Theorem 8.3.3) is intimately related to our randomized block model. Use the data in Example 8.3.4 where the 10 individuals are the blocks (extraneous factor) and before (x) and after (y) values are the two populations to be compared and show that the F statistic for testing the hypothesis that the two populations are the same is equal to $(-13.33)^2 = 177.59$, the square of the T value.

9. The standard T test for comparing the means of two normal populations (Theorem 8.3.1) actually gives the same test as the completely randomized model for two populations. Use the data of Example 8.3.1, and the completely randomized model to evaluate the F statistic for testing the two means are equal; you should find the F statistic equals $(-1.345)^2 = 1.81$, the square of the observed T statistic.

*10. (a) Let Y_{ij}, $i = 1, 2$, $j = 1, 2, \ldots, n_i$ be observed values from a completely randomized model. The F statistic for testing $H_0: \tau_1 = \tau_2 = 0$ is

$$F = \sum n_i(\overline{Y}_{i.} - \overline{Y}_{..})^2 \Big/ \frac{\sum\sum(Y_{ij} - \overline{Y}_{i.})^2}{\sum(n_i - 1)}$$

Show that $F = T^2$, where T is the T statistic of Theorem 8.3.1.

(b) Repeat (a) with the randomized block model ($k = 2$ populations, n observations from each) and show that the F statistic for testing $H_0: \tau_1 = \tau_2 = 0$ equals the square of the paired T statistic of Theorem 8.3.3 for testing $H_0: \mu_D = 0$.

9.4 Summary

Regression line through the origin, equal variances:

$$Y_i = bx_i + e_i, \qquad E[e_i] = E[e_ie_j] = 0, \qquad E[e_i^2] = \sigma^2$$

$$\hat{b} = \frac{\sum x_i y_i}{\sum x_i^2}, \qquad s^2 = \frac{1}{n-1}\sum(y_i - \hat{b}x_i)^2, \qquad s_{\hat{B}}^2 = \frac{s^2}{\sum x_i^2}$$

General straight line regression, equal variances:

$$Y_i = a + bx_i + e_i, \qquad E[e_i] = E[e_i e_j] = 0, \qquad E[e_i^2] = \sigma^2$$

$$\hat{b} = \frac{\sum (x_i - \bar{x})(y_i - \bar{y})}{\sum (x_i - \bar{x})^2}, \qquad \hat{a} = \bar{y} - \hat{b}\bar{x},$$

$$s^2 = \frac{1}{n-2} \sum (y_i - \hat{a} - \hat{b}x_i)^2, \qquad s_{\hat{B}}^2 = \frac{s^2}{\sum (x_i - \bar{x})^2},$$

$$s_{\hat{A}}^2 = \frac{s^2 \sum x_i^2}{n \sum (x_i - \bar{x})^2}, \qquad \text{Cov}[\hat{A}, \hat{B}] = \frac{-\bar{x}\sigma^2}{\sum (x_i - \bar{x})^2}.$$

Least squares estimators are best linear unbiased. With $\text{Var}[Y_i] = \sigma^2 c_i^2$, the best linear unbiased estimators come from minimizing

$$\sum \frac{(Y_i - g(x_i))^2}{c_i^2}.$$

With normality, the least squares estimators for a and b are also maximum likelihood estimators.

With normality, $(\hat{B} - b)/S_{\hat{B}}$ and $(\hat{A} - a)/S_{\hat{A}}$ both have the T distribution with $n - 2$ degrees of freedom.

With normality,

$$\hat{Y}_{x^*} = \hat{A} + \hat{B}x^*, \qquad S_{x^*}^2 = \left(\frac{1}{n} + \frac{(x^* - \bar{x})^2}{\sum (x_i - \bar{x})^2} \right) S^2, \quad S_{x^*+e}^2 = S_{x^*}^2 + S^2,$$

$$\frac{\hat{Y}_{x^*} - (a + bx^*)}{S_{x^*}}$$

has the T distribution with $n - 2$ degrees of freedom, as does $(\hat{Y}_{x^*} - Y)/S_{x^*+e}$, where Y is a future observed value with $x = x^*$.

Completely randomized model:

$$Y_{ij} = \mu + \tau_i + e_{ij}, \qquad i = 1, 2, \ldots, k, \qquad j = 1, 2, \ldots, n_i,$$

where μ and the τ_i's are unknown, the e_{ij}'s are independent, normal, $0, \sigma^2$. Analysis of variance given in Table 9.3.2.

Randomized block model:

$$Y_{ij} = \mu + \tau_i + \beta_j + e_{ij}, \qquad i = 1, 2, \ldots, k, \qquad j = 1, 2, \ldots, b$$

where μ, the τ_i's and β_j's are unknown, the e_{ij}'s are normal and independent, with mean 0, and variance σ^2. Analysis of variance given in Table 9.3.4. Assumes no interaction between populations and the extraneous factor.

10

Nonparametric Methods

In discussing tests of hypotheses and confidence intervals, we have in every instance assumed a random sample of a random variable whose probability law had a known form (normal, exponential, binomial, etc.). The hypotheses we considered specified values for parameters of the probability law of the random variable and the confidence intervals we examined gave interval estimates for the parameters. These techniques are frequently called *parametric* because of this; the procedures are designed to elicit information about specific parameters, taking advantage of the known probability law. It is, of course, possible that the assumed form for the probability law is incorrect, in which case the resulting inferences would be suspect. Most of the techniques we have discussed are fairly *robust*, meaning they can be expected to perform fairly well even if the actual probability law is not exactly the same as the one assumed. We unfortunately do not have the space to treat this important topic.

Instead, we will in this chapter investigate a number of *nonparametric* procedures, those that are frequently designed to make inferences about quantities that are not specifically parameters of an assumed probability law. These procedures generally make fewer assumptions about the underlying probability law and thus can be expected to perform quite well over a spectrum of possible distributions. This generality is, of course, gotten for a price: if a nonparametric procedure could be used to test a hypothesis, say, and a parametric procedure could be used to test the *same* hypothesis (say, one assuming a sample from a normal population), the parametric procedure is in general more sensitive, as long as the sample actually came from the assumed probability law. The nonparametric procedure, however, might well be preferred if one is not completely sure he or she has assumed the correct probability law.

10.1   Inferences About Quantiles

Recall that the 100pth quantile of a continuous probability law is the number t_p such that

$$F(t_p) = p,$$

where F is the distribution function. The median, $t_{.5}$, is frequently used as a measure of location and the interquartile range, $t_{.75} - t_{.25}$, or the interdecile range, $t_{.9} - t_{.1}$, are measures of spread for the probability law. In this section we will study some simple inference procedures for the quantiles of any continuous distribution.

Let $X_1, X_2, \ldots, X_n$ be a random sample of a continuous random variable X with distribution function $F(x)$, and let $X_{(1)}, X_{(2)}, \ldots, X_{(n)}$ represent the order statistics (ordered sample values, see Section 5.7). Then, as we saw in Section 6.2, $U_{(j)} = F(X_{(j)}), j = 1, 2, \ldots, n$, are the order statistics for a uniform $(0, 1)$ random variable; $U_{(j)}$ has a beta probability law with parameters $a = j, b = n - j + 1$, and thus

$$E[U_{(j)}] = \frac{j}{n + 1} :$$

that is,

$$E[F_X(X_{(j)})] = \frac{j}{n + 1}.$$

Then with $p = j/(n + 1)$, it seems natural to use $X_{(j)}$ as an estimator for t_p, the pth quantile, no matter what the continuous probability law may be. If $j/(n + 1) < p < (j + 1)/(n + 1)$, a natural estimator for t_p would be a linear combination of $X_{(j)}$ and $X_{(j+1)}$.

It is also easy to find confidence intervals for t_p using the order statistics. Repeating the reasoning used in Section 6.2, the event $\{X_{(j)} \leq t_p\}$ occurs if and only if j or more of the sample values are less than or equal to t_p. But each individual sample value is less than or equal to t_p with probability p, so the sample values can be thought of as independent Bernoulli trials, each with probability p of success. That is,

$$P(X_{(j)} \leq t_p) = \sum_{i=j}^{n} \binom{n}{i} p^i (1 - p)^{n-i}$$

a simple binomial computation. Thus $X_{(j)}$ itself serves as a lower confidence limit for t_p, with confidence coefficient

$$\sum_{i=j}^{n} \binom{n}{i} p^i (1 - p)^{n-i}.$$

If $k > j$,

$$\{X_{(j)} \le t_p < X_{(k)}\} \cup \{X_{(k)} \le t_p\} = \{X_{(j)} \le t_p\}$$

and the two events on the left are mutually exclusive. Thus for $k > j$,

$$P(X_{(j)} \le t_p < X_{(k)}) = P(X_{(j)} \le t_p) - P(X_{(k)} \le t_p)$$

$$= \sum_{i=j}^{n} \binom{n}{i} p^i (1 - p)^{n-i} - \sum_{i=k}^{n} \binom{n}{i} p^i (1 - p)^{n-i}$$

$$= \sum_{i=j}^{k-1} \binom{n}{i} p^i (1 - p)^{n-i}$$

and the interval $[X_{(j)}, X_{(k)})$ is a two-sided confidence interval for t_p with confidence coefficient

$$\sum_{i=j}^{k-1} \binom{n}{i} p^i (1 - p)^{n-i}.$$

EXAMPLE 10.1.1. In Example 7.3.4 the observed times to failure for 10 devices were (in ranked order) 37.6, 109.0, 129.9, 188.1, 409.5, 499.0, 529.5, 582.4, 607.5, 1947.0, and assuming these are a random sample of an exponential random variable, the observed 95% confidence limits for $1/\lambda$, the mean, were found to be (320.5, 925.9), an interval of length roughly 600 hours. For an exponential random variable,

$$F\left(\frac{1}{\lambda}\right) = 1 - e^{-\lambda/\lambda} = 1 - e^{-1} = .632,$$

so the mean $1/\lambda$ is the 63.2th quantile. The interval from the first-order statistic $x_{(1)} = 37.6$ to the 9th, $x_{(9)} = 607.5$, is shorter than 600 hours: the probability this interval covers $t_{.632} = 1/\lambda$ is

$$\sum_{i=1}^{8} \binom{10}{i} (.632)^i (.378)^{10-i} = .931.$$

Also

$$P(X_{(4)} \le t_{.632}) = \sum_{i=4}^{10} \binom{10}{i} (.632)^i (.368)^{10-i} = .966,$$

so the probability is .966 that $t_{.632}$ is larger than $x_{(4)} = 188.1$. ∎

To test hypotheses about t_p, the $100p$th quantile of a continuous distribution, based on a random sample of n observations from the population, we could convert the preceding confidence intervals into tests. That is, suppose we wanted to test $H_0 : t_p = t^*$ versus $H_1 : t_p \neq t^*$, where t^* is some specified number. If $[X_{(j)}, X_{(k)})$ is a $100(1 - \alpha)\%$ confidence interval for t_p, we accept H_0 if $X_{(j)} \leq t^* < X_{(k)}$ and reject H_0 otherwise: this test has $P(\text{Type I error}) = \alpha$.

This conversion of confidence intervals for t_p into tests actually is very simple and is equivalent to what is frequently called the *sign test*. As previously stated we will reject $H_0 : t_p = t^*$ if $t^* < X_{(j)}$ (which says that $j - 1$ or fewer of the X_i's are smaller than t^*) or if $t^* \geq X_{(k)}$ (which says that k or more of the X_i's are smaller than t^*). That is, the rule for accepting or rejecting $H_0 : t_p = t^*$ hinges on the value for Y, the number of X_i's that are smaller than the hypothesized value t^* (equivalently, the number of $t^* - X_1, t^* - X_2, \ldots, t^* - X_n$, which are positive, the reason it is called the sign test). If in fact $t_p = t^*$, the probability is p that each sample value does not exceed t^* and Y, the number of these in the sample of n, then is binomial, n, p. We reject H_0 if $Y \leq c$ or $Y \geq d$, where

$$P(Y \leq c) + P(Y \geq d) = \sum_{i=0}^{c} \binom{n}{i} p^i (1 - p)^{n-i} + \sum_{i=d}^{n} \binom{n}{i} p^i (1 - p)^{n-i}$$

$$= \alpha.$$

The sign test for $H_0 : t_p = t^*$ bases the decision to accept or reject H_0 on just the number of the $(t^* - X_i)$'s that are positive, ignoring their magnitudes. When $p = .5$ so we are testing a hypothesis about the median of the distribution (and the density function is symmetric about $t_{.5}$), Wilcoxon suggested a modification to the sign test, making it a more sensitive test. Again, let $X_1, X_2, \ldots, X_n$ be a random sample of a continuous random variable whose density function is symmetric, $f(t_{.5} + x) = f(t_{.5} - x)$, so in fact $\mu = t_{.5}$. We want to test $H_0 : t_{.5} = t^*$ versus $H_1 : t_{.5} \neq t^*$, where again t^* is some specified value. Also, we compute $t^* - X_1, t^* - X_2, \ldots, t^* - X_n$, as in the sign test, and then rank the absolute values of these differences in order of magnitude; let R_i be the rank of $|t^* - X_i|$ in this ordering. For example, with $n = 4$, suppose our observed values are 15.2, 8.1, 10.7, 12.2, respectively, and we want to test $H_0 : t_{.5} = 9$, say. Then the observed differences $t^* - X_i$ are $-6.2, .9, -1.7, -3.2$ and the ranks (from smallest to largest) of the absolute values are 4, 1, 2, 3, respectively, so $R_1 = 4$, $R_2 = 1$, $R_3 = 2$, $R_4 = 3$. Because we are sampling from a continuous population, the probability that any two absolute differences, $|t^* - X_i|$ and $|t^* - X_j|$, are exactly equal is zero so this possibility is ignored in the theory of the test. In practice, of course, ties could occur; it is recommended that the average of the ranks be assigned to those that are tied.

The quantities $R_1, R_2, ..., R_n$ are just a listing of the integers $1, 2, ..., n$ in some order; in fact, $R_1, R_2, ..., R_n$ is a random permutation of $1, 2, ..., n$. Now define

$$V_i = -1 \quad \text{if} \quad t^* - X_i < 0$$
$$= 1 \quad \text{if} \quad t^* - X_i > 0;$$

again, because of the continuity, no other cases need to be considered. If $H_0 : t_{.5} = t^*$ is true, $V_1, V_2, ..., V_n$ are independent, $P(V_i = 1) = P(V_i = -1) = \frac{1}{2}$ so $E[V_i] = 0$, $\text{Var}[V_i] = 1$ for each i. The Wilcoxon statistic for testing H_0 is

$$W = \sum_{i=1}^{n} R_i V_i;$$

this test is called the *signed rank* test.

The sum $W = \sum R_i V_i$ includes each of the integers 1 through n (exactly once) and each integer is multiplied by 1 or -1, determined by the V_i's, which are independent. Also, because of the symmetry of the density function, any deviation from $t_{.5}$ of a given magnitude is equally likely to be positive or negative, so the distribution for W is the same as the distribution of $\sum_{i=1}^{n} C_i$ where the C_i's are independent and

$$P(C_i = i) = P(C_i = -i) = \frac{1}{2}.$$

Thus

$$E[W] = \sum E[C_i] = 0$$

$$\text{Var}[W] = \sum_{i=1}^{n} E[C_i^2] = \sum_{i=1}^{n} i^2 = \frac{n(n+1)(2n+1)}{6}.$$

For small n it is possible (although tedious) to work out the probability distribution for W, assuming $H_0 : t_{.5} = t^*$ is true. The distribution is symmetric with a maximum at zero and is quite flat for small values of n. (Tables of the distribution for W, for small n, are available in many texts, including *Nonparametric Statistical Methods Based on Ranks*, by E. L. Lehmann, Holden-Day, 1975.) It can be shown that the distribution for $W/\sqrt{\text{Var}[W]}$ converges to the standard normal as $n \to \infty$, so a normal approximation for $P(W \leq w)$ is quite good for large n. Certainly, with $n \geq 20$ the normal approximation should be quite adequate for most purposes. In testing $H_0 : t_{.5} = t^*$ versus $H_1 : t_{.5} \neq t^*$, with the signed rank test, we would reject H_0 if $W \leq w_1$ (because this happens with large negative values of $t^* - X_i$, indicating $t_{.5}$ may be greater than t^*) or if $W \geq w_2$ (because this happens with large positive values of $t^* - X_i$, indicating $t_{.5}$ may be smaller than t^*), where $P(W \leq w_1) + P(W \geq w_2) = \alpha$, assuming H_0 is true.

EXAMPLE 10.1.2. A new synthetic fiber was tested for breaking strength: each of 20 sample filaments was stretched until a break occurred. The pressures needed (in pounds) to break the 20 fibers were respectively:

> 156.0, 255.5, 132.0, 246.7, 867.9, 86.4, 610.4, 125.7,
> 150.4, 117.6, 201.9, 207.2, 189.8, 585.8, 153.1, 565.4,
> 511.0, 567.0, 222.3, 141.5

It was expected that this fiber would have a median breaking strength of at least 400 pounds. Let us first use the sign test to test $H_0 : t_{.5} \geq 400$ versus $H_1 : t_{.5} < 400$. The test statistic for the sign test is simply the number Y of the $(400 - X_i)$ values that are positive—just the number of observed values smaller than 400; we easily find $y = 14$. We would want to reject H_0 for large values of Y, because large Y values indicate a large number of observations below the hypothesized median of 400 pounds, which in turn makes it appear the true median is below 400. With H_0 assumed true, Y is binomial with parameters $n = 20$, $p = \frac{1}{2}$, and we find $P(Y \geq 14) = .0577$. Thus we would reject H_0 with $\alpha = .0577$ using the sign test. Let us also use the signed rank test to test $H_0 : t_{.5} \geq 400$ versus $H_1 : t_{.5} < 400$. To do so, we must compute the $400 - X_i$ values, given as follows, in the same order as the original data:

> 244.0, 144.5, 268.0, 153.3, -467.9, 313.6, -210.4, 274.3,
> 249.6, 282.4, 198.1, 192.8, 210.2, -185.8, 246.9, -165.4,
> -111.0, -167.0, 177.7, 258.5

and then compute the ranks of the absolute values of these differences. In the same order, the ranks of the absolute values (the R_i's) are:

> 12, 2, 16, 3, 20, 19, 11, 17, 14, 18, 9, 8, 10, 7, 13,
> 4, 1, 5, 6, 15

respectively. The sum of the ranks of the negative values is 48 and the sum of the ranks of the positive values is 162, so the observed value for the Wilcoxon signed rank statistic is $w = 162 - 48 = 114$. Since $n = 20$ is "large," the Wilcoxon signed rank statistic is approximately normal with mean 0 and variance $n(n + 1)(2n + 1)/6 = 2870$. Again, we should reject $H_0 : t_{.5} \geq 400$ for large values of W (which are consistent with $t_{.5} < 400$); using the normal approximation, we find

$$P(W \geq 114) = P\left(\frac{W}{\sqrt{2870}} \geq \frac{114}{\sqrt{2870}} \right)$$

$$\doteq 1 - N(2.13) = .0166$$

so we would reject H_0 with $\alpha = .0166$. Notice that these same data seem more extreme (reject with a smaller α) using the signed rank test than with the sign test; this will frequently be the case and is what was meant by the earlier comment that the signed rank test is designed to be more sensitive. ∎

The sign test can be used to test hypotheses about the pth quantile, t_p, for any p, for continuous distributions. The signed rank test is for use specifically for testing hypotheses about $t_{.5}$, the median of a symmetric continuous distribution. In testing $H_0 : t_{.75} = t^*$ versus $H_1 : t_{.75} \neq t^*$, for example, we could still define the Wilcoxon statistic $W = \sum R_i V_i$, but its distribution is no longer easily worked out, nor independent of the underlying symmetric continuous probability law; thus we cannot use its value as a nonparametric test of the value of t_p, for $p \neq \frac{1}{2}$.

Both the sign test and signed rank test can be used for any cases in which the data can be transformed into variables whose density is hypothesized to be symmetric about $t_{.5}$. The following example discusses an application similar to the paired T test, which we discussed in Section 7.4.

EXAMPLE 10.1.3. To test whether or not some substance has an effect on the time needed for an individual to accomplish some task, suppose $n = 10$ individuals participate in an experiment. Each individual is timed performing the task initially, yielding $X_1, X_2, ..., X_{10}$. Then the same individuals consume the substance and again are timed in performing the task, giving $Y_1, Y_2, ..., Y_{10}$. The hypothesis that the substance has no effect implies the distribution for each X_i is the same as for the corresponding Y_i; thus the difference, $X_i - Y_i$, then has mean 0 and is symmetric, if the hypothesis is true. Furthermore, if we assume the distributions for the differences, $X_i - Y_i$, are the same for all 10 people and let $t_{.5}$ represent the median difference, we could use either the sign or signed rank test for $H_0 : t_{.5} \geq 0$ versus $H_1 : t_{.5} < 0$ (assuming we feel it is possible the substance may slow the reaction time). If for each pair we find $X_i - Y_i < 0$, so that each person's reaction time is slower after the substance than before, the number of positive differences is 0 and, with the sign test, we would reject H_0 with probability $P(Y \leq 0) = 1/2^{10} = .000977$. With this same assumed outcome (all differences negative) the sign of each rank is negative and the value of the Wilcoxon statistic is

$$w = - \sum_{i=1}^{10} i = -55,$$

its smallest possible value; since the only way we can observe $w = -55$ is for each V_i to equal -1, $P(W \leq -55) = 1/2^{10} = .000977$, the same significance level as the sign test.

Now suppose exactly nine of the differences, $X_i - Y_i$, are negative, and one is positive. With the sign test we would now reject $H_0 : t_{.5} \geq 0$ with probability

$$P(Y \leq 1) = \frac{1 + 10}{2^{10}} = .0107.$$

If we were to use Wilcoxon's signed rank test, there are 10 possible values for w, depending on the magnitude of the single positive difference relative to the others. If r_i is the rank of the single positive difference, it is easy to see that $w = -55 + 2r_i$, so w can vary from -53 to -35, depending on the value of r_i. By enumerating all the possible values for w (with $n = 10$) the values in Table 10.1.1 result (this is a partial listing of the distribution function for W, with $n = 10$ and H_0 true). Thus if the single positive differ-

Table 10.1.1

w	-55	-53	-51	-49	-47	-45	-43	-41	-39	-37	-35
$P(W \leq w)$ $\times 2^{10}$	1	2	3	5	7	10	14	19	25	33	43

ence were the smallest in magnitude we would reject H_0 with probability

$$P(W \leq -53) = \frac{2}{2^{10}} = .001954,$$

whereas if the single positive difference were largest in magnitude (of the 10) we would observe $w = -35$ and reject H_0 with probability

$$P(W \leq -35) = \frac{43}{2^{10}} = .0411.$$

It is not always the case that the signed rank significance level is smaller than the significance level of the sign test.

EXERCISE 10.1

1. A fuse manufacturing firm is interested in the distribution of times it takes their fuses to burn (and ignite the attached charge). The burning times of 15 fuses, in decimilliminutes, were:
1.0038, 1.0139, 1.0026, 1.0234, 1.0120, 0.9911, 1.0195, 1.0169, 1.0090, 1.0083, 1.0079, 0.9923, 0.9936, 0.9988, 1.0302.

(a) Rank these times in order of magnitude and estimate the quartiles of the distribution $t_{.25}, t_{.50}, t_{.75}$.

(b) Estimate the time, ζ, which 90 percent of the fuse burning times should exceed.

2. In his study of a certain species of ant, a zoologist visited a field inhabited by the species. In one section of the field he found 7 anthills; thus there are $\binom{7}{2} = 21$ distinct pairs which can be selected from these 7. He measured the distances (in meters) between all possible pairs of anthills, giving the following values.

> 3.95, 10.8, 9.57, 3.01, 19.35, 12.67, 1.05, 12.2,
> 6.86, 4.6, 3.33, 2.49, 3.39, 7.2, 3.96, 4.46, 4.54,
> 1.95, 5.79, 4.65, 3.8.

(a) What would you estimate as the median distance between anthills for this species?

(b) The six smallest of these values give the shortest observed distances between anthills. If an anthill claims territorial prerogative over all land within a circle, centered at the hill, with radius equal to half the distance to its nearest neighbor, what is your estimate of the median prerogative area for an anthill (area $= \pi r^2$)?

3. A college broadjumper made 20 jumps in competition in a large track meet. The distances he jumped on these 20 attempts are: (in meters)

> 7.13, 8.32, 8.20, 8.08, 8.39, 8.24, 8.38, 8.13,
> 8.27, 8.11, 8.61, 8.17, 7.96, 8.09, 7.88, 8.75,
> 8.41, 8.05, 8.21, 8.03.

(a) Estimate the distance d, his jump should exceed with probability .95.

(b) Estimate the median distance he could jump on this day.

4. The times of flight (in minutes) for a Boeing 747, from the time the wheels left the ground at San Francisco International until they touched down at Dulles International in Washington, D.C., for 15 successive flights are:

> 256.4, 252.0, 259.6, 258.9, 261.8, 251.2, 267.2,
> 252.7, 259.5, 260.1, 253.3, 255.3, 255.0, 251.0, 261.5.

(a) Estimate the median for this distribution.

(b) The airline involved wants its published flight time for this flight to be a value that they can meet at least 80% of the time. Based on the preceding observations, what would you recommend as the published flight time?

5. The times taken by 25 persons to complete a written California driver's license examination follow (in minutes).

> 10.72, 3.82, 4.42, 4.8, 11.91, 12.86, 8.16, 7.16,
> 8.38, 8.13, 12.41, 4.01, 12.7, 6.22, 7.88, 2.55,
> 3.55, 10.12, 8.82, 5.57, 3.58, 14.36, 2.7, 3.19, 14.03.

 (a) Estimate the 30th and 70th quantiles of the distribution of time required to complete this exam.

 (b) Based on these data, what is your estimate of the probability the next person to take this exam will require more than 10 minutes to complete it?

6. Evaluate a lower 95% (approximately) confidence limit for $t_{.9}$, the 90th quantile for the fuse burning times, based on the data in Exercise 10.1.1.

7. Using the data in Exercise 10.1.5, would you accept $H_0 : t_{.5} = 5$ versus $H_1 : t_{.5} \neq 5$ with $\alpha = .1$ (approximately)?

8. Compute the two-sided (approximate) 96% confidence limits for the median time of flight from San Francisco to Dulles, based on the data in Exercise 10.1.4.

9. A friend of the college broad jumper, mentioned in Exercise 10.1.3, claims this jumper has probability of at least .8 of jumping 8.5 meters on each of his attempts. Using the data in that exercise, would you accept this statement?

10. Use both the sign test and the signed rank test for $H_0 : t_{.5} = 8$ versus $H_1 : t_{.5} \neq 8$, with $\alpha = .05$, for the data given in Exercise 10.1.3.

11. Repeat Exercise 10.1.10 for the data in Exercise 10.1.5.

12. If t_p is the pth quantile of a normal distribution,

$$p = F(t_p) = N\left(\frac{t_p - \mu}{\sigma}\right)$$

 so $t_p = \mu + \sigma z_p$, where z_p is the pth quantile of the standard normal distribution.

 (a) Suggest a method of estimating t_p for a sample from a normal distribution.

 (b) Assume the flight times in Exercise 10.1.4 are observations from a normal population and use the procedure in (a) to answer Exercise 10.1.4 (b).

13. (a) Express the pth quantile, t_p, of an exponential distribution as a function of λ.

 (b) Use the data from Exercise 10.1.5, assuming these values came from an exponential distribution and estimate the 30th and 70th quantiles of the distribution.

10.2 The Run Test and the Wilcoxon–Mann–Whitney Test

In the last section we examined some procedures for estimating quantiles of continuous distributions and for testing hypotheses about continuous distributions. In this section we will continue to discuss samples from continuous distributions and two additional frequently used nonparametric tests that can be employed for a variety of hypotheses. We will begin with the run test.

Let us first discuss the possible arrangements of $m + n$ symbols, where m are of one kind (say, 0's) and the remaining n are of a second kind (say, 1's). The total number of arrangements possible is, of course, $\binom{m + n}{m}$, since any m of the $m + n$ positions could be chosen for the 0's, and then the 1's are placed in the remaining n positions. For example, with $m = 4, n = 7$, there are $\binom{11}{4} = 330$ possible arrangements, such as

$$0\ 0\ 0\ 0\ 1\ 1\ 1\ 1\ 1\ 1\ 1$$
$$0\ 1\ 0\ 1\ 1\ 1\ 0\ 1\ 1\ 0\ 1$$

and so on. For various purposes, as we will see, it is useful to know the number of *runs* in any such arrangement, where a run is defined to be a sequence (however long) of positions occupied by the same symbol. For example, the arrangement

$$1\ 1\ 1\ 1\ 1\ 1\ 0\ 0\ 0\ 0$$

has two runs, the arrangement

$$0\ 1\ 0\ 1\ 1\ 0\ 1\ 1\ 1\ 0\ 1$$

has eight runs, and so on, as marked by the underlining. With $m + n$ symbols placed in a row at random, each possible distinct arrangement has probability $1 / \binom{m + n}{m}$, and each such arrangement results in some number, R, of runs. Let us now derive the probability distribution for R, the total number of runs in a random arrangement. Clearly, the smallest possible number of runs is 2, given by one run of 0's followed by one run of 1's, or vice versa. The largest possible number of runs is given by alternating single 0's and single 1's; this number is $2m + 1$, if $m < n$, and is $2m$, if $m = n$. Any integer in between these two could occur as the number of runs, so the range of R is $\{2, 3, \ldots, 2m + 1\}$ or $\{2, 3, \ldots, 2m\}$, depending on whether $m < n$ or $m = n$.

Consider any odd observed value $r = 2k + 1$: this number of runs occurs in any arrangement that contains k runs of 0's and $k + 1$ runs of 1's, or vice versa. To get k runs of 0's, consider the m 0's lying in a row; there are $m - 1$ possible interior positions between pairs of 0's, and we can choose any $k - 1$ of these to receive dividers, delineating the k runs of 0's. This operation can be done in $\binom{m-1}{k-1}$ ways. Similarly, there are $\binom{n-1}{k}$ ways of splitting the n 1's into exactly $k + 1$ runs. These k runs of 0's can be merged with the $k + 1$ runs of 1's, then, in $\binom{m-1}{k-1}\binom{n-1}{k}$ ways, where the dividers of the 0's are replaced by the runs of 1's in all possible ways. Similarly, we will observe $r = 2k + 1$ runs if there are $k + 1$ runs of 0's and k runs of 1's; the number of arrangements yielding this outcome is $\binom{m-1}{k}\binom{n-1}{k-1}$, and we have

$$P(R = 2k + 1) = \frac{\binom{m-1}{k-1}\binom{n-1}{k} + \binom{m-1}{k}\binom{n-1}{k-1}}{\binom{m+n}{m}}$$

as the value of the probability function for R for odd observed values.

We will observe $r = 2k$, an even number of runs only if there are k runs of 0's and k runs of 1's. Using the preceding reasoning again, there are $\binom{m-1}{k-1}\binom{n-1}{k-1}$ arrangements containing k runs of 0's and k runs of 1's, reading from left to right, each starting with a run of 0's; clearly each of these could also be read from right to left, giving an equal number of arrangements that start with a run of 1's. Thus there are $2\binom{m-1}{k-1}\binom{n-1}{k-1}$ arrangements containing $2k$ runs and the probability function for R is

$$P(R = 2k) = \frac{2\binom{m-1}{k-1}\binom{n-1}{k-1}}{\binom{m+n}{m}}$$

$$P(R = 2k + 1) = \frac{\binom{m-1}{k-1}\binom{n-1}{k} + \binom{m-1}{k}\binom{n-1}{k-1}}{\binom{m+n}{m}},$$

where $k = 1, 2, ..., m \leq n$. It can be shown that

$$E[R] = \frac{2mn}{m + n} + 1,$$

$$\text{Var}[R] = \frac{2mn(2mn - m - n)}{(m + n)^2 (m + n - 1)}.$$

We will soon discuss how the distribution for R can be used in defining certain nonparametric tests.

The exact distribution for R can, of course, be evaluated from the preceding formulas, a chore that becomes more tedious as m and n get larger. Fortunately, it has been shown that as $m \rightarrow \infty$, $n \rightarrow \infty$, with $(m/n) \rightarrow \zeta$, the distribution for

$$\left(R - \frac{2m}{1 + \zeta} \right) \bigg/ \sqrt{\frac{4\zeta m}{(1 + \zeta)^3}}$$

tends to the standard normal. This suggests that, replacing ζ by m/n, the distribution for $\left(R - \dfrac{2mn}{m + n} \right) \bigg/ \dfrac{2mn}{(m + n)^{3/2}}$ should be approximately standard normal, for large m and n. Let us discuss using the normal approximation to find unusually small or unusually large values for R, the number of runs. Let t_p, where $p < .5$, be the largest integer such that $P(R \leq t_p) \leq p$; we might call t_p the pth quantile for the distribution for R. From the previous normal approximation, then

$$P\left(R \leq t_p \right) \doteq N\left(\left(t_p - \frac{2mn}{m + n} \right) \bigg/ \frac{2mn}{(m + n)^{3/2}} \right)$$

giving

$$t_p \doteq \frac{2mn}{m + n}\left(1 + \frac{z_p}{\sqrt{m + n}} \right),$$

where $N(z_p) = p$. It will also be convenient to let u_q be the smallest integer such that $P(R \geq u_q) \leq q$, where $q < .5$. Thus

$$1 - q \leq P(R < u_q) = P(R \leq u_q - 1)$$

$$\doteq N\left(\left(u_q - 1 - \frac{2mn}{m + n} \right) \bigg/ \frac{2mn}{(m + n)^{3/2}} \right)$$

giving

$$u_q \doteq 1 + \frac{2mn}{m + n}\left(1 + \frac{z_{1-q}}{\sqrt{m + n}} \right),$$

where $N(z_{1-q}) = 1 - q$. If both m and n are at least 20, these approximations for t_p and u_q, rounded to the nearest integer, are very good. In fact, if both m and n are at least 10, these approximations are still quite good and, where they are in error by 1, they are generally conservative. (For example, with $m = n = 10$, the exact value for $t_{.005}$ is 5, whereas the approximation gives 4; the exact value for $u_{.005}$ is 16, whereas the approximation gives 17.) We will employ this normal approximation to evaluate t_p and u_q for both m and n at least 10. For smaller values of m and n, the probability law for R is available from published tables or a hand-held calculator can be employed without much difficulty to evaluate this exact distribution. (A table of the exact distribution for R, for small m and n, can be found in *Practical Non-parametric Statistics*, 2nd ed., by W. J. Conover, Wiley, 1980.)

A major use for the distribution of R is in testing hypotheses of randomness of sample values. Suppose $X_1, X_2, \ldots, X_n$ is assumed to be a random sample of a random variable X; the observed sample values are $x_1, x_2, \ldots, x_n$ and let m_e represent the sample median (the middle sample value if n is odd, any number between the two middle values if n is even). In the order in which the observations occurred, replace x_i by 0 if $x_i < m_e$ and replace x_i by 1 if $x_i \geq m_e$. This then generates a sequence of 0's and 1's, which, if we have a random sample, should be in a random order. We would doubt the randomness of the sample if we observe either too few runs (0 0 0 0 1 1 1 1 1 would indicate an increasing trend) or too many runs (1 0 1 0 1 0 1 0 seems to indicate a cyclic effect). Thus to test the hypothesis that our sample values occurred in a random order, we would reject the hypothesis if the observed number of runs $r \leq t_p$ or $r \geq u_q$; our probability of type I error then is $\alpha = p + q$.

EXAMPLE 10.2.1.   In Example 10.1.2 are given the observed breaking strengths of 20 pieces of synthetic fiber. The median of these observed values is $\frac{1}{2}(201.9 + 207.2) = 204.6$; replacing each value smaller than the median by 0 and each value larger than the median by 1 gives the following sequence (in the order of the given observations):

$$0\ 1\ 0\ 1\ 1\ 0\ 1\ 0\ 0\ 0\ 0\ 1\ 0\ 1\ 0\ 1\ 1\ 1\ 1\ 0$$

This sequence contains $r = 13$ runs; to test the randomness of these observations, with $\alpha = .1$, we compute

$$t_{.05} = \frac{2(10)(10)}{10 + 10}\left(1 - \frac{1.64}{\sqrt{10 + 10}}\right) = 6.33 \sim 6$$

$$u_{.05} = 1 + \frac{2(10)(10)}{10 + 10}\left(1 + \frac{1.64}{\sqrt{10 + 10}}\right) = 14.67 \sim 15.$$

Since the observed $r = 13$ runs lies between these two values, we accept the hypothesis of randomness of these observations. ∎

The run test gives a simple way to test the hypothesis that two independent random samples came from the same population, although the next statistic we will discuss, the Wilcoxon–Mann–Whitney statistic, gives a more sensitive test for this type of hypothesis. Suppose $X_1, X_2, ..., X_n$ is a random sample of a continuous random variable X and $Y_1, Y_2, ..., Y_m$ is an independent random sample of a continuous random variable Y. We want to test the hypothesis that the distribution of X values is the same as the distribution of Y values, $H_0: F_X(t) = F_Y(t)$ for all t, versus the alternative that H_0 is false. To do so we can put all $m + n$ values together, rank these combined sample values in order of magnitude, and then replace the X values by 1's, the Y values by 0's, generating a sequence of n 1's and m 0's. If in fact H_0 is true, we would expect the 0's and 1's to be quite thoroughly intermixed, giving a large value for R, the number of runs. If the median for X is smaller than the median for Y, say, we would expect most of the 1's to occur early in the ranked sequence, leading to a fairly small number of runs. Or if the two medians are equal, but the interquartile range for X exceeds that of Y, we would expect most of the 0's in the middle, with 1's at both ends, again giving a relatively small number of runs. Thus we will reject $H_0: F_X(t) = F_Y(t)$ for all t if we find $R \leq t_p$; our probability of type I error is $\alpha = p$.

EXAMPLE 10.2.2. Assume $n = 13$ public school students in grade 6 took a standard mathematics achievement test, receiving the following scores: 578, 274, 344, 294, 617, 339, 270, 563, 336, 323, 440, 351, 384. We will assume these are a random sample of the achievement scores of all public school 6th graders in this school district (X population). Also, assume $m = 11$ private school students in grade 6 in the same area took the same achievement test; their scores were 545, 474, 512, 406, 477, 603, 471, 429, 417, 398, 523. We assume these are a random sample of all private school 6th grade achievement scores for this same area (Y population). Let us use the run statistic to test $H_0: F_X(t) = F_Y(t)$ for all t, the two populations of achievement scores are the same. To do so we must combine the two samples and rank them in magnitude, giving

270, 274, 294, 323, 336, 339, 344, 351, 384, 398, 406,
417, 429, 440, 471, 474, 477, 512, 523, 545, 563, 578,
603, 617.

Now we replace each X value by 1 and each Y value by 0, giving

1 1 1 1 1 1 1 1 1 0 0 0 0 1 0
0 0 0 0 0 1 1 0 1.

The observed number of runs is $r = 7$. Using the normal approximation

$$t_{.05} \doteq \frac{2(11)(13)}{11 + 13}\left(1 - \frac{1.64}{\sqrt{11 + 13}}\right) = 7.927 \cong 8$$

(again, actually the exact value) and since we have observed 8 or fewer runs in the combined sample, we would reject H_0 with $\alpha = .05$. It would appear from these samples that the private school students appear to score higher on the test, because the 0's are pretty well congregated toward the end of the combined ranked sample. ∎

Now let us discuss a more sensitive statistic that is frequently used to test that two independent samples came from the same continuous distribution, the Wilcoxon–Mann–Whitney test, also frequently called the *rank-sum test*. Let $X_1, X_2, \ldots, X_n$ be a random sample of a continuous random variable X and let $Y_1, Y_2, \ldots, Y_m$ be an independent random sample of a continuous random variable Y. As in the run test, the two samples are combined and ranked in order of magnitude from smallest to largest. Let W_X denote the sum of the ranks of the n X values and let W_Y be the sum of the ranks of the m Y values. Clearly,

$$W_X + W_Y = \sum_{i=1}^{m+n} i = (m + n)(m + n + 1)/2.$$

Wilcoxon suggested using W_Y (or equivalently W_X) as the test statistic for $H_0 : F_X(t) = F_Y(t)$ for all t. For small values of m and n it is not difficult (but it is tedious) to work out the exact distribution for W_Y under the assumption H_0 is true. For example, with $n = 4$, $m = 3$, there are $\binom{7}{3} = 35$ different collections of 3 ranks that might be assigned to the Y values, all of which are equally likely to occur if H_0 is true. These 35 possible assignments are listed in Table 10.2.1, along with the resulting value for W_Y.

Table 10.2.1

Ranks	W_Y	Ranks	W_Y	Ranks	W_Y	Ranks	W_Y	Ranks	W_Y
123	6	136	10	167	14	247	13	356	14
124	7	137	11	234	9	256	13	357	15
125	8	145	10	235	10	257	14	367	16
126	9	146	11	236	11	267	15	456	15
127	10	147	12	237	12	345	12	457	16
134	8	156	12	245	11	346	13	467	17
135	9	157	13	246	12	347	14	567	18

From this complete tabulation we can evaluate the probability distribution for W_Y, given in Table 10.2.2.

Table 10.2.2

w	6	7	8	9	10	11	12	13
$P(W_Y = w)$	.029	.029	.057	.086	.114	.114	.143	.114
$P(W_Y \le w)$	.029	.057	.114	.200	.314	.429	.571	.686

w	14	15	16	17	18
$P(W_Y = w)$	.114	.086	.057	.029	.029
$P(W_Y \le w)$	.800	.886	.943	.971	1.000

Granted this distribution for W_Y with $m = 3$, $n = 4$, we can see that $P(W_Y \le 6) + P(W_Y \ge 18) = .058$; thus if we had samples of $m = 3$ Y values and $n = 4$ X values and found the sum of the Y ranks equal to either 6 or 18, we would reject $H_0 : F_X(t) = F_Y(t)$ for all t, with $\alpha = .058$. (The books cited earlier by Lehmann (1975) and Conover (1980) give tables of the exact distribution for W_Y, for small m and n.) As m and n get larger, of course, the evaluation of the exact distribution for W_Y gets increasingly tedious. It can be shown (we will not do so) that

$$E[W_Y] = \frac{m(m + n + 1)}{2}$$

$$\text{Var}[W_Y] = \frac{mn(m + n + 1)}{12} ;$$

the distribution for W_Y is, in every case, symmetric about its mean value

$$P\left(W_Y = \frac{m(m + n + 1)}{2} + \delta \right) = P\left(W_Y = \frac{m(m + n + 1)}{2} - \delta \right)$$

for all possible δ, because for any possible selection of Y ranks, $R_1, R_2, \ldots, R_m$, there corresponds a set of Y ranks $R'_1, R'_2, \ldots, R'_m$ symmetrically displaced about $m(m + n + 1)/2$ (the "middle" rank) that has the same probability of occurrence.

Because of the symmetry of the distribution for W_Y, it is perhaps not too surprising that as m and n tend to infinity, the probability law for

$$\frac{W_Y - m(m + n + 1)/2}{\sqrt{mn(m + n + 1)/12}}$$

tends to the standard normal. The normal approximation (with continuity correction) is very good as long as both m and n are at least 10. Thus, as with the run test, let $t_p (p < .5)$ be the largest integer such that $P(W_Y \leq t_p)$ $\leq p$ and let u_q be the smallest integer such that $P(W_Y \geq u_q) \leq q$; the normal approximation with continuity correction then gives

$$t_p \doteq \frac{m(m + n + 1)}{2} + z_p \sqrt{mn(m + n + 1)/12} - \frac{1}{2}$$

$$u_q \doteq \frac{m(m + n + 1)}{2} + z_{1-q} \sqrt{mn(m + n + 1)/12} + \frac{1}{2},$$

where $N(z_p) = p$.

EXAMPLE 10.2.3.   Let us use the achievement test scores for public school (X) and private school (Y) students, given in Example 10.2.2, to illustrate the computation of the Wilcoxon statistic. Again, then, we want to test $H_0 : F_X(t) = F_Y(t)$ for all t, with $\alpha = .05$. The two sets of sample values were combined and ranked in that example, from which we see that the Y ranks (the 0 values) are 10, 11, 12, 13, 15, 16, 17, 18, 19, 20, 23, giving $w_Y = 174$. If H_0 is true, the mean of W_Y is $11(25)/2 = 137.5$ and its variance is 297.92. We would reject H_0 if either $w_Y \leq t_{.025}$ or $w_Y \geq u_{.025}$; because the observed value is greater than $E[W_Y]$, we need only evaluate

$$u_{.025} = 137.5 + 1.96 \sqrt{297.92 + \frac{1}{2}}$$

$$= 171.83 \sim 172.$$

Because $174 > 172$, we would again reject H_0 with $\alpha = .05$. ■

The Wilcoxon statistic is especially sensitive to cases in which the medians (or means) of the X and Y populations are not equal, because this generally causes the X and Y values not to be particularly well mixed, which in turn produces small or large values for W_Y. If, however, the two medians are equal, but the Y population interquartile range, say, is larger than that of the X population, we would expect the Y values generally to receive both the smaller and the larger ranks in the combined sample. But then W_Y, the sum of the Y ranks, would equal a moderate value, because summing high and low values produces a sum in the middle, and we would not expect to reject equality of the populations. Thus the Wilcoxon statistic would not be expected to be very sensitive to this type of difference in the populations. Several adaptations to the Wilcoxon statistic have been suggested for this case; we do not have the space to discuss these here.

At about the same time that Wilcoxon suggested using W_Y to test equality of two distributions, Mann and Whitney independently suggested the use of a statistic M_{XY} for the same purpose. Given independent random samples of n X values and m Y values, Mann and Whitney suggested forming the nm pairs (X_i, Y_j) and using the value of

$$M_{XY} = \text{number of these pairs with } X_i < Y_j$$

to test $H_0: F_X(t) = F_Y(t)$ for all t. But this statistic is actually just a linear function of W_Y and thus the two procedures are equivalent. To see that this is true, let $R_1 < R_2 < \cdots < R_m$ be the Y ranks in the combined sample. Then the smallest Y value exceeds $R_1 - 1$ X values, the second smallest exceeds $R_2 - 2$ X values, and so on. Thus

$$M_{XY} = \sum_{i=1}^{m} (R_i - i) = \sum R_i - \sum i$$

$$= W_Y - \frac{m(m+1)}{2}$$

so M_{XY} is a linear function of W_Y, and the test has come to be known as the Wilcoxon–Mann–Whitney test.

EXERCISE 10.2

1. An even number, k, of observations has been selected from a continuous distribution. Those $m = k/2$ observations that are smaller than the sample median are replaced by 0's while the remaining $n = k/2$ are replaced by 1's, giving an observed number of runs equal to r. What is the approximate observed significance level in testing the randomness of the sample?

2. Apply the run test to the college broad jumping data in Exercise 10.1.3 to test the randomness of the observations.

3. Test the randomness of the drivers exam times in Exercise 10.1.5 using the run statistic.

4. With $m = 4$, $n = 5$ evaluate the exact distribution for R, the number of runs.

5. Ten alkaline batteries of brand A were used in the same radio, one after the other. The observed lifetimes of the 10 were (in hours):
   51.4, 81.4, 87.2, 69.6, 90.8, 85.1, 73.9, 85.7, 68, 77.2.

Ten alkaline batteries of brand B were also tested to failure in the same radio. Their times to failure were:
76.7, 90.9, 87.3, 96.7, 76.5, 82.7, 79.6, 99, 82.4, 78.2.
Use the run statistic to test that the distributions of lifetimes are the same for these two brands with $\alpha = .1$.

6. Use the Wilcoxon–Mann–Whitney statistic to test equality of battery lifetime, using the data of Exercise 10.2.5.

7. Twelve cars of brand Z were selected at random (all one owner) from the used car lots in a large city. The number of miles registered on the odometers of the 12 were:
36623, 34358, 34222, 35397, 34503, 34833,
35934, 34999, 35584, 34092, 35034, 36719.
Ten cars of brand T were also selected at random (one owner) from the same used car lots. The number of miles registered on their odometers were:
34775, 31971, 31162, 29870, 38238,
33649, 29824, 32532, 33424, 27443.
Use the Wilcoxon–Mann–Whitney statistic to test the hypothesis that the number of miles driven until first trade-in is the same for these two car brands.

8. Describe a normal approximation for the distribution of M_{XY}, the Mann–Whitney statistic.

*9. (a) Another test that is sometimes used to test $H_0: F_X(t) = F_Y(t)$ for all t is called the *median* test. The two independent samples from two continuous populations are combined and the median of the combined samples, m^*, is determined. Then U, the number of Y values that are greater than m^* is determined. Assuming H_0 true evaluate $P(U = u)$, $u = 0, 1, 2, \dots, m$. (Assume n X values, $m \leq n$ Y values.)
   (b) Describe how you would use the value for U to test H_0.

*10. Use the median test (see Exercise 10.2.9) for the data given in Exercise 10.2.7.

10.3 Chi-Square Goodness of Fit Tests

Recall from Section 5.7 that a multinomial trial is a simple experiment with $k \geq 3$ different possible outcomes, a generalization of a Bernoulli trial. The probabilities of success for the k different outcomes are $p_1, p_2, \dots,$ p_k, where $\sum p_i = 1$. If n independent multinomial trials are performed, each with the same probabilities, $p_1, p_2, \dots, p_k$, and we define $X_i =$ number of times outcome i is observed in the n trials, $i = 1, 2, \dots, k$, then $(X_1, X_2, \dots,$

X_k) is called a multinomial random vector with parameters $n, p_1, p_2, \dots, p_k$. The probability function for $(X_1, X_2, \dots, X_k)$ is

$$p_{X_1, X_2, \dots, X_k}(x_1, x_2, \dots, x_k) = \frac{n!}{x_1! \, x_2! \dots x_k!} p_1^{x_1} p_2^{x_2} \cdots p_k^{x_k},$$

where $x_i = 0, 1, 2, \dots, n, \; i = 1, 2, \dots, k$ and $\sum x_i = n$. Each X_i individually is a binomial random variable with parameters n and p_i. A rather wide range of testing procedures follows from the following basic result, which we will not be able to prove.

Theorem 10.3.1. If $(X_1, X_2, \dots, X_k)$ is a multinomial random variable with parameters $n, p_1, p_2, \dots, p_k$, then the distribution function of the random variable

$$U = \sum_{i=1}^{k} \frac{(X_i - np_i)^2}{np_i}$$

approaches the χ^2 distribution function with $k - 1$ degrees of freedom as $n \to \infty$. That is,

$$\lim_{n \to \infty} F_U(t) = F_{\chi^2}(t), \qquad \text{for every } t,$$

where $F_{\chi^2}(t)$ is the χ^2 distribution function with $k - 1$ degrees of freedom. ∎

This result is directly and immediately useful in testing whether an observed vector $(x_1, x_2, \dots, x_k)$ has some specified multinomial distribution. That is, suppose n independent multinomial trials are observed and we want to test that the multinomial trials have specified probabilities for the various outcomes; call these specified values $p_1^0, p_2^0, \dots, p_k^0$ and we want to test $H_0 : p_1 = p_1^0, p_2 = p_2^0, \dots, p_k = p_k^0$ versus $H_1 : H_0$ is false. Once we have observed $(x_1, x_2, \dots, x_k)$, we can compute the observed value

$$u = \sum_{i=1}^{k} \frac{(x_i - np_i^0)^2}{np_i^0}$$

and if H_0 is true, we should find $u \leq \chi_{1-\alpha}^2$ ($k - 1$ degrees of freedom) unless an event with probability α has occurred; thus if we reject H_0 when we find $u > \chi_{1-\alpha}^2$, our probability of type I error is α. This test is very reasonable heuristically; if H_0 is in fact true, then $E[X_i] = np_i^0$ for each i and u should be small (since each $x_i - np_i^0$ should be small). If, however, $E[X_i] \neq np_i^0$ for some two or more X_i's, we would expect a large value for u and should reject H_0, which is just what is previously suggested.

The result given in Theorem 10.3.1 really has its basis in the Central

Limit Theorem. Many rules have been suggested for how large n should be before one could expect the χ^2 limiting distribution for U to be sufficiently accurate. Certainly if $np_i \geq 5$ for all i, which implies $n \geq 5k$ where k is the number of different outcomes for each trial, the χ^2 distribution should give a very good approximation to the exact distribution for U. If $k \geq 5$, it is possible to let one of the np_i be as small as 1 (requiring the others to be 5 or more) and still the distribution for U is pretty well approximated by the χ^2 with $k-1$ degrees of freedom.

EXAMPLE 10.3.1. Suppose the same die was rolled 200 times and the number of spots uppermost was observed for each roll. Each roll of the die is a multinomial trial with $k = 6$ possible outcomes, X_i = number of times face i occurred, $i = 1, 2, \ldots, 6$. The observed multinomial vector, for the $n = 200$ rolls, was $(33, 28, 39, 30, 38, 32)$; thus face 1 occurred on 33 rolls, face 2 on 28 rolls, and so on. We will use this observed vector, and Theorem 10.3.1, to test the hypothesis that this die is fair. If the die is fair, then the probability of occurrence for each face is $p_i = \frac{1}{6}$, $i = 1, 2, \ldots, 6$, so we will test $H_0 : p_1 = p_2 = \cdots = p_6 = \frac{1}{6}$. The expected number of times each face should occur is $np_i^0 = 200(\frac{1}{6}) = 33\frac{1}{3}$, $i = 1, 2, \ldots, 6$. The observed value for U is

$$
\begin{aligned}
u &= \sum_{i=1}^{6} \frac{(x_i - np_i^0)^2}{np_i^0} \\
&= \frac{(33 - 33\frac{1}{3})^2}{33\frac{1}{3}} + \frac{(28 - 33\frac{1}{3})^2}{33\frac{1}{3}} + \cdots + \frac{(32 - 33\frac{1}{3})^2}{33\frac{1}{3}} \\
&= 2.86.
\end{aligned}
$$

From the χ^2 table with $k - 1 = 5$ degrees of freedom we see $P(U \geq 2.86) \doteq .74$ so we would accept H_0 (unless we wanted to have a probability of being wrong equal to .74). ∎

Let $X_1, X_2, \ldots, X_n$ be a random sample of a random variable X. It is easy to use Theorem 10.3.1 to test the hypothesis that the probability law for X is of any completely specified form. Let $F_X(t)$ represent a specific distribution function with no unknown parameters and let R_X be the range of the random variable with this distribution function (the values for which the probability function or density function is positive). If $D_1, D_2, \ldots, D_k$ is any partition of R_X ($D_1 \cup D_2 \cup \cdots \cup D_k = R_X$ and $D_i \cap D_j = \varnothing$ for all $i \neq j$), we can use the sample values $X_1, X_2, \ldots, X_n$ to define a multinomial random variable $(Y_1, Y_2, \ldots, Y_k)$. Since $D_1, D_2, \ldots, D_k$ is a partition of R_X, each X_i must belong to exactly one of $D_1, D_2, \ldots, D_k$; that is, we can look at each sample value X_i as defining a multinomial trial. Now we simply define Y_j to be the number of X_i values that belong to D_j, $j = 1, 2, ..., k$, and $(Y_1, Y_2, ..., Y_k)$ then is a multinomial random vector with parameters $n, p_1, p_2, ..., p_k$ where $p_j = P(X \in D_j)$, $j = 1, 2, ..., k$. This establishes the following theorem.

Theorem 10.3.2. Let $X_1, X_2, ..., X_n$ be a random sample of a random variable X whose distribution function is completely specified and whose range is R_X. If $D_1, D_2, ..., D_k$ is any partition of R_X and Y_j is the number of X_i's that belong to D_j, then $(Y_1, Y_2, ..., Y_k)$ is a multinomial random vector with parameters $n, p_j = P(X \in D_j), j = 1, 2, ..., k$. ∎

We can then immediately use Theorem 10.3.1 to test that a random sample $X_1, X_2, ..., X_n$ was selected from any specific distribution function $F_X(t)$. The following two examples illustrate this.

EXAMPLE 10.3.2. Over two seasons a professional baseball player was at bat exactly 4 times in each of 200 games; his historical batting average was .300. Let X_i equal the number of hits he made in game $i, i = 1, 2, ..., 200$; the range for each X_i then is $\{0, 1, 2, 3, 4\}$, and we will assume $X_1, X_2, ..., X_{200}$ is a random sample of a random variable X. A natural partition of R_X then is to let the D_j's be single-element subsets of R_X: $D_1 = \{0\}$, $D_2 = \{1\}$, $D_3 = \{2\}$, $D_4 = \{3\}$, $D_5 = \{4\}$. Letting Y_j be the number of X_i's that belong to D_j, the observed Y_j's are 73, 82, 38, 7, 0, respectively; thus in 73 games he got 0 hits in 4 times at bat, in 82 games he got exactly 1 hit in 4 times at bat, and so on. Let us use these data to test the hypothesis that X is a binomial random variable with parameters $n = 4$, $p = .3$. Then the hypothesized values for the probabilities of the various outcomes occurring are:

$$p_1^0 = P(X \in D_1) = P(X = 0) = (.7)^4 = .2401$$
$$p_2^0 = P(X \in D_2) = P(X = 1) = 4(.7)^3 (.3) = .4116$$
$$p_3^0 = P(X \in D_3) = P(X = 2) = 6(.7)^2 (.3)^2 = .2646$$
$$p_4^0 = P(X \in D_4) = P(X = 3) = 4(.7)(.3)^3 = .0756$$
$$p_5^0 = P(X \in D_5) = P(X = 4) = (.3)^4 = .0081$$

and from these, we compute the expected values of the Y_j's as $200p_j^0$, giving 48.02, 82.32, 52.92, 15.12, 1.62, respectively. Since $k = 5$ and all of these exceed 5 except the last (which exceeds 1), the χ^2 distribution should give a good approximation for the distribution for U. The observed value for u is

$$\sum_{j=1}^{k} \frac{(y_j - 200p_j^0)^2}{200p_j^0} = \frac{(73 - 48.02)^2}{48.02} + \cdots + \frac{(0 - 1.62)^2}{1.62}$$

$$= 23.2,$$

which exceeds 14.9, the 99.5th quantile of the χ^2 distribution with 4 degrees of freedom so we would reject H_0 based on these data.   ∎

EXAMPLE 10.3.3. The manufacturer of an electron tube claims the times to failure for one of his products is an exponential random variable with parameter $\lambda = .005$. Four hundred of these tubes were placed on test and their times to failure observed. Their observed times to failure are summarized in Table 10.3.1.

Table 10.3.1

Time to Failure t	Number
$t \le 26.7$	36
$26.7 < t \le 57.5$	45
$57.5 < t \le 94.0$	45
$94.0 < t \le 138.6$	50
$138.6 < t \le 196.2$	65
$196.2 < t \le 277.3$	41
$277.3 < t \le 415.9$	58
$415.9 < t$	60

Let us use these data to test the hypothesis that $X_1, X_2, \ldots, X_{400}$ is a random sample of an exponential random variable with $\lambda = .005$. The table actually has already partitioned R_X into eight pieces ($D_1 = \{t : t \le 26.7\}$, $D_2 = \{t : 26.7 < t \le 57.5\}$, etc.) and the values in the column labeled number are in fact the observed Y_j's. We still require the p_j^0 values, given by $p_j^0 = P(X \in D_j)$. Thus

$$p_1^0 = \int_0^{26.7} .005\, e^{-.005x} dx = 1 - e^{-26.7(.005)}$$
$$= .125$$
$$p_2^0 = \int_{26.7}^{57.5} .005\, e^{-.005x} dx = e^{-26.7(.005)} - e^{-57.5(.005)}$$
$$= .125,$$

and so on. You can easily verify that each of the p_i^0 values is .125; the reason for this is subsequently discussed. Then the expected number in each of these intervals is $200(.125) = 50$, so the observed value for U is

$$u = \frac{(36 - 50)^2}{50} + \frac{(45 - 50)^2}{50} + \cdots + \frac{(60 - 50)^2}{50}$$

$$= 14.32.$$

Since $\chi^2_{.95} = 14.1$, with 7 degrees of freedom, we would reject H_0 with $\alpha = .05$. ∎

In using Theorem 10.3.2 to test hypotheses about $F_X(t)$, the choice of the sets in the partition, $D_1, D_2, \ldots, D_k$, and their number k, is not specified. For small sample sizes n, it is wise to use $k \leq n/5$, so that we can generally have expected numbers np_j^0 at least as large as 5 (to make the χ^2 approximation valid). For large sample sizes n, the larger k is taken to be the more computational labor that is involved, but with modern computing equipment this is no real deterrent. The larger k is, the better the chance of detecting (relatively small) deviations from the hypothesized distribution function. Once k is chosen, this still leaves the choice of boundaries for $D_1, D_2, \ldots, D_k$ (especially when R_X is continuous). There is some evidence that the power of the test is generally improved if we choose these boundaries (if possible) so that the p_j^0 values are approximately equal. In Example 10.3.3, the D_j's were specifically set so that all the p_j^0's would equal $1/k = .125$.

The result given in Theorem 10.3.2 is useful in testing that a random sample was selected from a population whose distribution function is completely specified, with no unknown parameters. In many applications we may want to test that sample values were selected from a normal population, for example, without specifiying the values for μ and σ^2; in short, it is frequently desired to test that a random sample was selected from a particular type of distribution (normal, exponential, binomial, Poisson, etc.) without specifying values for the parameters of the distribution. Fortunately, a simple modification is all that is needed to accomplish such a test, with the parameter values unspecified. This is presented in the following theorem, again without proof.

Theorem 10.3.3. Let $X_1, X_2, \ldots, X_n$ be a random sample of a random variable X whose distribution function $F_X(t)$ contains s unknown parameters and whose range is R_X; it is also assumed that $F_X(t)$ satisfies certain regularity conditions. Let $D_1, D_2, \ldots, D_k$ be a partition of R_X and let $(Y_1, Y_2, \ldots, Y_k)$ be the corresponding multinomial vector with parameters n, p_1, $p_2, \ldots, p_k$, where $p_j = P(X \in D_j)$. If $\hat{P}_1, \hat{P}_2, \ldots, \hat{P}_k$ are the maximum likelihood estimators for $p_1, p_2, \ldots, p_k$ (determined from $(Y_1, Y_2, \ldots, Y_k)$), then the distribution function for

$$U = \sum_{i=1}^{k} \frac{(Y_i - n\hat{P}_i)^2}{n\hat{P}_i}$$

converges to the distribution function of a χ^2 random variable with $k - 1 - s$ degrees of freedom as $n \to \infty$.   ∎

It is rather remarkable that we only have to use the maximum likelihood estimators for the p_i's (actually several other estimators could also be employed), compute U in the same way and then subtract one degree of freedom for every parameter estimated. The use of the theorem then is exactly the same, in testing that a random sample was selected from a given type of distribution, whose parameters do not have to be specified.

EXAMPLE 10.3.4.   In Example 10.3.2 the numbers of hits made by a professional baseball player, for 200 games, were used to test (and reject) the hypothesis that X, the number of hits in 4 times at bat, was binomial with $n = 4$, $p = .3$. It may be that these observed data are consistent with a binomial distribution, but we used the wrong value for p, or it may be that the data are not consistent with any binomial distribution. Let us use the same data to test H_0: X is binomial with $n = 4$, where p is unspecified. As discussed in Theorem 10.3.3, we must first determine the maximum likelihood estimator for p, from the observed Y_j's; in a sense, this picks out the "best possible" binomial distribution to compare with the observed data. The observed Y_j's were $(73, 82, 38, 7, 0)$ respectively. If we assume X is binomial, $n = 4$, p unspecified, the values for the multinomial p_i's are:

$$p_1 = P(X = 0) = \binom{4}{0} p^0 (1 - p)^{4-0} = (1 - p)^4$$

$$p_2 = P(X = 1) = 4p(1 - p)^3$$
$$p_3 = P(X = 2) = 6p^2(1 - p)^2$$
$$p_4 = P(X = 3) = 4p^3(1 - p)$$
$$p_5 = P(X = 4) = p^4 ;$$

each of $p_1, p_2, \ldots, p_5$ is a simple function of the binomial p. The multinomial likelihood function is

$$L(p) = \frac{200!}{73!\,82!\,38!\,7!\,0!} \left[(1 - p)^4\right]^{73} \left[4p(1 - p)^3\right]^{82} \cdots \left[p^4\right]^0$$

$$= c p^{82 + 2(38) + 3(7) + 4(0)} (1 - p)^{4(73) + 3(82) + 2(38) + 1(7) + 0(0)}$$

$$= c p^{179}(1 - p)^{621},$$

where c is a constant. The maximizing value for p (since $L(p)$ is proportional to the binomial likelihood function), then is

$$\hat{p} = \frac{179}{179 + 621} = .224.$$

With this estimate for the binomial p we then compute estimates for the multinomial p_j's

$$\hat{p}_1 = (.776)^4 = .363$$
$$\hat{p}_2 = 4(.224)(.776)^3 = .419$$
$$\hat{p}_3 = .181, \qquad \hat{p}_4 = .035, \qquad \hat{p}_5 = .003$$

and the estimates for the $E[Y_j]$ values

$$200\hat{p}_1 = 72.6, \qquad 200\hat{p}_2 = 83.8, \qquad 200\hat{p}_3 = 36.2, \qquad 200\hat{p}_4 = 7.0,$$
$$200\hat{p}_5 = 0.6.$$

Because the last of the expected numbers is smaller than 1, we combine the last two parts of the partition, letting $D_4 = \{x : x = 3 \text{ or } 4\}$. Thus our observed data for the four outcomes are $(73, 82, 38, 7)$ and the expected numbers are $(72.6, 83.8, 36.2, 7.6)$. The observed value for U is

$$u = \frac{(73 - 72.6)^2}{72.6} + \cdots + \frac{(7 - 7.6)^2}{7.6}$$

$$= .178.$$

Since with $(s = 1)$ $4 - 1 - 1 = 2$ degrees of freedom, $\chi^2_{.75} = 2.77$ we would accept H_0: X is binomial; it appears this player's hitting ability in these games is not up to his historical average.   ∎

The estimation of the parameter(s) of $F_X(t)$, from the multinomial values $(Y_1, Y_2, \ldots, Y_k)$, is frequently not so simple as it was in Example 10.3.4. Let us discuss the exponential case (like Example 10.3.3). Assume $X_1, X_2, \ldots, X_n$ is a random sample of a random variable X and we want to test H_0: $F_X(t) = 1 - e^{-\lambda t}$ for all t; that is, that X is an exponential random variable with parameter λ (unspecified). To do so, we partition $R_X = \{x : x > 0\}$ into, say, $k = 8$ pieces, defined by

$$D_1 = \{x : 0 < x \le t_1\}$$
$$D_{j+1} = \{x : t_j < x \le t_{j+1}\}, \qquad j = 1, 2, \ldots, 6$$
$$D_8 = \{x : x > t_7\},$$

where $t_1 < t_2 < \cdots < t_7$ are fixed constants. Then the multinomial parameters are:

$$p_1 = F_X(t_1) = 1 - e^{-\lambda t_1}$$
$$p_{j+1} = F_X(t_{j+1}) - F_X(t_j) = e^{-\lambda t_j} - e^{-\lambda t_{j+1}},$$
$$j = 1, 2, \ldots, 6$$
$$p_8 = 1 - F_X(t_7) = e^{-\lambda t_7}.$$

The likelihood function, given the observed multinomial vector $(y_1, y_2, \ldots, y_8)$, then is

$$L(\lambda) = \frac{(\sum y_i)!}{\prod y_j!} (1 - e^{-\lambda t_1})^{y_1} (e^{-\lambda t_1} - e^{-\lambda t_2})^{y_2} \cdots (e^{-\lambda t_7})^{y_8}.$$

Determining the value of λ that maximizes $L(\lambda)$ is not a trivial task; there is no simple closed form solution for the maximizing λ, although its value could be determined numerically. If one does not have access to a simple way of determining the value $\hat{\lambda}$ that maximizes $L(\lambda)$, how could the test of H_0 be accomplished? A natural procedure would be to use the original data, $x_1, x_2, \ldots, x_n$ to compute $\bar{x}$ (or at least to approximate the value for $\bar{x}$) and then use $\hat{\lambda} = 1/\bar{x}$, the maximum likelihood estimate for the exponential case, to compute $\hat{p}_1, \hat{p}_2, \ldots, \hat{p}_8$, the estimates of the multinomial parameters; in turn, we could then evaluate

$$u = \sum_{j=1}^{k} \frac{(y_j - n\hat{p}_j)^2}{n\hat{p}_j}.$$

Chernoff and Lehmann (*Annals of Mathematical Statistics*, 1954) showed that if this procedure is followed, then the distribution function for U is bounded by $F_1(t)$ and $F_2(t)$, where $F_1(t)$ is the χ^2 distribution function with $k - 1 - s$ degrees of freedom ($s = 1$ in this case) and $F_2(t)$ is the χ^2 distribution with $k - 1$ degrees of freedom, as $n \to \infty$. Thus for any observed value u,

$$1 - F_1(u) \le P(U \ge u) \le 1 - F_2(u),$$

as $n \to \infty$, so we can at least get a good approximation for $\alpha = P(\text{type I error})$.

EXAMPLE 10.3.5.   Let us illustrate the preceding discussion by using the data from Example 10.3.3 to test $H_0 : F_X(t) = 1 - e^{-\lambda t}$ for some value of λ. Because we do not have the original x_i's we cannot in fact evaluate $\bar{x}$ exactly. What we can do, though, is to get a pretty good approximation for $\bar{x}$ by using the data summarized in Table 10.3.1. There were 36 items whose times to failure fell between 0 and 26.7; the midpoint of this interval is 13.35. The sum of these 36 times should be approximately equal to $36(13.35) = 480.6$; similarly, the sum of the 45 failure times that fell between 26.7 and 57.5 should be approximately equal to $45((26.7 + 57.5)/2) = 1894.50$, and so on. We can easily approximate the sums of the failure times for each of the first seven intervals in this way. The last interval, though, defined by $t > 415.9$, is unbounded. What is a reasonable way to approximate the sum of the times in this interval? One rational procedure is: We are testing whether the observed data seem consistent with an exponential probability law. For the exponential case the density function for X given $X > a$, is

$$f(x) = \frac{\lambda e^{-\lambda x}}{1 - F_X(a)} = \frac{\lambda e^{-\lambda x}}{e^{-\lambda a}} = \lambda e^{-\lambda(x-a)}, \qquad x > a.$$

The mean of this density is

$$E[X|X > a] = e^{\lambda a} \int_a^\infty \lambda x e^{-\lambda x} \, dx$$

$$= e^{\lambda a} \frac{(\lambda a + 1)}{\lambda} e^{-\lambda a} = a + \frac{1}{\lambda}$$

(recall the memoryless property). Now we know that 340 failure times were less than or equal to 415.9; thus we could use 415.9 as the estimate of the $\frac{340}{400} = 85$th quantile of the distribution. For the exponential

$$\int_0^{415.9} \lambda e^{-\lambda x} \, dx = 1 - e^{-415.9\lambda} = .85$$

implies

$$\lambda = -\frac{1}{415.9} \ln .15 = .00456$$

or $1/\lambda = 219.2$. Thus we might take $a + 1/\lambda = 415.9 + 219.2 = 635.1$ as a reasonable average for the times that exceed 415.9 and $60(635.1) = 38106$ as an approximation to the sum of 60 such times. We summarize much of this in Table 10.3.2.

Table 10.3.2

Time to Failure	Number	Approximate sum
$t \leq 26.7$	36	480.6
$26.7 < t \leq 57.5$	45	1894.5
$57.5 < t \leq 94.0$	45	3408.8
$94.0 < t \leq 138.6$	50	5815.0
$138.6 < t \leq 196.2$	65	10881.0
$196.2 < t \leq 277.3$	41	9706.8
$277.3 < t \leq 415.9$	58	20102.8
$t > 415.9$	60	38106.0

Thus $\sum x_i \doteq 90395.5$, $\bar{x} \doteq 225.99$, $\hat{\lambda} \doteq \frac{1}{225.99}$. With this value we can now compute the $\hat{p}_j$'s and the estimated $n\hat{p}_j$'s as given in Table 10.3.3.

Table 10.3.3

Time to Failure	Observed	$\hat{p}_j$	Expected
$t \leq 26.7$	36	.1114	44.56
$26.7 < t \leq 57.5$	45	.1132	45.28
$57.5 < t \leq 94.0$	45	.1156	46.24
$94.0 < t \leq 138.6$	50	.1182	47.28
$138.6 < t \leq 196.2$	65	.1218	48.72
$196.2 < t \leq 277.3$	41	.1266	50.64
$277.3 < t \leq 415.9$	58	.1344	53.76
$t > 415.9$	60	.1588	63.52

The observed value for U is

$$u = \frac{(36 - 44.56)^2}{44.56} + \cdots + \frac{(60 - 63.52)^2}{63.52} = 9.64;$$

since the observed value for u (9.64) is about the 85th quantile for the χ^2 distribution with $k - 1 - s = 6$ degrees of freedom and is about the 78th quantile for the χ^2 distribution with $k - 1 = 7$ degrees of freedom, $.15 \leq P(U \geq 9.64) \leq .22$, so we would accept H_0 (for any $\alpha \leq .15$). ■

EXERCISE 10.3

1. A college track man put the shot 100 times in practice in a week. The distances he threw it (measured in feet) are recorded in the following table.

Distance y	Frequency x
$y \leq 61$	12
$61 < y \leq 63$	20
$63 < y \leq 65$	40
$65 < y \leq 67$	25
$y > 67$	3

With $\alpha = .01$, test the hypothesis that the distance, Y, that he can put the shot is a normal random variable with $\mu = 63$ feet, $\sigma = 2$ feet.

2. The number of misprints per page in a printed book is frequently taken to be a Poisson random variable, because they are presumably independent and, if the whole book were done by the same typesetter, the rate should be constant from the first page to the last. A misprint count was made for 100 pages of a recent novel with the numbers of misprints found as:

Number of Misprints	Number of Pages
0	65
1	25
2	8
3	2
total	100

Test the hypothesis that the number of misprints per page is a Poisson random variable with $\mu = .4$, using $\alpha = .1$.

3. It would seem reasonable to assume the final digit of a randomly selected telephone number is equally likely to be 0 or 1 or 2 or $\cdots$ or 9. Select a page at random from your telephone book and count the number of 0's, 1's, 2's, ..., 9's that occur as last digits. Then test the hypothesis that the last digit is equally likely to be 0 or 1 or $\cdots$ or 9, with $\alpha = .1$. (You should get a sample of size $n = 350$ or so since this is roughly the number of phone numbers listed on a single page.)

4. A pair of dice was rolled 500 times. The sums that occurred were as recorded in the following table. Test the hypothesis that the dice are fair, with $\alpha = .05$

Sum	Frequency
2, 3, or 4	74
5 or 6	120
7	83
8 or 9	135
10, 11, or 12	88

5. A "random" generator was used to generate 1000 numbers on the interval (0, 1). The numbers generated are as follows:

Interval	(0, .1]	(.1, .2]	(.2, .3]	(.3, .4]	(.4, .5]
Number generated	114	100	99	98	111

Interval	(.5, .6]	(.6, .7]	(.7, .8]	(.8, .9]	(.9, 1)
Number generated	104	106	95	92	81

Would you accept the hypothesis that this generator works as it should, based on this sample?

6. A newly designed IQ test was given to a "random sample" of 300 students aged 15. The scores made by these 300 students are summarized as follows.

Score x	Number
$x \leq 70$	5
$70 < x \leq 80$	13
$80 < x \leq 90$	41
$90 < x \leq 100$	59
$100 < x \leq 110$	92
$110 < x \leq 120$	55
$120 < x \leq 130$	18
$x > 130$	17

The test was expected to produce scores that were normally distributed with $\mu = 100, \sigma = 15$.

(a) Would you accept the hypothesis that these scores are a random sample from a normal population with $\mu = 100, \sigma = 15$?

(b) Would you accept the hypothesis that these scores are a random sample from a normal population? (The smallest score was 48 and the largest was 142.)

7. In a recent 72-hour holiday period in the United States there was a total of 288 fatal auto accidents. The number of fatal accidents, per hour, during this period was as follows.

Number per Hour	Number of Hours
0 or 1	6
2	11
3	15
4	14
5	12
6	8
7 or more	6

Test the hypothesis that the number of accidents per hour, during such a holiday weekend, is a Poisson random variable.

*8. A 1972 government report gave the following distribution for the number of days that hospitalized people spent in the hospital in 1971.

Number of Days	Number of Cases
1	89
2	152
3	105
4—5	165
6—9	221
10—14	124
15—30	106
31 and over	38

Test the hypothesis that these data were selected from an exponential distribution with $\lambda = \frac{1}{8.691}$, with $\alpha = .10$.

9. Between January 1, 1965 and February 9, 1971, inclusive, a period of 2231 days, a total of 163 earthquakes were recorded somewhere in the world, each of magnitude 4 or more on the Richter scale. Knowing the calendar date and time at which each quake occurred, one can compute the number of days between successive quakes and the 163 quakes give a total of 162 differences, for days between earthquakes. The following table summarizes the number of days between quakes over this period.

Number of Days	Frequency
0–4	50
5–9	31
10–14	26
15–19	17
20–24	10
25–29	8
30–34	6
35–39	6
40 or more	8[†]

Test the hypothesis that these values are a random sample of an exponential random variable, with $\alpha = .05$.

10. Test the hypothesis that the data in Table 6.1.2 summarizes a random sample from a normal population, with $\alpha = .05$.

10.4 Contingency Tables

In many cases the results of multinomial trials can be classified according to two (or more) criteria. For example, a voter in a presidential election could be classified according to the person for whom he or she will vote, and according to his party affiliation. A household selected at random can be classified according to the annual income of the head of the household, as well as his or her racial background. A consumer of a given product can be classified according to his or her age bracket and according to his or her brand preference. Each fire in a large city can be classified according to the district in which it was located and according to whether or not it appeared to be arson-caused.

When multinomial trials can be classified according to two criteria, it is frequently of interest to test whether the two criteria are independent. As we will see, Theorem 10.3.3 can be directly employed for this purpose. We assume we have n independent multinomial trials; each trial results in one of $k = rc$ different outcomes. To distinguish the different levels of the two criteria of classification, we will use a double subscripting for the components of the multinomial random vector and for the multinomial probabilities. Thus Y_{ij} represents the number of trials whose outcomes fell in level i, of criteria 1, and level j, of criteria 2, $i = 1, 2, \ldots, r, j = 1, 2, \ldots, c$; p_{ij} is the probability of level i criteria 1, level j criteria 2 occurring on each trial.

[†] These 8 values were 40, 43, 44, 49, 58, 60, 81, and 109 days.

Thus with $r = 2, j = 3$, for example, there are $2 \cdot 3 = 6$ different possible outcomes for each trial, and we can picture the multinomial vector as a two-dimensional array (called a *contingency table*).

$$
\begin{array}{ccc}
Y_{11} & Y_{12} & Y_{13} \\
Y_{21} & Y_{22} & Y_{23}
\end{array}
$$

and we can do the same for the multinomial probabilities:

$$
\begin{array}{ccc}
p_{11} & p_{12} & p_{13} \\
p_{21} & p_{22} & p_{23}
\end{array}
$$

The rows represent levels of criteria 1 and the columns represent the levels of criteria 2. Just as with the probability function for a two-dimensional discrete random vector, we get the total probability of observing an outcome at level i, criteria 1, by summing over the levels of criteria 2;

$$ p_{i.} = \sum_{j} p_{ij} = P(\text{outcome in level } i, \text{criteria 1}), $$

$i = 1, 2, \ldots, r$ and we get the total probability of observing an outcome at level j, criteria 2, by summing over the levels of criteria 1:

$$ p_{.j} = \sum_{i} p_{ij} = P(\text{outcome in level } j, \text{criteria 2}), $$

$j = 1, 2, \ldots, c$. These summations correspond to summing across a row or down a column in the two-dimensional array of p_{ij}; they occur as totals at the margins of the rows and columns.

If the two criteria of classification are in fact independent, then it must be true that

$$ p_{ij} = p_{i.} p_{.j}, $$

the probability of the outcome falling in cell (i, j) of the contingency table is given by the product of the probabilities of falling in row i, and in row j, for all (i, j). Thus testing independence of the two criteria in a contingency table means we want to test $H_0: p_{ij} = p_{i.} p_{.j}$ for all (i, j) versus $H_1: H_0$ is false.

As already mentioned, we can employ Theorem 10.3.3 to design a test for H_0. Let Y_{ij}, $i = 1, 2, \ldots, r$, $j = 1, 2, \ldots, c$ represent the components of the multinomial random vector with parameters n and p_{ij}. Then if H_0 is true, the maximum likelihood estimator for $p_{i.}$ is $\hat{P}_{i.} = Y_{i.}/n$ and the maximum likelihood estimator for $p_{.j}$ is $\hat{P}_{.j} = Y_{.j}/n$ (see Exercise 7.1.21) so the maximum likelihood estimate of the probability outcome (i, j) occurs is

$$ \hat{P}_{ij} = \hat{P}_{i.} \hat{P}_{.j} = \frac{Y_{i.} Y_{.j}}{n^2} $$

and the estimator for $E[Y_{ij}]$ is

$$n\hat{P}_{ij} = \frac{Y_{i.}\,Y_{.j}}{n} = E_{ij}.$$

Then from Theorem 10.3.3, if $H_0: p_{ij} = p_{i.}\,p_{.j}$ is true, the distribution function for

$$U = \sum_i \sum_j \frac{(Y_{ij} - E_{ij})^2}{E_{ij}}$$

approaches that of a χ^2 random variable with $rc - 1 - [(r-1) + (c-1)]$ $= (r-1)(c-1)$ degrees of freedom as $n \to \infty$. The value for s is $(r-1) + (c-1)$ since we only need to estimate $p_{1.}, p_{2.}, \ldots, p_{r-1.}$ ($\sum_i p_{i.} = 1$) and $p_{.1}, p_{.2}, \ldots, p_{.c-1}$ ($\sum_j p_{.j} = 1$), if the two variables are independent. When H_0 is not true we would expect rather large disparities between the observed values, Y_{ij}, and the expected values, E_{ij}, meaning the value for U should be large in this case. Thus we reject H_0 if we find $U \geq \chi^2_{1-\alpha}$, selected from the χ^2 distribution with $(r-1)(c-1)$ degrees of freedom.

EXAMPLE 10.4.1. Contingency tables are frequently employed in social science research. A random sample of 1397 American adults, taken in the spring of 1975, provided the data given in Table 10.4.1.

Table 10.4.1

Amount of Confidence in Large Companies

		Great Deal	Only Some	Hardly Any	Total
Race	White	269	699	274	1242
	Black	16	100	39	155
	Total	285	799	313	1397

One can use a contingency table test of the hypothesis that these two variables, race and amount of confidence in large companies, are independent; if independence is accepted, this would imply that both races have the same distribution for the amount of confidence in large companies. The expected number in cell (i, j), if H_0 is true, is given by $e_{ij} = y_{i.}y_{.j}/n$. These quantities are displayed in Table 10.4.2.

Table 10.4.2

Expected Numbers If Independent

	Great Deal	Only Some	Hardly Any
White	253.38	710.35	278.27
Black	31.62	88.65	34.73

The observed value for U is

$$u = \sum_i \sum_j \frac{(y_{ij} - e_{ij})^2}{e_{ij}}$$

$$= \frac{(269 - 253.38)^2}{253.38} + \cdots + \frac{(39 - 34.73)^2}{34.73}$$

$$= 10.9,$$

which we compare with the quantiles of the χ^2 distribution with $2(1) = 2$ degrees of freedom. Since $\chi^2_{.995} = 10.6$, we would reject independence with $\alpha = .005$. It would appear from these data that blacks were, in general, more skeptical about large companies, because a greater proportion of blacks had hardly any confidence, and a smaller proportion of blacks had a great deal of confidence in them. ■

The chi-square statistic can be used to test independence of categorical variables, as in Example 10.4.1. A categorical variable is one whose levels are categories, non numerical quantities that do not have the properties of real numbers. In this case we are in fact comparing how the counts are distributed over the levels of variable 2, for the various levels of variable 1 (or vice versa). If these counts are distributed over the variable 2 levels in the same way, for all levels for variable 1, then the two categorical variables are independent; if they are distributed sufficiently differently in the sample, we will conclude the two variables are not independent. It is, of course, not necessary to restrict the use of the test to only categorical variables. It can as well be applied to numerical variables, in which case we are in fact testing the independence of random variables (testing whether the two components of a two dimensional random vector are independent). This is illustrated in the following example.

EXAMPLE 10.4.2. Two hundred 15-year-old students each took two aptitude tests, one a measure of mathematical aptitude and the other a measure of musical aptitude. The two hundred vectors of scores are summarized in Table 10.4.3.

Table 10.4.3

Math Score X

	$x \leq 290$	$290 < x \leq 320$	$320 < x \leq 350$	$x > 350$	Totals
$z \leq 275$	16	12	6	1	35
Music $\quad 275 < z \leq 300$	19	31	13	5	68
Score $\quad 300 < z \leq 325$	5	26	24	17	72
$Z \quad z > 325$	1	6	9	9	25
Total	41	75	52	32	200

Again, if the two variables X (math score) and Z (music score) are independent the expected number in cell (i, j) is $e_{ij} = y_{i.}y_{.j}/n$. These values are given in Table 10.4.4.

Table 10.4.4

	$x \leq 290$	$290 < x \leq 320$	$320 < x \leq 350$	$x > 350$
$z \leq 275$	7.175	13.125	9.100	5.600
$275 < z \leq 300$	13.540	25.500	17.680	10.880
$300 < z \leq 325$	14.760	27.000	18.720	11.520
$z > 325$	5.125	9.375	6.500	4.000

The observed value for U is

$$u = \sum_i \sum_j \frac{(x_{ij} - e_{ij})^2}{e_{ij}} = 45.6,$$

which considerably exceeds $\chi^2_{.995} = 23.6$, with 9 degrees of freedom. If these 200 vectors are a random sample from the population of (X, Z) values for all 15 year olds, it would appear that X and Z are not independent random variables. ∎

This contingency table statistic can also be used to test that several different populations all have the same probability law (this is frequently called a test of homogeneity for the populations). Suppose we are given independent random samples from m populations; the size of sample selected from population i is n_i. Let $X_{ij}, i = 1, 2, \ldots, m, j = 1, 2, \ldots, n_i$ represent the sample values and let $D_1, D_2, ..., D_k$ be a partition of R_X, the common range for all the populations. Define Y_{ir} to be the number of X_{ij} values that belong to D_r; for example, Y_{11} is the number of values from sample 1 that belong to D_1, Y_{13} is the number of values from sample 1 that belong to D_3, Y_{21} is the number of values from sample 2 that belong to D_1, and so on. The counts from sample i, $(Y_{i1}, Y_{i2}, ..., Y_{ik})$, define a multinomial vector with parameters n_1 and $p_{i1}, p_{i2}, ..., p_{ik}$, for $i = 1, 2, ..., m$; because the original samples are independent, these multinomial vectors are independent. Now if all the samples came from populations with the same probability law (the hypothesis H_0), it follows that $p_{11} = p_{21} = \cdots = p_{m1}, p_{12} = p_{22} = ... = p_{m2}$, and so on. That is, the probability that a sample value belongs to D_r is the same (call this probability p_r) regardless of which population it was selected from. But then we easily see that the maximum likelihood estimators for $p_1, p_2, ..., p_k$ are given by

$$\hat{P}_r = \frac{Y_{.r}}{\sum m_j}, \qquad r = 1, 2, ..., k,$$

the overall proportion of all the observations that belong to D_r. The maximum likelihood estimate for the expected number of values from sample i that belong to D_r (that is, the estimate of $E[Y_{ir}]$) then is $e_{ir} = m_i y_{.r}/\sum m_j$, if H_0 is true. (Again, with the Y_{ir} values displayed in an $m \times k$ array, this is simply the product of the totals for row i and column r, divided by the total sample size $\sum m_j$.)

For any fixed i,

$$U_i = \sum_r \frac{(Y_{ir} - n_i p_{ir})^2}{n_i p_{ir}}$$

is approximately a χ^2 random variable with $k - 1$ degrees of freedom, and because of the independence between samples,

$$U = \sum_i U_i = \sum_i \sum_r \frac{(Y_{ir} - n_i p_{ir})^2}{n_i p_{ir}}$$

then is approximately χ^2 with $m(k - 1)$ degrees of freedom. To compute the estimate of the expected values, the e_{ir}'s, we had to estimate $p_1, p_2, ..., p_k$ so we "lose" $s = k - 1$ degrees of freedom (since $p_1 + p_2 + \cdots + p_k = 1$) and from Theorem 10.3.3,

$$U = \sum_{i=1}^{m} \sum_{r=1}^{n_i} \frac{(Y_{ir} - E_{ir})^2}{E_{ir}}$$

is approximately a χ^2 random variable with $m(k - 1) - (k - 1) =$

$(m - 1)(k - 1)$ degrees of freedom. Again, we would reject H_0 if we find $U \geq \chi^2_{1-\alpha}$ to have P(type I error) $= \alpha$(approximately). This test of homogeneity is illustrated in the following example.

EXAMPLE 10.4.3. Each of 3 different manufacturers (call them G, W, S) produces 60-watt light bulbs and each of the 3 claims the same average light output and the same average lifetime. Random samples of $m_1 = 50$ bulbs from G, $m_2 = 60$ bulbs from W and $m_3 = 50$ bulbs from S were placed on test and left on until they burned out. The observed lifetimes for the 160 bulbs tested are summarized in Table 10.4.5.

Table 10.4.5

Bulb Counts Y_{ir}

Observed Time to Failure x

	$x \leq 700$	$700 < x \leq 750$	$750 < x \leq 800$	$x > 800$	Totals
Manufacturer G	2	20	25	3	50
Manufacturer W	7	18	22	13	60
Manufacturer S	2	15	21	12	50
Totals	11	53	68	28	160

Under the assumption that the bulb lifetimes have the same probability law, for all 3 manufacturers, the column totals divided by $\sum m_i = 160$ give the estimates for p_1, p_2, p_3, p_4:

$$\hat{p}_1 = \tfrac{11}{160}, \qquad \hat{p}_2 = \tfrac{53}{160}, \qquad \hat{p}_3 = \tfrac{68}{160}, \qquad \hat{p}_4 = \tfrac{28}{160}.$$

From these we then compute the expected numbers for each cell, the e_{ir} values given in Table 10.4.6.

Table 10.4.6

Expected Number e_{ir}

	$x \leq 700$	$700 < x \leq 750$	$750 < x \leq 800$	$x > 800$
Manufacturer G	3.4375	16.5625	21.25	8.75
Manufacturer W	4.125	19.875	25.5	10.5
Manufacturer S	3.4375	16.5625	21.25	8.75

Then the observed value for U is

$$u = \sum_i \sum_r \frac{(y_{ir} - e_{ir})^2}{e_{ir}} = 10.97.$$

With $2(3) = 6$ degrees of freedom, $\chi^2_{.9} = 10.6$, $\chi^2_{.95} = 12.6$, so we would accept H_0 with $\alpha = .05$ and reject H_0 with $\alpha = .10$. ∎

EXERCISE 10.4

1. A sample of 2000 medical records was examined and the following data resulted.

	Died of Cancer of Intestines	Died of All Other Causes	Totals
Smokers	22	1178	1200
Nonsmokers	26	774	800
Total	48	1952	2000

Assume that these results were the outcome of a random sample from a certain population and test that the two classifications are independent, with $\alpha = .05$.

2. Records of 10,000 auto accidents were examined to determine the degree of injury to the driver and whether or not he was using a seat belt. The data are summarized as follows.

	Seat Belt	No Seat Belt	Total
Minor injuries	2500	1500	4000
Major injuries	450	4550	5000
Death	50	950	1000
Total	3000	7000	10,000

Test the hypothesis that severity of injury to the driver is independent of whether the driver wears a seat belt, with $\alpha = .01$.

3. A random sample of 4000 individuals (all male of the same age) yielded the following data.

Annual Income

Highest Education Attained	Less Than 5000	5000 to 15,000	More Than 15,000	Total
Grade school	350	35	15	400
High school	100	850	50	1000
College	40	1200	760	2000
Graduate	10	415	175	600
Total	500	2500	1000	4000

Test the hypothesis that annual salary (for males of this age) is independent of education attained, with $\alpha = .1$.

4. If random samples, of size 1000, were selected from the U.S. population in 1900, 1930, and 1965, and the ages of the people selected were recorded, the resulting data would give a table as follows.

Age

	Under 14	14 to 19	20 to 24	25 to 44	45 to 64	65 and over	Total
1900	322	120	97	282	138	41	1000
1930	273	113	89	295	175	55	1000
1965	289	106	70	242	200	93	1000

Given these data, would you accept the hypothesis that the distributions of ages in the U.S. population was the same for all 3 years?

5. A second independent sample of people's attitudes toward large companies, like the one in Example 10.4.1, was also taken in the spring of 1975. The results of this sample are summarized as follows.

Amount of Confidence

	Great Deal	Only Some	Hardly Any	Totals
White	232	641	371	1244
Black	34	81	72	187
Totals	266	722	443	1431

Using these results, test the independence of the two categories with $\alpha = .05$.

*6. Are the data in Example 10.4.1 and Exercise 10.4.5 consistent?

7. Given that we have random samples from each of two Bernoulli distributions, we can use the contingency table approach discussed in this section to test the hypothesis that the two Bernoulli probabilities, p_1 and p_2 are equal. Specifically, assume we have a random sample of size n_1 from a Bernoulli population with parameter p_1 and an independent random sample of size n_2 from a second Bernoulli population with parameter p_2. Then the observations can be arranged in a 2×2 table as follows.

Number of

	Successes	Failures	Total
Population 1	X_1	$n_1 - X_1$	n_1
Population 2	X_2	$n_2 - X_2$	n_2

X_1 then is a binomial random variable with parameters n_1, p_1 and X_2 is an independent binomial random variable with parameters n_2, p_2.

(a) Define

$$Z = \frac{X_1/n_1 - X_2/n_2}{\sqrt{\left(\dfrac{X_1 + X_2}{n_1 + n_2}\right)\left(1 - \dfrac{X_1 + X_2}{n_1 + n_2}\right)\left(\dfrac{1}{n_1} + \dfrac{1}{n_2}\right)}}$$

and explain why Z is approximately a standard normal random variable if in fact $p_1 = p_2$.

(b) Show that $Z^2 = U$, the χ^2 statistic used to test independence in a 2×2 contingency table, if $n_1 = n_2$.

10.5 Summary

Nonparametric procedures: Procedures designed to perform well with a relatively weak set of assumptions.

Confidence interval for t_p, $100p$th quantile: $[X_{(j)}, X_{(k)})$ where confidence coefficient is

$$\sum_{i=j}^{k-1} \binom{n}{i} p^i (1-p)^{n-i}.$$

Sign test: Used to test that $t_p = t^*$; test statistic is number of observations that exceed t^*.

Signed rank test: Used to test hypotheses about median for a symmetric continuous probability law; test statistic depends on both ranks and magnitudes.

Run test: Statistic is defined by number of runs, used to test randomness of observations, and for other purposes.

Wilcoxon–Mann–Whitney test: Used to test that two independent samples came from the same continuous distribution.

χ^2 Goodness-of-fit test: Used to test that sample observations came from specified probability law, with known or unknown parameters.

Contingency tables: Summaries of multinomial trials whose outcomes have been classified according to two (or more) criteria; χ^2 test for independence of two criteria.

Tests of homogeneity: Tests that samples from 2 or more populations have the same probability law; χ^2 statistic used for large samples.

11

Bayesian Methods

In Chapters 7 and 8 we discussed the definition and use of some classical methods for estimating parameters and testing hypotheses. In this chapter we will look at techniques based on Bayes theorem (Theorem 2.6.1). A prime difference between these techniques and the classical ones already discussed is the interpretation of the probabilities used. In essentially everything we have studied so far, the probabilities used were interpreted in a frequency sense; they were referring to an experiment that could be repeated an indefinite number of times, and if the probability of occurrence of an event A was .3, we meant that A would be expected to occur in about 30 percent of these repetitions. This type of interpretation of probability is called objective or frequentist; the numbers called probabilities are measuring relative frequencies of occurrence in repetitions of the basic experiment.

In common English usage, however, probability is frequently used in another, more subjective sense. For example, we have all heard such statements as "it probably will rain tomorrow" or "the chances are 3 out of 5 the Yankees will win the pennant this year" or "most likely the bank robbery was an inside job." In each of these cases the individual making the statement is using his own experience and knowledge as the basis for the statement and is not referring to some experiment that can or will be repeated an indefinite number of times. These are all examples of uses of subjective probabilities, ones that base their validity strictly on the beliefs of the individuals making them. Thus a subjective probability is measuring a person's "degree of belief" in a proposition, which is not necessarily the same as its long-term frequency of occurrence if, indeed, he is referring to something that could be repeated.

The Bayesian techniques that we will study in this chapter all make use of subjective probabilities measuring degrees of belief about the value or values of unknown parameters. These subjective probabilities are used to define what is called the prior distribution for the parameter. Thus when using Bayesian methods, we will act as though an unknown parameter is a random variable and has a known prior distribution (prior to taking a sample); this prior distribution summarizes our subjective degree of belief about the unknown value of the parameter. If we are fairly certain of the parameter's value, we will choose a prior with a small variance; if we are less certain about its value we will choose a prior with a larger variance. After the prior is specified the sample values are observed and used to compute what is called the posterior distribution of the parameter. This posterior distribution is made up of both the subjective prior information (our degree of belief) about the parameter and the objective sample information. The posterior distribution then is used to construct an estimator of the unknown parameter or to make interval statements about the unknown parameter.

11.1 Prior and Posterior Distributions

In Chapter 7 we examined both the method of moments and maximum likelihood criteria for using a sample to estimate the value of an unknown parameter θ. We assumed that the probability law of a random variable X, which we could sample, was dependent on the unknown value of θ; then the sample information was used to construct a guess or estimate of the value of θ. Neither of these two methods called for any information for estimating θ other than the sample values that occurred. Moreover, if such extraneous information were available, it would not have been possible to make use of it. In many situations additional information is available about the value of the unknown parameter; if this information can be used to construct a prior distribution for the parameter θ, then the Bayesian methods discussed in this chapter may be used to estimate the unknown value of θ.

DEFINITION 11.1.1. The *prior distribution* of a parameter θ is a probability function or probability density function expressing our degree of belief about the value of θ, prior to observing a sample of a random variable X whose distribution function depends on θ. ∎

It should be stressed that the prior distribution for a parameter θ is making use of additional assumed information above and beyond anything to be observed from a random sample of X. Since this information is expressed as a probability function or density function, we will let Θ be the symbol for the corresponding random variable and then θ would represent its observed value (which would be the same as the true unknown value). Some examples should clarify this notation.

EXAMPLE 11.1.1. Suppose that we have a brand new 50-cent piece and are interested in estimating θ, the probability of getting a head with this coin on a single flip. We know that θ must lie between zero and 1. If we are not willing to assume any values in this interval are more likely than others, we could then reasonably assume a uniform prior for θ on the interval $(0, 1)$. This would be written

$$f_\Theta(\theta) = 1, \qquad 0 < \theta < 1$$
$$= 0, \qquad \text{otherwise.}$$

This would, in a sense, correspond to an assumption of total ignorance; we feel that all possible values of θ are equally likely. On the other hand, we might feel justified in assuming that θ must certainly lie in the interval from .4 to .6, since the coin would appear to be quite symmetric, and that each of these values is equally likely to occur. Our prior for θ then would be

$$f_\Theta(\theta) = 5, \qquad .4 < \theta < .6$$
$$= 0, \qquad \text{otherwise.}$$

This is a stronger assumption, because we have ruled out values of θ below .4 and above .6. Or we might again reason that the value of θ must certainly lie in the interval from .4 to .6 but that only .4, .5, or .6 are possible values for θ with .5 being twice as likely. Then we would be assuming the discrete prior

$$p_\Theta(\theta) = \tfrac{1}{4}, \qquad \text{at } \theta = .4 \text{ or } .6$$
$$= \tfrac{1}{2}, \qquad \text{at } \theta = .5$$
$$= 0, \qquad \text{otherwise.} \qquad \blacksquare$$

It should be noted that the priors mentioned in Example 11.1.1 were quite arbitrary and dependent on the sort of assumptions one is willing to make regarding the unknown value of θ; many other possible assumed priors could be mentioned. As we will see, the final answer arrived at in using a Bayes' technique is generally dependent on the particular prior assumed, so the assumption should not be taken lightly. Furthermore, two different people observing the same sample values might very well arrive at different estimates of the value of the same parameter because they assumed different prior information.

In some cases the prior information for the value of a parameter may take a rather natural form. The following example illustrates this.

EXAMPLE 11.1.2. Suppose a lot of 1000 items is received from a supplier; the lot contains θ (unknown) defective items. Also suppose we have dealt with this supplier many times in the past and, from this past experience, we have found that 5 percent of the items he supplies are defective. If we make the assumption each item he produces has probability .05 of being defective, and defectives occur independently, the natural prior to use for θ, the number of defectives in the lot, is a binomial distribution with $n = 1000$, $p = .05$. Thus we would assume the prior for θ to be

$$p_\Theta(\theta) = \binom{1000}{\theta}(.05)^\theta (.95)^{1000-\theta}, \qquad \theta = 0, 1, \ldots, 1000. \quad \blacksquare$$

With the foregoing discussion in mind regarding an assumed prior for an unknown parameter θ, let us switch our attention to the probability density function for a continuous random variable X that depends on θ. We previously denoted this density simply by $f_X(x)$; this notation does not express the fact that $f_X(x)$ actually is dependent on the value of θ. While discussing Bayesian techniques in this chapter, we will use a conditional probability notation for such a density to stress the fact that $f_X(x)$ can only be used to compute probabilities for a given value of θ. Thus we will write $f_{X|\Theta}(x|\theta)$ for the density of X. (If X is discrete, we will analogously use $p_{X|\Theta}(x|\theta)$ for its probability function in the context of Bayesian methods.) Thus if X is an exponential random variable with parameter θ, we will write

$$\begin{aligned}f_{X|\Theta}(x|\theta) &= \theta e^{-\theta x}, \qquad x > 0,\ \theta > 0 \\ &= 0, \qquad\qquad \text{otherwise}\end{aligned}$$

to stress the conditional dependence of the density on θ. Or if X is binomial with parameters n and θ, we will denote its probability function by

$$p_{X|\Theta}(x|\theta) = \binom{n}{x}\theta^x (1-\theta)^{n-x}, \qquad x = 0, 1, \ldots, n, \qquad 0 \le \theta \le 1,$$

$$= 0, \qquad\qquad\qquad \text{otherwise}.$$

Generally, the parameter n is known, so it plays no special role in this conditional notation.

We are in essence acting as though the parameter of the probability law for X is itself a random variable when we assume a prior distribution for such a parameter; thus the density and probability functions that we are used to are actually conditional densities or probability functions. The notation is meant to acknowledge this fact.

Our notation for the density function of the elements of a random sample of n observations of X will also be changed accordingly. Thus if $X_1, X_2, \ldots, X_n$ is a random sample of an exponential random variable with parameter θ the joint density for the sample will be denoted by

$$f_{\mathbf{X}|\Theta}(\mathbf{x}|\theta) = \theta^n e^{-\theta \sum x_i}.$$

Given a prior density for θ, $f_{\Theta}(\theta)$, and the conditional density of the elements of a sample, $f_{\mathbf{X}|\Theta}(\mathbf{x}|\theta)$, the joint (unconditional) density for the sample and the parameter, is simply the product of these two functions:

$$f_{\mathbf{X},\Theta}(\mathbf{x}, \theta) = f_{\mathbf{X}|\Theta}(\mathbf{x}|\theta) f_{\Theta}(\theta).$$

(The reader might like to review conditional densities in Chapter 5.) Then the marginal density of the sample values, which is independent of θ, is given by the integral of the joint density over the range of Θ. Thus

$$f_{\mathbf{X}}(\mathbf{x}) = \int_{\text{Range of } \Theta} f_{\mathbf{X},\Theta}(\mathbf{x}, \theta)\, d\theta.$$

This will be referred to as the *marginal of the sample*. The posterior density for the parameter θ is defined as follows.

DEFINITION 11.1.2.   The *posterior density* for θ is the conditional density of Θ, given the sample values. Thus

$$f_{\Theta|\mathbf{X}}(\theta|\mathbf{x}) = \frac{f_{\mathbf{X},\Theta}(\mathbf{x}, \theta)}{f_{\mathbf{X}}(\mathbf{x})}. \qquad \blacksquare$$

This posterior density thus is simply the conditional density of Θ, given the sample values; the prior density expresses our degree of belief of the location of the value of θ prior to sampling, and the posterior density expresses our degree of belief of the location of θ, given the results of the sample. Let us consider some examples of these manipulations.

EXAMPLE 11.1.3.   Let us return to the coin of Example 11.1.1 and, based on the result of a single flip, see how the posterior of θ changes for the three priors presented there. Thus if we flip this 50-cent piece one time and let $X = 1$ if we get a head and $X = 0$ if we get a tail, the probability function for the sample is

$$
\begin{aligned}
p_{X|\Theta}(x|\theta) &= 1 - \theta, &&\text{at } x = 0 \\
&= \theta, &&\text{at } x = 1.
\end{aligned}
$$

If we assume our prior to be

$$f_{\Theta}(\theta) = 1, \qquad 0 < \theta < 1,$$

then the joint density function for X and Θ is

$$f_{X,\Theta}(x,\theta) = 1 - \theta, \qquad \text{for } x = 0, \quad 0 < \theta < 1$$
$$= \theta, \qquad \text{for } x = 1, \quad 0 < \theta < 1.$$

(Note that this is actually a mixture, corresponding to the joint function of a discrete and a continuous random variable; we will use f to denote the joint function in such cases.) The marginal for X then is

$$p_X(x) = \int_0^1 (1 - \theta)d\theta = \tfrac{1}{2}, \qquad \text{for} \qquad x = 0$$

$$= \int_0^1 \theta\, d\theta = \tfrac{1}{2}, \qquad \text{for} \qquad x = 1.$$

The posterior for θ then is

$$f_{\Theta|X}(\theta|x) = 2(1 - \theta), \qquad \text{for} \quad x = 0, \quad 0 < \theta < 1$$
$$= 2\theta \qquad \text{for} \quad x = 1, \quad 0 < \theta < 1.$$

Prior to flipping the coin we felt that the probability that θ exceeded $\tfrac{1}{2}$ was $\tfrac{1}{2}$; after we flip the coin and get a head ($x = 1$), the same probability is

$$\int_{1/2}^1 2\theta\, d\theta = \tfrac{3}{4}.$$

Or if we get a tail ($x = 0$), this same probability is

$$\int_{1/2}^1 2(1 - \theta)\, d\theta = \tfrac{1}{4}.$$

Thus our posterior probability that $p > \tfrac{1}{2}$ is either $\tfrac{3}{4}$ or $\tfrac{1}{4}$, depending on whether we get a head when the coin is flipped. On the other hand, if we take

$$f_\Theta(\theta) = 5, \qquad .4 < \theta < .6$$

as our prior for θ, then the marginal for X is

$$p_X(x) = \int_{.4}^{.6} 5(1 - \theta)d\theta = .5, \qquad \text{for} \qquad x = 0$$

$$= \int_{.4}^{.6} 5\theta\, d\theta = .5, \qquad \text{for} \qquad x = 1.$$

Notice that the marginal for X is the same in this case as it was earlier. The posterior for θ now is

$$
\begin{aligned}
f_{\Theta|X}(\theta|x) &= 10(1 - \theta), & \text{for} & \quad x = 0, & .4 < \theta < .6 \\
&= 10\theta, & \text{for} & \quad x = 1, & .4 < \theta < .6.
\end{aligned}
$$

Similarly, if we use the discrete prior

$$
\begin{aligned}
p_{\Theta}(\theta) &= \tfrac{1}{4}, & \text{for} & \quad \theta = .4 \text{ or } .6 \\
&= \tfrac{1}{2}, & \text{for} & \quad \theta = .5,
\end{aligned}
$$

then the posterior for θ is discrete as well and we find

$$
\begin{aligned}
p_{\Theta|X}(\theta|x = 0) &= .3, & \text{at } \theta = .4 \\
&= .5, & \text{at } \theta = .5 \\
&= .2, & \text{at } \theta = .6
\end{aligned}
$$

and

$$
\begin{aligned}
p_{\Theta|X}(\theta|x = 1) &= .2, & \text{at } \theta = .4 \\
&= .5, & \text{at } \theta = .5 \\
&= .3, & \text{at } \theta = .6.
\end{aligned}
$$ ∎

EXAMPLE 11.1.4.   Let us continue the discussion of Example 11.1.2. A lot of 1000 items is received, of which θ are defective; the prior probability function for θ is

$$
p_{\Theta}(\theta) = \binom{1000}{\theta}(.05)^{\theta}(.95)^{1000-\theta}, \qquad \theta = 0, 1, \ldots, 1000.
$$

Suppose we select a random sample of 10 items from this lot, and let X be the number of defectives in the sample. Then the distribution for X, given $\Theta = \theta$, is hypergeometric,

$$
p_{X|\Theta}(x|\theta) = \frac{\binom{\theta}{x}\binom{1000-\theta}{10-x}}{\binom{1000}{10}}, \qquad x = 0, 1, \ldots, 10,
$$

and the joint probability function for X and Θ is

$$
p_{X,\Theta}(x, \theta) = \frac{\binom{\theta}{x}\binom{1000-\theta}{10-x}}{\binom{1000}{10}}\binom{1000}{\theta}(.05)^{\theta}(.95)^{1000-\theta},
$$

for $x = 0, 1, \ldots, 10$, $\theta = x, x + 1, \ldots, 990 + x$. Granted x defectives (and $10 - x$ good items) were found in the sample of 10, the smallest possible value for θ is x and the largest is $1000 - (10 - x) = 990 + x$, giving the range quoted for θ. By writing out the factorials and canceling common factors, one can easily see that

$$\frac{\binom{\theta}{x}\binom{1000 - \theta}{10 - x}}{\binom{1000}{10}}\binom{1000}{\theta} = \binom{10}{x}\binom{990}{\theta - x}$$

so the joint probability function for X and Θ is

$$p_{X,\Theta}(x, \theta) = \binom{10}{x}\binom{990}{\theta - x}(.05)^{\theta}(.95)^{1000 - \theta},$$

with the ranges for x and θ as previously given. Then the marginal probability function for X is

$$p_X(x) = \sum_{\theta = x}^{990 + x}\binom{10}{x}\binom{990}{\theta - x}(.05)^{\theta}(.95)^{1000 - \theta}$$

$$= \binom{10}{x}(.05)^x(.95)^{10 - x}\sum_{\theta - x = 0}^{990}\binom{990}{\theta - x}(.05)^{\theta - x}(.95)^{990 - (\theta - x)}$$

$$= \binom{10}{x}(.05)^x(.95)^{10 - x}, \qquad x = 0, 1, \ldots, 10;$$

that is, X is binomial, $n = 10$, $p = .05$. The posterior probability function for Θ then is

$$p_{\Theta|X}(\theta|x) = \frac{p_{X,\Theta}(x, \theta)}{p_X(x)}$$

$$= \binom{990}{\theta - x}(.05)^{\theta - x}(.95)^{990 - (\theta - x)}, \qquad \theta = x, x + 1, \ldots, 990 + x,$$

a binomial probability law with $n = 990$, $p = .05$, shifted x units to the right. ∎

EXERCISE 11.1

1. Assume the probability that an item coming off an assembly line is defective in some respect is a constant θ. Suppose we select 1 item at

random from the assembly line (constituting a sample of 1 observation of a Bernoulli random variable with parameter θ) and we assume a uniform prior density for θ on the interval $(0, 1)$, What is the posterior for θ, if the selected item is defective? If it is not defective?

2. Repeat Exercise 11.1.1, assuming θ is uniform on (a, b) where $a \geq 0$, $b \leq 1$.

*3. A new psychological test has been developed to measure "intelligence." Assume the score an individual will make on the test is a normal random variable with unknown mean μ and known variance of 10. Prior information on μ (from similar tests already in use) would lead us to assume a normal prior density for μ with mean 100 and variance 5. Given a random sample of 20 individual scores on the test, compute the posterior density for μ.

*4. If in Exercise 11.1.3 we assume μ has a uniform prior on the interval $(90, 110)$ (rather than the normal prior used there), show that the posterior for μ is

$$\frac{1}{\sqrt{\pi}} \frac{1}{N\left(\dfrac{110 - \bar{x}}{1/\sqrt{2}}\right) - N\left(\dfrac{90 - \bar{x}}{1/\sqrt{2}}\right)} e^{-(\bar{x} - \mu)^2}, \quad 90 < \mu < 110,$$

where $\bar{x}$ is the observed mean of the 20 scores.

5. Assume the length of nylon string that can be extruded by a machine with no break occuring is an exponential random variable X with parameter λ. Also, assume an exponential prior density on λ with parameter $\frac{1}{1000}$. Given a sample of 1 observation on X (and we find $x = 1100$ feet), compute the posterior density for λ.

6. The time it will take a certain track star to run a 100-yard dash is a uniform random variable on the interval $(\alpha, 10.5)$ measured in seconds. Assume α is equally likely to be 9.5 or 9.7 seconds and compute the posterior for α, given a sample of 1 time of 10 seconds.

7. A coin is flipped 1 time. The probability θ of getting a head is equally likely to be $\frac{1}{4}$, $\frac{1}{2}$, or $\frac{3}{4}$. Compute the posterior for θ, given that we get a a tail on the single flip. Compute the posterior for θ, given that we get a head on the single flip.

8. Assume we repeatedly receive shipments of 50 items from the same supplier and that, in the past, 5 percent of the items received have been defective. What would you use as a prior for D, the unknown number of defectives in a shipment of 50 items just received?

9. In the state of California 52 percent of all registered voters are Democrats, 42 percent are Republicans and the remaining 6 percent are independent or some other party. How would you use this information

to construct a prior distribution for θ, the proportion of votes that the Republican nominee for governor will receive?

10. Assume, in Exercise 11.1.8, a random sample of 3 items is selected from the shipment of 50 and none of the 3 is defective. What is the posterior distribution for D, the number of defective items in the shipment?

11.2 Bayesian Estimators

Suppose a random variable X has density function $f_{X|\Theta}(x|\theta)$, the parameter θ has a prior density $f_\Theta(\theta)$, and, given a random sample $\mathbf{X} = (X_1, X_2, \ldots, X_n)$, the posterior density for θ is $f_{\Theta|\mathbf{X}}(\theta|\mathbf{x})$. We have seen how $\mathbf{X}$ can be used to estimate θ, with either the method of moments or maximum likelihood (or least squares) approaches, none of which make use of the prior or posterior densities for θ. Can the prior and/or posterior information be used in some way to generate an estimate of θ, given we have observed $\mathbf{X} = \mathbf{x}$?

Bayesian estimation procedures can be motivated in a number of ways. One of the most frequently employed actually derives from decision theory, which we do not have the space to explore in any detail. This approach suggests the following rationale in terms of estimating an unknown parameter. Let $X_1, X_2, \ldots, X_n$ be a random sample of a random variable X with density function $f_{\mathbf{X}|\Theta}(\mathbf{x}|\theta)$. (Although we will explicitly discuss the continuous, density function situation, the approach can be obviously modified for the discrete case.) We assume a prior, $f_\Theta(\theta)$, and a derived posterior density, $f_{\Theta|\mathbf{X}}(\theta|\mathbf{x})$, given the sample values $\mathbf{x}$. After the sample values are known, the posterior density summarizes the prior and sample information about θ. Let g represent an estimate of θ, which in general would be a function of $\mathbf{x}$, the observed sample values. Decision theory postulates the existence of a *loss function* $l(\theta, g)$, which, for every possible true value θ and estimate g, expresses the loss suffered; we would, of course, like to find the g that minimizes the loss suffered. With the Bayesian approach, we act as though the unknown parameter is a random variable Θ, which, in turn, would imply that for any given estimate g the loss, $l(\Theta, g)$, then is also a random variable. Given the sample values $\mathbf{x}$ and the posterior $f_{\Theta|\mathbf{X}}(\theta|\mathbf{x})$, the Bayesian approach says the estimate g should be chosen to minimize the *expected* loss $E[l(\Theta, g)|\mathbf{x}]$. This is summarized in the following definition.

DEFINITION 11.2.1. The *Bayesian estimate* of an unknown parameter θ is the value g that minimizes $E[l(\Theta, g)|\mathbf{x}]$, the posterior expected value for the loss function. ∎

With this approach, the Bayesian estimate depends (quite heavily) on the assumed form for the loss function $l(\theta, g)$. By far the most commonly assumed loss function is

$$l(\theta, g) = (\theta - g)^2,$$

called a squared error loss function. Note that with this loss function the smallest possible loss is 0, which occurs only if we choose $g = \theta$; the bigger the difference between the estimate, g, and θ, the larger the loss (actually, the square of the error that, of course, gets large quickly with increasing $|\theta - g|$). Using this loss function (which we always will unless specified otherwise), we find that the Bayesian estimate for θ then is the value that minimizes

$$E[l(\Theta, g)|\underline{x}] = E[(\Theta - g)^2|\underline{x}];$$

but as seen in Exercise 3.3.9, the constant g that minimizes the square of the distance between a random variable Θ and g is the mean of the probability law for Θ. That is, the Bayesian estimate is the mean of the posterior for Θ, as given in the following theorem.

Theorem 11.2.1. With a squared error loss function the Bayesian estimate θ^* of an unknown parameter θ is the mean of the posterior distribution. ∎

The following examples illustrate the Bayesian estimates for some cases we discussed earlier.

EXAMPLE 11.2.1. In Example 11.1.3 we used three different priors for θ, the probability of getting a head on flipping a coin, and for each we derived the posterior distribution. Let us now compute the mean value for each posterior, which would be the Bayesian estimate, using the corresponding prior. For the posterior

$$\begin{aligned} f_{\Theta|X}(\theta|x) &= 2(1 - \theta), & x = 0, && 0 < \theta < 1 \\ &= 2\theta, & x = 1, && 0 < \theta < 1 \end{aligned}$$

we have

$$\begin{aligned} \theta^* &= \tfrac{1}{3}, & \text{if} && x = 0 \\ &= \tfrac{2}{3}, & \text{if} && x = 1, \end{aligned}$$

since these are the respective means for the two possible x values we might observe in the sample. For the posterior

$$\begin{aligned} f_{\Theta|X}(\theta|x) &= 10(1 - \theta), & x = 0, && .4 < \theta < .6 \\ &= 10\theta, & x = 1, && .4 < \theta < .6 \end{aligned}$$

we find

$$\theta^* = .4933, \quad \text{if} \quad x = 0.$$
$$= .5067, \quad \text{if} \quad x = 1.$$

For the discrete posterior

$$p_{\Theta|X}(\theta|x = 0) = .3, \quad \text{at} \quad \theta = .4$$
$$= .5, \quad \text{at} \quad \theta = .5$$
$$= .2, \quad \text{at} \quad \theta = .6$$

and

$$p_{\Theta|X}(\theta|x = 1) = .2, \quad \text{at} \quad \theta = .4$$
$$= .5, \quad \text{at} \quad \theta = .5$$
$$= .3, \quad \text{at} \quad \theta = .6,$$

we find

$$\theta^* = .49, \quad \text{if} \quad x = 0$$
$$= .51, \quad \text{if} \quad x = 1.$$

Both the maximum likelihood and method of moments estimates would be 0 if $x = 0$, 1 if $x = 1$, so in this case the Bayesian estimates seem to give more sensible results. ∎

EXAMPLE 11.2.2.   In Example 11.1.4 we found the posterior density for the number of defectives, θ, in a lot of 1000, given that $X = x$ were found in a random sample of 10, to be

$$p_{\Theta|X}(\theta|x) = \binom{990}{\theta - x}(.05)^{\theta - x}(.95)^{990 - (\theta - x)}, \qquad \theta = x, x + 1, \ldots, 990 + x,$$

where $x = 0, 1, 2, \ldots, 10$. Notice then that $\Theta - x$ is binomial, $n = 990$, $p = .05$, so

$$E[\Theta - x|x] = E[\Theta|x] - x = 990(.05) = 49.50$$

from which it follows that the Bayesian estimate for θ is

$$\theta^* = E[\Theta|x] = 49.50 + x,$$

where x is the number of defectives in the sample of 10. It is interesting to reconsider the Bayesian setting. For the prior we have assumed Θ is binomial, $n = 1000$, $p = .05$, and thus we expected the lot to contain $1000(.05) = 50$ defectives before we took our sample of 10. If in the 10 selected there are $x = 0$ defectives our Bayesian estimate has dropped to $\theta^* = 49.5 + 0 = 49.5$. If in the sample there were $x > 0$ defectives the Bayesian estimate has increased to $49.5 + x > 50$. Because, if there are 50 defectives in the lot, the most likely outcome is $x = 0$, this estimate does make sense. Note that with a sample of 10 from the lot the possible values for θ^* are $49.5, 50.5, \ldots,$ 59.5, all kept fairly close to the prior guess of 50. ∎

As the examples and exercises of Section 11.1 may have indicated, the computation of the posterior density can involve mathematical tedium (or even more serious difficulties). Therefore people using Bayesian procedures have a tendency to choose their prior densities carefully, matching them in a sense with the density for the sample values to minimize the chore of finding the posterior density. Table 11.2.1 presents a collection of sample densities and frequently employed priors, together with the resulting posteriors. The parameters of the prior can be selected in each given case to match the desired prior information. Notice that in essentially all cases the sample density, prior and posterior, are all the same functional form; this is described by saying a *conjugate* prior is used. Conjugate priors tend to minimize mathematical difficulties in the analyses.

EXAMPLE 11.2.3.   Assume each item coming off a production line either is or is not defective, so we can call each item a Bernoulli trial. Assume the trials are independent with $P(\text{defective}) = \theta$ for each trial. If we select n items from the production line, then, and let

$$X_i = 1 \text{ if item } i \text{ is defective}$$
$$X_i = 0 \text{ if item } i \text{ is not defective,}$$

then $X_1, X_2, \ldots, X_n$ is a random sample of a Bernoulli random variable X with parameter θ. We see from line 5 in Table 11.2.1 (where p is used instead of θ) that the conjugate prior for θ is a beta density with parameters a and b. Recall that the mean and variance of a beta random variable are $a/(a + b)$ and $ab/(a + b)^2 (a + b + 1)$, respectively. If we are to use a beta prior for θ we must specify values for a and b, or equivalently prior values for $E[\Theta]$ and $\text{Var}[\Theta]$ from which we can solve for a and b. For example, for our production line case suppose our prior information suggests $E[\Theta] = .01$ and $\text{Var}[\Theta] = .0001$ (so $\sigma = .01$); the larger we take $\text{Var}[\Theta]$, the less sure we are of our prior information, in a sense (there are consistency requirements here too, if we are to use a beta prior, since the mean and variance are not independently varying, but we do not have the space to discuss these). Thus we determine a and b by solving

$$\frac{a}{a + b} = .01 \qquad \frac{ab}{(a + b)^2 (a + b + 1)} = .0001$$

simultaneously. This gives $b = 99(98)/100 = 97.02$, $a = b/99 = .98$ as the

Table 11.2.1 Priors and posteriors, Given a Random Sample of Size n

Line	Density of Sample (given value of unknown parameter)	Prior Density for Unknown Parameter	Posterior Density for Unknown Parameter
1	$\left(\dfrac{1}{2\pi\sigma_X^2}\right)^{n/2} e^{-\Sigma(x_i-\mu)^2/2\sigma_X^2}$ (σ_X^2 known)		
2	$\left(\dfrac{1}{2\pi\sigma^2}\right)^{n/2} e^{-\Sigma(x_i-\mu_X)^2/2\sigma^2}$ (μ_X known)	$\dfrac{1}{m!}[m\sigma_0^2]^{m+1}\left(\dfrac{1}{\sigma^2}\right)^{m+2} e^{-m\sigma_0^2/\sigma^2}$	$\dfrac{1}{\left(m+\dfrac{n}{2}\right)!}\,b^{n/2+m+1}\left(\dfrac{1}{\sigma^2}\right)^{n/2+m+2} e^{-b/\sigma^2}$, $\quad b = m\sigma_0^2 + \dfrac{1}{2}\sum(x_i-\mu_X)^2$
3	$\lambda^n e^{-\lambda\Sigma x_i}$	$\dfrac{1}{m!}\left(\dfrac{m+1}{\lambda_0}\right)^{m+1}\lambda^m e^{-(m+1)\lambda/\lambda_0}$	$\dfrac{\{n\bar{x}+[(m+1)/\lambda_0]\}^{n+m+1}}{\Gamma(n+m+1)}\lambda^{n+m} e^{-\lambda[n\bar{x}+(m+1)/\lambda_0]}$
4	$\dfrac{\lambda^{\Sigma x_i}}{\prod x_i!}e^{-n\lambda}$	$\dfrac{1}{m!}\left(\dfrac{m+1}{\lambda_0}\right)^{m+1}\lambda^m e^{-(m+1)\lambda/\lambda_0}$	$\dfrac{[n+(m+1)/\lambda_0]^{c+1}}{c!}\lambda^c e^{-[n+(m+1)/\lambda_0]\lambda}$, $\quad c = m + \sum x_i$
5	$p^{\Sigma x_i}(1-p)^{n-\Sigma x_i}$	$\dfrac{\Gamma(a+b)}{\Gamma(a)\Gamma(b)}p^{a-1}(1-p)^{b-1}$	$\dfrac{\Gamma(a+b+n)}{\Gamma(a+\sum x_i)\Gamma(b+n-\sum x_i)}p^{a+\sum x_i-1}(1-p)^{b+n-\Sigma x_i-1}$
6	$p^n(1-p)^{\Sigma x_i - n}$	$\dfrac{\Gamma(a+b)}{\Gamma(a)\Gamma(b)}p^{a-1}(1-p)^{b-1}$	$\dfrac{\Gamma(a+b+\sum x_i)}{\Gamma(a+n)\Gamma(b+\sum x_i-n)}p^{a+n-1}(1-p)^{b+\Sigma x_i-n-1}$

values for the parameters of the prior density for Θ. Now if we observe $\sum X_i = \sum x_i$ from the sample we see that the posterior for Θ is again a beta density with parameters $(a + \sum x_i), (b + n - \sum x_i)$. Thus the Bayesian estimate for θ is the mean of this posterior density

$$\theta^* = \frac{a + \sum x_i}{(a + \sum x_i) + (b + n - \sum x_i)} = \frac{a + \sum x_i}{a + b + n}$$

or

$$\theta^* = \frac{\sum x_i + .98}{n + 97.02}$$

for the preceding parameter values. Both the method of moments and maximum likelihood estimates give $\bar{x} = \sum x_i/n$ as the estimate for this case. If in the prior we had chosen $a = \sum x_i, b = n - \sum x_i$ (so $E[\Theta] = \bar{x}$, $\text{Var}[\Theta] = \bar{x}(1 - \bar{x})/(n + 1)$ were the prior mean and variance) the Bayesian estimate would have been

$$\theta^* = \frac{\sum x_i + \sum x_i}{\sum x_i + (n - \sum x_i) + n} = \frac{\sum x_i}{n},$$

the same as maximum likelihood. ∎

The Bayesian and maximum likelihood estimates may be equal, but in general they are not. It can be shown they will, in general, differ by an amount that is small compared with $1/\sqrt{n}$, where n is the sample size. For any finite sample size the Bayesian estimate is "shaded" toward the prior mean, the best guess for θ before any sample values were taken. This effect disappears as n increases indefinitely.

EXAMPLE 11.2.4. Suppose the time to failure for some device is assumed to be an exponential random variable with parameter λ. From line 3 of Table 11.2.1 the suggested conjugate prior for λ is a gamma density with parameters $n = m + 1$, $(m + 1)/\lambda_0$ where m and λ_0 are known. With this parametrization for the gamma density

$$E[\Lambda] = \frac{(m + 1)}{(m + 1)/\lambda_0} = \lambda_0$$

$$\text{Var}[\Lambda] = \frac{(m + 1)}{((m + 1)/\lambda_0)^2} = \frac{\lambda_0^2}{m + 1},$$

so prior knowledge for $E[\Lambda]$ and $\text{Var}[\Lambda]$ can again easily be used to specify values for λ_0 and m. Suppose we feel the mean time to failure for this device should be about 500 hours; since $E[X] = 1/\lambda$, this is equivalent to feeling λ is about $1/500 = .02$. We will take this as the value for the prior mean, $\lambda_0 = .02$. If we are not too certain about this information we should choose $\text{Var}[\Lambda]$ to be large, or equivalently we should take m to be small. We will arbitrarily choose $m = 0$ (making the prior actually exponential). If from a random sample of n values of X we observe $\sum X_i = \sum x_i$, the posterior for Λ (Table 11.2.1) then is a gamma density with parameters $n + 1, n\bar{x} + 1/\lambda_0 = n\bar{x} + 500$ and whose mean is

$$\lambda^* = \frac{n + 1}{n\bar{x} + 500}$$

the Bayesian estimate for λ. The Bayesian estimate for $E[X]$, the mean time to failure then is

$$\frac{1}{\lambda^*} = \frac{n\bar{x} + 500}{n + 1};$$

both maximum likelihood and the method of moments give $\bar{X}$ as the estimate for $E[X]$. If it should happen that $\bar{x} = 500$ (the assumed prior mean for X) then again λ^* is the same as the maximum likelihood estimate. ∎

If, in fact, we are absolutely certain that the value of a parameter θ is θ_0, then the prior for θ would be

$$\begin{aligned} p_\Theta(\theta) &= 1, \quad \text{for } \theta = \theta_0 \\ &= 0, \quad \text{otherwise.} \end{aligned}$$

In such a case we would certainly not want to take a sample to estimate θ, because we would already be willing to assume that θ is known. However, an interesting property of Bayesian estimators is that the sample values cannot change our minds about such a strong assumption, no matter what happens in the sample. To see this, assume we have a continuous random variable X whose density depends on θ; the density function for a random sample of n observations of X then is

$$f_{X|\Theta}(\mathbf{x}|\theta)$$

and the joint density of $\mathbf{X}$ and Θ is

$$\begin{aligned} f_{X,\Theta}(\mathbf{x}, \theta) &= f_{X|\Theta}(\mathbf{x}|\theta), \quad \text{for} \quad \theta = \theta_0 \\ &= 0, \quad \text{otherwise,} \end{aligned}$$

and the marginal for $\mathbf{X}$ is

$$f_X(\mathbf{x}) = f_{X|\Theta}(\mathbf{x}|\theta_0)$$

since there is only one value of θ (θ_0) to average over. Then the posterior for θ is

$$f_{\Theta|X}(\theta|x) = \frac{f_{X,\Theta}(x,\theta)}{f_X(x)}$$

$$= 1, \quad \text{for} \quad \theta = \theta_0$$
$$= 0, \quad \text{otherwise.}$$

Thus the posterior probability that $\Theta = \theta_0$ is 1 and we are still absolutely certain that $\Theta = \theta_0$, whether or not the sample results are consistent with this value.

EXERCISE 11.2

1. Evaluate the expected values for the 3 Bayesian estimators in Example 11.2.1. Are any of them unbiased?
2. (a) What is the Bayesian estimate of the fraction defective in the lot discussed in Example 11.2.2.?
   (b) Is the Bayesian estimator for θ in Example 11.2.2 unbiased? What is its variance and mean square error?
3. Let θ be the fraction defective produced by a production line, as in Example 11.2.3 and assume a uniform $(0, 1)$ prior for θ.
   (a) Evaluate the Bayesian estimate for θ.
   (b) Is this Bayesian estimator unbiased? Compute its mean square error.
4. Assume individuals in the population form independent Bernoulli trials where $p = P(\text{person has hypoglycemia})$. Persons are tested at random until $r = 15$ are found who have hypoglycemia. Let $X = x$ be the observed number of persons required. From previous surveys of the population it is felt that p is quite certainly about .10 so for a prior it was desired to have $E[P] = .1$, $\text{Var}[P] = .01$. With the prior density suggested in Table 11.2.1, what is the Bayesian estimate for p?
*5. At a large industrial concern, assume the number of acts of sabotage that are committed successfully, in a year, is a binomial random variable with parameters n (unknown) and $p = .2$. The parameter n represents the total number of acts attempted in the year; only those that are successful are really known to have occurred. As a prior distribution for n, use

$$P(N = n) = \frac{15^n}{n!} e^{-15}, \quad n = 0, 1, 2, \ldots;$$

that is, the prior is Poisson $\mu = 15$. With a sample of one observation of X (number of successful acts in 1 year) what is the Bayesian estimate for n, the number attempted that year?

*6. The number of orders, per week, that a wholesaler receives for a given item is a Poisson random variable X with parameter λ. From previous experience it is felt that a good prior mean for λ is 10.

    (a) Assume an exponential prior density for λ and evaluate the Bayesian estimate for λ, granted that 80 orders were received in a 10-week period.

    (b) What does the estimate become if there were 80 orders in a 5-week period?

7. Rather than a squared-error loss function, assume an absolute error loss function, $l(\theta, g) = |\theta - g|$, where the parameter θ has a continuous prior $f_\Theta(\theta)$ (and a continuous posterior $f_{\Theta|X}(\theta|x)$). Show that the Bayesian estimate for θ is the median of the posterior. (See Exercise 6.1.12).

8. Rework Example 11.2.3 using the absolute error loss function described in Exercise 11.2.7, that is, find θ^* with this loss function.

9. Rework Example 11.2.4 using the absolute error loss function described in Exercise 11.2.7.

*10. Verify the posterior density functions given in Table 11.2.1.

11.3 Bayesian Intervals

Confidence intervals for unknown parameters have known probabilities of covering the parameter values, over repeated samples of the same size. Given a random sample of a random variable, the confidence interval can be evaluated and, in a sense, we are $100(1 - \alpha)\%$ sure the observed confidence limits cover the true unknown parameter value. Very similar manipulations can be accomplished with the Bayesian approach. Suppose we are given a random sample of a random variable X whose probability law depends on an unknown parameter θ. The parameter θ has a prior density $f_\Theta(\theta)$; once the sample values $x_1, x_2, ..., x_n$, are known, we can compute the posterior density $f_{\Theta|X}(\theta|x)$, which summarizes all the current information about θ, both prior and sample. Then, if $c_1 < c_2$ are two constants such that

$$P(c_1 \le \theta \le c_2 | x) = 1 - \alpha,$$

we are $100(1 - \alpha)\%$ sure that (c_1, c_2) includes θ, given the sample values. We will call such an interval (c_1, c_2) a $100(1 - \alpha)\%$ *Bayesian interval* for θ.

EXAMPLE 11.3.1. Let us use the scenario and computations from Example 11.2.3 to examine a Bayesian interval for θ, the proportion of defective items coming off the assembly line. The prior used for θ was a beta density with parameters a and b, $\sum x_i$ defectives were observed in n

items from the assembly line and the posterior density for θ is beta with parameters $a + \sum x_i$, $b + n - \sum x_i$. Then a $100(1 - \alpha)\%$ Bayesian interval for θ is determined by

$$P(c_1 < \Theta < c_2 | x) =$$

$$\int_{c_1}^{c_2} \frac{\Gamma(a + b + n)}{\Gamma(a + \sum x_i)\Gamma(b + n - \sum x_i)} \theta^{a + \sum x_i - 1} (1 - \theta)^{b + n - \sum x_i - 1} d\theta$$

$$= \int_{c_1}^{c_2} f_{\Theta | \mathbf{x}}(\theta | \mathbf{x}) d\theta = 1 - \alpha.$$

Recall from our discussion of the beta probability law in Section 4.6 (actually, the discussion of the rth largest of n independent uniform $(0, 1)$ random variables) that

$$\int_0^t \frac{\Gamma(n + 1)}{\Gamma(r)\Gamma(n - r + 1)} y^{r - 1}(1 - y)^{n - r} dy = 1 - \sum_{k=0}^{r-1} \binom{n}{k} t^k (1 - t)^{n - k}.$$

Then, with the correspondence $a + \sum x_i = r$, $b + n - \sum x_i = n - r + 1$, $y = \theta, t = c_1$ or c_2, we can write

$$\int_{c_1}^{c_2} f_{\Theta | \mathbf{x}}(\theta | \mathbf{x}) d\theta = \int_0^{c_2} f_{\Theta | \mathbf{x}}(\theta | \mathbf{x}) d\theta - \int_0^{c_1} f_{\Theta | \mathbf{x}}(\theta | \mathbf{x}) d\theta$$

$$= \left(1 - \sum_{k=0}^{a + \sum x_i - 1} \binom{a + b + n}{k} c_2^k (1 - c_2)^{a + b + n - k} \right)$$

$$- \left(1 - \sum_{k=0}^{a + \sum x_i - 1} \binom{a + b + n}{k} c_1^k (1 - c_1)^{a + b + n - k} \right)$$

$$= \sum_{k=0}^{a + \sum x_i - 1} \binom{a + b + n}{k} c_1^k (1 - c_1)^{n - k}$$

$$- \sum_{k=0}^{a + \sum x_i - 1} \binom{a + b + n}{k} c_2^k (1 - c_2)^{n - k},$$

as long as a and b are integers. That is, the Bayesian limits (c_1, c_2) actually are given by finding two binomial probability laws, one with parameters

$a + b + n$, c_1, the other with $a + b + n$, c_2, such that the difference in the two distribution functions at $t = a + \sum x_i - 1$ is equal to $1 - \alpha$. Granted an algorithm for a hand-held calculator for evaluating binomial distribution functions, it can be used to find c_1 and c_2 for the given values of a, b, n, and $\sum x_i$. It is interesting to compare this Bayesian interval with the confidence interval for p (now called θ) derived in Section 7.3. That $100(1 - \alpha)\%$ interval is (p_1, p_2); our earlier formulas for determining p_1 and p_2 are easily seen to be equivalent to

$$\sum_{k=0}^{\Sigma x_i - 1} \binom{n}{k} p_1^k (1 - p_1)^{n-k} - \sum_{k=0}^{\Sigma x_i} \binom{n}{k} p_2^k (1 - p_2)^{n-k} = 1 - \alpha,$$

the same general type of equation that determines the Bayesian limits (c_1, c_2). A major difference is the Bayesian interval corresponds to $n + a + b$ trials (versus n for the confidence interval) that result in $a + \sum x_i$ successes (versus $\sum x_i$ for the confidence interval). ∎

The normal distribution provides a good approximation to the binomial, for large n. Thus this normal approximation can be used to approximate the Bayesian limits for the Bernoulli θ as discussed in Example 11.3.1. If Y is binomial, n, p, the normal approximation to the binomial distribution function is given by

$$P(Y \le y) \doteq N\left(\frac{y - np}{\sqrt{npq}}\right).$$

The use of this approximation is illustrated in the following example.

EXAMPLE 11.3.2.   Suppose a and b are integers and $a + b + n$ is large. Then, using the normal approximation, we have

$$\sum_{k=0}^{a + \Sigma x_i - 1} \binom{a + b + n}{k} c_1^k (1 - c_1)^{a+b+n-k}$$

$$\doteq N\left(\frac{a + \sum x_i - 1 - (a + b + n)c_1}{\sqrt{(a + b + n)c_1(1 - c_1)}}\right)$$

$$\sum_{k=0}^{a + \Sigma x_i - 1} \binom{a + b + n}{k} c_2^k (1 - c_2)^{a+b+n-k}$$

$$\doteq N\left(\frac{a + \sum x_i - 1 - (a + b + n)c_2}{\sqrt{(a + b + n)c_2(1 - c_2)}}\right)$$

so we would want c_1 and c_2 such that

$$N\left(\frac{a + \sum x_i - 1 - (a + b + n)c_1}{\sqrt{(a + b + n)c_1(1 - c_1)}}\right)$$

$$- N\left(\frac{a + \sum x_i - 1 - (a + b + n)c_2}{\sqrt{(a + b + n)c_2(1 - c_2)}}\right) = 1 - \alpha.$$

We can satisfy this equation by solving

$$\frac{a + \sum x_i - 1 - (a + b + n)c_1}{\sqrt{(a + b + n)c_1(1 - c_1)}} = z_{1-\alpha/2}$$

$$\frac{a + \sum x_i - 1 - (a + b + n)c_2}{\sqrt{(a + b + n)c_2(1 - c_2)}} = z_{\alpha/2} = -z_{1-\alpha/2}$$

for c_1 and c_2. The algebra involved is just like that used with the normal approximation to the binomial confidence interval discussed in Section 7.3; ignoring terms in $z_{1-\alpha/2}^2/(a + b + n)$ the solutions are

$$c_1 = \theta^* - z_{1-\alpha/2}\sqrt{\frac{\theta^*(1 - \theta^*)}{a + b + n}}$$

$$c_2 = \theta^* + z_{1-\alpha/2}\sqrt{\frac{\theta^*(1 - \theta^*)}{a + b + n}},$$

where

$$\theta^* = \frac{\sum x_i - 1 + a}{a + b + n}$$

is essentially the Bayesian estimate for θ. The confidence limits using the same large sample approximation are

$$p_1 = \hat{p} - z_{1-\alpha/2}\sqrt{\hat{p}(1 - \hat{p})/n}$$

$$p_2 = \hat{p} + z_{1-\alpha/2}\sqrt{\hat{p}(1 - \hat{p})/n},$$

where $\hat{p} = \sum x_i/n$ is the maximum likelihood estimate. The length and location of these Bayesian intervals depends on both the prior mean and variance. To compare the Bayesian and confidence intervals, Table 11.3.1 presents the 90% confidence intervals for p that result from a sample of size $n = 100$ giving $\sum x_i = 5$ and giving $\sum x_i = 10$ defective items. Four Bayesian

Table 11.3.1

		Estimate	90% Limits
	Confidence interval	.05	(.0141, .0859)
	Bayesian (a)	.05	(.0173, .0827)
$\sum x_i = 5$	(b)	.05	(.0246, .0754)
	(c)	.0545	(.0189, .0902)
	(d)	.0667	(.0332, .1002)
	Confidence interval	.10	(.0506, .1494)
	Bayesian (a)	.0917	(.0484, .1359)
$\sum x_i = 10$	(b)	.075	(.0444, .1056)
	(c)	.10	(.0529, .1471)
	(d)	.10	(.0597, .1403)

limits are given for each of these cases, depending on the prior mean and variance assumed.

(a) $a = 1,$ $b = 19,$ $E[\Theta] = .05,$ $\text{Var}[\Theta] = .002262$
(b) $a = 5,$ $b = 95,$ $E[\Theta] = .05,$ $\text{Var}[\Theta] = .000470$
(c) $a = 1,$ $b = 9,$ $E[\Theta] = .10,$ $\text{Var}[\Theta] = .008182$
(d) $a = 5,$ $b = 45,$ $E[\Theta] = .10,$ $\text{Var}[\Theta] = .001768$

Thus as has been mentioned, the Bayesian estimate is shaded toward the prior mean and agrees with maximum likelihood when the prior mean equals $\sum x_i/n$. The length of the Bayesian interval decreases with smaller prior variances. In a sense, the Bayesian estimate (and interval) augments the observations with $a + b$ more trials, a of which are successes; the difference between the Bayesian values, and the classical approach, disappears as n increases (as long as the prior for Θ allows the true θ as a possible "observed" value). ∎

EXAMPLE 11.3.3. In Example 11.2.4, the time to failure for a device was assumed to be exponential with parameter λ. The conjugate prior for λ (Table 11.2.1) is a gamma density with mean λ_0 and variance $\lambda_0^2/(m + 1)$. If $X_1, X_2, \ldots, X_n$ is a random sample of n failure times the posterior for λ is again gamma with parameters $n + m + 1$, $n\bar{x} + (m + 1)/\lambda_0$ so the posterior mean and variance are

$$\frac{\lambda_0(n + m + 1)}{(m + 1 + n\lambda_0\bar{x})}, \qquad \frac{\lambda_0^2(n + m + 1)}{(m + 1 + n\lambda_0\bar{x})^2},$$

respectively. The Bayesian estimate for $\mu = 1/\lambda$, the mean time to failure, then is

$$\frac{1}{\lambda_0}\left(\frac{m+1}{n+m+1}\right) + \bar{x}\left(\frac{n}{n+m+1}\right),$$

a weighted mean of $1/\lambda_0$ and $\bar{x}$, the maximum likelihood estimate for $1/\lambda$; again, the Bayesian estimate for the mean time to failure is "shaded" toward $1/\lambda_0$, the prior estimate. A $100(1-\alpha)\%$ Bayesian interval for the mean time to failure μ then can be computed from the posterior density.

$$P\left(\frac{1}{c_2} \leq \Lambda \leq \frac{1}{c_1}\middle|\mathbf{x}\right) = P\left(c_1 \leq \frac{1}{\Lambda} \leq c_2\middle|\mathbf{x}\right)$$

$$= \int_{1/c_2}^{1/c_1} \frac{a^{n+m+1}}{\Gamma(n+m+1)} \lambda^{n+m} e^{-a\lambda} \, d\lambda$$

$$= \frac{1}{2^{n+m+1}\Gamma(n+m+1)} \int_{2a/c_2}^{2a/c_1} v^{n+m} e^{-v/2} \, dv,$$

where $a = n\bar{x} + (m+1)/\lambda_0$ and the last integral comes from the change of variable $v = 2a\lambda$. But after this change of variable the integrand is a χ^2 density with $2(m+n+1)$ degrees of freedom (as long as m is an integer). Thus with $2(m+n+1)$ degrees of freedom, the area between $\chi^2_{\alpha/2}$ and $\chi^2_{1-\alpha/2}$ is $1-\alpha$ so we can take

$$\frac{2a}{c_2} = \chi^2_{\alpha/2}, \qquad \frac{2a}{c_1} = \chi^2_{1-\alpha/2}$$

and the $100(1-\alpha)\%$ Bayesian limits for μ, the mean time to failure, are

$$c_1 = \frac{2a}{\chi^2_{1-\alpha/2}} = \frac{2n\bar{x} + 2(m+1)/\lambda_0}{\chi^2_{1-\alpha/2}}$$

$$c_2 = \frac{2a}{\chi^2_{\alpha/2}} = \frac{2n\bar{x} + 2(m+1)/\lambda_0}{\chi^2_{\alpha/2}}.$$

Recall from Chapter 7 that the maximum likelihood estimate for $\mu = 1/\lambda$ is $\bar{x}$, the sample mean; the $100(1-\alpha)\%$ confidence limits for μ are

$$\frac{2n\bar{x}}{\chi^2_{1-\alpha/2}}, \qquad \frac{2n\bar{x}}{\chi^2_{\alpha/2}},$$

where these χ^2 quantiles come from the χ^2 density with $2n$ degrees of freedom. Again, then, the Bayesian limits are very similar to the confidence limits for λ;

in the Bayesian case the degrees of freedom are increased by $2(m + 1)$ and the sample total of the failure times is increased by $(m + 1)/\lambda_0$, as though $m + 1$ additional observations were made, each of which equaled $1/\lambda_0$. Ninety percent confidence intervals and Bayesian intervals are compared in Table 11.3.2, for a sample of size $n = 10$ of an exponential random variable with mean $\mu = 1/\lambda$. Two cases are considered, one in which $\sum x_i = 995$ and the other in which $\sum x_i = 5104$. Again, two different Bayesian priors are used, each with two different prior variances.

(a) $\lambda_0 = .01\ \ = E[\Lambda]$; $m = 0,$ $\mathrm{Var}[\Lambda] = .0001$
(b) $\lambda_0 = .01\ \ = E[\Lambda],$ $m = 10,$ $\mathrm{Var}[\Lambda] = .0000091$
(c) $\lambda_0 = .002 = E[\Lambda],$ $m = 0,$ $\mathrm{Var}[\Lambda] = .000004$
(d) $\lambda_0 = .002 = E[\Lambda],$ $m = 10,$ $\mathrm{Var}[\Lambda] = .00000036$

Table 11.3.2

		Estimate	90% limits
	Confidence interval	99.5	(63.4, 182.6)
	Bayes (a)	99.5	(64.6, 178.0)
$\sum x_i = 995$	(b)	99.8	(72.1, 149.1)
	(c)	135.9	(88.2, 243.1)
	(d)	309.3	(223.6, 462.3)
	Confidence interval	510.4	(325.1, 936.5)
	Bayes (a)	473.1	(307.0, 846.2)
	(b)	295.4	(213.6, 441.6)
$\sum x_i = 5104$	(c)	509.5	(330.6, 911.2)
	(d)	505.0	(365.0, 754.7)

∎

Bayesian procedures can also be used in testing hypotheses. Recall that a simple hypothesis is one that completely specifies the probability law for a random variable X. Suppose X is a random variable whose probability law depends on a parameter θ and based on a random sample of X, we want to test $H_0: \theta = \theta_0$ versus $H_1: \theta = \theta_1$, where θ_0 and θ_1 are specified. In the Bayesian sense, then, there are only two possible values for θ, θ_0 and θ_1, so the prior for θ is discrete; let

$$p_0 = P(\Theta = \theta_0), \qquad p_1 = P(\Theta = \theta_1).$$

The posterior probability function for Θ, given $\mathbf{X} = \mathbf{x}$, is

$$P(\Theta = \theta_0 | \mathbf{x}) = \frac{p_0 f_{\mathbf{X}|\Theta}(\mathbf{x}|\theta_0)}{p_0 f_{\mathbf{X}|\Theta}(\mathbf{x}|\theta_0) + p_1 f_{\mathbf{X}|\Theta}(\mathbf{x}|\theta_1)}$$

$$P(\Theta = \theta_1 | \mathbf{x}) = \frac{p_1 f_{\mathbf{X}}(\mathbf{x}|\theta_1)}{p_0 f_{\mathbf{X}}(\mathbf{x}|\theta_0) + p_1 f_{\mathbf{X}}(\mathbf{x}|\theta_1)}$$

It would seem rational to accept H_0 if the posterior probability $\Theta = \theta_0$ exceeds the posterior probability $\Theta = \theta_1$, that is, if

$$\frac{P(\Theta = \theta_0 | \mathbf{x})}{P(\Theta = \theta_1 | \mathbf{x})} \geq 1,$$

or

$$\frac{f_{\mathbf{X}}(\mathbf{x}|\theta_0)}{f_{\mathbf{X}}(\mathbf{x}|\theta_1)} \geq \frac{p_1}{p_0}$$

and to reject H_0 otherwise. That is, the natural Bayesian acceptance region is determined by the value of

$$\frac{f_{\mathbf{X}}(\mathbf{x}|\theta_0)}{f_{\mathbf{X}}(\mathbf{x}|\theta_1)},$$

the ratio of the likelihood of the sample values, given θ_0, versus the likelihood given $\Theta = \theta_1$. But this is exactly the same as the Neyman–Pearson best test (except that k is actually specified by p_1/p_0, rather than by α). Thus the Bayesian procedure leads to the same way of partitioning the sample space for X into acceptance versus rejection regions as the Neyman–Pearson approach, for simple H_0 versus simple H_1.

In testing composite hypotheses, a natural Bayesian procedure is given by using the $100(1 - \alpha)\%$ Bayesian interval for θ as the acceptance region for the test, as we have discussed with confidence intervals. Because the confidence interval depends on the prior used for θ, so does the test. For example, by referring to Table 11.3.1, with a sample of $n = 100$ observations of a Bernoulli random variable, observing $\sum x_i = 5$ would lead us to reject $H_0: \theta = .02$ if our prior was specified by (b) $E[\Theta] = .05$, $\text{Var}[\Theta] = .000470$ or (d) $E[\Theta] = .10$, $\text{Var}[\Theta] = .001768$, although we would accept H_0 for the priors in (a) and (c) (and for the classical non-Bayesian approach).

EXERCISE 11.3

1. If in a random sample of $n = 50$ independent Bernoulli trials we observe $\sum x_i = 20$ successes, evaluate the 95% Bayesian limits for $\theta = P(\text{success})$ assuming a uniform $(0, 1)$ prior for θ. Compare these with the 95% confidence limits for θ.

*2. In $n = 2$ repeated independent Bernoulli trials with $\theta = P(\text{success})$, one success was observed.
   (a) Assume θ is uniform on $(0, .2)$ and find the posterior for θ.
   (b) Can you evaluate a 90% Bayesian interval for θ?

3. For a random sample of $n = 20$ exponential random variables with parameter λ, it was found that $\sum x_i = 1982$.
   (a) Evaluate 95% confidence limits for $\mu = 1/\lambda$.
   (b) Assume a gamma prior for λ with mean $\lambda_0 = .05$ and variance .0005. Evalute 95% Bayesian limits for $\mu = 1/\lambda$.

4. Reread Exercise 11.2.4. The observed value for X was $x = 125$.
   (a) With prior parameters $a = 1, b = 8$, evaluate the Bayesian estimate for p.
   (b) Compute 90% Bayesian limits for p.

*5. For the case discussed in Exercise 11.2.6, assume the same prior information and evaluate 90% Bayesian limits for λ for the two cases cases given, (a) and (b).

6. For the two cases mentioned in Exercise 11.3.5, evaluate 90% confidence limits for λ.

7. A random sample of $n = 10$ exponential random variables gave $\sum x_i = 995$.
   (a) With $\alpha = .1$, would you accept $H_0 : \mu = 70$?
   (b) Assuming a gamma prior for λ with $\lambda_0 = .01$, $m = 20$, would you accept $H_0 : \mu = 70$?

8. A Las Vegas casino claims that one person in eight leaves their casino as a winner. Of 100 patrons leaving the casino, it was found that $\sum x_i = 10$ were winners. Would you accept $H_0 : p = .125$ (versus $H_1 : p \neq .125$ with $\alpha = .1$), based on this sample?

9. Assuming a prior for p with $a = 1$, $b = 10$, and the data given in Exercise 11.3.8, would you accept $H_0 : p = .125$ (versus $H_1 : p \neq .125$) with $\alpha = .1$?

11.4 Summary

Prior distribution for parameter θ: Density or probability function expressing prior belief about value for θ.

Posterior distribution for parameter θ: Conditional density or probability function for θ, given the values of a random sample.

Loss function $l(\theta, g)$: Function expressing loss suffered if true parameter value is θ and g is used as an estimate for θ.

Bayesian estimate: Estimate that minimizes the expected value of the loss function, with respect to the posterior probability law. Posterior mean with a squared error loss function.

Posteriors resulting from conjugate priors: Table 11.2.1.

Bayesian interval: Interval that has given posterior probability.

Appendix

Table 1   Standard Normal Distribution Function

$$N(t) = \int_{-\infty}^{t} \frac{1}{\sqrt{2\pi}} e^{-z^2/2}\, dz$$

t	0	1	2	3	4	5	6	7	8	9
−3.	.0013	.0013	.0013	.0012	.0012	.0011	.0011	.0011	.0010	.0010
−2.9	.0019	.0018	.0017	.0017	.0016	.0016	.0015	.0015	.0014	.0014
−2.8	.0026	.0025	.0024	.0023	.0023	.0022	.0021	.0021	.0020	.0019
−2.7	.0035	.0034	.0033	.0032	.0031	.0030	.0029	.0028	.0027	.0026
−2.6	.0047	.0045	.0044	.0043	.0041	.0040	.0039	.0038	.0037	.0036
−2.5	.0062	.0060	.0059	.0057	.0055	.0054	.0052	.0051	.0049	.0048
−2.4	.0082	.0080	.0078	.0075	.0073	.0071	.0069	.0068	.0066	.0064
−2.3	.0107	.0104	.0102	.0099	.0096	.0094	.0091	.0089	.0087	.0084
−2.2	.0139	.0136	.0132	.0129	.0125	.0122	.0119	.0116	.0113	.0110
−2.1	.0179	.0174	.0170	.0166	.0162	.0158	.0154	.0150	.0146	.0143
−2.0	.0227	.0222	.0217	.0212	.0207	.0202	.0197	.0192	.0188	.0183
−1.9	.0287	.0281	.0274	.0268	.0262	.0256	.0250	.0244	.0239	.0233
−1.8	.0359	.0351	.0344	.0336	.0329	.0322	.0314	.0307	.0300	.0294
−1.7	.0446	.0436	.0427	.0418	.0409	.0401	.0392	.0384	.0375	.0367
−1.6	.0548	.0537	.0526	.0516	.0505	.0495	.0485	.0475	.0465	.0455
−1.5	.0668	.0655	.0643	.0630	.0618	.0606	.0594	.0582	.0571	.0559
−1.4	.0808	.0793	.0778	.0764	.0749	.0735	.0721	.0708	.0694	.0681
−1.3	.0968	.0951	.0934	.0918	.0901	.0885	.0869	.0853	.0838	.0823
−1.2	.1151	.1131	.1112	.1093	.1075	.1056·	.1038	.1020	.1003	.0985
−1.1	.1357	.1335	.1314	.1292	.1271	.1251	.1230	.1210	.1190	.1170
−1.0	.1587	.1562	.1539	.1515	.1492	.1469	.1446	.1423	.1401	.1379
−.9	.1841	.1814	.1788	.1762	.1736	.1711	.1685	.1660	.1635	.1611
−.8	.2119	.2090	.2061	.2033	.2005	.1977	.1949	.1921	.1894	.1867
−.7	.2420	.2389	.2358	.2326	.2297	.2266	.2236	.2206	.2177	.2148
−.6	.2743	.2709	.2676	.2643	.2611	.2578	.2546	.2514	.2483	.2451
−.5	.3085	.3050	.3015	.2981	.2946	.2912	.2877	.2843	.2810	.2776
−.4	.3446	.3409	.3372	.3336	.3300	.3264	.3228	.3192	.3156	.3121
−.3	.3821	.3783	.3745	.3707	.3669	.3632	.3594	.3557	.3520	.3483
−.2	.4207	.4168	.4129	.4090	.4052	.4013	.3974	.3936	.3897	.3859
−.1	.4602	.4562	.4522	.4483	.4443	.4404	.4364	.4325	.4286	.4247
−.0	.5000	.4960	.4920	.4880	.4840	.4801	.4761	.4721	.4681	.4641

Table 1 Standard Normal Distribution Function (*continued*)

t	0	1	2	3	4	5	6	7	8	9
.0	.5000	.5040	.5080	.5120	.5160	.5199	.5239	.5279	.5319	.5359
.1	.5398	.5438	.5478	.5517	.5557	.5596	.5636	.5675	.5714	.5753
.2	.5793	.5832	.5871	.5910	.5948	.5987	.6026	.6064	.6103	.6141
.3	.6179	.6217	.6255	.6293	.6331	.6368	.6406	.6443	.6480	.6517
.4	.6554	.6591	.6628	.6664	.6700	.6736	.6772	.6808	.6844	.6879
.5	.6915	.6950	.6985	.7019	.7054	.7088	.7123	.7157	.7190	.7224
.6	.7257	.7291	.7324	.7357	.7389	.7422	.7454	.7486	.7517	.7549
.7	.7580	.7611	.7642	.7673	.7704	.7734	.7764	.7794	.7823	.7852
.8	.7881	.7910	.7939	.7967	.7995	.8023	.8051	.8079	.8106	.8133
.9	.8159	.8186	.8212	.8238	.8264	.8289	.8315	.8340	.8365	.8389
1.0	.8413	.8438	.8461	.8485	.8508	.8531	.8554	.8577	.8599	.8621
1.1	.8643	.8665	.8686	.8708	.8729	.8749	.8770	.8790	.8810	.8830
1.2	.8849	.8869	.8888	.8907	.8925	.8944	.8962	.8980	.8997	.9015
1.3	.9032	.9049	.9066	.9082	.9099	.9115	.9131	.9147	.9162	.9177
1.4	.9192	.9207	.9222	.9236	.9251	.9265	.9279	.9292	.9306	.9319
1.5	.9332	.9345	.9357	.9370	.9382	.9394	.9406	.9418	.9429	.9441
1.6	.9452	.9463	.9474	.9484	.9495	.9505	.9515	.9525	.9535	.9545
1.7	.9554	.9564	.9573	.9582	.9591	.9599	.9608	.9616	.9625	.9633
1.8	.9641	.9649	.9656	.9664	.9671	.9678	.9686	.9693	.9700	.9706
1.9	.9713	.9719	.9726	.9732	.9738	.9744	.9750	.9756	.9761	.9767
2.0	.9773	.9778	.9783	.9788	.9793	.9798	.9803	.9808	.9812	.9817
2.1	.9821	.9826	.9830	.9834	.9838	.9842	.9846	.9850	.9854	.9857
2.2	.9861	.9864	.9868	.9871	.9875	.9878	.9881	.9884	.9887	.9890
2.3	.9893	.9896	.9898	.9901	.9904	.9906	.9909	.9911	.9913	.9916
2.4	.9918	.9920	.9922	.9925	.9927	.9929	.9931	.9932	.9934	.9936
2.5	.9938	.9940	.9941	.9943	.9945	.9946	.9948	.9949	.9951	.9952
2.6	.9953	.9955	.9956	.9957	.9959	.9960	.9961	.9962	.9963	.9964
2.7	.9965	.9966	.9967	.9968	.9969	.9970	.9971	.9972	.9973	.9974
2.8	.9974	.9975	.9976	.9977	.9977	.9978	.9979	.9979	.9980	.9981
2.9	.9981	.9982	.9982	.9983	.9984	.9984	.9985	.9985	.9986	.9986
3.	.9987	.9987	.9987	.9988	.9988	.9989	.9989	.9989	.9990	.9990

Table 2 χ^2 Distribution Function Quantiles, t_k

$$F_{\chi^2}(t_k) = \int_0^{t_k} \frac{1}{2^{d/2}\Gamma(d/2)} x^{(d-2)/2} e^{-x/2}\, dx = k$$

Degrees of freedom = d

d	.005	.010	.025	.050	.100	.250	.500	.750	.900	.950	.975	.990	.995
1	.0000393	.000157	.000982	.00393	.0158	.102	.455	1.32	2.71	3.84	5.02	6.63	7.88
2	.0100	.0201	.0506	.103	.211	.575	1.39	2.77	4.61	5.99	7.38	9.21	10.6
3	.0717	.115	.216	.352	.584	1.21	2.37	4.11	6.25	7.81	9.35	11.3	12.8
4	.207	.297	.484	.711	1.06	1.92	3.36	5.39	7.78	9.49	11.1	13.3	14.9
5	.412	.554	.831	1.15	1.61	2.67	4.35	6.63	9.24	11.1	12.8	15.1	16.7
6	.676	.872	1.24	1.64	2.20	3.45	5.35	7.84	10.6	12.6	14.4	16.8	18.5
7	.989	1.24	1.69	2.17	2.83	4.25	6.35	9.04	12.0	14.1	16.0	18.5	20.3
8	1.34	1.65	2.18	2.73	3.49	5.07	7.34	10.2	13.4	15.5	17.5	20.1	22.0
9	1.73	2.09	2.70	3.33	4.17	5.90	8.34	11.4	14.7	16.9	19.0	21.7	23.6
10	2.16	2.56	3.25	3.94	4.87	6.74	9.34	12.5	16.0	18.3	20.5	23.2	25.2
11	2.60	3.05	3.82	4.57	5.58	7.58	10.3	13.7	17.3	19.7	21.9	24.7	26.8
12	3.07	3.57	4.40	5.23	6.30	8.44	11.3	14.8	18.5	21.0	23.3	26.2	28.3
13	3.57	4.11	5.01	5.89	7.04	9.30	12.3	16.0	19.8	22.4	24.7	27.7	29.8
14	4.07	4.66	5.63	6.57	7.79	10.2	13.3	17.1	21.1	23.7	26.1	29.1	31.3
15	4.60	5.23	6.26	7.26	8.55	11.0	14.3	18.2	22.3	25.0	27.5	30.6	32.8

Table 2 χ^2 Distribution Function Quantiles, t_k (continued)

k													
16	5.14	5.81	6.91	7.96	9.31	11.9	15.3	19.4	23.5	26.3	28.8	32.0	34.3
17	5.70	6.41	7.56	8.67	10.1	12.8	16.3	20.5	24.8	27.6	30.2	33.4	35.7
18	6.26	7.01	8.23	9.39	10.9	13.7	17.3	21.6	26.0	28.9	31.5	34.8	37.2
19	6.84	7.63	8.91	10.1	11.7	14.6	18.3	22.7	27.2	30.1	32.9	36.2	38.6
20	7.43	8.26	9.59	10.9	12.4	15.5	19.3	23.8	28.4	31.4	34.2	37.6	40.0
21	8.03	8.90	10.3	11.6	13.2	16.3	20.3	24.9	29.6	32.7	35.5	38.9	41.4
22	8.64	9.54	11.0	12.3	14.0	17.2	21.3	26.0	30.8	33.9	36.8	40.3	42.8
23	9.26	10.2	11.7	13.1	14.8	18.1	22.3	27.1	32.0	35.2	38.1	41.6	44.2
24	9.89	10.9	12.4	13.8	15.7	19.0	23.3	28.2	33.2	36.4	39.4	43.0	45.6
25	10.5	11.5	13.1	14.6	16.5	19.9	24.3	29.3	34.4	37.7	40.6	44.3	46.9
26	11.2	12.2	13.8	15.4	17.3	20.8	25.3	30.4	35.6	38.9	41.9	45.6	48.3
27	11.8	12.9	14.6	16.2	18.1	21.7	26.3	31.5	36.7	40.1	43.2	47.0	49.6
28	12.5	13.6	15.3	16.9	18.9	22.7	27.3	32.6	37.9	41.3	44.5	48.3	51.0
29	13.1	14.3	16.0	17.7	19.8	23.6	28.3	33.7	39.1	42.6	45.7	49.6	52.3
30	13.8	15.0	16.8	18.5	20.6	24.5	29.3	34.8	40.3	43.8	47.0	50.9	53.7

Abridged with permission from Catherine M. Thompson, "Tables of percentage points of the incomplete Beta function and of the chi-square distribution," *Biometrika*, **32**, 1941.

Table 3 *T* Distribution Function Quantiles

$$F_T(t_k) = \int_{-\infty}^{t_k} \frac{\Gamma((d+1)/2)}{\sqrt{d\pi}\,\Gamma(d/2)}(1 + u^2/d)^{-(d+1)/2}\,du = k$$

d	.60	.75	.90	.95	.975	.99	.995	.9995
1	.325	1.000	3.078	6.314	12.706	31.821	63.657	636.619
2	.289	.816	1.886	2.920	4.303	6.965	9.925	31.598
3	.277	.765	1.638	2.353	3.182	4.541	5.841	12.941
4	.271	.741	1.533	2.132	2.776	3.747	4.604	8.610
5	.267	.727	1.476	2.015	2.571	3.365	4.032	6.859
6	.265	.718	1.440	1.943	2.447	3.143	3.707	5.959
7	.263	.711	1.415	1.895	2.365	2.998	3.499	5.405
8	.262	.706	1.397	1.860	2.306	2.896	3.355	5.041
9	.261	.703	1.383	1.833	2.262	2.821	3.250	4.781
10	.260	.700	1.372	1.812	2.228	2.764	3.169	4.587
11	.260	.697	1.363	1.796	2.201	2.718	3.106	4.437
12	.259	.695	1.356	1.782	2.179	2.681	3.055	4.318
13	.259	.694	1.350	1.771	2.160	2.650	3.012	4.221
14	.258	.692	1.345	1.761	2.145	2.624	2.977	4.140
15	.258	.691	1.341	1.753	2.131	2.602	2.947	4.073
16	.258	.690	1.337	1.746	2.120	2.583	2.921	4.015
17	.257	.689	1.333	1.740	2.110	2.567	2.898	3.965
18	.257	.688	1.330	1.734	2.101	2.552	2.878	3.922
19	.257	.688	1.328	1.729	2.093	2.539	2.861	3.883
20	.257	.687	1.325	1.725	2.086	2.528	2.845	3.850
21	.257	.686	1.323	1.721	2.080	2.518	2.831	3.819
22	.256	.686	1.321	1.717	2.074	2.508	2.819	3.792
23	.256	.685	1.319	1.714	2.069	2.500	2.807	3.767
24	.256	.685	1.318	1.711	2.064	2.492	2.797	3.745
25	.256	.684	1.316	1.708	2.060	2.485	2.787	3.725
26	.256	.684	1.315	1.706	2.056	2.479	2.779	3.707
27	.256	.684	1.314	1.703	2.052	2.473	2.771	3.690
28	.256	.683	1.313	1.701	2.048	2.467	2.763	3.674
29	.256	.683	1.311	1.699	2.045	2.462	2.756	3.659
30	.256	.683	1.310	1.697	2.042	2.457	2.750	3.646
40	.255	.681	1.303	1.684	2.021	2.423	2.704	3.551
60	.254	.679	1.296	1.671	2.000	2.390	2.660	3.460
120	.254	.677	1.289	1.658	1.980	2.358	2.617	3.373
∞	.253	.674	1.282	1.645	1.960	2.326	2.576	3.291

Degrees of freedom = d, k

Abridged with permission from R. A. Fisher and Frank Yates, *Statistical Tables*, Oliver and Boyd, Ltd., Edinburgh, 1938.

Table 4 F Distribution Function Quantiles, t_k

$$\int_0^{t_k} \frac{\Gamma\left(\dfrac{d_1+d_2}{2}\right)}{\Gamma\left(\dfrac{d_1}{2}\right)\Gamma\left(\dfrac{d_2}{2}\right)} \left(\frac{d_1}{d_2}\right)^{d_1/2}\; \frac{x^{d_1/2-1}}{\left(1+\dfrac{d_1}{d_2}x\right)^{(d_1+d_2)/2}}\,dx = k$$

$$k = .5$$

	1	2	3	4	5	6	7	8	9	10	12	14	16	18	20	30	40	∞
1	1.000	1.500	1.709	1.823	1.894	1.942	1.977	2.004	2.025	2.042	2.067	2.086	2.100	2.110	2.119	2.145	2.158	2.198
2	0.667	1.000	1.135	1.207	1.252	1.282	1.305	1.321	1.334	1.345	1.361	1.372	1.381	1.388	1.393	1.410	1.418	1.443
3	0.585	0.881	1.000	1.063	1.102	1.129	1.148	1.163	1.174	1.183	1.197	1.207	1.215	1.220	1.225	1.239	1.246	1.268
4	0.549	0.828	0.941	1.000	1.037	1.062	1.080	1.093	1.104	1.113	1.126	1.135	1.142	1.147	1.152	1.165	1.172	1.192
5	0.528	0.799	0.907	0.965	1.000	1.024	1.041	1.055	1.065	1.073	1.086	1.094	1.101	1.106	1.111	1.123	1.130	1.149
6	0.515	0.780	0.886	0.942	0.977	1.000	1.017	1.030	1.040	1.050	1.060	1.069	1.075	1.080	1.085	1.097	1.103	1.122
7	0.506	0.767	0.871	0.926	0.960	0.983	1.000	1.013	1.022	1.030	1.042	1.051	1.057	1.062	1.066	1.079	1.085	1.103
8	0.499	0.757	0.860	0.915	0.948	0.971	0.988	1.000	1.010	1.018	1.029	1.038	1.044	1.049	1.053	1.065	1.071	1.089
9	0.494	0.749	0.852	0.906	0.939	0.962	0.978	0.990	1.000	1.008	1.019	1.028	1.034	1.039	1.043	1.055	1.061	1.079
10	0.490	0.743	0.845	0.899	0.932	0.954	0.971	0.983	0.992	1.000	1.012	1.020	1.026	1.031	1.035	1.047	1.053	1.071
11	0.486	0.739	0.840	0.893	0.926	0.948	0.964	0.977	0.986	0.994	1.005	1.013	1.020	1.025	1.028	1.040	1.046	1.064
12	0.484	0.735	0.835	0.888	0.921	0.943	0.959	0.972	0.981	0.989	1.000	1.008	1.014	1.019	1.023	1.035	1.041	1.058
13	0.481	0.731	0.832	0.885	0.917	0.939	0.955	0.967	0.977	0.984	0.996	1.004	1.010	1.015	1.019	1.030	1.036	1.054
14	0.479	0.729	0.828	0.881	0.914	0.936	0.952	0.964	0.973	0.981	0.992	1.000	1.006	1.011	1.015	1.026	1.032	1.050
15	0.478	0.726	0.826	0.878	0.911	0.933	0.948	0.960	0.970	0.977	0.989	0.997	1.003	1.007	1.011	1.023	1.029	1.046
16	0.476	0.724	0.823	0.876	0.908	0.930	0.946	0.958	0.967	0.975	0.986	0.994	1.000	1.005	1.009	1.020	1.026	1.043
17	0.475	0.722	0.821	0.874	0.906	0.928	0.943	0.955	0.965	0.972	0.983	0.992	0.998	1.002	1.006	1.017	1.023	1.040
18	0.474	0.721	0.819	0.872	0.904	0.926	0.941	0.953	0.962	0.970	0.981	0.989	0.995	1.000	1.004	1.015	1.021	1.038
19	0.473	0.719	0.818	0.870	0.902	0.924	0.939	0.951	0.961	0.968	0.979	0.987	0.993	0.998	1.002	1.013	1.019	1.036
20	0.472	0.718	0.816	0.868	0.900	0.922	0.938	0.950	0.959	0.966	0.977	0.986	0.992	0.997	1.000	1.011	1.017	1.034
22	0.470	0.715	0.814	0.866	0.898	0.919	0.935	0.947	0.956	0.963	0.974	0.982	0.989	0.993	0.997	1.008	1.014	1.031
24	0.469	0.714	0.812	0.863	0.895	0.917	0.932	0.944	0.953	0.961	0.972	0.980	0.986	0.991	0.994	1.006	1.011	1.028
26	0.468	0.712	0.810	0.861	0.893	0.915	0.930	0.942	0.951	0.959	0.970	0.978	0.984	0.989	0.992	1.004	1.009	1.026
28	0.467	0.711	0.808	0.860	0.892	0.913	0.929	0.940	0.950	0.957	0.968	0.976	0.982	0.987	0.990	1.002	1.007	1.024
30	0.466	0.709	0.807	0.858	0.890	0.912	0.927	0.939	0.948	0.955	0.966	0.974	0.980	0.985	0.989	1.000	1.006	1.023
40	0.463	0.705	0.802	0.854	0.885	0.907	0.922	0.934	0.943	0.950	0.961	0.969	0.975	0.980	0.983	0.994	1.000	1.017
∞	0.455	0.693	0.789	0.839	0.870	0.891	0.907	0.918	0.927	0.934	0.945	0.953	0.959	0.963	0.967	0.978	0.983	1.000

Table 4 F Distribution Function Quantiles, t_k (continued)

$$k = .75$$

									d_1									
d_2	1	2	3	4	5	6	7	8	9	10	12	14	16	18	20	30	40	∞
1	5.828	7.500	8.200	8.581	8.820	8.983	9.102	9.192	9.263	9.320	9.406	9.468	9.515	9.552	9.581	9.670	9.714	9.849
2	2.571	3.000	3.153	3.232	3.280	3.312	3.335	3.353	3.366	3.377	3.393	3.405	3.414	3.421	3.426	3.443	3.451	3.476
3	2.024	2.280	2.356	2.390	2.410	2.422	2.430	2.436	2.441	2.445	2.450	2.454	2.456	2.459	2.460	2.465	2.467	2.474
4	1.807	2.000	2.047	2.064	2.072	2.077	2.079	2.080	2.081	2.082	2.083	2.083	2.083	2.083	2.083	2.082	2.082	2.081
5	1.692	1.853	1.884	1.893	1.895	1.894	1.894	1.892	1.891	1.890	1.888	1.886	1.884	1.883	1.882	1.878	1.876	1.869
6	1.621	1.762	1.784	1.787	1.785	1.782	1.779	1.776	1.773	1.771	1.767	1.764	1.761	1.759	1.757	1.751	1.748	1.737
7	1.573	1.701	1.717	1.716	1.711	1.706	1.701	1.697	1.693	1.690	1.684	1.680	1.676	1.674	1.671	1.664	1.659	1.645
8	1.538	1.657	1.668	1.664	1.658	1.651	1.645	1.640	1.635	1.631	1.624	1.619	1.615	1.612	1.609	1.600	1.595	1.578
9	1.512	1.624	1.632	1.625	1.617	1.609	1.602	1.596	1.591	1.586	1.579	1.573	1.568	1.564	1.561	1.551	1.545	1.526
10	1.492	1.598	1.603	1.595	1.585	1.577	1.569	1.562	1.556	1.551	1.543	1.537	1.531	1.527	1.524	1.512	1.506	1.484
11	1.475	1.577	1.580	1.570	1.560	1.550	1.542	1.535	1.528	1.523	1.514	1.507	1.501	1.497	1.493	1.480	1.474	1.450
12	1.461	1.560	1.561	1.550	1.539	1.529	1.520	1.512	1.505	1.500	1.490	1.483	1.477	1.472	1.468	1.454	1.447	1.422
13	1.450	1.545	1.545	1.534	1.521	1.511	1.501	1.493	1.486	1.480	1.470	1.462	1.456	1.451	1.447	1.432	1.425	1.398
14	1.440	1.533	1.532	1.519	1.507	1.495	1.485	1.477	1.470	1.463	1.453	1.445	1.438	1.433	1.428	1.414	1.406	1.377
15	1.432	1.523	1.520	1.507	1.494	1.482	1.472	1.463	1.456	1.449	1.438	1.430	1.423	1.417	1.413	1.397	1.389	1.359
16	1.425	1.514	1.510	1.497	1.483	1.471	1.460	1.451	1.443	1.437	1.426	1.417	1.410	1.404	1.399	1.383	1.374	1.343
17	1.419	1.506	1.502	1.487	1.473	1.461	1.450	1.441	1.433	1.426	1.414	1.405	1.398	1.392	1.387	1.370	1.361	1.329
18	1.413	1.499	1.494	1.479	1.464	1.452	1.441	1.431	1.423	1.416	1.404	1.395	1.388	1.381	1.376	1.359	1.350	1.316
19	1.408	1.493	1.487	1.472	1.457	1.444	1.432	1.423	1.415	1.407	1.395	1.386	1.378	1.372	1.367	1.349	1.339	1.305
20	1.404	1.487	1.481	1.465	1.450	1.437	1.425	1.415	1.407	1.400	1.387	1.381	1.370	1.363	1.358	1.340	1.330	1.294
22	1.396	1.477	1.470	1.454	1.438	1.424	1.413	1.402	1.394	1.386	1.374	1.366	1.360	1.349	1.343	1.325	1.314	1.276
24	1.390	1.470	1.462	1.445	1.428	1.414	1.402	1.392	1.383	1.375	1.362	1.353	1.347	1.336	1.331	1.311	1.300	1.261
26	1.384	1.463	1.454	1.437	1.420	1.406	1.394	1.383	1.374	1.366	1.352	1.343	1.336	1.331	1.320	1.300	1.289	1.247
28	1.380	1.457	1.448	1.430	1.413	1.399	1.386	1.375	1.366	1.358	1.344	1.334	1.326	1.321	1.311	1.291	1.279	1.236
30	1.376	1.452	1.443	1.424	1.407	1.392	1.380	1.368	1.359	1.351	1.337	1.327	1.318	1.313	1.303	1.282	1.270	1.226
40	1.363	1.436	1.424	1.404	1.386	1.371	1.357	1.346	1.335	1.327	1.312	1.300	1.291	1.284	1.276	1.253	1.240	1.188
∞	1.323	1.386	1.369	1.346	1.325	1.307	1.291	1.277	1.265	1.255	1.237	1.222	1.211	1.200	1.191	1.160	1.140	1.000

Table 4 F Distribution Function Quantiles, t_k (continued)

$k = .90$

d_2 \ d_1	1	2	3	4	5	6	7	8	9	10	12	14	16	18	20	30	40	∞
1	39.86	49.50	53.59	55.83	57.24	58.20	58.91	59.44	59.86	60.20	60.71	61.07	61.35	61.57	61.74	62.27	62.53	63.33
2	8.526	9.000	9.162	9.243	9.293	9.326	9.349	9.367	9.380	9.393	9.408	9.420	9.429	9.436	9.441	9.458	9.466	9.491
3	5.538	5.462	5.391	5.343	5.309	5.285	5.266	5.252	5.240	5.230	5.216	5.205	5.196	5.190	5.184	5.168	5.160	5.134
4	4.545	4.325	4.191	4.107	4.051	4.010	3.979	3.955	3.936	3.920	3.896	3.878	3.864	3.853	3.844	3.817	3.804	3.761
5	4.060	3.780	3.620	3.520	3.453	3.404	3.368	3.339	3.316	3.297	3.268	3.247	3.230	3.217	3.207	3.174	3.157	3.105
6	3.776	3.463	3.289	3.181	3.108	3.055	3.014	2.983	2.958	2.937	2.905	2.881	2.863	2.848	2.836	2.800	2.781	2.722
7	3.589	3.257	3.074	2.960	2.883	2.827	2.785	2.752	2.725	2.702	2.668	2.643	2.623	2.607	2.595	2.556	2.535	2.471
8	3.458	3.113	2.924	2.806	2.727	2.668	2.624	2.589	2.561	2.538	2.502	2.475	2.455	2.438	2.425	2.383	2.361	2.293
9	3.360	3.006	2.813	2.693	2.611	2.551	2.505	2.469	2.440	2.416	2.379	2.351	2.330	2.312	2.298	2.255	2.232	2.159
10	3.285	2.924	2.728	2.605	2.522	2.461	2.414	2.377	2.347	2.323	2.284	2.255	2.233	2.215	2.201	2.155	2.132	2.055
11	3.225	2.860	2.660	2.536	2.451	2.389	2.342	2.304	2.274	2.248	2.209	2.179	2.156	2.138	2.123	2.076	2.052	1.972
12	3.176	2.807	2.606	2.480	2.394	2.331	2.283	2.245	2.213	2.189	2.147	2.117	2.094	2.075	2.060	2.012	1.986	1.904
13	3.136	2.763	2.560	2.434	2.347	2.283	2.234	2.195	2.164	2.138	2.097	2.066	2.042	2.023	2.007	1.958	1.931	1.846
14	3.102	2.726	2.522	2.395	2.307	2.243	2.193	2.154	2.122	2.095	2.054	2.022	1.998	1.979	1.962	1.912	1.885	1.797
15	3.073	2.695	2.490	2.361	2.273	2.208	2.158	2.118	2.086	2.059	2.017	1.985	1.961	1.941	1.924	1.873	1.845	1.755
16	3.048	2.668	2.462	2.333	2.244	2.178	2.128	2.088	2.055	2.028	1.985	1.953	1.928	1.908	1.891	1.839	1.811	1.718
17	3.026	2.645	2.437	2.308	2.218	2.152	2.102	2.061	2.028	2.001	1.958	1.925	1.900	1.879	1.862	1.809	1.780	1.686
18	3.007	2.624	2.416	2.286	2.196	2.130	2.078	2.038	2.005	1.977	1.933	1.900	1.875	1.854	1.837	1.783	1.754	1.657
19	2.990	2.606	2.397	2.266	2.176	2.109	2.058	2.017	1.984	1.956	1.912	1.878	1.852	1.831	1.814	1.759	1.730	1.631
20	2.975	2.589	2.380	2.249	2.158	2.091	2.040	1.998	1.965	1.937	1.892	1.859	1.833	1.811	1.794	1.738	1.708	1.607
22	2.949	2.561	2.351	2.219	2.128	2.060	2.008	1.967	1.933	1.904	1.859	1.825	1.798	1.777	1.759	1.702	1.671	1.567
24	2.927	2.538	2.327	2.195	2.103	2.035	1.983	1.941	1.906	1.878	1.832	1.797	1.770	1.748	1.730	1.672	1.641	1.533
26	2.909	2.519	2.308	2.175	2.082	2.014	1.961	1.919	1.884	1.855	1.809	1.774	1.747	1.724	1.706	1.647	1.615	1.504
28	2.894	2.503	2.291	2.157	2.064	1.996	1.943	1.900	1.865	1.836	1.790	1.765	1.726	1.704	1.685	1.625	1.592	1.478
30	2.881	2.489	2.276	2.142	2.049	1.980	1.927	1.884	1.849	1.820	1.773	1.744	1.709	1.686	1.667	1.607	1.573	1.456
40	2.835	2.440	2.226	2.091	1.997	1.927	1.872	1.829	1.793	1.763	1.715	1.687	1.665	1.649	1.605	1.541	1.506	1.377
∞	2.705	2.303	2.084	1.945	1.847	1.774	1.717	1.670	1.632	1.599	1.546	1.505	1.471	1.444	1.421	1.342	1.295	1.000

Table 4 F Distribution Function Quantiles, t_k (continued)

$k = .95$

d_1

d_2	1	2	3	4	5	6	7	8	9	10	12	14	16	18	20	30	40	∞
1	161.4	199.5	215.7	224.6	230.2	234.0	236.8	238.9	240.5	241.9	243.9	245.4	246.5	247.3	248.0	250.1	251.1	254.3
2	18.51	19.00	19.16	19.25	19.30	19.33	19.35	19.37	19.39	19.40	19.41	19.42	19.43	19.44	19.45	19.46	19.47	19.50
3	10.12	9.552	9.277	9.117	9.013	8.941	8.887	8.845	8.812	8.786	8.745	8.715	8.692	8.675	8.660	8.617	8.594	8.526
4	7.709	6.944	6.591	6.388	6.256	6.163	6.094	6.041	5.999	5.964	5.912	5.873	5.844	5.821	5.803	5.746	5.717	5.628
5	6.608	5.786	5.409	5.192	5.050	4.950	4.876	4.818	4.772	4.735	4.678	4.636	4.604	4.579	4.558	4.496	4.464	4.365
6	5.991	5.143	4.757	4.534	4.387	4.284	4.207	4.147	4.099	4.060	4.000	3.956	3.922	3.896	3.874	3.808	3.774	3.669
7	5.592	4.737	4.347	4.120	3.972	3.866	3.787	3.726	3.677	3.637	3.575	3.529	3.494	3.467	3.445	3.376	3.340	3.223
8	5.318	4.459	4.066	3.838	3.688	3.581	3.500	3.438	3.388	3.347	3.284	3.237	3.202	3.173	3.150	3.079	3.043	2.928
9	5.117	4.263	3.863	3.633	3.482	3.374	3.293	3.230	3.179	3.137	3.073	3.025	2.989	2.960	2.936	2.864	2.826	2.707
10	4.965	4.105	3.708	3.478	3.326	3.217	3.135	3.072	3.020	2.978	2.913	2.865	2.828	2.798	2.774	2.700	2.661	2.538
11	4.844	3.984	3.602	3.357	3.204	3.095	3.012	2.948	2.896	2.854	2.788	2.739	2.701	2.671	2.646	2.570	2.531	2.404
12	4.747	3.886	3.498	3.259	3.106	2.996	2.913	2.849	2.796	2.753	2.687	2.637	2.599	2.568	2.544	2.466	2.426	2.296
13	4.667	3.806	3.415	3.203	3.025	2.915	2.832	2.767	2.714	2.671	2.604	2.554	2.515	2.484	2.459	2.380	2.339	2.206
14	4.600	3.739	3.347	3.127	2.958	2.848	2.764	2.699	2.646	2.602	2.534	2.484	2.445	2.413	2.388	2.308	2.266	2.131
15	4.543	3.682	3.289	3.066	2.937	2.790	2.707	2.641	2.588	2.544	2.475	2.424	2.385	2.353	2.328	2.247	2.204	2.066
16	4.494	3.634	3.240	3.014	2.877	2.741	2.657	2.591	2.538	2.494	2.425	2.373	2.333	2.302	2.276	2.194	2.151	2.010
17	4.451	3.592	3.198	2.969	2.827	2.748	2.614	2.548	2.494	2.450	2.381	2.329	2.289	2.257	2.230	2.148	2.104	1.960
18	4.414	3.555	3.160	2.931	2.785	2.697	2.577	2.510	2.477	2.412	2.342	2.290	2.250	2.217	2.191	2.107	2.063	1.917
19	4.381	3.522	3.127	2.898	2.749	2.655	2.609	2.477	2.423	2.378	2.308	2.256	2.215	2.182	2.155	2.071	2.026	1.878
20	4.351	3.493	3.098	2.866	2.718	2.619	2.563	2.448	2.393	2.348	2.278	2.225	2.184	2.151	2.124	2.039	1.994	1.843
22	4.301	3.443	3.049	2.817	2.665	2.561	2.494	2.397	2.343	2.297	2.226	2.173	2.131	2.098	2.071	1.984	1.938	1.783
24	4.260	3.403	3.009	2.776	2.623	2.508	2.442	2.355	2.301	2.256	2.183	2.130	2.088	2.054	2.027	1.939	1.892	1.733
26	4.225	3.369	2.975	2.743	2.587	2.474	2.402	2.321	2.265	2.221	2.167	2.094	2.052	2.018	1.990	1.901	1.853	1.691
28	4.196	3.340	2.947	2.714	2.558	2.445	2.359	2.291	2.236	2.190	2.120	2.064	2.021	1.987	1.959	1.869	1.820	1.654
30	4.171	3.316	2.922	2.690	2.534	2.421	2.334	2.266	2.211	3.000	2.092	2.042	1.995	1.960	1.932	1.841	1.792	1.622
40	4.085	3.232	2.839	2.606	2.449	2.336	2.249	2.180	2.124	2.077	2.003	1.948	1.906	1.874	1.851	1.744	1.693	1.509
∞	3.842	2.996	2.605	2.372	2.214	2.099	2.010	1.938	1.880	1.831	1.752	1.692	1.643	1.604	1.571	1.459	1.394	1.000

Table 4 F Distribution Function Quantiles, t_k (continued)

$k = .99$

d_1

d_2	1	2	3	4	5	6	7	8	9	10	12	14	16	18	20	30	40	∞
1	4052.	5000.	5403.	5625.	5764.	5859.	5928.	5981.	6022.	6056.	6106.	6143.	6170.	6192.	6209.	6261.	6227.	6366.
2	98.50	99.00	99.17	99.25	99.30	99.33	99.36	99.37	99.39	99.40	99.42	99.43	99.44	99.44	99.45	99.47	99.47	99.50
3	34.12	30.82	29.46	28.71	28.24	27.91	27.67	27.49	27.35	27.23	27.05	26.92	26.83	26.75	26.69	26.51	26.41	26.12
4	21.20	18.00	16.69	15.98	15.52	15.21	14.98	14.80	14.66	14.55	14.37	14.25	14.15	14.08	14.02	13.84	13.75	13.46
5	16.26	13.27	12.06	11.39	10.97	10.67	10.46	10.29	10.16	10.05	9.888	9.770	9.680	9.610	9.553	9.379	9.291	9.020
6	13.75	10.93	9.780	9.148	8.746	8.466	8.260	8.102	7.976	7.874	7.718	7.605	7.519	7.451	7.396	7.229	7.143	6.880
7	12.25	9.547	8.451	7.847	7.460	7.191	6.993	6.840	6.719	6.620	6.469	6.359	6.275	6.209	6.155	5.992	5.908	5.650
8	11.26	8.649	7.591	7.006	6.632	6.371	6.178	6.029	5.911	5.814	5.667	5.559	5.477	5.412	5.359	5.198	5.116	4.859
9	10.56	8.022	6.992	6.422	6.057	5.802	5.613	5.467	5.351	5.257	5.111	5.005	4.924	4.860	4.808	4.649	4.567	5.111
10	10.04	7.559	6.552	5.994	5.636	5.386	5.200	5.057	4.942	4.849	4.706	4.601	4.520	4.457	4.405	4.247	4.165	4.706
11	9.646	7.206	6.217	5.668	5.316	5.069	4.886	4.744	4.632	4.539	4.397	4.293	4.213	4.150	4.099	3.941	3.860	4.397
12	9.380	6.927	5.953	5.412	5.064	4.821	4.639	4.499	4.388	4.296	4.155	4.052	3.972	3.909	3.858	3.701	3.619	4.155
13	9.074	6.701	5.739	5.205	4.862	4.620	4.441	4.302	4.191	4.100	3.960	3.857	3.778	3.716	3.665	3.507	3.425	3.165
14	8.862	6.515	5.564	5.035	4.695	4.456	4.278	4.140	4.030	3.939	3.800	3.698	3.619	3.556	3.505	3.348	3.266	3.004
15	8.683	6.359	5.417	4.893	4.556	4.318	4.142	4.004	3.895	3.805	3.666	3.564	3.485	3.423	3.372	3.214	3.132	2.868
16	8.531	6.226	5.292	4.773	4.437	4.202	4.026	3.890	3.780	3.691	3.553	3.451	3.372	3.310	3.259	3.101	3.018	2.753
17	8.400	6.112	5.185	4.669	4.336	4.102	3.927	3.791	3.682	3.593	3.455	3.353	3.275	3.212	3.162	3.003	2.920	2.653
18	8.285	6.013	5.092	4.579	4.248	4.015	3.841	3.705	3.597	3.508	3.371	3.269	3.190	3.128	3.077	2.919	2.835	2.566
19	8.185	5.926	5.010	4.500	4.171	3.939	3.765	3.631	3.523	3.434	3.297	3.195	3.117	3.054	3.003	2.844	2.761	2.489
20	8.096	5.849	4.938	4.431	4.103	3.871	3.699	3.564	3.457	3.368	3.231	3.130	3.051	2.989	2.938	2.778	2.695	2.421
22	7.945	5.719	4.817	4.313	3.988	3.758	3.587	3.453	3.346	3.258	3.121	3.019	2.941	2.879	2.827	2.667	2.583	2.306
24	7.823	5.614	4.718	4.218	3.895	3.667	3.496	3.363	3.256	3.168	3.032	2.930	2.852	2.789	2.738	2.577	2.492	2.211
26	7.721	5.526	4.637	4.140	3.818	3.591	3.421	3.288	3.182	3.094	2.958	2.857	2.778	2.715	2.664	2.503	2.417	2.131
28	7.636	5.453	4.568	4.074	3.754	3.528	3.358	3.226	3.120	3.032	2.896	2.795	2.716	2.653	2.602	2.440	2.353	2.064
30	7.562	5.390	4.510	4.018	3.699	3.473	3.305	3.173	3.067	2.981	2.843	2.742	2.663	2.600	2.549	2.386	2.299	2.006
40	7.314	5.178	4.313	3.828	3.514	3.291	3.124	2.993	2.888	2.801	2.665	2.566	2.494	2.421	2.369	2.203	2.114	1.805
∞	6.635	4.605	3.782	3.319	3.017	2.802	2.639	2.511	2.407	2.321	2.185	2.082	2.000	1.934	1.878	1.696	1.592	1.000

589

Answers to Exercises

Chapter 1

EXERCISE 1.1

3. $F = \{1, 2, ..., 10\}, \quad E = \{1, 2\}, \quad D = \{9, 10\}$

5. (a) False (b) False (c) True (d) True (e) True
    (f) True (g) False (h) True (i) True (j) False
    (k) False (l) False (m) False (n) False

6. (a) True (b) False (c) False (d) True (e) False
    (f) True (g) False

7. Yes, No, No

8. (a) Only if one carefully defines the term family.
    (b) Element

9. No

10. No

11. $\bar{A} = \{9, 10\}$, $\bar{B} = \{1, 5, 6, 7, 8, 9, 10\}$, $\bar{C} = \{2, 4, 6, 8, 10\}$, No

12. Yes

15. $\bar{U} = \varnothing$, $\varnothing = U$

EXERCISE 1.2

2. $A \cup B = \{x: 0 \leq x \leq 1\}$, $A \cup C = \{0, \frac{1}{2}, 1\}$, $B \cup C = B$
   $A \cap B = \emptyset = A \cap C$, $B \cap C = C$

4. $F \subset E$

5. $F \subset E$

6. (a) $A \cup B = B$
   (b) $A \cap B = A$
   (c) $A \cup C = \{0, 1, \ldots, 10\}$
   (d) $A \cap C = \{1, 2, \ldots, 6\}$
   (e) $A \cup D = \{0, 1, \ldots, 10, 20, 30\}$
   (f) $A \cap D = \{10\}$
   (g) $B \cup C = \{x: x = 0$ or $1 \leq x \leq 10\}$
   (h) $B \cap C = \{1, 2, \ldots, 6\}$
   (i) $B \cup D = \{x: x = 0, 10, 20$ 30 or $1 \leq x \leq 10\}$

   (j) $B \cap D = \{10\}$
   (k) $C \cup D = \{0, 1, \ldots, 6, 10, 20, 30\}$
   (l) $C \cap D = \{0\}$
   (m) $A \cup B \cup C = B \cup C$
   (n) $A \cap (B \cup C) = A$
   (o) $A \cup (B \cap C) = A$
   (p) $A \cap B \cap C = A \cap C$
   (q) $C \cup (A \cap D) = \{0, 1, \ldots, 6, 10\}$
   (r) $(A \cup B) \cap (C \cup B) = B$

8. Safecrackers in prison in United States.

9. Sociologists in the United States.

10. (a) False   (b) True   (c) False

14. $A \times B = \{(1, 2), (1, 1), (2, 2), (2, 1)\}$
    $A \times C = \{(1, 10), (1, 12), (2, 10), (2, 12)\}$
    $B \times C = A \times C$
    $B \times A = A \times B$
    $C \times A = \{(10, 1), (12, 1), (10, 2), (12, 2)\}$
    $C \times B = C \times A$
    $A \times A = A \times B = B \times B$
    $C \times C = \{(10, 10), (10, 12), (12, 10), (12, 12)\}$
    $A \times B \times C = \{(1, 2, 10), (1, 2, 12), (1, 1, 10), (1, 1, 12), (2, 2, 10), (2, 2, 12),$
    $(2, 1, 10), (2, 1, 12)\}$
    $C \times B \times A = \{(10, 1, 2), (12, 1, 2), (10, 1, 1), (12, 1, 1), (10, 2, 2), (12, 2, 2),$
    $(10, 2, 1), (12, 2, 1)\}$
    $C \times A \times B = C \times B \times A$

15. $A = B$ or A or B is null.

16. No.

17. No.

EXERCISE 1.3

2. $R_g = \{2, 3, ..., 12\}$

3. $R_h = \{0, 1, 2, 3\}$.   One, 3

4. $R_d = \{-5, -4, ..., 4, 5\}$

5. $R_a = \{0, 1, ..., 5\}$

6. $R_p = \{0, \frac{1}{5}, \frac{2}{5}, \frac{3}{5}, \frac{4}{5}, 1\}$

7. $R_q = \{2, 1, \frac{1}{2}, \frac{1}{3}, \frac{1}{4}\}$

Chapter 2

EXERCISE 2.1

1. $S = \{1, 2, 3, ..., 10\}$

2. $S = S_1 \times S_1$, where $S_1 = \{1, 2, ..., 10\}$

3. $S = \{(x, y): x = 1, 2, ..., 10; y = 1, 2, ..., 10; x \neq y\}$

4. $A = \{7, 8, 9, 10\}$ $B = \{1, 2, ..., 6\}$

5. $C = \{(x_1, x_2): x_1 = 7, 8, 9, 10; x_2 = 1, 2, ..., 10\}$
   $D = \{(x_1, x_2): x_1 = 1, 2, ..., 10; x_2 = 7, 8, 9, 10\}$
   $E = \{(x_1, x_2): x_1 = 7, 8, 9, 10; x_2 = 7, 8, 9, 10\}$
   Yes

6. $S = \{(x, y): x = a, b, ..., e; y = a, b, ..., e\}$

7. $S = \{(x, y): x = \text{bald, brown, black}; y = \text{blue, brown}\}$
   $A = \{(\text{bald}, y): y = \text{blue, brown}\}$
   $B = \{(x, \text{blue}): x = \text{bald, brown, black}\}$
   $C = \{(\text{brown, brown})\}$

8. $S = \{(x, y): x = 1, 2, 3; y = 1, 2, 3; x \neq y\}$
   $A = \{(1, y): y = 2, 3\}$
   $B = \{(x, 1): x = 2, 3\}$
   $C = \{(3, 2), (2, 3)\}$

9. $S = \{(x_1, x_2, x_3): x_i = r, b_1, g, b_2; i = 1, 2, 3\}$
   $A = \{(r, r, r)\}$
   $B = \{(r, g, b_1), (r, b_1, g), (b_1, r, g), (b_1, g, r), (g, r, b_1), (g, b_1, r)\}$
   $C = \{(x_1, x_2, x_3): x_i = r, b_1, g, b_2; i = 1, 2, 3; x_1 \neq x_2 \neq x_3\}$
   $D = \varnothing$

10. $S = \{(x_1, x_2, x_3, x_4): x_i = 1, 2, 3; i = 1, 2, 3, 4\}$
     $A = \{(1, 1, 1, 1)\}$
     $B = \{(1,1,2,2), (1,2,1,2), (1,2,2,1), (2,2,1,1), (2,1,2,1), (2,1,1,2)\}$
     $C = \{(x_1, x_2, x_3, x_4): x_i = 1, 2, 3; \ i = 1, 2, 3, 4; \ 3 \ \text{different numbers occur}\}$

11. $S = \{(x_1, x_2, x_3): x_i = 1, 2, \ldots, 9; i = 1, 2, 3; x_1 \neq x_2 \neq x_3\}$
     $A = \{(5, x_2, x_3): x_i = 1, 2, \ldots, 9; i = 2, 3; x_2 \neq 5, x_3 \neq 5, x_2 \neq x_3\}$
     $B = \{(x_1, x_2, 5): x_i = 1, 2, \ldots, 9; i = 1, 2; x_1 \neq 5, x_2 \neq 5, x_1 \neq x_2\}$
     $C = \{(x_1, x_2, x_3): x_i = 1, 2, \ldots, 9; i = 1, 2, 3; x_1 \neq x_2 \neq x_3, x_1 \neq 5,$
     $\qquad\qquad x_2 \neq 5, x_3 \neq 5\}$

12. $S = \{(x_1, x_2): x_1 \leq 1600, x_2 \leq 1600\}$
     $A = \{(x_1, x_2): x_1 \leq 1000, x_2 \leq 1000\}$
     $B = \{(x_1, x_2): 1000 \leq x_1 \leq 1600, 1000 \leq x_2 \leq 1600\}$
     $C = \{(x_1, x_2): x_1 = 1000 \ \text{and} \ 1000 \leq x_2 \leq 1600 \ \text{or} \ 1000 \leq x_1 \leq 1600$
     $\qquad\qquad \text{and} \ x_2 = 1000\}$
     $D = \{(x_1, x_2): x_1 \leq x_2 = 1000 \ \text{or} \ x_2 \leq x_1 = 1000\}$

13. $S = \{0, 1, 2, \ldots, 30\}$
     $A = \{5, 6, \ldots, 30\}$
     $B = \{0, 1, 2, \ldots, 5\}$
     $C = \{5\}$

14. $S = S_1 \times S_1$
     $A = \{(x_1, x_2): x_1 = 5, 6, \ldots, 30, x_2 = 0, 1, \ldots, 30\}$
     $B = \{(x_1, x_2): x_1 = 0, 1, \ldots, 30, x_2 = 5, 6, \ldots, 30\}$
     $C = A \cap B$

15. $S = \{(x_1, x_2, \ldots, x_{50}): x_i = 0, 1; i = 1, 2, \ldots, 50\}$
     $A = \{(1, 1, \ldots, 1)\}$
     $B = \{(0, 0, \ldots, 0)\}$
     $C = \{(0, 0, 1, 1, \ldots, 1)\}$

EXERCISE 2.2

1. (a) $\frac{1}{3}$ (b) $\frac{2}{3}$ (c) $\frac{2}{3}$ (d) $\frac{1}{3}$ (e) 1   (f) $\frac{2}{3}$

2. (a) $\frac{3}{10}$ (b) $\frac{7}{10}$ (c) $\frac{1}{2}$ (d) $\frac{3}{10}$ (e) 1   (f) $\frac{1}{2}$

3. (a) 0   (b) 1   (c) 0   (d) 0   (e) 1   (f) 0, No

4. $A =$	$\varnothing$	$\{a\}$	$\{b\}$	S
$P(A) =$	0	$\frac{1}{10}$	$\frac{9}{10}$	1
$P(A) =$	0	$\frac{1}{2}$	$\frac{1}{2}$	1
$P(A) =$	0	$\frac{4}{5}$	$\frac{1}{5}$	1

7. (a) $\frac{1}{2}$ (b) $\frac{1}{2}$ (c) $\frac{1}{3}$ (d) $\frac{1}{3}$ (e) $\frac{2}{3}$ (f) $\frac{1}{6}$ (g) $\frac{1}{6}$ (h) $\frac{5}{6}$

8. (a) $\frac{7}{12}$ (b) $\frac{11}{12}$ (c) $\frac{1}{12}$ (d) $\frac{1}{4}$ (e) $\frac{5}{12}$ (f) $\frac{3}{4}$

9. No

10. No

11. No

12. (a) .1 (b) .7 (c) .2

13. Yes

14. (a) .8 (b) .5 (c) .9 (d) .6

15. (a) .8 (b) .1 (c) .1

EXERCISE 2.3

1. (a) $\frac{4}{7}$ (b) $\frac{11}{21}$ (c) $\frac{2}{3}$

2. $\frac{3}{4}, \frac{1}{4}$

3. $\frac{3}{8}, \frac{3}{16}, \frac{1}{16}$

4. $\frac{1}{2}$

5. $\frac{26}{52}, \frac{13}{52}, \frac{4}{52}, \frac{1}{52}$

6. $\frac{1}{5}, \frac{1}{5}, \frac{1}{25}, \frac{5}{25}$

7. $\frac{1}{36}, \frac{2}{26}, \frac{3}{36}, \frac{4}{36}, \frac{5}{36}, \frac{6}{36}, \frac{5}{36}, \frac{4}{36}, \frac{3}{36}, \frac{2}{36}, \frac{1}{36}$

8. $\frac{1}{8}, \frac{3}{8}, \frac{3}{8}, \frac{1}{8},$

9. $\frac{18}{40}, \frac{3}{20}, \frac{7}{40}, \frac{15}{40}, \frac{17}{40}, \frac{1}{10}$

10. $P(\text{Doug wins}) = \frac{8}{27},$ $P(\text{Joe wins}) = \frac{2}{27},$ $P(\text{Hugh wins}) = \frac{13}{27},$ $P(\text{Ray wins}) = \frac{4}{27}$

11. $P(\text{Doug loses first}) = \frac{9}{27}$, $P(\text{Joe loses first}) = \frac{18}{27}$, $P(\text{Hugh loses first}) = \frac{9}{27}$, $P(\text{Ray loses first}) = \frac{18}{27}$

12. $\frac{1}{8}, \frac{1}{8}$

EXERCISE 2.4

1. $3!$

2. $3!$

3. $6!$

4. 3^3

5. 2^4

6. $_4P_3$

7. 2^3

8. No

9. $\begin{pmatrix} 28 \\ 5 \end{pmatrix}$

10. $\begin{pmatrix} 8 \\ 2 \end{pmatrix}$, 168 games

11. $\begin{pmatrix} 10 \\ 3 \end{pmatrix}$

12. $\begin{pmatrix} 20 \\ 5 \end{pmatrix}$

13. $\begin{pmatrix} 15 \\ 2 \end{pmatrix}$

14. $\begin{pmatrix} 15 \\ 3 \end{pmatrix}, \begin{pmatrix} 15 \\ k \end{pmatrix}$

15. $\begin{pmatrix} 9 \\ 3 \end{pmatrix}$

16. $\begin{pmatrix} 2 \\ 2 \end{pmatrix}\begin{pmatrix} 7 \\ 1 \end{pmatrix}, \begin{pmatrix} 7 \\ 3 \end{pmatrix}$

17. $\dbinom{2}{1}\dbinom{3}{1}\dbinom{4}{1}$

18. (a) $(26)^3 (10)^3$

    (b) 17,380,000

    (c) $\dbinom{6}{3} 26^3 10^3$

19. (a) $\dbinom{100}{10}$

    (b) $1.105 \times (10)^{13}$

    (c) $6.26 \times (10)^{12}$

22. (a) $\dbinom{8}{3}$

    (b) $\dbinom{4}{2}\dbinom{4}{1}$

    (c) 32

23. $\dfrac{20!}{8!7!5!}$

24. (a) $\dfrac{19!}{7!7!5!}$

    (b) $\dfrac{18!}{6!7!5!}$

25. $\dfrac{11!}{4!4!2!1!},\ \dfrac{9!}{4!3!2!}$

26. 176

EXERCISE 2.5

1. $\frac{5}{14},\ \frac{15}{28}$

2. $\frac{3}{28},\ \frac{1}{14}$

3. 52,488/90,000, 23,328/90,000

4. 9/90,000, $9 \cdot 9 \cdot 8 \cdot 7 \cdot 6/90,000$

5. $\binom{9}{2}\Big/\binom{10}{3}, \binom{8}{1}\Big/\binom{10}{3}$

6. $\binom{99}{1}\Big/\binom{100}{2}, 1\Big/\binom{100}{2}$

7. (a) $\frac{1}{9}$ (b) $\frac{4}{9}$

8. $\frac{2}{6}$

9. $\frac{1}{32}, \frac{6}{32}$

10. (a) $\binom{13}{2}\binom{4}{2}\binom{4}{2}\binom{44}{1}$ (b) $\binom{13}{1}\binom{12}{1}\binom{4}{3}\binom{4}{2}$

    (c) $\binom{4}{1}\binom{13}{5}$ (d) $\binom{4}{1}10$

11. $\dfrac{12\cdot 11\cdot 10\cdots(13-n)}{12^n}$

12. (a) $\frac{1}{6}$ (b) $\frac{2}{6}$

13. (a) $\binom{10}{3}\cdot 3!/10^3$ (b) $\binom{10}{1}\Big/10^3$

14. $.05, 0, \frac{9}{20}$

15. $\frac{36}{70}$

EXERCISE 2.6

1. $\frac{1}{4}, \frac{1}{4}, \frac{1}{4}, \frac{1}{4}, \frac{1}{3}, \frac{1}{3}, \frac{1}{3}, \frac{1}{3}$

2. All $\frac{1}{4}$

3. $\frac{1}{2}$

4. $\frac{3}{4}, \frac{3}{6}$

5. $\frac{5}{12}, 1-\frac{5}{12}$

6. $\frac{9}{36}, \frac{3}{4}$

7. $\binom{26}{5}\Big/\binom{52}{5}, \binom{13}{5}\Big/\binom{52}{5}$

8. $\frac{10}{30}$

9. $\frac{150}{189}$

10. $\frac{7}{10}$

11. .087

12. .199, $\frac{1}{199}$

13. .807

14. $\frac{6}{21}$, $\frac{5}{21}$, $\frac{4}{21}$, $\frac{3}{21}$, $\frac{2}{21}$, $\frac{1}{21}$

15. (a) $\frac{1}{3}$ (b) No

16. (a) .3024 (b) .216, .18, .1

17. (a) $\frac{1}{25}$ (b) $\frac{1}{10}$

18. .310, .496, .194

EXERCISE 2.7

1. Yes

2. Yes

3. No

4. Yes

5. No

8. $(.7)^{15}$, $14(.3)(.7)^{14}$

9. (a) .07 (b) .24 (c) .51

10. (b) $(.9)^{10}$ (c) $10(.1)(.9)^9 + (.9)^{10}$

11. (a) $\binom{990}{20} \Big/ \binom{1000}{20}$ (b) $\binom{990}{19}\binom{10}{1} \Big/ \binom{1000}{20}$

12. .8147

13. (a) .007 (b) .101

14. (a) .024 (b) .976 (c) .037

EXERCISE 2.8

1. (a) $\left(\frac{5}{6}\right)^9\left(\frac{1}{6}\right)$ (b) $\frac{1}{6} + \frac{5}{6}\left(\frac{1}{6}\right) + \frac{1}{6}\left(\frac{5}{6}\right)^2$ (c) $\frac{6}{11}$

2. (a) $\frac{11}{36}$ (b) $\frac{5}{11}$

3. (a) .01   (b) $\frac{1}{111}$

4. (a) $(.9)^4$ (b) $(.9)^{10}$

5. (a) $\frac{3}{10}$ (b) $\frac{2}{10}$

6. $\frac{1}{12}, \frac{1}{12}, \frac{5}{12}$

7. .707

8. $\frac{2}{3}$

9. (a) $\frac{3}{4}$ (b) $\frac{1}{5}$

10. (a) $\frac{1}{2}$ (b) $\frac{1}{3}$

11. (a) $n/2^{n-1}$ (b) $\dfrac{1}{1 - \left(1 - \dfrac{n}{2^{n-1}}\right)^2}$, no

13. $\frac{8}{25}, \frac{6}{25}$

14. .644, .958, .653

Chapter 3

EXERCISE 3.1

1. $p_Y(y) = \frac{1}{4}$, $y = 1, 2, 3, 4$

2.

$z =$	3	4	5	6	7
$p_Z(z) =$	$\frac{1}{6}$	$\frac{1}{6}$	$\frac{2}{6}$	$\frac{1}{6}$	$\frac{1}{6}$

3.

$z =$	2	3	4	5	6	7	8
$p_Z(z) =$	$\frac{1}{16}$	$\frac{2}{16}$	$\frac{3}{16}$	$\frac{4}{16}$	$\frac{3}{16}$	$\frac{2}{16}$	$\frac{1}{16}$

4.

$x =$	2	8	5	10	13	17	18	20	25	32
$p_X(x) =$	$\frac{1}{16}$	$\frac{1}{16}$	$\frac{2}{16}$	$\frac{2}{16}$	$\frac{2}{16}$	$\frac{2}{16}$	$\frac{1}{16}$	$\frac{2}{16}$	$\frac{2}{16}$	$\frac{1}{16}$

5.

$x =$	19	$19\frac{1}{2}$	20	$20\frac{1}{2}$	$21\frac{1}{2}$	22	$22\frac{1}{2}$	23	$23\frac{1}{2}$	25
$p_X(x) =$	$\frac{3}{45}$	$\frac{12}{45}$	$\frac{9}{45}$	$\frac{4}{45}$	$\frac{3}{45}$	$\frac{4}{45}$	$\frac{4}{45}$	$\frac{4}{45}$	$\frac{1}{45}$	$\frac{1}{45}$

6.

$x =$	1	2	3	4
$p_X(x) =$	$\frac{1}{4}$	$\frac{1}{4}$	$\frac{1}{4}$	$\frac{1}{4}$

7. $p_Y(y) = \dfrac{\dbinom{26}{y}\dbinom{26}{5-y}}{\dbinom{52}{5}}$, $y = 0, 1, 2, 3, 4, 5$

8. $p_Z(z) = \dfrac{\dbinom{13}{z}\dbinom{39}{5-z}}{\dbinom{52}{5}}$, $z = 0, 1, 2, 3, 4, 5$

9.

$m =$	1	2	3	4	5	6
$p_M(m) =$	$\frac{1}{36}$	$\frac{3}{36}$	$\frac{5}{36}$	$\frac{7}{36}$	$\frac{9}{11}$	$\frac{11}{36}$

10.

$w =$	1	2	3	4	5	6
$p_W(w) =$	$\frac{11}{36}$	$\frac{9}{36}$	$\frac{7}{36}$	$\frac{5}{36}$	$\frac{3}{36}$	$\frac{1}{36}$

11. $p_X(x) = \frac{1}{2}$, $x = 1, 2$

12. $p_X(x) = \dfrac{\dbinom{20}{x}\dbinom{80}{3-x}}{\dbinom{100}{3}}$, $x = 0, 1, 2, 3$

EXERCISE 3.2

5. $F_Z(t) = 0,\ t < 0$
$\qquad\quad = \frac{1}{3},\ 0 \le t < 1$
$\qquad\quad = \frac{2}{3},\ 1 \le t < 2$
$\qquad\quad = 1,\ 2 \le t$

6. $F_U(t) = 0,\ t < -3$
$\qquad\quad = \frac{1}{2},\ -3 \le t < 0$
$\qquad\quad = \frac{2}{3},\ 0 \le t < 4$
$\qquad\quad = 1,\ 4 \le t$

7. $\frac{1}{2}$

8. $\dfrac{\sqrt{3}-1}{2}$

10. No

12. $F_X(t) = 0,$ $t < 99$
   $\quad\quad\;\; = t - 99,$ $99 \le t \le 100$
   $\quad\quad\;\; = 1,$ $t > 100$

13. $F_Y(t) = 0,$ $t < 0$
   $\quad\quad\;\; = 1 - (1 - t)^2,\; 0 \le t \le 1$
   $\quad\quad\;\; = 1,$ $t > 1$

14. $F_Z(t) = 1 - e^{-10t},\; t \ge 0$

15. $F_X(t) = \frac{2}{25}(x - 69)^2,$ $69 \le t < 71.5$
   $\quad\quad\;\; = 1 - \frac{2}{25}(74 - t)^2,$ $71.5 \le t \le 74$
   $\quad\quad\;\; = 1,$ $t > 74$

16. $F_X(t) = 0,$ $x < 0$

   $\quad\quad\;\; = \frac{1}{2} - \dfrac{(15 - t)^2}{450},$ $0 \le t \le 15$

   $\quad\quad\;\; = \frac{1}{2} + \dfrac{(15 - t)^2}{450},$ $15 < t \le 30$

   $\quad\quad\;\; = 1,$ $t > 30$

EXERCISE 3.3

1. (a) $7\frac{1}{2}$ (b) 85 (c) $\frac{15}{64}$ (d) $69\frac{1}{2}$ (e) $\sigma_X^2 = 28\frac{3}{4}$, $\sigma_X = 5.362$

2. (a) 0   (b) $\frac{1}{3}$ (c) 2   (d) 7   (e) $\frac{1}{3}$ (f) .577

3. (a) $\frac{1}{3}$ (b) $\frac{1}{6}$ (c) $106\frac{5}{6}$ (d) 2   (e) $\frac{1}{18}$ (f) .236

5. (a) 10 (b) 20 (c) 55   (d) 99 (e) 47 (f) 80
   (g) 50

6. $1.649, .833, e - 1, \dfrac{(e - 1)(3 - e)}{2}$

7. $\log_2 1.5,\ \log_2 1.4,\ 2 - (\ln 2)^{-1},\ .285$

9. $68\frac{1}{2}$

10. $-\$.70,\ -\$70,\ 899.91,\ 89100$

11. $7,\ 5\frac{5}{6}$

12. $3.22,\ 42.15$

13. (a) geometric, $p = \frac{1}{2}$ (b) does not exist
    (c) no amount large enough

14. 21

15. 1

16. 2, $2\frac{2}{3}$

EXERCISE 3.4

1. $(5 + e^t)/6$, $\frac{1}{6}, \frac{1}{6}, \frac{1}{6}$

2. $\dfrac{e^{40t} - e^{30t}}{10t}$

3. $\dfrac{a}{a - t}, \dfrac{1}{a}, \dfrac{1}{a^2}$

4. $\dfrac{(1 + t)^3}{8}$, $\frac{3}{2}, \frac{3}{4}$

5. $\dfrac{\sum t^i}{n}$

6. $2(e^t - 1)$, $\mu_X = \sigma_X^2 = 2$

7. $p_X(k) = \dfrac{1}{k!}$, $k = 1, 2, 3, \ldots$

8. $e^{2(t-1)}$

9. $\dfrac{1}{(1 - t)^2}$, $-2 \ln(1 - t)$

10. $p(k) = \dbinom{n}{k} p^k q^{n-k}$, $k = 0, 1, 2, \ldots, n$

11. $(q + pe^t)^n$, $n \ln(q + pe^t)$

12. $\displaystyle\sum_{j=1}^{n} \dfrac{e^{jt}}{n}$

13. $\dfrac{e^{tb} - e^{ta}}{t(b - a)}$

14. $t^b \, \psi_X(t^a)$

EXERCISE 3.5

1. $1 - F_X\left(\dfrac{t-a}{b}\right) + p_X\left(\dfrac{t-a}{b}\right)$

2. $b = 2$

4. $F_Y(t) = 0, \qquad\qquad t < 13$

$\qquad = \dfrac{t-13}{4}, \quad 13 \le t \le 17$

$\qquad = 1, \qquad\qquad t > 17$

$f_Y(t) = \tfrac{1}{4}, \qquad\qquad 13 < t < 17$

5. $F_Z(t) = 0, \qquad\qquad t < -1$

$\qquad = (t+1)^3, \quad -1 \le t \le 0$

$\qquad = 1, \qquad\qquad t > 0$

$f_Z(t) = 3(t+1)^2, \; -1 < t < 0$

6. $F_U(t) = 0, \quad t < -120$

$\qquad = \tfrac{1}{4}, \quad -120 \le t < -50$

$\qquad = \tfrac{3}{4}, \quad -50 \le t < 20$

$\qquad = 1, \quad t \ge 20$

$p_U(u) = \tfrac{1}{4}, \quad u = -120, \; 20$

$\qquad = \tfrac{1}{2}, \quad u = -50$

7. $F_X(t) = 1 - e^{-(t+7)/2}, \quad t \ge -7$

$\qquad = 0, \qquad\qquad\quad t < -7$

$f_X(t) = \tfrac{1}{2}e^{-(t+7)/2}, \qquad t > -7$

8. $F_Y(t) = F_X(t^2), \; t > 0$

9. $f_Y(t) = 2t f_X(t^2), \; t > 0$

10. $F_Y(t) = t, \quad 0 < t < 1$

$f_Y(t) = 1, \quad 0 < t < 1$

11. (a) $F_X(t) = 0, \qquad\qquad\quad t < -1$

$\qquad = \dfrac{(t+1)^2}{8}, \qquad -1 \le t \le 1$

$\qquad = 1 - \dfrac{(3-t)^2}{8}, \quad 1 < t \le 3$

(b) $F_Y(t) = 0,$ ⠀⠀⠀⠀⠀⠀⠀$t < 0$

⠀⠀⠀⠀$= 1,$ ⠀⠀⠀⠀⠀⠀⠀$t > 3$

⠀⠀⠀⠀$= 1 - \dfrac{(3-t)^2}{8},$ ⠀$1 < t \le 3$

(c) $f_Y(t) = \frac{1}{2},$ ⠀⠀⠀⠀⠀$0 < t < 1$

⠀⠀⠀⠀$= \left(\dfrac{3-t}{4}\right),$ ⠀$1 < t \le 3$

12. $F_Z(t) = 1 - \dfrac{(3-\sqrt{t})^2}{8},$ ⠀$1 < t \le 9$

⠀⠀$f_X(t) = \dfrac{1}{2\sqrt{t}}$ ⠀⠀⠀⠀$0 < t < 1$

⠀⠀⠀⠀$= \dfrac{3-\sqrt{t}}{8\sqrt{t}},$ ⠀⠀$1 \le t < 9$

13. $P(X = k) = pq^k, k = 0, 1, 2, \ldots,$ where $p = 1 - e^{-1}, q = 1 - p$

Chapter 4

EXERCISE 4.1

1. $5/6, 25/36, .723, .196$
2. $1.6, 1.789, .218, .087, .069$
3. $.246, 5$
4. $.349, 1$
5. binomial, $n = 7, p = .2, .2097, .9996$
6. $p = \frac{1}{3}, n = 18$
7. 99, no
8. $-.0965, -.965$
9. $.5981, .1962$
10. (a) $.590$ (b) $.00001$
11. (a) 1 ⠀⠀(b) $.358$

12. (a) 3.58 (b) .000035
13. binomial, $n = 10$, $p = \frac{1}{2}$, $11/2^9$
14. (a) linear function of a binomial random variable (b) 700
15. No
16. .2, .8, 1

EXERCISE 4.2

1. $\frac{3}{4}$, $\frac{7}{8}$
2. geometric, $p = \frac{18}{38} = .474$, 2.111
3. geometric, $p = \frac{2}{38}$, 19, 684
4. geometric, $p = .1$, 10
5. $5
6. $\dfrac{pe^t}{1 - qe^t}$, $t < -\ln q$, $\left(\dfrac{pe^t}{1 - qe^t} \right)^r$
7. (a) .9919 (b) .9163 (c) .7692
8. (a) geometric, $p = .05$ (b) 20
9. negative binomial, $r = 5$, $p = .05$, 100
10. $P(X = k) = (.3)^k (.7)$, $k = 0, 1, 2, \ldots$
11. negative binomial, $r = 3$, $p = .001$
12. (a) $\dfrac{p}{1 - q^2}$ (b) $\frac{1}{3}$ (c) no

EXERCISE 4.3

1. $P(X = k) = \dfrac{\dbinom{3}{k} \dbinom{2}{2 - k}}{\dbinom{5}{2}}$, $k = 0, 1, 2$

2. $P(X = k) = \dfrac{\dbinom{26}{k}\dbinom{26}{13-k}}{\dbinom{52}{13}}$, $k = 0, 1, 2, \ldots, 13$, 6.5, 2.49

3. $P(X = k) = \dfrac{\dbinom{8}{k}\dbinom{2}{5-k}}{\dbinom{10}{5}}$, $k = 0, 1, \ldots, 5$

$P(Y = k) = \dfrac{\dbinom{8}{5-k}\dbinom{2}{k}}{\dbinom{10}{5}}$, $k = 0, 1, \ldots, 5$

4. $\frac{22}{35}$

5. .9810, .1609

6. $P(X = k) = \dfrac{\dbinom{10}{k}\dbinom{35}{10-k}}{\dbinom{45}{10}}$, $k = 0, 1, \ldots, 10$, 2.22, 1.37

7.

x	1	2	3	4	5	6
$p_X(x)$	$\frac{6}{21}$	$\frac{5}{21}$	$\frac{4}{21}$	$\frac{3}{21}$	$\frac{2}{21}$	$\frac{1}{21}$

8.

y	2	3	4	5	6	7
$p_Y(y)$	$\frac{1}{21}$	$\frac{2}{21}$	$\frac{3}{21}$	$\frac{4}{21}$	$\frac{5}{21}$	$\frac{6}{21}$

10. hypergeometric; $m = 100$, $w = 20$, $b = 80$, $n = 5$

EXERCISE 4.4

1. .3679, .2642

2. .8825, .9692, .8825

3. .6321, .2526

4. 6, 2.4495, .6065

5. (a) .9990 (b) .9753
6. (a) $1000m^2$ (b) $2303m^2$
8. .6065, .3033, .0902
9. .9841, .9841
10. .0498, .2240, 149.4
11. (a) .1804, .3233, .8571 (b) .0025
12. (a) X is Poisson, $\mu = \frac{1}{2}$ (b) .6065 (c) .0003
13. .9950

EXERCISE 4.5

1. $f(x) = \frac{1}{2}$, $0 < x < 1$, $2 < x < 3$, $\frac{3}{2}$, 1.041

2. $f_Y(y) = \dfrac{1}{4\sqrt{y}}$, $0 < y < 1$, $4 < y < 9$

3. $f_Y(y) = \dfrac{2y}{3}$, $1 < y < 2$

4. $f_W(w) = \dfrac{1}{2\sqrt{w}}$, $1 < w < 4$, same

5. $\ln 2$

6. .2865, .5654, 0

7. $p_Y(y) = \frac{1}{10}$, $y = 0, 1, \dots, 9$

8. $\dfrac{\sqrt{3}}{2}$

9. $\ln \dfrac{1}{q}$

10. 2

11. Erlang, $r = 3$, λ

12. (a) Geometric, $p = 1 - e^{-\lambda}$

13. (a) $\frac{1}{10}$ (b) $\frac{1}{10}$ (c) $e^{-10/4}$

14. (a) .3162 (b) .459

15. $\left(1 + \dfrac{\lambda b}{\lambda a + 1}\right) e^{-\lambda b}$, $(\lambda b + 1) e^{-\lambda b}$

16. $1 - \dfrac{b}{10 - a}$, $a + b < 10$, No

EXERCISE 4.6

1. (a) $\frac{1}{2}$, $\frac{1}{12}$

   (b)

$r =$	1	2	3	4	5
$E[Y_r] =$	$\frac{1}{6}$	$\frac{1}{3}$	$\frac{1}{2}$	$\frac{2}{3}$	$\frac{5}{6}$
$\mathrm{Var}[Y_r] =$	$\frac{5}{294}$	$\frac{8}{294}$	$\frac{9}{294}$	$\frac{8}{294}$	$\frac{5}{294}$

2. $f(y) = 30y^2(1 - y)^2$

3. .3085, .9987

4. .3085, .8413, .1574

5. For $t \geq 0$, $F_U(t) = 2N(t) - 1$, $f_U(t) = (\sqrt{2/\pi})\, e^{-t^2/2}$, $t > 0$

6. .3174, .3830

7. 0, 1.34

8. μ, 1.34σ

12. .449, 1.221

13. (9.51, 10.49)

14. (a) 183.6 (b) 212.8 cm

15. (a) 664 (b) 567, 433 (c) 80%

17. (a) $\sigma = 10.78$, $\mu = 36.21$ (b) 33.52

18. $t_k = z^2_{(k+1)/2}$

19. $t_k = z_{(k+1)/2}$

Chapter 5

EXERCISE 5.1

1. (a)

$$x_1 =$$

	0	1	2
0	$\frac{1}{16}$	$\frac{1}{8}$	$\frac{1}{16}$
$x_2 = 1$	$\frac{1}{8}$	$\frac{1}{4}$	$\frac{1}{8}$
2	$\frac{1}{16}$	$\frac{1}{8}$	$\frac{1}{16}$

(b)

$$x_1 =$$

	0	1	2
0	$\frac{81}{256}$	$\frac{27}{128}$	$\frac{9}{256}$
$x_2 = 1$	$\frac{27}{128}$	$\frac{9}{64}$	$\frac{3}{128}$
2	$\frac{9}{256}$	$\frac{3}{128}$	$\frac{1}{256}$

2. (a)

$$y_1 =$$

	0	1	2	3
0	$\frac{1}{16}$	$\frac{1}{16}$	0	0
1	$\frac{1}{16}$	$\frac{3}{16}$	$\frac{1}{8}$	0
$y_2 = 2$	0	$\frac{1}{8}$	$\frac{3}{16}$	$\frac{1}{16}$
3	0	0	$\frac{1}{16}$	$\frac{1}{16}$

(b)

$$y_1 =$$

	0	1	2	3
0	$\frac{81}{256}$	$\frac{27}{256}$	0	0
1	$\frac{27}{256}$	$\frac{63}{256}$	$\frac{18}{256}$	0
$y_2 = 2$	0	$\frac{18}{256}$	$\frac{15}{256}$	$\frac{3}{256}$
3	0	0	$\frac{3}{256}$	$\frac{1}{256}$

3. In Exercise 5.1.1,

$x =$	0	1	2
$p(x) =$	$\frac{1}{4}$	$\frac{1}{2}$	$\frac{1}{4}$

$y =$	0	1	2
$p(y) =$	$\frac{9}{16}$	$\frac{3}{8}$	$\frac{1}{16}$

In Exercise 5.1.2,

$x =$	0	1	2	3
$p(x) =$	$\frac{1}{8}$	$\frac{3}{8}$	$\frac{3}{8}$	$\frac{1}{8}$

$y =$	0	1	2	3
$p(y) =$	$\frac{27}{64}$	$\frac{27}{64}$	$\frac{9}{64}$	$\frac{1}{64}$

4. $\dfrac{2}{\ln 2}$

5. $\dfrac{9}{40}$

6.

	$x_1 =$	
	0	1
$x_2 = \quad 0$	$\frac{1}{2} - a$	a
1	a	$\frac{1}{2} - a$

where $0 < a \leq \frac{1}{2}$

7. $f(x) = 1,\ 0 < x < 1,\ f(y) = -\ln y,\ 0 < y < 1$

8. (a) $\dfrac{1}{\sqrt{2}}$ (b) $f(x) = \dfrac{1}{\sqrt{2}},\ -\dfrac{1}{\sqrt{2}} < x < \dfrac{1}{\sqrt{2}}$ for both

9. (a) 1   (b) $\begin{aligned} f(x) &= 1 + x,\ -1 < x < 0 \\ &= 1 - x,\quad 0 < x < 1 \end{aligned}\Bigg\}$ for both

10. $f_{X_1}(x_1) = \lambda e^{-\lambda x_1},\ x_1 > 0,\ f_{X_2}(x_2) = \lambda^2 x_2 e^{-\lambda x_2},\ x_2 > 0$

11. $f(x_1, x_2) = \dfrac{1}{2\pi},\qquad\qquad 0 < x_1^2 + x_2^2 < 1$

$\qquad = \dfrac{1}{64.8\pi} - \dfrac{\sqrt{x_1^2 + x_2^2}}{648\pi},\ 1 < x_1^2 + x_2^2 \leq 100$

12. $\frac{1}{6}, \frac{2}{9}, \frac{1}{9}$

13. $p_{X_1, X_2}(x_1, x_2) = p^2 q^{x_2 - 2},\ x_1 = 1, 2, 3, \ldots,$
$\qquad\qquad\qquad\qquad x_2 = x_1 + 1, x_1 + 2, \ldots$
$p_{X_1}(x_1) = pq^{x_1 - 1},\ x_1 = 1, 2, 3, \ldots,$
$p_{X_2}(x_2) = \begin{pmatrix} x_2 - 1 \\ 1 \end{pmatrix} p^2 q^{x_2 - 2},\ x_2 = 2, 3, \ldots,$

14. (a) $1 - 2e^{-1}$ (b) e^{-1}

EXERCISE 5.2

3.

Given $y_1 =$	$y_2 =$ 0	1	2	3
0	$\frac{1}{2}$	$\frac{1}{2}$	0	0
1	$\frac{1}{6}$	$\frac{1}{2}$	$\frac{1}{3}$	0
2	0	$\frac{1}{3}$	$\frac{1}{2}$	$\frac{1}{6}$
3	0	0	$\frac{1}{2}$	$\frac{1}{2}$

Given $y_1 =$	$y_2 =$ 0	1	2	3
0	$\frac{3}{4}$	$\frac{1}{4}$	0	0
1	$\frac{3}{12}$	$\frac{7}{12}$	$\frac{2}{12}$	0
2	0	$\frac{6}{12}$	$\frac{5}{12}$	$\frac{1}{12}$
3	0	0	$\frac{3}{4}$	$\frac{1}{4}$

4. $2x,\ 0 < x < 1,\ \dfrac{1}{y \ln 2},\ 1 < y < 2$

7. $p_{X_2|X_1}(x_2 \mid 100) = \dfrac{\dbinom{100}{x_2}\dbinom{100}{8 - x_2}}{\dbinom{1200}{x_2}},\ x_2 = 0, 1, \ldots, 8$

$$p_{X_1, X_2} = \dbinom{8}{x_2}\dbinom{1192}{x_1 - x_2}(.01)^{x_1}(.99)^{1200 - x_1}\quad \text{for}$$

$$x_2 = 0, 1, \ldots, 8,$$
$$x_1 = x_2, x_2 + 1, \ldots, 1200$$
X_2 is binomial, $n = 8$, $p = .01$

8. $p_{X_2}(x_2) = \dfrac{\dbinom{20}{x_2}\dbinom{980}{12 - x_2}}{\dbinom{1000}{12}},\ x_2 = 0, 1, \ldots, 12$

9. $f_{Y|X}(y \mid \frac{1}{2}) = 2,\ 0 < y < \frac{1}{2},\ f_{X|Y}(x \mid \frac{1}{2}) = \dfrac{1}{(2 - \sqrt{2})\sqrt{x}},\ \frac{1}{2} < x < 1$

10. $f_{X,Y}(x, y) = \dfrac{3x + y}{4} e^{-x-y},\ x > 0,\ y > 0$

$f_{X|Y}(x \mid y) = \dfrac{3x + y}{3 + y} e^{-x},\ x > 0,\ y > 0$

11.

$x_2 =$	$x_1 =$ 0	1
0	$\frac{1}{10}$	$\frac{3}{10}$
1	$\frac{3}{10}$	$\frac{3}{10}$

12.

	$x_3 =$ 0		$x_3 =$ 1	
$x_2 =$	$x_1 =$ 0	1	$x_1 =$ 0	1
0	0	$\frac{1}{10}$	$\frac{1}{10}$	$\frac{1}{5}$
1	$\frac{1}{10}$	$\frac{1}{5}$	$\frac{1}{5}$	$\frac{1}{10}$

13. $P(Y = y) = \dfrac{1}{y!} e^{-1}, \; y = 0, 1, 2, \ldots$

14. (a) $p_{X_1, X_2}(x_1, x_2) = \frac{1}{100}, \; x_1 = 0, 1, \ldots, 9, \; x_2 = 0, 1, \ldots, 9$
     (b) .34

15. .249

EXERCISE 5.3

1. $\frac{1}{4}, \frac{1}{2}, 0$

2. $\dfrac{n + y}{2}, \dfrac{x + 1}{2}$

3. (a) 0   (b) $\dfrac{(n + 2)(n - 1)}{36}$

4. $\rho = \dfrac{-3}{\sqrt{31(23)}}$

5. $\frac{12}{350}, \dfrac{24}{\sqrt{(48)(37)}}$

6. 60 months

7. No

8. 0

EXERCISE 5.4

1. (a) 7, 116 (b) 3, 116 (c) 13, 356 (d) 0, 800
2. 1080, 17.32
3. (a) 25, 8.06 (b) 5, 13.60 (c) 0, 36.88
4. (a) 84 (b) 164 (c) 36 (d) 280
5. (a) -75 (b) -220 (c) 500
6. (a) $1 - 2.5e^{-1.5}$ (b) $3.5e^{-2.5}$
7. (a) .1251 (b) .9582
8. (a) .3465 (b) .7392
9. (a) Binomial, $n = 15$, $p = .55$

   (b) $P(W = 15 - y, Y = y) = \binom{15}{y}(.45)^y(.55)^{15-y}$,

   $y = 0, 1, \ldots, 15$

   (c) $\rho = -1$
10. (a) 14, 3.74 (b) .3585
11. (a) .3679   (b) 1.1774
12. (a) .8413   (b) $\frac{1}{2}$
13. (a) .2483   (b) .8944 (c) .8888
14. (a) $a_1 = -\frac{2}{5}a_2, a_2$ is arbitrary

    (b) $\dfrac{a_1}{a_2} = -\dfrac{b_2 + 5\rho b_1}{25b_1 + 5\rho b_2}$
15. v, $2v$

EXERCISE 5.5

5. 2000, 1000
6. 728
7. 250
8. 90

9. 10

10. $n = 80$

EXERCISE 5.6

1. .8643, .6939

2. 1, 2, n, 1

3. 0

4. .3300

5. .5558

6. .0262

7. 24

8. 41

9. 65.31, 69.02, 74.28, 78.74, 106.97, 111.70, 118.38, 123.98

10. 34.87, 40.15, 43.44

11. .8708, 1, .8708

12. .0159, .0160, .0118

14. .9332

15. (a) negative binomial, $r = 50$, $p = .05$ (b) .9265

EXERCISE 5.7

1. (a) $P(X_1 = x, \ X_2 = n - x) = \binom{n}{x} p^x (1 - p)^{n-x}$,

    $x = 0, 1, \ldots, n$
    (b) -1

2. (a) multinomial, n, .4, .3, .2, .1   (b) .0063   (c) .0051

3. (a) $\binom{6}{1}\left(\frac{1}{6}\right)^4 = .00077$ (b) $\dfrac{2(5!)}{6^5}$

4. $P(X_1 = x_1, X_2 = x_2, \ldots, X_c = x_c) = \dfrac{\binom{k_1}{x_1}\binom{k_2}{x_2}\cdots\binom{k_c}{x_c}}{\binom{m}{n}}$

for all $0 \le x_i \le n$ such that $\sum x_i = n$

5. $P(X_A = x_1, X_B = x_2, X_C = x_3, X_D = x_4)$

$$= \dfrac{\binom{45,000}{x_1}\binom{35,000}{x_2}\binom{15,000}{x_3}\binom{5,000}{x_4}}{\binom{100,000}{1,000}}$$

6. (a) .8413, .1587   (b) .2810

7. (a) .1587   (b) .9773   (c) .9826

8. (a) .9406   (b) $a = 1206.20$

9. $a = 2.08\sigma$

10. $a = \frac{1}{2}t_{.95} = 1.066$

11. essentially .99

12. $\frac{1}{2}$

13. (a) $n(1 - u)^{n-1}, \; 0 < u < 1$ (b) $nu^{n-1}, \; 0 < u < 1$

(c) $1 - \dfrac{1}{2^n}, \dfrac{1}{2^n}$

14. $\displaystyle\sum_{j=k}^{n}\binom{n}{j}\dfrac{1}{2^n}$

15. (a) χ^2 random variable with 4 degrees of freedom

Chapter 6

EXERCISE 6.1

1. (a) $\bar{x} = 406.48, \; s^2 = 20498.593, \; s = 143.173$

2. (a) $\bar{x} = 744.15, \; s^2 = 11996.555, \; s = 109.529$

3. (a) $\bar{x} = 8.132$, $s^2 = .32727$, $s = .5721$
4. (a) $\bar{x} = 52.60$, $s^2 = 788.766$, $s = 28.085$
5. (a) $\bar{x} = 122.667$, $s^2 = 74.738$, $s = 8.645$
6. 425, 296
7. 120, 12
8. 52.5, 52
10. (a) 60% (b) 29.1, 5

EXERCISE 6.2

1. (a) .3830 (b) .8684
2. (a) .3935 (b) .6321
3. (a) .25   (b) .5   (c).1
4. (a)

		$x_1 =$			
		3.10	3.85	4.50	6.50
	3.10	$\frac{2}{15}$	$\frac{2}{15}$	$\frac{4}{45}$	$\frac{2}{45}$
$x_2 =$	3.85	$\frac{2}{15}$	$\frac{1}{15}$	$\frac{1}{15}$	$\frac{1}{30}$
	4.50	$\frac{4}{45}$	$\frac{1}{15}$	$\frac{1}{45}$	$\frac{1}{45}$
	6.50	$\frac{2}{45}$	$\frac{1}{30}$	$\frac{1}{45}$	0

(b) $\frac{2}{15}$ (c) $\frac{2}{45}$

5. (a)

		$x_1 =$			
		3.10	3.85	4.50	6.50
	3.10	.16	.12	.08	.04
$x_2 =$	3.85	.12	.09	.06	.03
	4.50	.08	.06	.04	.02
	6.50	.04	.03	.02	.01

(b) .16 (c) .05

7. (a) .0668 (b) .3585 (c) .4991
8. (a) 2.0794 (b) .6931
9. $(13.03, \infty)$, $(0, 284.74)$, $(10.01, 584.87)$

10. (a) .0571, .2266 (b) .5, .8

11. No, $E[S] = \sigma \dfrac{\Gamma\left(\dfrac{n}{2}\right)}{\Gamma\left(\dfrac{n-1}{2}\right)} \sqrt{\dfrac{2}{n-1}}$

12. $n = 68$

13. .99

14. (a) gamma, parameters n and λ
    (b) χ^2 with $2n$ degrees of freedom

15. $E[\overline{X}] = \mu$, $\text{Var}[\overline{X}] = \dfrac{\sigma^2}{n}\left(\dfrac{M-n}{M-1}\right)$, $E[S^2] = \dfrac{\sigma^2 M}{M-1}$

16. (a) $(.5)^n$ (b) $(.5)^n$ (c) $1 - (.5)^{n-1}$

17. (a) $X_{((n+1)/2)}$ (b) $\frac{1}{2}X_{(1)} + \frac{1}{2}X_{(n)}$, No

Chapter 7

EXERCISE 7.1

1. .62

2. 3, 1.50

3. $63\frac{1}{3}$

4. $\tilde{p} = \dfrac{1}{\bar{x}}$

5. .19

6. $\tilde{p} = \bar{x}$

7. $\tilde{p} = \dfrac{\bar{x}}{n}$

8. $\tilde{\sigma}^2 = \dfrac{1}{n}\sum x_i^2 - 25$

9. $\tilde{p} = \overline{X} + 1 - \dfrac{M_2}{\overline{X}}$, $\tilde{n} = \dfrac{\overline{X}}{\tilde{p}} = \dfrac{\overline{X}^2}{\overline{X}(1+\overline{X}) - M_2}$

10. $\hat{\sigma}^2 = \dfrac{1}{n} \sum (x_i - 10)^2$

11. $\hat{\mu} = 15, \ \hat{\sigma}^2 = 8.25$

12. $\hat{\mu} = \bar{x}$

13. $\hat{\mu} = 1.5$

14. $\hat{p} = \dfrac{1}{\bar{x}}$

15. $\hat{p} = \dfrac{\bar{x}}{n}$

16. $\tilde{\mu} = .41$

17. $\hat{\mu} = .41$

18. $\hat{\lambda} = .0036$

19. $\tilde{\mu}_1 = \bar{x}_1, \ \tilde{\mu}_2 = \bar{x}_2, \ \tilde{\sigma}_1^2 = \dfrac{1}{n} \sum x_{i1}^2 - \bar{x}_1^2, \ \tilde{\sigma}_2^2 = \dfrac{1}{n} \sum x_{i2}^2 - \bar{x}_2^2,$

$\tilde{\rho} = \dfrac{\sum (x_{i1} - \bar{x}_1)(x_{i2} - \bar{x}_2)}{\sqrt{\sum (x_{i1} - \bar{x}_1)^2 \sum (x_{i2} - \bar{x}_2)^2}}$, maximum likelihood is the same

20. $.476$

EXERCISE 7.2

1. $\frac{1}{6}$

2. $\dfrac{1}{n-1}$

3. $a = \dfrac{1}{2n}$

4. $b = \dfrac{1}{n-1}, \ a = 2$

5. $a = \frac{2}{3}$

6. $\dfrac{npq + p^2}{(n+1)^2}$

7. $a = \dfrac{n}{n+1}$

8. $\dfrac{pq}{n}$, $\overline{X}$

9. $\dfrac{q}{np^2}$, $\overline{X}$

10. $\dfrac{1}{n\lambda^2}$, $\overline{X}$

11. $\dfrac{1}{r\lambda^2 n}$, $\dfrac{\overline{X}}{r}$

12. (a) the minimum value $X_{(1)}$ (b) $\dfrac{1}{2n}$, No

    (c) $X_{(1)} - \dfrac{1}{n}$ is unbiased for α, variance $\dfrac{1}{n^2}$

13. (b) $a = \dfrac{\sigma_2^2}{\sigma_1^2 + \sigma_2^2} = \dfrac{\dfrac{1}{\sigma_1^2}}{\dfrac{1}{\sigma_1^2} + \dfrac{1}{\sigma_2^2}}$

14. $E[\hat{\Gamma}] = \dfrac{n\gamma}{n+1}$, $MSE[\hat{\Gamma}] = \dfrac{2\gamma^2}{(n+2)(n+1)}$, $E[2\overline{X}] = \gamma$,

    $\dfrac{4\gamma^2}{12n} = MSE[2\overline{X}]$, $MSE[\hat{\Gamma}] \leq MSE[2\overline{X}]$, both estimators are consistent

16. The bound for unbiased estimators for μ then is $\sigma^2/n = \mathrm{Var}[\overline{X}]$ and for unbiased estimators for σ^2 is $2\sigma^4/n$. No

17. $a = \dfrac{1}{n+1}$

18. normal, $\mu = p$, $\sigma^2 = \dfrac{p^2 q}{40}$

19. normal μ, μ^2/n

EXERCISE 7.3

1. (a) $\bar{x} = 4.328, s = .3610$ (b) (4.119, 4.537)   (c) (22.04, 24.28)

2. (.062, .434), (.248, .659)

3. 4.119, 24.28

4. 1.69 and 6.45

5. $50.70 to $193.50

6. (.0194, .0405)

7. (a) 247   (b) .5557

8. 477.16, 679.08

9. (a) .862 σ (b) .80

10. .372σ (b) .997

11. (a) gamma with parameters $2n$ and λ (b) $k = 2\lambda$
     (c) confidence limits for λ are $\chi^2_{\alpha/2}/2Y$, $(\chi^2_{1-\alpha/2}/2Y)$. The confidence limits for $\mu = 2/\lambda$ are twice the reciprocals of these values

12. 1.97, 5.09

13. $\bar{X} \pm z_{1-\alpha/2} \sqrt{\bar{X}/n}$

14. (2.36, 4.18)

15. (2.42, 4.32)

EXERCISE 7.4

1. The maximizing values are the same as for the individual samples treated separately.

2. $\hat{\mu}_X = \dfrac{\sum x_i + 4\sum y_i}{n + 8m}$, $\hat{\sigma}^2 = \dfrac{1}{m+n}\left(\sum(x_i - \hat{\mu}_X)^2 + 2\sum(y_i - 2\hat{\mu}_X)^2\right)$

3. $\hat{\lambda} = .2002$

4. .340, 3.057

5. (.059, 3.990)

6. .352

7. .610, 48.02

8. $(-66.4, 23.5)$

9. $(-.135, -.004)$. Since this 90 percent interval does not include 0, we can be 90 percent sure that $p_1 - p_2$ does not equal zero; that is, p_1 and p_2 are unequal.

Chapter 8

EXERCISE 8.1

1. $\alpha = .0039$, $\beta = .684$

2. $\alpha = .0002$, $\beta = .49$

3. $\sum \ln x_i \geq -n \ln 2 - \ln k$

4. Reject H_0, accept H_0

5. $\sum (x_i - 5)^2 \geq c$, $\sum (x_i - 5)^2 \leq c$

6. Accept H_0

7. $\sum x_i \leq c$

8. Accept H_0, $\beta = .333$

9. $\sum (x_i - 100.2)^2 \geq c$

10. $n = 231$, $c = 173.01$, reject H_0.

11. $n \sim 135$, $c = 1.716$, accept H_0.

12. $\alpha \doteq .1$, $\beta \doteq .37$

13. $\alpha = .1335$, $\beta = .2514$

EXERCISE 8.2

1. Reject H_0.

2. Reject H_0.

3. Accept H_0.

4. Accept H_0.

5. (a) $\mu = 1000 \lambda$ (b) reject H_0.

6. Accept H_0.

7. Accept H_0.

10. Accept H_0.

11. Accept H_0.

12. Accept H_0.

13. Reject H_0.

14. Accept H_0.

EXERCISE 8.3

1. Accept H_0.

2. Reject H_0: $\sigma_X^2 = \sigma_Y^2$, Still accept H_0: $\mu_D \geq \mu_T$.

3. Reject H_0.

4. Reject H_0: $\sigma_X^2 = \sigma_Y^2$. Still reject H_0: $\mu_X = \mu_Y$.

5. Accept H_0.

6. Accept H_0.

7. Reject H_0 for any $\alpha \geq .006$.

8. With the repaired T test, reject with any $\alpha \geq .004$. If samples were independent, reject with $\alpha \geq .26$.

9. The hypothesis of equality would be rejected for $\alpha \geq .2$.

10. .80

11. .4

12. Accept H_0.

13. Reject H_0.

Chapter 9

EXERCISE 9.1

1. $\hat{b} = .006784$, $\sum (y_i - \hat{b}x_i) = .540483$, $s^2 = .060054$. The sum of squares of the residuals for the model including the Y intercept a are smaller.

2. $\hat{b} = .0102$, $\hat{a} = 11.0052$, $s^2 = 4.9666$

$\hat{y}_i$	15.91	21.28	27.87	31.33	37.17	41.90	47.45	51.99	56.99	62.12
$\hat{e}_i$	$-.09$	.28	$-.13$	.33	3.17	-5.10	.45	.99	-1.01	1.12

3. (a) $\hat{b} = .9788$, $s^2 = .1765$
   (b) $\hat{b} = .9823$, $\hat{a} = -.2467$, $s^2 = .1823$

5. $\hat{b} = .9778$

9. $\sum (x_i - \bar{x})^2 = 0$

10. $\hat{B} = \dfrac{1}{n} \sum \dfrac{Y_i}{x_i}$

11. (a) No. (b) Yes, the case in which $\text{Var}[Y_i] = \sigma^2 x_i$.

12. $\sigma^2 \left(\dfrac{1}{n} + \dfrac{(x^* - \bar{x})^2}{\sum (x_i - \bar{x})^2} \right)$, smallest for $x^* = \bar{x}$

13. (a) $\hat{b} = 196.97$, $s^2 = 15803797.4$, $s_{\hat{b}} = 27.98$, $60\hat{b} = 11818.26$
    (b) $\hat{b} = 187.25$, $s^2 = 472439.60$, $s_{\hat{b}} = 32.96$, $60\hat{b} = 11,235$
    (c) $\hat{b} = 168.47$, $s^2 = 130.91$, $s_{\hat{b}} = 37.79$, $60\hat{b} = 10108$

14. $\hat{\Lambda} = \dfrac{\bar{Y}}{\bar{x}}$

EXERCISE 9.2

1. 95% limits for a are (7.46, 14.55), 95% limits for b are (.0091, .0113), 95% confidence limits for σ^2 are (2.27, 18.23).

2. 2.397

3. (a) $\hat{b} = .1287$, $\hat{a} = 377.22$, $s^2 = 1608.38$
   (b) Reject H_0 with any $\alpha \geq .6$

4. We would reject $H_0 : b = 0$ for any $\alpha \geq .06$.

5. (a) 20.71 (b) 11.43

6. Reject $H_0 : b \geq 1$.

7. (a) $\hat{b} = (\bar{z} - \bar{y})/(x_2 - x_1)$, $\hat{a} = x_1(\bar{y} - \bar{z})/(x_2 - x_1)$ (b) No

8. (a) 69.43 (b) (69.022, 69.838)

10. (a) $y_i = x_i + e_i$
    (b) $\hat{b} = 1.045$, $\hat{a} = -.090$, $s = .116$, $s_{\hat{a}} = .086$, $s_{\hat{b}} = .172$

EXERCISE 9.3

(a) $\hat{\mu} = 12.05$, $\hat{\tau}_1 = -2.25$, $\hat{\tau}_2 = .45$, $\hat{\tau}_3 = 1.73$, $s^2 = 4.59$
(b) Reject H_0.

2. (a) $\hat{\mu} = 50.30$, $\hat{\tau}_1 = .03$, $\hat{\tau}_2 = .20$, $\hat{\tau}_3 = -.30$, $s^2 = 2.32$
(b) Accept H_0.

3. (a) $\hat{\mu} = 91.88$, $\hat{\tau}_1 = -8.38$, $\hat{\tau}_2 = -6.88$, $\hat{\tau}_3 = -5.13$, $\hat{\tau}_4 = 20.38$,
$s^2 = 147.54$
(b) Reject equality for any $\alpha \geq .02$.

4. (a) $\hat{\mu} = 14.18$, $\hat{\beta}_1 = -2.66$, $\hat{\beta}_2 = -1.31$, $\hat{\beta}_3 = 1.32$, $\hat{\beta}_4 = 2.62$,
$\hat{\beta}_5 = -.78$, $\hat{\beta}_6 = .82$, $\hat{\tau}_1 = -2.63$, $\hat{\tau}_2 = -4.38$, $\hat{\tau}_3 = 3.40$,
$\hat{\tau}_4 = 3.62$, $s^2 = 4.24$
(b) Reject equality of chemicals for any $\alpha \geq .01$.

5. Reject equality for $\alpha \geq .04$.

Chapter 10

EXERCISE 10.1

1. (a) .9911   .9923   .9936   .9988   1.0026   1.0038   1.0079
1.0083   1.0090   1.0120   1.0139   1.0169   1.0195   1.0234
1.0302
$\hat{t}_{.5} = 1.0083$, $\hat{t}_{.25} = .9988$, $\hat{t}_{.75} = 1.0169$
(b) $\hat{\zeta} = .9918$

2. (a) 4.54 meters   (b) $(2.75)^2 \pi$

3. (a) 7.17 meters   (b) 8.19

4. (a) 256.4   (b) 252.14

5. (a) 4.34, 10.24   (b) .37

6. $x_{(12)} = 1.0169$

7. Accept $H_0: t_{.5} = 5$

8. 252.7, 260.1

9. Yes.

10. Reject with either.

11. Accept with either.

12. (a) $\hat{t}_p = \bar{x} + \hat{\sigma} z_p$ (b) $\hat{t}_{.8} = 253.22$

13. (a) $t_p = -\dfrac{1}{\lambda} \ln(1 - p)$ (b) 2.74, 9.25

EXERCISE 10.2

1. For $r < n + 1$ the observed significance level is
$2N\left(\dfrac{(r - n)\sqrt{2}}{\sqrt{n}}\right)$; for $r > n + 1$ the observed significance level
is $2\left(1 - N\left(\dfrac{(r - 1 - n)\sqrt{2}}{\sqrt{n}}\right)\right)$.

2. Accept the randomness hypothesis.

3. Accept the randomness.

4.

$r =$	2	3	4	5	6	7	8	9
$P(R = r) =$	$\frac{2}{126}$	$\frac{7}{126}$	$\frac{24}{126}$	$\frac{30}{126}$	$\frac{36}{126}$	$\frac{18}{126}$	$\frac{8}{126}$	$\frac{1}{126}$

5. Accept the hypothesis.

6. Accept equality with $\alpha < .16$.

7. Reject H_0 for any $\alpha \geq .0042$.

8. M_{XY} is approximately normal with $\mu = mn/2$,
$\sigma = \sqrt{(mn)(m + n + 1)/12}$.

10. Reject $H_0 : F_Z(t) = F_T(t)$ for all t for any $\alpha \geq .03$.

EXERCISE 10.3

1. Reject H_0.

2. Accept H_0.

3. Accept the hypothesis.

4. Accept the hypothesis.

5. Accept the hypothesis.

6. (a) Reject. (b) If we reject H_0, $.05 \leq \alpha \leq .12$, based on these data.

7. Accept the hypothesis.

8. Reject.

10. Reject H_0.

EXERCISE 10.4

1. Reject independence.

2. Reject independence.

3. Reject independence.

4. The 3 age distributions are not the same.

5. Reject independence.

6. Reject consistency.

Chapter 11

EXERCISE 11.1

1. $f(\theta \mid X = 1) = 2\theta \qquad\qquad 0 < \theta < 1$
    $f(\theta \mid X = 0) = 2(1 - \theta), \quad 0 < \theta < 1$

2. $f(\theta \mid X = 1) = \dfrac{2\theta}{b^2 - a^2}, \qquad\qquad a < \theta < b$

    $f(\theta \mid X = 0) = \dfrac{2(1 - \theta)}{1 - (1 - b + a)^2}, \quad a < \theta < b$

3. Normal, mean $\frac{10}{11}\bar{x} + \frac{100}{11}$, variance $\frac{5}{11}$

5. $f(\lambda \mid x) = (x + a)^2 \, \lambda e^{-\lambda(x + a)}, \; \lambda > 0, \, a = \frac{1}{100}$

6. $P(\alpha = 9.5 \mid x) = 1 \quad$ for $\quad 9.5 < x \leq 9.7$
    $\qquad\qquad\quad\; = \frac{4}{9} \quad$ for $\quad 9.7 < x < 10.5$
    $P(\alpha = 9.7 \mid x) = 0 \quad$ for $\quad 9.5 < x \leq 9.7$
    $\qquad\qquad\quad\; = \frac{5}{9} \quad$ for $\quad 9.7 < x < 10.5$

7.

$\theta =$	$\frac{1}{4}$	$\frac{1}{2}$	$\frac{3}{4}$
$P(\Theta = \theta) =$	$\frac{1}{2}$	$\frac{1}{3}$	$\frac{1}{6}$

is the posterior for Θ given $X = 0$.

$\theta =$	$\frac{1}{4}$	$\frac{1}{2}$	$\frac{3}{4}$
$P(\Theta = \theta) =$	$\frac{1}{6}$	$\frac{1}{3}$	$\frac{1}{2}$

is the posterior for Θ given $X = 1$.

8. D is binomial, $n = 50$, $p = .05$.

10. $P(D = d \mid X = 0) = \binom{47}{d}(.05)^d(.95)^{47-d}$, $d = 0, 1, ..., 47$

EXERCISE 11.2

1. No

2. (a) $.0495 + \dfrac{X}{1000}$

    (b) No, $\text{Var}[49.50 + X] = 10\dfrac{\theta}{1000}\dfrac{(1000 - \theta)}{1000}\left(\dfrac{1000 - 10}{999}\right)$,

    $MSE[49.50 + X] = \text{Var}[X] + (49.5 - .99\theta)^2$

3. (a) $\dfrac{\sum X_i + 1}{n + 2}$

    (b) No, $MSE\left[\dfrac{\sum X_i + 1}{n + 2}\right] = \dfrac{n\theta(1 - \theta) + (2\theta - 1)^2}{(n + 2)^2}$

4. $p^* = \dfrac{15.8}{8 + x}$

5. $n^* = 12 + x$

6. (a) $\lambda^* = \dfrac{y + 2}{n + .1} = 8.119$ (b) $\lambda^* = \dfrac{80 + 2}{5 + .1}$

EXERCISE 11.3

1. The approximate 95% Bayesian limits are (.253, .517).
    The approximate 95% confidence limits are (.264, .536).

2. (a) $f(\theta \mid 1) = 57.7\theta(1 - \theta)$, $0 < \theta < .2$ (b) $(.042, .194)$

3. (a) For λ, $(.006, .013)$ and for μ, $(67.4, 165.2)$ (b) $(58.7, 130.5)$

4. (a) $\frac{16}{134}$ (b) .070, .159

5. 4.35, 13.02 (b) 7.63, 30.63

6. (a) 5.10, 14.7 (b) 8.74, 40.6

7. (a) Accept H_0. (b) Accept H_0.

8. Accept H_0.

9. Accept H_0.

Index